Climate Change and Human Responses

Vertebrate Paleobiology and Paleoanthropology Series

Edited by

Eric Delson
Vertebrate Paleontology, American Museum of Natural History
New York, NY 10024, USA
delson@amnh.org

Eric J. Sargis
Anthropology, Yale University
New Haven, CT 06520, USA
eric.sargis@yale.edu

Focal topics for volumes in the series will include systematic paleontology of all vertebrates (from agnathans to humans), phylogeny reconstruction, functional morphology, Paleolithic archaeology, taphonomy, geochronology, historical biogeography, and biostratigraphy. Other fields (e.g., paleoclimatology, paleoecology, ancient DNA, total organismal community structure) may be considered if the volume theme emphasizes paleobiology (or archaeology). Fields such as modeling of physical processes, genetic methodology, nonvertebrates or neontology are out of our scope.

Volumes in the series may either be monographic treatments (including unpublished but fully revised dissertations) or edited collections, especially those focusing on problem-oriented issues, with multidisciplinary coverage where possible.

Editorial Advisory Board
Ross D.E. MacPhee (American Museum of Natural History), **Peter Makovicky** (The Field Museum), **Sally McBrearty** (University of Connecticut), **Jin Meng** (American Museum of Natural History), **Tom Plummer** (Queens College/CUNY).

More information about this series at http://www.springer.com/series/6978

Climate Change and Human Responses

A Zooarchaeological Perspective

Edited by

Gregory G. Monks

Department of Anthropology, University of Manitoba, Winnipeg, MB, Canada

Springer

Editor
Gregory G. Monks
Department of Anthropology
University of Manitoba
Winnipeg, MB
Canada

ISSN 1877-9077   ISSN 1877-9085 (electronic)
Vertebrate Paleobiology and Paleoanthropology Series
ISBN 978-94-024-1105-8   ISBN 978-94-024-1106-5 (eBook)
DOI 10.1007/978-94-024-1106-5

Library of Congress Control Number: 2016957497

© Springer Science+Business Media Dordrecht 2017
This work is subject to copyright. All rights are reserved by the Publisher, whether the whole or part of the material is concerned, specifically the rights of translation, reprinting, reuse of illustrations, recitation, broadcasting, reproduction on microfilms or in any other physical way, and transmission or information storage and retrieval, electronic adaptation, computer software, or by similar or dissimilar methodology now known or hereafter developed.
The use of general descriptive names, registered names, trademarks, service marks, etc. in this publication does not imply, even in the absence of a specific statement, that such names are exempt from the relevant protective laws and regulations and therefore free for general use.
The publisher, the authors and the editors are safe to assume that the advice and information in this book are believed to be true and accurate at the date of publication. Neither the publisher nor the authors or the editors give a warranty, express or implied, with respect to the material contained herein or for any errors or omissions that may have been made.

Cover Illustration: This Inuit hunter on melting sea ice is a metaphor that captures the sense of this volume. Photo courtesy of National Geographic Society.

Printed on acid-free paper

This Springer imprint is published by Springer Nature
The registered company is Springer Science+Business Media B.V.
The registered company address is: Van Godewijckstraat 30, 3311 GX Dordrecht, The Netherlands

To Janet, with gratitude, and to Denise, Reed, Abigail, Benjamin and Eden with love

Foreword: The Essential Past

Life was much simpler when I was a student. Past climate change was a matter of oscillations from warm to cold and back again, at the rather gradual pace that the palynological record seemed to show. Animals responded by latitudinal shifts of distribution (reindeer) or by going extinct (mammoths) (Evans 1975; West 1968; Zeuner 1959, 1961). And people did much the same, adjusting their distribution and lifeways as climate and environmental change dictated, at least until the 'revolution' of the Neolithic allowed man (*sic*) to make himself (Childe 1936; Clark and Piggott 1970). Anthropogenic climate change had yet to hit the agenda, and even Marine Oxygen Isotope Stages were little more than a rumour (Shackleton et al. 1977). Palaeozoology was something rather *outré* practised by a few palaeontologists, whilst field zoology was carried out in the real world. Those two research communities seldom met, let alone communicated.

Today's students of palaeozoology and zooarchaeology (collectively the 'palaeo record', for brevity) have a far more complex knowledge base with which to contend. Proxy records in marine oozes and polar ice show that global climates have swung from one state to another with remarkable rapidity, inserting brief temperate interstadials into our beloved ice ages, and acute cold snaps even into the Holocene (Thomas et al. 2007). Our understanding of the response of animal communities to those climate changes is still only partial, though we have learned that community disequilibrium may have been significant (Wolverton et al. 2009), and that some species successfully habitat-tracked, whilst others did not (Dalén et al. 2007). The timing and rate of extinction events has been re-thought, with mammoths surviving in Siberia as the pyramids were under construction in Egypt (Guthrie 2004), and the palaeo record has forced a re-think of biogeographical assumptions, for example regarding the past distribution of elephants in Island Southeast Asia (Cranbrook et al. 2007). And rapid climate change has become a topic for the present and future, not only for the past, as atmospheric and climate data accumulate to confirm Arrhenius's theories about global warmth (Arrhenius 1896).

In many respects we live in exciting times. Methodological advances and refinements allow us to ask questions that would have been unanswerable twenty years ago. Advances in the use of the stable isotope record stratified into datable sediments, shells and skeletal tissues have simply revolutionized our investigations of climate changes, animal migration and transhumance, and diet (Gil et al. 2011; Schwarz et al. 2010). Our chronologies have become more refined. Individually characterized volcanic tephras provide marker-horizons on a regional scale (Lane et al. 2011). Dendrochronological sequences have been pushed further back, to the Pleistocene/Holocene boundary (Kromer 2009), and developed for new parts of the world (Gou et al. 2008; Turney et al. 2010). Even that old workhorse, radiocarbon dating, has been given a new lease of life as AMS dating and new sample preparation methods have squeezed ± errors down to a few decades even on dates in the Late Pleistocene (Higham et al. 2006). Rather pleasingly, the other advance in radiocarbon dating, Bayesian calibration, uses a statistical technique first published in 1763 (Bayes and Price 1763; Bronk Ramsey 2009), reminding us that old tools can be put to new uses. We now have better resolution on dates back to MIS3 than ever before. Our resolution has also been enhanced with regard to taxonomy. We are no longer restricted to asking whether a particular species survived a climate change event, but whether the same genomic clades of that species did so (Hornsby and Matocq 2012). For at least some species, the response to climate change can now be studied at the population level, and some of the soft-tissue phenotypic attributes of those animals can be inferred (Pruvost et al. 2011).

What impact have all of these advances had on the dialogue between neo- and palaeozoology? On the palaeo side of that uneasy relationship, we have been accustomed to think in terms of whole organisms, populations of individuals and communities of species. People in the past acted as populations within vertebrate communities, and it was that relationship that underpinned much of our research. The individualistic, selfish politics of the 1980s bred a trend towards the individual within archaeology, one aspect of the 'post-processual critique' (e.g., Earle and Preucel 1987), steering us away from the generalizations of processual archaeology that lent themselves so well to comparison and integration with the generalizations of community ecology. Meanwhile, technical developments within the biological sciences focussed interest at the sub-organism level, to details of biomolecular variation and, of course, to genomic research (Strasser 2008), with particular consequences for studies of the evolutionary record (e.g., Hackett et al. 2008). Biomolecular research is expensive and quickly came to attract a high proportion of available research funds, thus drawing in more bright young researchers in a positive feedback loop. This has somewhat changed the dialogue between palaeo and neozoology.

There have been remarkable advances in our understanding of ancient populations of animals, and even extinct species (e.g., Edwards et al. 2010), with neozoologists quarrying the palaeo record for DNA and other biomolecules to give time-depth to the present-day research questions, and meanwhile reading a palaeo record from the present-day genomic archive (e.g., Leonardi et al. 2012; Searle et al. 2009). Certainly, the palaeo record *is* an archive, and a precious and complex one at that. But more than that, it has its own research questions and agenda. It is a common criticism of the palaeo record that it lacks the precision of real-time neozoology. Even with advances in dating methods, our chronological resolution still cannot match that of a field experiment, and our data are often patchy in their temporal and geographical distribution. Furthermore, we are limited in our capacity for replication: if there are only two known assemblages of MIS3 date from a given region, then we cannot simply increase our inadequate sample size by doing more field research. However, the value of the palaeo record has to be seen as clearly as its weaknesses. As Lyman (2011) points out with exemplary precision, it is *precisely* the longer perspective of the palaeo record that makes it valuable as a source for conservation research, encompassing evolutionary timescales and past periods of rapid climate change.

One issue that afflicts both neo- and palaeozoology is the shifting baseline syndrome (Papworth et al. 2009; Pauly 1995). When studies of today's fauna seek to quantify a change in distribution or abundance, we have to make comparisons with some baseline: numbers of Monks' Porpoise have fallen 80% since 1990, for example. Why 1990? Is that when systematic records began, or when the current population fluctuation seems to have begun? If the latter, do we have really good data to show that the 1990 population was typical of the preceding few centuries? Hardly ever, and more often the chosen baseline is defined by the previous high (or, less often, low) peak in a rather short run of records, or by the beginning of that time-series of data. When do good records of animal distribution and abundance actually begin; that is, what is the baseline against which we assess the current state of the world's fauna? Even a very generous estimate would put that baseline somewhere late in the nineteenth century, acknowledging the pioneering, but region-specific, work of Humboldt and Wallace (e.g., Von Humboldt 1849; Wallace 1895). A more realistic global baseline would be well into the twentieth century, long after the consequences of human population growth and industrialization had wrought substantial changes to most global biomes. It is little wonder, then, that the palaeozoology and zooarchaeology records often produce surprises, such as tigers on Palawan (Ochoa and Piper 2013).

These baseline issues are a cause for concern. Neozoology alone will consistently understate changes in terrestrial faunas on anything other than very short (i.e., decadal) timescales. For example, without the palaeo record, the devastating reduction in avian biodiversity throughout the Pacific over the last millennium or so would not be apparent (Steadman 1995). We would note the extinctions and extirpations of the last 50–100 years, and probably bemoan the impact that 'westernization' seems to have had on this island Eden. We would certainly not appreciate the damage that had already been done before the first Europeans sailed into the region (Boyer 2008).

Looking forward, too, there is a challenge for palaeozoology. Despite the potential for understatement of the human impact on faunas, we *are* aware of, and concerned about, that impact, and one reaction has been to try to repair the damage. Restoration ecology seeks to go one step beyond mitigating the worst consequences of human impact by 'restoring' habitat patches and larger environments through carefully planned reintroduction of some species and extirpation of others (for an overview, see Suding 2011). But what is the baseline state towards which that ecology is being 'restored'? If restoration ecology is to have a firm evidential basis, a high-quality palaeo record is essential. The alternative is to limit ourselves to restoring damaged and degraded environments only to a previous state that happens to be our first detailed record of that environment. It is quite likely that record will be of an environment that was already impacted, perhaps degraded, and quite possibly in a state of disequilibrium. Without a good long-term record, restoration ecology may be rebuilding communities that are inherently unstable and of low resilience. The same applies to National Parks and other attempts at conservation of landscapes and their biotas. The wild places that John Muir and others sought to preserve may have been dynamic systems already undergoing processes of rapid change, so that at least some of our conservation efforts may be seeking to impose stability on inherently unstable systems.

At its most intrusive, restoration ecology grades into rewilding, with the objective of returning some environments to their supposed condition before human perturbation and management became dominant (Brown et al. 2011; Meyer 2010). That issue in itself begs a number of questions. All species cause some degree of disturbance to the ecosystem in which they live. How dominant does the human impact have to be in order for the ecology to be 'wild' no longer? To put that another way, could rewilding accept restoration to a biota that includes co-existing humans? Rewilding proposals include replacing extinct species with their closest extant ecological equivalents. Applying these principles to the Great Plains of North America, some have suggested trying to put back the megafauna that greeted the first Beringian human migrants. As mammoths, *Smilodon* and *Arctodus* have, regrettably, quit the scene, the proposed rewilding would populate the Plains with elephants, lions and bears (oh my!) (Donlan 2005; Donlan et al. 2006). Given that many contemporary Americans find it unacceptable to share space even with coyotes and bobcats, one can only suppose that this new megafauna would be extirpated even more quickly than its predecessor.

The point, of course, is that the palaeo record must be more than just a species list, a wish-list of potential reintroductees with which to wage the green campaign. Within Europe, various EU Species and Habitats Directives build on the 1979 Bern Convention on the Conservation of European Wildlife and Natural Habitats to effectively oblige signatory states to engage in restorative reintroduction. On what criteria? There has to be evidence that a species was extirpated from the region concerned as a direct or indirect consequence of human activity. This treaty obligation has justified the reintroduction of red kite *Milvus milvus* to England, supported by historical records of the persecution of that species, and zooarchaeological records of its formerly wide range. What about species for which there is only the palaeo record, their extirpation pre-dating reliable documentary records? Survival of lynx *Lynx lynx* into Anglo-Saxon times makes it likely that this species was lost to England through habitat destruction and, possibly, direct persecution. A case for the reintroduction of lynx to

England can be, and has been, made (Hetherington et al. 2008). How far into the past can we extend that principle? That lugubrious cervid the elk *Alces alces* (i.e., moose in North America, not the American elk *Cervus elaphus*) survived into the Holocene in England, and was certainly hunted by Mesolithic hunter-gatherers (Kitchener 2010). However, it would be difficult to argue that elk were extirpated through human predation. More likely, habitat changes and fragmentation following early Holocene warming progressively produced then replaced the wet woodland environments in which elk could flourish. Flooding of the North Sea basin, isolating British elk from continental populations, may have contributed to their vulnerability. Given that effective population size can now be estimated from genetic diversity in palaeofaunal assemblages (Hofreiter and Stewart 2009), that possibility could be investigated if enough well-dated elk remains could be assembled and sampled.

There is another challenge for the palaeo record, one that relates directly to modern concerns: how far back in time do conservation efforts extend? National Parks and species directives are phenomena of the last century or so. However, if the impacts of climate change and human activities had a discernible effect on the distribution or numbers of a species deeper into the past, would prehistoric peoples have taken note and responded? At its simplest level, of course, that may only mean prey-switching or moving home-range to match the movements of a favoured prey population. We can certainly detect those responses in the palaeo record and, at their simplest, see them in terms of logistic optimisation. Crabs are running out – eat more clams. Does that give prehistoric peoples due credit for their intelligence and forward planning? The motivation might equally have been 'crabs are running out – stop eating crabs so they can recover'. Mobile foragers might simply move to a stretch of coast where crabs were still abundant, but what of more sedentary foragers, as in the Pacific Northwest? Butler and Campbell (2004) make the point that the region shows appreciable stability in settlement and subsistence: to what extent does that stability reflect care in the use, and therefore the conservation, of animal resources? Our conservation efforts today depend upon written records that chart past abundance and distribution, however fallibly. Non-literate societies had the memories and knowledge of older generations as their shifting baseline. It seems to be a common human cross-cultural trait that with grey hair comes the mindset that the past was different, usually better (e.g., see the beginning of this foreword!). The grandparent generation may have been critical in enabling a forager group to detect the consequences of their impacts or of some local change in weather patterns or more general climate, and to plan their response.

In a similar way, neozoology needs the 'grandparent' memory of the palaeo record, and archaeology needs its use of that record to be informed by sound zoology and ecology. We need to keep the inter-disciplinary conversations going, to keep an integration of ideas and evidence from the short- to the very long-term, through collaborative projects and through publications that offer new ideas and information to all researchers concerned with the response of human and other animal populations to each other and to their climatic context. This volume provides just such a conversation, demonstrating how far our palaeo studies have come since the days of Zeuner, and how much precious synergy results from taking a truly ecological perspective on the human past.

Terry O'Connor
PALAEO and Department of Archaeology
University of York
UK

References

Arrhenius, S. (1896). On the influence of carbonic acid in the air upon the temperature of the ground. *London, Edinburgh, and Dublin Philosophical Magazine and Journal of Science (fifth series), 41*, 237–275.

Bayes, T., & Price, R. (1763). An essay towards solving a problem in the doctrine of chance. *Philosophical Transactions of the Royal Society of London, 53*, 370–418.

Boyer, A. G. (2008). Extinction patterns in the avifauna of the Hawaiian Islands. *Diversity and Distributions, 14*(3), 509–517.

Bronk Ramsey, C. (2009). Bayesian analysis of radiocarbon dates. *Radiocarbon, 51*(1), 337–360.

Brown, C., McMorran, R., & Price, M. F. (2011). Rewilding – a new paradigm for nature conservation in Scotland? *Scottish Geographical Journal, 127*(4), 288–314.

Butler, V. L., & Campbell, S. K. (2004). Resource intensification and resource depression in the Pacific Northwest of North America: A zooarchaeological review. *Journal of World Prehistory, 18*(4), 327–405.

Childe, V. G. (1936). *Man makes himself*. London: Watts.

Clark, J. G., & Piggott, S. (1970). *Prehistoric societies*. London: Penguin Books.

Cranbrook, Earl of, Payne, J., & Leh, C. M. U. (2007). Origin of the elephants *Elephas maximus* L. of Borneo. *Sarawak Museum Journal, 63*(84), 95–124.

Dalén, L., Nyström, V., Valdiosera, C., Germonpré, M., Sablin, M., Turner, E., et al. (2007). Ancient DNA reveals lack of postglacial habitat tracking in the arctic fox. *Proceedings of the National Academy of Sciences, 104*(16), 6726–6729.

Donlan, J. (2005). Re-wilding North America. *Nature, 436*, 913–914.

Donlan, J., Berger, J., Bock, C. E., Bock, J. H., Burney, D. A., Estes, J. A., et al. (2006). Pleistocene rewilding: An optimistic agenda for twenty-first century conservation. *The American Naturalist, 168*(5), 660–681.

Earle, T. K., & Preucel, R. W. (1987). Processual archaeology and the radical critique. *Current Anthropology, 28*(4), 501–538.

Edwards, C. J., Magee, D. A., Park, S. D. E., McGettigan, P. A., & Lohan, A. J. (2010), A complete mitochondrial genome sequence from a Mesolithic wild Aurochs (*Bos primigenius*). *PLoS ONE, 5*(2), e9255. doi: 10.1371/journal.pone.0009255.

Evans, J. G. (1975). *The environment of early man in the British Isles*. London: Paul Elek Books.

Gil, A. F., Neme, G. A., & Tykot, R. (2011). Stable isotopes and human diet in central Western Argentina. *Journal of Archaeological Science, 38*(7), 1395–1404.

Gou, X., Peng, J., Chen, F., Yang, M., Levia, D. F., & Li J. (2008). A dendrochronological analysis of maximum summer half-year temperature variation over the past 700 years on the northeastern Tibetan plateau. *Theoretical and Applied Climatology, 93*(3–4), 195–206.

Guthrie, R. D. (2004). Radiocarbon evidence of mid-Holocene mammoths stranded on an Alaskan Bering Sea island. *Nature, 429* (17 June 2004), 746–749.

Hackett, S. J., Kimball, R. T., Reddy, S., Bowie, R. C. K., Braun, E. L., Chojnowski, J. L., et al. (2008). A phylogenomic study of birds reveals their evolutionary history. *Science, 320*(5884), 1763–1768.

Hetherington, D. A., Miller, D. R., Macleod, C. D., & Gorman, M. L. (2008). A potential habitat network for the Eurasian lynx *Lynx lynx* in Scotland. *Mammal Review, 38*(4), 285–303.

Higham, T. F. G., Jacobi, R. M., & Bronk Ramsey, C. (2006). AMS radiocarbon dating of ancient bone using ultrafiltration. *Radiocarbon, 48*(2), 179–185.

Hofreiter, M., & Stewart, J. (2009). Ecological change, range fluctuations and population dynamics during the Pleistocene. *Current Biology, 19*(14), 584–594.

Hornsby, A. D., & Matocq, M. D. (2012). Differential regional response of the bushy-tailed woodrat (*Neotoma cinerea*) to late Quaternary climate change. *Journal of Biogeography, 39*(2), 28–305.

Kitchener, A. C. (2010). The Elk. In T. O'Connor & N. Sykes (Eds.) *Extinctions and invasions: A social history of British fauna* (pp. 36–42). Oxford: Windgather Press.

Kromer, B. (2009). Radiocarbon and dendrochronology. *Dendrochronologia, 27*(1), 15–19.

Lane, C. S., Andrič, M., Cullen, V. L., & Blockley, S. P. (2011). The occurrence of distal Icelandic and Italian tephra in the Lateglacial of Lake Bled, Slovenia. *Quaternary Science Reviews, 30*(9–10), 1013–1018.

Leonardi, M., Gerbault, P., Thomas, M. G., & Burger, J. (2012). The evolution of lactase persistence in Europe. A synthesis of archaeological and genetic evidence. *International Dairy Journal, 22*(2), 88–97.

Louys, J. (2012). The future of mammals in southeast Asia: Conservation insights from the fossil record. In J. Louys (Ed.) *Palaeontology in Ecology and Conservation* (pp. 227–238). Berlin: Springer Verlag.

Lyman, R.L. (2011). A warrant for applied palaeozoology. *Biological Reviews*, Early View, doi: 10.1111/j.1469-185X.2011.00207.x.

Maschner, H. D. G. (2012). Archaeology of the Northwest Coast. In T. Pauketat (Ed.), *The Oxford handbook of North American archaeology* (pp. 160–172). Oxford: Oxford University Press.

Meyer, T. (2010). Rewilding Germany. *International Journal of Wilderness, 16*(3), 8–12.

Ochoa, J., & Piper, P. J. (2017). Holocene large mammal extinctions in Palawan Island, Philippines. In G. G. Monks (Ed.), *Climate change and human responses: A zooarchaeological perspective*. Dordrecht: Springer.

Papworth, S. K., Rist, J., Coad, L., & Milner-Gulland, E. J. (2009). Evidence for shifting baseline syndrome in conservation. *Conservation Letters, 2*(2), 93–100.

Pauly, D. (1995). Anecdotes and the shifting baseline syndrome of fisheries. *Trends in Ecology and Evolution, 10*(10), 430.

Pruvost, M., Bellone, R., Benecke, N., Sandoval-Castellanos, E., Cieslaka, M., Kuznetsova, T., et al. (2011). Genotypes of predomestic horses match phenotypes painted in Paleolithic works of cave art. *Proceedings of the National Academy of Sciences of the USA, 108*(46), 18626–18630. 15 November 2011.

Schwarz, H. P., White, C. D., & Longstaff, F. J. (2010). Stable and radiogenic isotopes in biological archaeology: Some applications. *Isoscapes, 3*, 335–356.

Searle, J. B., Jones, C. S., Gündüz, I., Scascitelli, M., Jones, E. P., Herman, J. S., et al. (2009). Of mice and (Viking?) men: Phylogeography of British and Irish house mice. *Proceedings of the Royal Society B, 276* (1655), 201–207.

Shackleton, N. J., Lamb, H. H., Worssam, B. C., Hodgson, J. M., Lord, A. R., Shotton, F. W., et al. (1977). The oxygen isotope stratigraphic record of the late pleistocene. *Philosophical Transactions of the Royal Society of London B, 280*(972), 169–182.

Steadman, D. W. (1995). Prehistoric extinctions of Pacific island birds: Zoogeography meets zooarchaeology. *Science, 267*(5201), 1123–1131.

Strasser, B. J. (2008). GenBank – natural history in the 21st century? *Science, 322*(5901), 537–538.

Suding, K. N. (2011). Towards an era of restoration in ecology: Successes, failures, and opportunities ahead. *Annual Review of Ecology, Evolution and Systematics, 42*, 465–487.

Thomas, E. R., Wolff, E. W., Mulvaney, R., Steffensen, J. P., Johnsen, S. J., Arrowsmith, et al. (2007). The 8.2 ka event from Greenland ice cores. *Quaternary Science Reviews, 26*(1–2), 70–81.

Turney, C. S. M., Fifield, L. K., Hogg, A. G., Palmer, J. G., Hughen, K., Baillie, M. G. L., et al. (2010). The potential of New Zealand kauri (*Agathis australis*) for testing the synchronicity of abrupt climate change during the Last Glacial Interval (60,000–11,700 years ago). *Quaternary Science Reviews, 29*(27–8), 3677–3682.

Von Humboldt, A. (1849). *Aspects of nature, indifferent lands and different climates with scientific elucidations*. London: John Murray.

Wallace, A. R. (1895). *Island life or the phenomena and causes of insular faunas and floras*. (2nd ed.). London: Macmillan & Co.

West, R. G. (1968). *Pleistocene geology and biology with especial reference to the British Isles*. Harlow: Longmans.

Wolverton, S., Lyman, R. L., Kennedy, J. H., & La Point, T. W. (2009). The Terminal Pleistocene extinctions in North America, hypermorphic evolution, and the dynamic equilibrium model. *Journal of Ethnobiology, 29*(1), 28–63.

Zeuner, F. E. (1959). *The Pleistocene period: Its climate, chronology and faunal successions*. London: Hutchinson Scientific and Technical.

Zeuner, F. E. (1961). Faunal evidence for Pleistocene climates, *Annals of the New York Academy of Sciences, 95*, 502–507

Conference Organizers' Preface

This publication is one of the volumes of the proceedings of the 11th International Conference of the International Council for Archaeozoology (ICAZ), which was held in Paris (France) 23rd–28th August 2010. ICAZ was founded in the early 1970s and ever since has acted as the main international organization for the study of animal remains from archaeological sites. The International Conferences of ICAZ are held every four years, with the Paris meeting – the largest ever – following those in Hungary (Budapest), the Netherlands (Groningen), Poland (Szczecin), England (London), France (Bordeaux), USA (Washington, DC), Germany (Constance), Canada (Victoria), England (Durham) and Mexico (Mexico City). The next meeting was held in Argentina in 2014. The Paris conference – attended by some 720 delegates from 56 countries – was organized as one general and thirty thematic sessions, which attracted, in addition to archaeozoologists (zooarchaeologists), scholars from related disciplines such as bone chemistry, genetics, morphometry anthropology, archaeobotany, and mainstream archaeology. This conference was also marked by the involvement in the international archaeozoological community of increasing numbers of individuals from countries of Latin America and of South and East Asia.

As nearly 800 papers were presented at the Paris conference in the form of either oral or poster presentations, it was not possible to organize a comprehensive publication of the proceedings. It was left up to the session organizers to decide if the proceedings of their session would be published and to choose the form such a publication would take. A comprehensive list of publications of the 11th ICAZ International Conference is regularly updated and posted on the ICAZ web site (http://alexandriaarchive.org/icaz/pdf/ICAZ-Paris-publications.pdf).

The conference organizers take this opportunity to thank the Muséum national d'Histoire naturelle, the Université Pierre et Marie Curie, the Centre national de la Recherche scientifique and the ICAZ Executive Committee for their support during the organization of the conference, and all session organizers – some of them now being book editors – for all their hard work. The conference would not have met with such success without the help of the Alpha Visa Congrès Company, which was in charge of conference management. Further financial help came from the following sources: La Région Île-de-France, the Bioarch European network (French CNRS; Natural History Museum Brussels; Universities of Durham, Aberdeen, Basel and Munich), the LeCHE Marie Curie International Training Network (granted by the European Council), the Institute of Ecology and Environment of the CNRS, the Institut National de Recherche en Archéologie Préventive (INRAP), the European-Chinese Cooperation project (ERA-NET Co-Reach), the Centre National Interprofessionnel de l'Économie Laitière (CNIEL) and its Observatory for Food Habits (OCHA), the Ville de Paris, the Société des Amis du Muséum, the French Embassies in Beijing and Moscow, the laboratory of "Archaeozoology-Archaeobotany" (UMR7209, CNRS-MNHN), the School of Forensics of Lancaster, English Heritage and private donors.

Jean-Denis Vigne, Christine Lefèvre and Marylène Patou-Mathis
Organizers of the 11th ICAZ International Conference

Volume Editor's Preface

The publication of this volume has been severely delayed by unforeseen circumstances. Readers should not judge authors harshly for events that were beyond their control. All except the final chapter were completed by August 2012, although minor substantive updates have been made since then in several cases. The availability of the final chapter due to a change of authorship and unforeseen events has been the source of delay.

Many hands have touched this volume, and it is appropriate to thank them here. The contributors, first and foremost, have laboured long and hard to prepare their papers for publication here. Each has also acted as a referee for one other paper in the volume, and they have, in reading all papers for useful cross-references for their own papers, provided constructive criticism to their colleagues. I take this opportunity to thank them all for their patience, forbearance and hard work. Two external reviewers per chapter are also to be thanked most gratefully for their time, effort and expertise in offering timely and constructive criticism that improved each submission. Confidentiality prevents me from naming you, but the two dozen of you know who you are, and I say "thank you" most sincerely for your help. I thank Terry O'Connor for agreeing to write the Foreword to this volume, and I thank Daniel Sandweiss for stepping in and writing the final chapter. I also thank the VERT series editors, Eric Delson and Eric Sargis, for agreeing to take on this volume and for guiding me through the fine points of bringing it to publication readiness. I further thank Margaret Deignan at Springer Verlag for her help and encouragement with the publication of this volume. Finally, I thank my wife, Janet, for facilitating in so many ways my work on this volume.

Gregory G. Monks

Contents

1 **Introduction: Why This Volume?** 1
Gregory G. Monks

Part I The Pleistocene – Holocene Transition

2 **The Southern Levant During the Last Glacial and Zooarchaeological Evidence for the Effects of Climate-Forcing on Hominin Population Dynamics** 7
Miriam Belmaker

3 **Quaternary Mammals, People, and Climate Change: A View from Southern North America** 27
Ismael Ferrusquía-Villafranca, Joaquín Arroyo-Cabrales, Eileen Johnson, José Ruiz-González, Enrique Martínez-Hernández, Jorge Gama-Castro, Patricia de Anda-Hurtado and Oscar J. Polaco

4 **Holocene Large Mammal Extinctions in Palawan Island, Philippines** 69
Janine Ochoa and Philip J. Piper

5 **Human Response to Climate Change in the Northern Adriatic During the Late Pleistocene and Early Holocene** 87
Suzanne E. Pilaar Birch and Preston T. Miracle

Part II The Early – Mid-Holocene

6 **Early to Middle Holocene Climatic Change and the Use of Animal Resources by Highland Hunter-Gatherers of the South-Central Andes** 103
Hugo D. Yacobaccio, Marcelo Morales and Celeste Samec

7 **Climate Change at the Holocene Thermal Maximum and Its Impact on Wild Game Populations in South Scandinavia** 123
Ola Magnell

Part III The Recent Holocene

8 **Oxygen Isotope Seasonality Determinations of Marsh Clam Shells from Prehistoric Shell Middens in Nicaragua** 139
André C. Colonese, Ignacio Clemente, Ermengol Gassiot and José Antonio López-Sáez

9 **Climatic Changes and Hunter-Gatherer Populations: Archaeozoological Trends in Southern Patagonia** 153
Diego Rindel, Rafael Goñi, Juan Bautista Belardi and Tirso Bourlot

10	**Evidence of Changing Climate and Subsistence Strategies Among the Nuu-chah-nulth of Canada's West Coast** Gregory G. Monks	173
11	**Biometry and Climate Change in Norse Greenland: The Effect of Climate on the Size and Shape of Domestic Mammals** Julia E.M. Cussans	197

Part IV Overview and Retrospective

12	**Zooarchaeology in the 21st Century: Comments on the Contributions** Daniel H. Sandweiss	219

Index 227

Contributors

Joaquín Arroyo-Cabrales Laboratorio de Arqueozoología, Subdirección de Laboratorios y Apoyo Académico Instituto Nacional de Antropología e Historia, Centro Histórico, México, D. F., Mexico

Juan Bautista Belardi UNPA-CONICET, Río Gallegos, Santa Cruz, Argentina

Miriam Belmaker Department of Anthropology, Henry Kendall College of Arts and Sciences, Harwell Hall, 2nd Floor, University of Tulsa, Tulsa, OK, USA

Tirso Bourlot CONICET-INAPL, UBA, Capital Federal, Argentina

Ignacio Clemente Department of Archaeology and Anthropology (IMF-CSIC), AGREST (Generalitat de Catalunya), Barcelona, Spain

André C. Colonese BioArCh Department of Archaeology, University of York, York, UK

Julia E.M. Cussans Archaeological Solutions, Bury St. Edmunds, Suffolk, UK

Patricia de Anda-Hurtado Instituto de Geología, Universidad Nacional, Autónoma de México, Ciudad Universitaria, Coyoacan, Mexico, D. F., Mexico

Ismael Ferrusquía-Villafranca Instituto de Geología, Universidad Nacional, Autónoma de México, Ciudad Universitaria, Coyoacan, Mexico, D. F., Mexico

Jorge Gama-Castro Instituto de Geología, Universidad Nacional, Autónoma de México, Ciudad Universitaria, Coyoacan, Mexico, D. F., Mexico

Ermengol Gassiot Department of Prehistory, Universitat Autònoma de Barcelona Edifici B, Barcelona, Spain

Rafael Goñi CONICET-INAPL, UBA, Capital Federal, Argentina

Eileen Johnson Museum of Texas Tech University, Lubbock, TX, USA

José Antonio López-Sáez Archaeobiology Group (CCHS-CSIC), Madrid, Spain

Ola Magnell Statens Historiska Museer, Lund, Sweden

Enrique Martínez-Hernández Instituto de Geología, Universidad Nacional, Autónoma de México, Ciudad Universitaria, Coyoacan, Mexico, D. F., Mexico

Preston T. Miracle Division of Archaeology, Department of Archaeology and Anthropology, University of Cambridge, Cambridge, UK

Gregory G. Monks Department of Anthropology, University of Manitoba, Winnipeg, MB, Canada

Marcelo Morales CONICET–Instituto de Arqueologia, Universidad de Buenos Aires, Buenos Aires, Argentina

Janine Ochoa Department of Anthropology, University of the Philippines, Diliman, Quezon City, Philippines

Suzanne E. Pilaar Birch Department of Anthropology and Department of Geography, University of Georgia, GA, USA

Philip J. Piper School of Archaeology and Anthropology, Australian National University, Canberra, ACT, Australia; Archaeological Studies Program, Palma Hall, University of the Philippines, Diliman, Quezon City, Philippines

Oscar J. Polaco Laboratorio de Arqueozoología, Instituto Nacional de Antropología e Historia, Centro Histórico, México, D. F., Mexico

Diego Rindel CONICET-INAPL, UBA, Capital Federal, Argentina

José Ruiz-González Instituto de Geología, Universidad Nacional, Autónoma de México, Ciudad Universitaria, Coyoacan, Mexico, D. F., Mexico

Celeste Samec CONICET–Instituto de Arqueologia, Universidad de Buenos Aires, Buenos Aires, Argentina

Daniel H. Sandweiss Department of Anthropology, University of Maine, Orono, ME, USA

Hugo D. Yacobaccio CONICET–Instituto de Arqueologia, Universidad de Buenos Aires, Buenos Aires, Argentina

Chapter 1
Introduction: Why This Volume?

Gregory G. Monks

Abstract Concerns about climate change have arisen in proportion to the recognized effects of human interaction with the animate and inanimate components of the earth and its atmosphere. Those concerns are the foundational rationale for this volume. Climate change has occurred at many scales over millions of years, but humans are relatively recent newcomers. Throughout the tenure of *Homo*, a gradual shift has occurred in terms of the impact of humans on terrestrial and aquatic plant and animal communities and, more recently, on the atmosphere. Archaeological research provides a chronological framework for the interactions between humans, plants, animals, and the environment. The chapters in this volume focus on past relationships between humans and climate as seen in the bone remains of animals with which humans interacted. Through understanding these past relationships, viable responses to current climate change and environmental variability issues can be developed.

Keywords Climate change • Humans • Animals • Environment • Shifting baseline • Anthropogenic processes

> "And God said, 'Let the earth bring forth living creatures of every kind: cattle and creeping things and wild animals of the earth of every kind.' And it was so.
> God made the wild animals of the earth of every kind, and the cattle of every kind, and everything that creeps upon the ground of every kind. And God saw that it was good.
> Then God said, 'Let us make humankind in our image, according to our likeness; and let them have dominion over the fish of the sea, and over the birds of the air, and over the cattle, and over all the wild animals of the earth, and over every creeping thing that creeps upon the earth.'"
> Genesis 1:24–26

> "And every living substance was destroyed which was upon the face of the ground, both man, and cattle, and the creeping things, and the fowl of the heavens: and they were destroyed from the earth: and Noah only remained alive, and they that were with him in the ark."
> Genesis 7:23

These quotations may be familiar to some readers. Whether or not one subscribes to the belief systems from which they are derived, the quotations can easily be seen to speak to the ancient and intimate relationship that has existed between humans and animals and to the ways in which humans have coped through time with the changing environments in which they have lived. The quotations assert an *ab origine* relationship between climate and environment, animals and humans that is the *raison d'être* for this volume.

The Wider Intellectual Context

Concern over human impacts on the natural environment has been growing for some time (e.g., Muir 1897; Grey Owl (A. S. Belaney) 1931; Carson 1962), but increasing attention to the dynamic linkages between natural and cultural processes has lately taken a leading role in public discussion. Atmospheric degradation as it affects human health and global environmental stability, for example, occupies considerable amounts of more and less informed discussion, both pro and con. One of the key questions in the current discussion of global climate change rests on the extent to which trends in proxy measures such as arctic ice degradation or polar bear mortality result from natural, long-term cyclical and/or periodic events in the earth's geologic history or from anthropogenic processes such as fossil fuel combustion. This question of natural versus anthropogenic processes is a very important one, not just in the discussion of current issues but also as it affects our future. Like it or not, we are integral, and highly affective, parts of the global ecosystem. Perhaps a

G.G. Monks (✉)
Department of Anthropology, University of Manitoba,
15 Chancellors Circle, Winnipeg, MB R3T 5V5, Canada
e-mail: Gregory.Monks@umanitoba.ca

focal question might be put: how do we identify which environmental effects are natural and which are anthropogenic?

The importance of these relatively recent kinds of questions has led many natural scientists to search for proxy indicators of past climate and environment. Ice core and sea sediment isotope analyses, dendrochronological and palynological investigations, and paleontology, for example, have all furnished information at varying levels of resolution about the climates and environments that have existed in the more or less deep past. There are three things to consider here, though; the "shifting baseline" problem, the time scale in relation to the level of resolution, and, most importantly, the effects of humans. These problems are all related. For example, many studies, such as those in fisheries management, begin with the present situation as a baseline and attempt to make decisions based on situations already heavily impacted by humans. Resolution may be very fine-grained, but sufficient time depth is unavailable (see Francis and Hare 1994) and human impact is not factored out. The intractability of the problem in separating these effects with an adequate level of chronological resolution is further exemplified by the debate over the causes of Pleistocene megafaunal extinctions worldwide.

Archaeologists, and latterly zooarchaeologists, continue to wrestle with the problem of human agency in these megafaunal extinctions, but that particular issue is indicative of larger contributions that zooarchaeologists can make, e.g., FAUNMAP in the United States and INPN in France. One of these contributions is inherent in archaeology itself; namely, time depth. The other of these contributions is empirical evidence of human interaction with, and exploitation of, the natural environment. Taken together, these two contributions recommend the archaeological record as a valuable source of chronological control over human-environment interactions. Zooarchaeology is valuably positioned to help address issues such as the "shifting baseline", time depth, and chronological control. That this potential is already being realized is evident from, for example, the emerging perspective of historical ecology (Crumley 1994), from zooarchaeological contributions to wildlife management policy (Lyman 2006) from ecotype distinctions within ringed seal populations (Crockford and Frederick 2011) and from anthropological concerns with environmental change (Miller et al. 2011).

It is useful to consider, as do the papers in this volume, the responses of past populations to climatic and environmental stability and change, and equally the effects of human actions on stability and change in climate and environment. Both perspectives are valuable because elements of those responses and effects may provide useful analogues, e.g., for wildlife resource managers or coastal urban planners, in the present discussion of changing climate. This is not to suggest that one should uncritically adopt ideas about what did or did not work so well in the past and what might or might not work so well in the present or the future. Clearly, cultures today are much more complex than they were 5,000 or 10,000 years ago, and their effects on the global environment are much more profound. If, however, one takes a step back and looks at the structures of the strategies and the outcomes of what people of the past did when they faced or caused climatic and environmental change, lessons are there to be learned.

The Objective of This Volume

The present volume is intended to show how archaeology, and zooarchaeology in particular, can contribute to the current discussion of climate change by presenting selected studies on the ways in which past human groups responded to climatic and environmental change. In particular, it shows how these responses can be seen in the animal remains that people left behind in their occupation sites. Many of these bones represent food remains while others represent commensal animals, so the environments in which these animals lived can be identified and human use of those environments can be understood. In the case of climatic change resulting in environmental change, these animal remains can indicate that a change has occurred, in climate, environment and human adaptation, and can indicate also the specific details of those changes. Regardless of the genesis of the animal remains (food refuse, commensal animals), the animals from which they came and the chemical properties of their skeletal elements can reveal properties of the climate and environment in which those animals lived.

This volume also has a larger objective. It is intended to bring two things to the attention of natural historians (ecologists, climatologists, zoologists, botanists). The first is that humans affected past ecosystems, so that the assumptions of pristine conditions may not apply. Human effects on the environment must be considered. The second is that archaeology, especially zooarchaeology and archaeobotany, can provide information with time depth about the past relationships between people and their environments in the context of climatic change. Sea cores and lake sediments are very useful, but the interface between those identified conditions and the people that faced them is found in animal and plant remains in archaeological sites. The intent of this volume, then, is to place it in a publication venue where natural historians are likely to see the information that zooarchaeology can provide to them so that they can engage in a closer dialogue with archaeologists in general, recognizing that humans have been part of the earth's ecosystem for several million years. During that time, there have been changes in that global ecosystem, just as there appears to be at present, so understanding what has happened in the past may be useful for dealing with what is happening in the present and what may happen in the future.

This volume is intended to contribute to the long-standing practice of inter-disciplinary archaeological research and to the more recent trend in applied zooarchaeology (e.g., Lyman and Cannon 2004; Lyman 2006; Maschner et al. 2008; Miller et al. 2011). Whether one's research effort in these various disciplines is directed at molecular properties of individuals, physical and behavioural properties of individuals, single taxa, predator-prey relations, or entire ecosystems [here I follow from Lyman's (2011: 27) discussion of the "individualistic hypothesis"], zooarchaeology has the potential to shed light on the human and non-human past under chronologically controlled conditions. This capacity for extensive chronological control of human-environment interactions implies the value to be gained all around by increased attention to zooarchaeological research by other natural science disciplines, and vice versa. It is to be hoped this volume will foster both greater understanding of the strengths and perspectives of each area of study and increased interaction between them.

The Papers in This Volume

The papers in this volume examine the geochemical, taxonomic and morphological characterization of the effects of climatic and environmental change on animals exploited by past human populations and, consequently, on human survival strategies in the face of such changes. The temporal scope of the studies ranges from the Late Pleistocene to the Late Holocene, and the geographic scope encompasses the Old World, the New World and the western Pacific. The diversity within the temporal and spatial ranges of these papers is united by the focus on zooarchaeological remains and by the analytic themes that are applied across the ranges of time and space. Similarly, the contributing authors present a world-wide diversity of backgrounds and perspectives. There are authors from South America, Mesoamerica, North America, the Philippines, northern Europe, southern Europe and the Middle East. The fact that so many archaeozoologists who come from so many different parts of the world and who study topics of such geographic and temporal diversity find it important to discuss the issue of climatic and environmental change and the ways humans have responded to it suggests that zooarchaeology has something worthwhile to contribute to the issues that continue to affect us as a global population today.

This volume has the following structure. This introductory chapter (Chap. 1) outlines the rationale for, and the organization of, the volume and highlights the issues and important contributions of each paper. The final chapter (Sandweiss 2017) synthesizes the substantive, methodological and conceptual contributions that emerge from the chapters and points to the lessons that we can take into our contemporary situation from the insights provided in the volume. Between these bookend chapters, there are three groups of papers organized in rough chronological order so as to capture some of the responses to global climatic changes that zooarchaeologists working in different areas have been able to examine.

Part I, the Pleistocene-Holocene transition group, consists of four papers that cover the southern Levant, eastern Mediterranean, Mexico and the Philippines. Chapter 2, by Miriam Belmaker (2017), examines macro and micro fauna, including ungulate tooth wear, to evaluate the hypothesis that climate forcing played a role in the shift in hominin taxa in the southern Levant during MIS 6-3). Chapter 3, by Ferrusquía-Villafranca et al. (2017), compares the Pleistocene and Early Holocene mammalian taxa from northern Mexico the southwestern United States and reviews the evidence for both climatic and anthropogenic effects on these animals. Chapter 4, by Ochoa and Piper (2017), traces changes in vegetation and land area on northern Palawan Island, Philippines, at the end of the Pleistocene and during the Early Holocene and the natural and human impacts on selected major fauna of the two epochs. Pilaar Birch and Miracle (2017: Chap. 5) examine a number of questions about human choices in the face of changing sea levels, land areas and fauna in the Adriatic islands at the Terminal Pleistocene and Early Holocene.

Part II, the Early and Middle Holocene group, is addressed by two papers covering the puna of the southern Andes and southern Scandinavia. The exploitation of camelids in the puna of the southern Andes during climatic changes from the earliest to the Middle Holocene is explored by Hugo Yacobaccio (2017: Chap. 6). The second paper in this section shifts attention from southern to northern hemisphere and from the New World to the Old. In Chap. 7 (2017), Ola Magnell addresses the same time period, early to Mid-Holocene, in southern Scandinavia in terms of changing climate, the effects of climate change on wild game populations, and human hunting strategies.

Part III, the Recent Holocene group, consists of four papers covering coastal Nicaragua, montane Patagonia, the northwest coast of North America and Norse Greenland. Chapter 8 by Colonese et al. (2017) uses stable isotope analysis of bivalve shells to explore climate shifts that resulted in varying freshwater inputs to estuarine environments in Nicaragua and human exploitation of these ecozones. In Chap. 9, Rindel et al. (2017) evaluate the role of climatic change in relation to exploited camelid species in mid-altitude southern Patagonia. Chapter 10 by Monks

(2017) considers evidence for climatic change and altered subsistence strategies, especially those focused on fish, between the Medieval Climatic Anomaly (MCA) and the Little Ice Age (LIA) on the west coast of Canada. Norse experiences of shifting climate in Greenland on either side of the MCA-LIA transition is the subject of the paper by Cussans (2017: Chap. 11). The paper shows that the size of sheep in both western and eastern Greenland settlements decreases as climatic conditions deteriorate at the MCA-LIA transition.

Part IV, Overview and Retrospective, consists of a single paper (Chap. 12) (Sandweiss 2017). In it, he provides a synthetic overview of the substantive, methodological and theoretical contributions contained in the chapters in this volume. As well, he points to areas of fruitful interaction between natural scientists, wildlife managers and conservationists who have an interest in historical population processes and archaeologists who can provide deep temporal perspectives on animal populations.

Acknowledgments Thanks to Terry O'Connor and Virginia Butler for their helpful comments on an earlier draft of this chapter. Any errors or oversights are solely those of the author.

References

Belmaker, M. (2017). The southern Levant during the last glacial and zooarchaeological evidence for the effects of climatic-forcing on hominin population dynamics. In G. G. Monks (Ed.), *Climate change and human responses: A zooarchaeological perspective* (pp. 7–25). Dordrecht: Springer.

Carson, R. (1962). *Silent spring*. New York: Houghton Mifflin.

Colonese, A., Clemente, I., Gassiot, E., & López-Sáez, J. (2017). Oxygen isotope seasonality determinations of marsh clam shells from prehistoric shell middens in Nicaragua. In G. G. Monks (Ed.), *Climate change and human responses: A zooarchaeological perspective* (pp. 139–152). Dordrecht: Springer.

Crockford, S., & Frederick, S. G. (2011). Neoglacial sea ice and life history flexibility in ringed and fur seals. In T. Braje & T. C. Rick (Eds.), *Human impacts on seals, sea lions, and sea otters* (pp. 65–91). Berkeley and Los Angeles: University of California Press.

Crumley, C. (Ed.). (1994). *Historical ecology: Cultural knowledge and changing landscapes*. Santa Fe: School of American Research Press.

Cussans, J. (2017). Biometry and climate change in Norse Greenland: The effect of climate on the size and shape of domestic mammals. In G. G. Monks (Ed.), *Climate change and human responses: A zooarchaeological perspective* (pp. 197–216). Dordrecht: Springer.

Ferrusquía-Villafranca, I., Arroyo-Cabrales, J., Johnson, E., Ruiz-González, J., Martínez-Hernández, E., Gama-Castro, J., et al. (2017). Quaternary mammals, people, and climate change: A view from southern North America. In G. G. Monks (Ed.), *Climate change and human responses: A zooarchaeological perspective* (pp. 27–67). Dordrecht: Springer.

Francis, R. C., & Hare, S. R. (1994). Decadal-scale regime shifts in the large marine ecosystems of the north-east Pacific: A case for historical science. *Fisheries Oceanography, 3*(4), 279–291.

Grey Owl (Belaney, A. S.). (1931). *The men of the last frontier*. London: Country Life.

Lyman, R. (2006). Paleozoology in the service of conservation biology. *Evolutionary Anthropology, 15*, 11–19.

Lyman, R. (2011). A history of paleoecological researh on sea otters, and pinnipeds of the eastern Pacific rim. In T. Braje & T. C. Rick (Eds.), *Human impacts on seals, sea lions and sea otters* (pp. 19–40). Berkeley and Los Angeles: University of California Press.

Lyman, R. L., & Cannon, K. P. (Eds.). (2004). *Zoology and conservation*. Salt Lake City: University of Utah Press.

Magnell, O. (2017). Climate at the Holocene thermal maximum and its impact on wild game populations in south Scandinavia. In G. G. Monks (Ed.), *Climate change and human responses: A zooarchaeological perspective* (pp. 123–135). Dordrecht: Springer.

Maschner, H., Betts, M., Reedy-Maschner, K., & Trites, A. (2008). A 4500-year time series of Pacific Cod (*Gadus macrocephalus*): Archaeology, regime shifts, and sustainable fisheries. *Fishery Bulletin, 106*(4), 386–394.

Miller, N., Moore, K. M., & Ryan, K. (Eds.). (2011). *Sustainable lifeways: Cultural persistence in an ever-changing environment*. Philadelphia: University of Pennsylvania Museum of Archaeology and Anthropology.

Monks, G. G. (2017). Evidence for changing climate and subsistence strategies among the Nuu-chah-nulth on Canada's west coast. In G. G. Monks (Ed.), *Climate change and human responses: A zooarchaeological perspective* (pp. 173–196). Dordrecht: Springer.

Muir, J. (1897). The American forests. *Atlantic Monthly, 80*(478), 145–157.

Ochoa, J., & Piper, P. (2017). Holocene large mammal extinctions in Palawan Island, Philippines. In G. G. Monks (Ed.), *Climate change and human responses: A zooarchaeological perspective* (pp. 69–86). Dordrecht: Springer.

Pilaar Birch, S., & Miracle, P. T. (2017). Human response to climate change in the northern Adriatic during the Late Pleistocene and Early Holocene. In G. G. Monks (Ed.), *Climate change and human responses: A zooarchaeological perspective* (pp. 87–100). Dordrecht: Springer.

Rindel, D., Goñi, R., Belardi, J. B., & Bourlot, T. (2017). Climatic changes and hunter-gatherer populations: Archaeozoological trends in southern Patagonia. In G. G. Monks (Ed.), *Climate change and human responses: A zooarchaeological perspective* (pp. 153–172). Dordrecht: Springer.

Sandweiss, D. (2017). Commentary. In G. G. Monks (Ed.), *Climate change and human responses: A zooarchaeological perspective* (pp. 219–225). Dordrecht: Springer.

Yacobaccio, H. D., Morales, M., & Samec, C. (2017). Early to Middle Holocene climatic change and the use of animal resources by highland hunter-gatherers of the south-central Andes. In G. G. Monks (Ed.), *Climate change and human responses: A zooarchaeological perspective* (pp. 103–121). Dordrecht: Springer.

Part I
The Pleistocene – Holocene Transition

Chapter 2
The Southern Levant During the Last Glacial and Zooarchaeological Evidence for the Effects of Climate-Forcing on Hominin Population Dynamics

Miriam Belmaker

Abstract Climate forcing has been suggested as a possible explanation for dispersal/extinction of hominins in the Southern Levant during the Middle Paleolithic (MP). Evidence from fauna has produced ambiguous results, suggesting that inter-site variation in Last Glacial faunas reflect spatial differences within the region. This study presents a multivariate approach to test the effect of climate change on mammalian communities during the Last Glacial in the Levant and analyzes the distribution of micro and macromammals from the site in the Levant spanning Marine Isotope Stage (MIS) 6-2 using non-metric multidimensional scaling (NMMDS). Results indicate that inter-site differences in faunal composition of Middle Paleolithic sites in the Levant do not reflect an abrupt climate change but are consistent with a spatial environmental mosaic within the Levant. This suggests that although hominin taxa show evidence of turnover during the Late Pleistocene in the Levant, we need to be more cautious about the role of climate forcing in the process.

Keywords Neanderthal extinction • Faunal response to climate change • Mammal community structure • NMMDS

Introduction

The past years have seen a growing interest in the role of climate in European Neanderthal population dynamics (Gamble et al. 2004; Stewart 2005; Finlayson et al. 2006; Finlayson and Carrion 2007; Tzedakis et al. 2007). Several hypotheses have been suggested for the extinction of the Neanderthals in Europe. Primarily, it has been suggested that modern humans outcompeted the Neanderthals when they arrived in the Europe around 40–45 ka (Bar-Yosef 2000; Kuhn et al. 2004; Mellars 2004, 2006). In contrast, it has been suggested that an increase in climatic fluctuations during the Last Glacial may have led to a fragmentation of Neanderthal habitats, leading to a decrease in the effective population size and finally to their extinction (Finlayson 2004; Finlayson et al. 2006; Finlayson and Carrion 2007). While this hypothesis has gained traction based on a wide range of paleoecological evidence in Europe, which points to an increase in amplitude, frequency and variability of climate change towards the end of the Last Glacial (Guiot et al. 1993; Gamble et al. 2004; Miracle et al. 2009), the paleoecological situation in the Southern Levant is more complex and suggests that a "one answer fits all" solution may not be applicable to the question of the effect of climate on large hominins in general and the Neanderthals in particular.

The Southern Levant is located in mid-latitudes and has a more temperate climate than Europe and more moderate climatic fluctuations (Enzel et al. 2008). It therefore provides us with a unique opportunity to study the possible effect of climate change during the Last Glacial on the Levantine Neanderthal population and their extinction from the region around 45 ka. Within this context it is of interest to test how environmental and climatic changes played out in this local arena in relation to the Levantine populations of Neanderthals.

Tchernov (1992) raised the hypothesis of the relationship between climate change and hominin taxa in the Levant. He suggested that faunal turnover of rodent taxa between MIS 5 and 4 and between MIS 4 and 3 was concurrent with observed shifts in hominin species. The shift between Anatomically Modern Humans (AMH) and Neanderthals (Valladas et al. 1987, 1999; Schwarcz et al. 1989; Valladas and Joron 1989; Solecki and Solecki 1993; Rink et al. 2003) coincided with a shift from a Saharo-Arabian rodent community to that dominated by a more Euro-Siberian rodent community. This shift is dated to the MIS 5/4 transition. The shift between Neanderthals and modern humans (Millard

M. Belmaker (✉)
Department of Anthropology, Henry Kendall College of Arts and Sciences, Harwell Hall, University of Tulsa, 800 S. Tucker Dr., Tulsa, OK 74104, USA
e-mail: miriam-belmaker@utulsa.edu

2008) coincided with a shift from a Euro-Siberian rodent community back to a Saharo-Arabian one, dated to the MIS 4/3 transition. This hypothesis was based on faunal data from two main sites: Qafzeh, dated to 100–90 ka, and Kebara, dated to ca. 65 ka. However, these two sites are located in two distinct regions in the Southern Levant, Qafzeh being further east than Kebara, which is located in Mount Carmel on the Coast. In contrast to this hypothesis, Jelinek suggested that the observed differences between the micromammal communities might result from the east-west differences in precipitation and an ecotone gradient in vegetation and microhabitat climate rather than temporal differences (Jelinek 1982).

Since the original development of these hypotheses, Shea (2008) proposed that the human population turnover in the Southern Levant was climatically driven. This was based on a wide array of paleoclimate proxies such as speleothem isotope data, ocean foraminifera, and pollen cores, which pointed to a paleoecological pattern of aridification throughout MIS 4, culminating in a cold and dry period known as the Heinrich 5 (H5) event. This temporal paleoclimatic pattern was interpreted as a regional decrease in environmental productivity, which led to the demise of the local Levantine Neanderthals similar to the model proposed for Europe (Shea 2008).

The aim of this paper is to reevaluate the evidence for an environmental shift, which occurred between Marine Oxygen Isotope Stage (MIS) 4 and 3 and which may have contributed to the extinction of the Levantine Neanderthals. This analysis will use data derived from Southern Levantine assemblages of small and large mammals dated to the Last Glacial. It is hypothesized that if the regional environmental mosaic is the driving force that controls the variance in fauna among sites during the Last Glacial, it will reveal a spatial patterning according to geographic provenance related to rainfall and/or annual temperature (Danin and Orshan 1990). On the other hand, if global climate change is the trigger for the change in faunal communities in the Southern Levant, we will see a greater similarity among sites of similar age than of similar region.

The Levant

The Levant is a unique biogeographic entity within Southwestern Asia. It lies at the crossroads of Africa and Eurasia and has more lush environments when compared to the alternative dispersal corridor on the southern fringes of the Arabian Peninsula (Thomas 1985; Tchernov 1988). The southern path through the margins of the Arabian Peninsula could be crossed at the Bab el-Mandeb straits during low sea level stands, and along the area that receives the summer Indian Ocean monsoon to be followed by a passageway through the Hormuz straits into the southern coast of Iran; a northern path could have been taken via the Nile river and the Sinai Peninsula.

The East Mediterranean region is located between the more temperate European climatic zone in the north and the hyper arid regions of the Saharo-Arabian desert belt in the south (Frumkin and Stein 2004). Southwestern Asia includes fauna from three biogeographic provinces: Palaearctic, Oriental and Ethiopian in different proportions depending on the environmental condition in each phytogeographic region (Harrison and Bates 1991). Many animal taxa are similar over two or more provinces, giving the entire Near East a coherent faunal community (Harrison and Bates 1991). Two main regions can be observed: the first includes the Mesic Mediterranean, Pontic and Iranian plateau provinces with Palaeartic taxa, and the second includes the Xeric Mediterranean and Arabian provinces with Ethiopian elements.

Methods

Data from herbivores from two groups of taxa differing in size, namely medium-sized herbivores and micromammals, were analyzed. Micromammals include taxa smaller than one kilogram (kg) (e.g., Andrews 1990). Confining the analysis to taxa of a similar size range (in the broad sense), and a community of trophically similar and sympatric species (Hubbell 2001), reduces the effect of sampling of rare species (such as primates and carnivores) as well as collection bias of smaller and larger taxa. The use of individual abundance data, i.e., Number of Identified Specimens or NISP, is particularly problematic in fossil analyses as it is mostly driven by taphonomic rather than paleoenvironmental factors (Behrensmeyer et al. 2000).

Only data for presence absence was used to maintain high ecological fidelity. Using relative abundance (i.e., those for which we could can we observe a change in the relative abundance of a species which can be related to climatic change) rests on our ability to remove one or more taphonomic biases, which may erroneously produce the appearance of change where one may not have existed. For example, given two assemblages, if one was accumulated by carnivores and the other not, the selectivity of prey by carnivores may increase the proportion of specific species in the fossil (death) assemblage compared to an assemblage accumulated by other factors. This would create the impression that an increase in species proportion had occurred between strata, when in actuality it did not.

Analysis of taphonomic factors and live-death comparisons has shown that presence-absence preserves the strongest fidelity between the living community and the death assemblage, followed by rank abundance (Kidwell and Flessa 1995, 1996; Roy et al. 1996; Behrensmeyer et al. 2000; Rogers and Kidwell 2000; Kidwell 2001, 2002, 2008; Kidwell et al. 2001; Kidwell and Holland 2002; Zohar et al. 2008; Tomasovych and Kidwell 2009; Terry 2010; Belmaker and Hovers 2011). Evidence for the fidelity of relative abundance for vertebrates such as micromammals (Terry 2010) is usually limited to recent death accumulation with time and space averaging of up to several hundreds rather than thousands of years.

Macromammal and micromammal data from archeological sites were retrieved from the literature. The macromammal sites, date and the reference from which they were retrieved are presented in Table 2.1 and the micromammal sites, date and the reference from which they were retrieved are presented in Table 2.2. Location of sites mentioned in the text are shown in Fig. 2.1a (Middle Paleolithic) and b (Upper Paleolithic). The strength of meta-analysis, such as this, which relies on data retrieved from the literature, is in the ability to amass large dataset, which provides robust statistical results. However, when using published literature, the difficulty is in controlling for similar data acquisition. Since method of excavation highly affects both species richness and abundance, we can expect fauna richness to differ among sites with different collection protocols. Total number of NISP from the sites ranged from 2 to over 8000 for macromammals and from 79 to over 28,000 for the micromammals.

In order to correct for sites that were under-sampled, only sites with five or more species recovered at the site were included. A minimum of five species per site was chosen as a cut off number as it was the median number of species per site in the assemblages. This procedure was adopted to avoid the problem of under-sampling due either to small sample size or to high selectivity, both of which would not represent the environment adequately. While it is possible that some sites with low number of species were not entirely representative of the environment, the main premise of this study was similarity or dissimilarity among site that was achieved by maintaining similar sampling strategies among sites while maintaining a large enough sample size sites over all.

Macromammal species used in the analysis included *Hippopotamus amphibius, Coelodonta antiquitatis, Bos primigenius, Cervus elaphus, Dama mesopotamica, Capreolus capreolus, Sus scrofa, Alcelaphus buselaphus, Equus asinus, Equus* cf. *mauritanicus, E. hemiones* and *E. hydruntinus*. Other species were identified at the genus level due to similarities between congeneric taxa; identification to the species level was not always possible when two species may be present at the same site. These included *Ovis/Capra/Ibex, Gazella* sp., *Equus* sp., and *Camelus* sp. Micromammal taxa included in this study were: *Suncus etruscus, Suncus murinus, Crocidura russula, Crocidura leucodon, Talpa chthonia (=T. davidiana), Sciurus anomalus, Myomimus qafzensis, Myomimus roachi (=Myomimus personatus), Allocricetulus magnus, Mesocricetus auratus, Cricetulus migratorius, Gerbillus dasyurus, Meriones tristrami, Spalax ehrenbergi, Ellobius fuscocapillus, Microtus guentheri, Arvicanthis ectos, Mastomys batei, Mus macedonicus, Apodemus mystacinus, Apodemus sylvaticus, Apodemus flavicollis, Arvicola terrestris, Acomys cahirinus, Rattus rattus*. Several taxa were only analyzed at the genus level due to several sympatric congeners and included: *Dryomys* sp., *Psammomys* sp., *Eliomys* sp., *Allactaga* sp., and *Jaculus* sp.

Data were derived from sites spanning MIS 6-2 throughout Southern Levant. Data were analyzed by two different independent variables. The first was *environment* and the second was *period*. Five environmental categories were chosen based on precipitation level (Danin and Orshan 1990) and are presented in Table 2.3.

In order to assign the variable *Period* to sites, I relied on cultural period attribution derived from the literature. The chronology of the Middle and Upper Paleolithic of the Levant has been based on the succession of the Levantine Mousterian lithic assemblages based on the three phase model proposed by Copeland and modeled after the three major phased in Tabun Cave: Tabun D, C and B. Subsequently named Early Middle Mousterian, Middle Middle Mousterian and Later Middle Mouserian (see Shea 2003 for details). Radiometric dates using ^{14}C, TL, ESR and U series that were applied to sites were able to confirm the basic model. These three stages also corresponded to MIS 7-6, MIS 4 and MIS 4/3 respectively (Wallace and Shea 2006 and references therein). Since many sites with faunal remains were assigned to the different lithic traditions but were not radiometrically dated, we used these traditions as markers of chronology, albeit relative, because it allowed a larger subset of sites and has been proven to span the geographic range of the southern Levant in this paper (Bar-Yosef 1992). Table 2.4 presents the list of cultural entities used and the estimate absolute dates assigned to them. Since we wanted to include several younger sites from the Upper Paleolithic, we continued with the scheme as well for the younger sites.

To analyze the data, non-metric multidimensional scaling (NMMDS) was used, which is a non-parametric version of principal coordinates analysis (PCA). PCA is used on parametric variables only, while NMMDS is based on a similarity/distance matrix, thus non-parametric. It attempts to approximate the ranks of the dissimilarities between sites based on species occurrences. NMMDS creates a configuration of points whose inter-point distances approximate a monotonic transformation of the original dissimilarities. The algorithm then attempts to place the data points in a two- or three-dimensional coordinate system such that the ranked differences are preserved. For example, if the original distance between points three and eight is the tenth largest of all distances between any two points, points three and eight will be

Table 2.1 List of sites used in macromammal analysis

#	Site	Culture	Environment	Reference for fauna
1	Ein Difla	Tabun D	Arid	Lindley and Clark (1987) in Shea (2003)
2	Bezez Cave B	Tabun D	Hyper Mesic Mediterranean	Garrard (1982) in Shea (2003)
3	Hayonim E > 405 bd	Tabun D	Mesic Mediterranean	Stiner and Tchernov (1998) in Shea (2003)
4	Tabun D	Tabun D	Mesic Mediterranean	Bar-Yosef (1989) in Shea (2003)
5	Rosh Ein Mor	Tabun D	Semi Arid	Tchernov (1986) in Shea (2003)
6	Ksar Akil 18-10	Tabun C	Hyper Mesic Mediterranean	Garrard (1980) in Rabinovich (2003)
7	Naamé	Tabun C	Hyper Mesic Mediterranean	Fleisch (1970) in Shea (2003)
8	Ras el Kelb A-O	Tabun C	Hyper Mesic Mediterranean	Garrard (1982) in Shea (2003)
9	Tabun C	Tabun C	Mesic Mediterranean	Bar-Yosef (1989) in Shea (2003)
10	Hayonim E < 405 bd	Tabun C	Mesic Mediterranean	Stiner and Tchernov (1998) in Shea (2003)
11	Qafzeh V–XV	Tabun C	Xeric Mediterranean	Rabinovich and Tchernov (1995) in Shea (2003)
12	Qafzeh XVI–XXIV	Tabun C	Xeric Mediterranean	Rabinovich and Tchernov (1995) in Shea (2003)
13	Dederiyeh 11	Tabun C-B	Semi Arid	Griggo (1998) in Shea (2003)
14	Dederiyeh 3	Tabun C-B	Semi Arid	Griggo (1998) in Shea (2003)
15	Douara III	Tabun B	Arid	Payne (1983) in Shea (2003)
16	Douara IV	Tabun B	Arid	Payne (1983) in Shea (2003)
17	Amud B	Tabun B	Mesic Mediterranean	Rabinovich and Hovers (2004)
18	Skhul B	Tabun B	Mesic Mediterranean	Bar-Yosef (1989) in Shea (2003)
19	Tabun B	Tabun B	Mesic Mediterranean	Bar-Yosef (1989) in Shea (2003)
20	Kebara F	Tabun B	Mesic Mediterranean	Davis (1982) in Shea (2003)
21	Umm el Tlel V2βa	Tabun B	Semi Arid	Griggo (1998) in Shea (2003)
22	Umm el Tlel VIIa	Tabun B	Semi Arid	Griggo (1998) in Shea (2003)
23	Umm el Tlel VI3 b'1	Tabun B	Semi Arid	Griggo (1998) in Shea (2003)
24	Far'ah II L.1	Tabun B	Semi Arid	Gilead and Grigson (1984) in Shea (2003)
25	Far'ah II L.2	Tabun B	Semi Arid	Gilead and Grigson (1984) in Shea (2003)
26	Abu Noshra I	MP-UP transition	Arid	Phillips (1988) in Rabinovich (2003)
27	Abu Noshra II	MP-UP transition	Arid	Philips (1988) in Rabinovich (2003)
28	Abu Halka IVd	MP-UP transition	Hyper Mesic Mediterranean	Garrard (1980) in Rabinovich (2003)
29	Antelias	MP-UP transition	Hyper Mesic Mediterranean	Garrard (1980) in Rabinovich (2003)
30	Abu Halka IVf	MP-UP transition	Hyper Mesic Mediterranean	Garrard (1980) in Rabinovich (2003)
31	Abu Halka IVe	MP-UP transition	Hyper Mesic Mediterranean	Garrard (1980) in Rabinovich (2003)
32	Ksar Akil 6-9	MP-UP transition	Hyper Mesic Mediterranean	Garrard (1980) in Rabinovich (2003)
33	Abu Halka IVc	MP-UP transition	Hyper Mesic Mediterranean	Garrard (1980) in Rabinovich (2003)
34	Ain Aqev (D31)	MP-UP transition	Semi Arid	Tchernov (1976) in Rabinovich (2003)
35	Shukbah D	MP-UP transition	Xeric Mediterranean	Garrod and Bate (1942) in Shea (2003)
36	El Wad E	Early Ahmarian	Mesic Mediterranean	Garrard (1980) in Rabinovich (2003)
37	Kebara III	Early Ahmarian	Mesic Mediterranean	Garrard (1980) in Rabinovich (2003)
38	Emireh	Early Ahmarian	Mesic Mediterranean	Bate (1927) in Rabinovich (2003)
39	El Wad F	Early Ahmarian	Mesic Mediterranean	Garrard (1980) in Rabinovich (2003)
40	Erq el Ahmar B	Early Ahmarian	Semi Arid	Vaufrey (1951) in Rabinovich (2003)
41	Erq el Ahmar C	Early Ahmarian	Semi Arid	Vaufrey (1951) in Rabinovich (2003)
42	Erq el Ahmar D	Early Ahmarian	Semi Arid	Vaufrey (1951) in Rabinovich (2003)
43	Erq el Ahmar E	Early Ahmarian	Semi Arid	Vaufrey (1951) in Rabinovich (2003)
44	Erq el Ahmar F	Early Ahmarian	Semi Arid	Vaufrey (1951) in Rabinovich (2003)
45	Qafzeh	Early Ahmarian	Xeric Mediterranean	Rabinovich (1998) in Rabinovich (2003)
46	Hayonim D3	Levantine Aurignacian	Mesic Mediterranean	Rabinovich 1998 in Rabinovich (2003)
47	Hayonim D1/2	Levantine Aurignacian	Mesic Mediterranean	Rabinovich (1998) in Rabinovich (2003)
48	Rakefet B-G 18-23 XVI	Levantine Aurignacian	Mesic Mediterranean	Garrard (1980) in Rabinovich (2003)
49	Rakefet B-G 18-23 XV	Levantine Aurignacian	Mesic Mediterranean	Garrard (1980) in Rabinovich (2003)
50	Hayonim D4	Levantine Aurignacian	Mesic Mediterranean	Rabinovich (1998) in Rabinovich (2003)
51	Sefunim 9-11	Levantine Aurignacian	Mesic Mediterranean	Tchernov (1984) in Rabinovich (2003)
52	Rakefet B-G 18-23 XIV	Levantine Aurignacian	Mesic Mediterranean	Garrard (1980) in Rabinovich (2003)

(continued)

Table 2.1 (continued)

#	Site	Culture	Environment	Reference for fauna
53	El Wad D	Levantine Aurignacian	Mesic Mediterranean	Garrard (1980) in Rabinovich (2003)
54	Yabrud II	Levantine Aurignacian	Semi Arid	Lehmann (1970) in Rabinovich (2003)
55	El Quesir D	Levantine Aurignacian	Semi Arid	Perrot (1955) in Rabinovich (2003)
56	El Quesir C	Levantine Aurignacian	Semi Arid	Perrot (1955) in Rabinovich (2003)
57	El Wad C	Atlitian	Mesic Mediterranean	Garrard (1980) in Rabinovich (2003)
58	Nahal Ein Gev I	Atlitian	Xeric Mediterranean	Davis (1982) in Rabinovich (2003)
59	Tor Hamar G	Nebekian	Arid	Klein (1995) in Rabinovich (2003)
60	Uwaynid 14	Nebekian	Arid	Garrard et al. (1988) in Rabinovich (2003)
61	Tor Hamar F	Nebekian	Arid	Klein (1995) in Rabinovich (2003)
62	Uwaynid 18	Nebekian	Arid	Garrard et al. (1988) in Rabinovich (2003)
63	Jilat 9	Nebekian	Arid	Garrard et al. (1988) in Rabinovich (2003)
64	Yabrud III	Nebekian	Semi Arid	Lehmann (1970) in Rabinovich (2003)
65	Fazael X	UP-Early Epipaleolithic	Arid	Davis (1982) in Rabinovich (2003)
66	Fazael XI	UP-Early Epipaleolithic	Arid	Davis (1982) in Rabinovich (2003)
67	Masraq e-Naj	UP-Early Epipaleolithic	Arid	Perrot (1955) in Rabinovich (2003)
68	El-Bezez A	UP-Early Epipaleolithic	Arid	Garrard (1980) in Rabinovich (2003)
69	WHS 784	UP-Early Epipaleolithic	Arid	Clark et al. (2000) in Rabinovich (2003)
70	Fazael IX	UP-Early Epipaleolithic	Arid	Davis (1982) in Rabinovich (2003)
71	WHS 618	UP-Early Epipaleolithic	Arid	Clark et al. (2000) in Rabinovich (2003)
72	Ksar Akil 25-19	UP-Early Epipaleolithic	Hyper Mesic Mediterranean	Garrard (1980) in Rabinovich (2003)
73	Ksar Akil 9-6	UP-Early Epipaleolithic	Hyper Mesic Mediterranean	Garrard (1980) in Rabinovich (2003)
74	Ohalo II	UP-Early Epipaleolithic	Xeric Mediterranean	Rabinovich (1998) in Rabinovich (2003)

Table 2.2 List of sites used in micromammal analysis

#	Site	Culture	Environment	References
1	Hayonim (Lower E)	Tabun D	Mesic Mediterranean	Tchernov (1998)
2	Tabun D	Tabun D	Mesic Mediterranean	Tchernov (1994)
3	Hayonim (Upper E)	Tabun C	Mesic Mediterranean	Tchernov (1994)
4	Tabun C	Tabun C	Mesic Mediterranean	Tchernov (1994)
5	Qafzeh (XVII–XXIII)	Tabun C	Xeric Mediterranean	Tchernov (1998)
6	Tabun B	Tabun B	Mesic Mediterranean	Tchernov (1994)
7	Kebara VI–XII	Tabun B	Mesic Mediterranean	Tchernov (1998)
8	Geula	Tabun B	Mesic Mediterranean	Heller (1970)
9	Amud	Tabun B	Mesic Mediterranean	Belmaker and Hovers (2011)
10	Qafzeh 2-9	UP	Xeric Mediterranean	Tchernov (1994)
11	Douara	Tabun B	Arid	Payne (1983)
12	Ksar Akil XXXIII–XXVIA	Tabun B	Hyper Mesic Mediterranean	Kersten (1992)
13	Kebara UP	UP	Mesic Mediterranean	Tchernov (1994)
14	Rakefet UP	UP	Mesic Mediterranean	Tchernov in Ronen (1984)
15	Rakefet Natufian	Natufian	Mesic Mediterranean	Nadel et al. (2008)
16	El Wad Natufian	Natufian	Mesic Mediterranean	Weissbrod et al. (2005)
17	Sefunim	UP	Mesic Mediterranean	Tchernov in Ronen (1984)
18	Ohalo II	UP	Xeric Mediterranean	Belmaker et al. (2001)

placed such that their Euclidean distance in the 2D plane or 3D space is still the tenth largest. It is important to note that NMMDS does not take absolute distances into account. Thus, large inter-point distances correspond to large dissimilarities, and small inter-point distances to small dissimilarities.

The degree of correspondence between the distance among the points implied by the NMMDS map and the raw data matrix is measured by the Kruskal statistics of *stress*. Stress is a measure of how well the solution recreates the dissimilarities. Smaller values indicate a better fit and are defined as $\sqrt{\dfrac{\sum_{i,j}(d_{ij}-x_{ij})^2}{\sum_{i,j}d_{ij}^2}}$ where d_{ij} is the association between i and j as measured by the similarity index and x_{ij} is the associations between i and j as predicted using distances on the Shepard's plot (i.e., by the regression). A stress value between 0.1 and

Table 2.3 Environmental categories and corresponding climatic parameters (After Danin and Orshan 1990)

Environment category	Mean annual precipitation (mm)	Lang's rain factor[a]	Emberger pluviothermic quotient[b]
Arid	<50	<7	<10
Semi-arid	50–300	13–9	40–10
Xeric Mediterranean	300–540	28–15	90–40
Mesic Mediterranean	540–780	29–37	122–90
Hyper Mesic Mediterranean	780–1200	>61	>141

[a]Lang's rain factor is calculated as $L = \frac{P}{T}$ where P is the mean annual precipitation and T is the mean annual temperature

[b]Emberger's pluviothermic quotient is calculated as $E = \frac{P \times 100}{(T_{mx} + T_{Mn}) + (T_{mx} - T_{Mn})}$ where P is the mean annual precipitation and T_{mx} is the average mean temperature of the warmest month and T_{mn} is the average mean temperature of the coldest month

Table 2.4 Culture categories

Time period	MIS	Culture	Absolute dates ka
Early Middle Paleolithic	MIS 6	Tabun D	250–130
Middle Middle Paleolithic	MIS 5	Tabun C	128–75
		Tabun C-B	
Late Middle Paleolithic	MIS 4/early MIS 3	Tabun B	75–47/45
Early Upper Paleolithic	MIS 3	MP-UP transition (Emiran)	47/45–40/38
	MIS 3	Levantine Aurignacian	36–38
	MIS 3	Early Ahmarian	38–25
Late Upper Paleolithic	MIS 3	Atlitian	25–20
	MIS 3	Nebekian	22–20
	MIS 2	UP-Early Epipaleolithic	20–15
Holocene	MIS 1	Natufian	12.5–10.2

0.15 is considered "good" and below 0.1 "excellent". Any value above 0.15 is considered unacceptable.

Different indices were used for the different groups of taxa (macromammals and micromammals) following Hausdorf and Hennig (2003). For micromammals the Kulczynski similarity index for binary data was used; this index is more appropriate for taxa that have home ranges of unequal sizes as it would be expected from smaller mammals that may have endemic populations. The index for similarity between sites j and k is calculated as $d_{jk} = \frac{\frac{M}{M+N_j} + \frac{M}{M+N_k}}{M+N}$, where M is the number of matches between site j and k and N_j is the number of species unique to site j and N_k is the number of species unique to site K.

For macromammals the Jaccard similarity index was used; this index is more appropriate for taxa that have home ranges of unequal size. The index is for similarity between sites j and k and is calculated as $d_{jk} = \frac{M}{M+N}$ where M is the number of matches between site j and k and N is the number of species unique to either site j or k.

Differences in the NMMDS scores between habitat and periods were analyzed using two-way ANOVA and post hoc Bonferroni corrections for multiple comparisons.

To visualize the results in a form that more closely resembles a PCA result, a PCA on the raw NMMDS results (as these are distributed normally) was applied. Doing so assured that the axes were uncorrelated with one another and therefore allowed for a more robust interpretation.

Statistical analyses used the statistical programs PAST 2.2, SPSS 18.0 and Aable for the Mac.

Results
Macromammals

Results from the NMMDS for the *environment* variable indicate that good representation was obtained (Kruskal's stress = 0.1486). Axes 1, 2 and 3 of the NMMDS account for 93.36% of the variance (R^2 = 0.429, 0.2445 and 0.2662 respectively; since NMMDS axis numbers are arbitrary, the percent of variance represented by the R^2 does not decrease with the increasing axis number).

NMMDS scores on Axis 1 (42.9% of the variance) and Axis 2 (24.45% of the variance) were significantly different between *period* categories ($F_{9,67}$ = 3.065, P = 0.005 and $F_{9,73}$ = 3.516, P = 0.001), while differences in scores between *period* categories on Axis 3, which accounts for 26.62% of the variance, were not significant ($F_{9,73}$ = 2.5, P = 0.016). NMMDS scores on Axis 1 (42.9% of the variance) and Axis 2 (24.45% of the variance) were highly significantly different between *environment* categories ($F_{5,73}$ = 11.4643, P < 0.001 and $F_{5,73}$ = 12.2599, P < 0.001), while differences in scores between *period* categories on Axis 3, which accounts for 26.62% of the variance were not significant ($F_{5,73}$ = 1.015,

$P = 0.416$). The effect of *period*environment* was not significant for Axis 1 (P value = 0.033) and 3 (P value = 0.936), but was significant for Axis 2 (P value < 0.001).

Bonferroni post hoc analysis for the variable *period* along the first axis shows that two of the pair-wise comparisons were significant: Tabun B vs. the Levantine Aurignacian (P value < 0.001) and Tabun B vs. the UP-Early Epipaleolithic transition (P value < 0.001). Along the second axis, pair-wise comparisons were significant: Tabun B vs. Nebekian (P value < 0.001) Tabun C vs. Nebekian (P value < 0.001), Tabun D vs. Nebekian (P value < 0.001) and Levantine Aurignacian vs. Nebekian (P value < 0.001). Since the Nebekian is limited to the arid regions of South Jordan, the significant difference between sites in the mesic Mediterranean, throughout the Last Glacial, and a specific time period with a local and limited geographic distribution is evident in the significant correlation along this axis between *period* and *environment*.

These results can be visualized in scatter plots. Figure 2.2 presents the NMMDS PCA for the sites when they were coded for *environment*.

Axis 1 explains the difference between all the Mediterranean habitats (hyper mesic, mesic and xeric Mediterranean and the arid ones (semi-arid and arid), while Axis 2 is less clear. However, it appears to explain the difference between the arid habitat and those with some rainfall. All the habitats that score above 0.0, including the hyper Mediterranean, mesic Mediterranean, and the semi-arid habitat have rainfall amounts of above 150 mm (mm) annually. A score below 0.0 includes arid habitats (albeit with some sites from semi-arid environments) and indicates little to no rainfall.

Figure 2.3 presents the NMMDS PCA for the sites when they were coded for *period*.

It shows that Tabun B sites have lower values than other sites along the first axis and that Nebekian sites have lower values along the second axis.

Since the previous tests included a wide range of habitats ranging from mesic Mediterranean to arid, one could argue that the faunal communities' differences among them are much greater than any changes we may expect in any given region, despite significant climate changes. Thus, in order to test if temporal climate changes had an impact on the local faunal community, a subset of the data confined only to the Mediterranean sites was retested. The Mediterranean sites (mesic and xeric) were chosen because they include the majority of Neanderthal sites in the region and allow for the testing of these hypotheses with an ample sample size.

Results for this test indicated there was no clear clustering among the sites that could be observed, and the Kruskal stress fell higher than the cutoff level of 0.15 (and even the very conservative value of 0.30), suggesting that there was too much noise in the dataset. In sum, results from the macromammals support the hypothesis that ecotones and mosaic habitat rather than a temporal shift accounts for the variance in large mammal distribution across sites in the Last Glacial of the Southern Levant.

Micromammals

Results from the NMMDS indicate that an excellent representation was obtained (Kruskal stress = 0.099). Axes 1, 2 and 3 of the NMMDS account for 100% of the variance (R^2 = 0.48, 0.28 and 0.32 respectively). NMMDS scores on Axes 1 and 2 were not significantly different between *period* categories ($F_{4,17} = 0.34$, $P > 0.5$ and $F_{4,17} = 0.40$, $P > 0.5$). However, differences in scores between periods on Axis 3, which accounts for 32% of the variance, were highly significant ($F_{4,17} = 7.414$, $P = 0.002$). Bonferroni post hoc analysis indicates that four of the pair-wise comparisons were significant: Tabun D vs. the Natufian (P value < 0.001), Tabun C vs. the Natufian (P value = 0.002), Tabun B vs. the Natufian (P value < 0.001 and the UP vs. the Natufian (P value = 0.003).

These results can be visualized in scatter plots. Figure 2.4 presents the NMMDS PCA for Axes 1 and 2 when the sites were coded for *period*. Along the first and second axes, there is no distinction according to time periods, similar to the results obtained for macromammals; however, along the third axis (Fig. 2.4b), there is a gradient that positively correlates with time. The earliest sites, i.e., MIS 6, have very negative values, all MIS 5-3 have near neutral values and MIS 2 sites have positive values. While there is a significant correlation with time, it is worth noting that there is no correlation among the sites spanning the MIS 4-3), which represent the time frame during which Neanderthals dispersed into the region and disappeared from it.

NMMDS scores on Axes 1 (48% of the variance) and 2 (28% of the variance) were significantly different between *environment* categories ($F_{4,17} = 6.82$, $P = 0.005$ and $F_{4,17} = 10.3$, $P \leq 0.001$), while, differences in scores between *period* categories on Axis 3 were not significant ($F_{4,17} = 0.015$, $P \geq 0.5$). Bonferroni post hoc analysis for the first axis indicates that only one of the pair-wise comparisons was significant: mesic vs. xeric Mediterranean

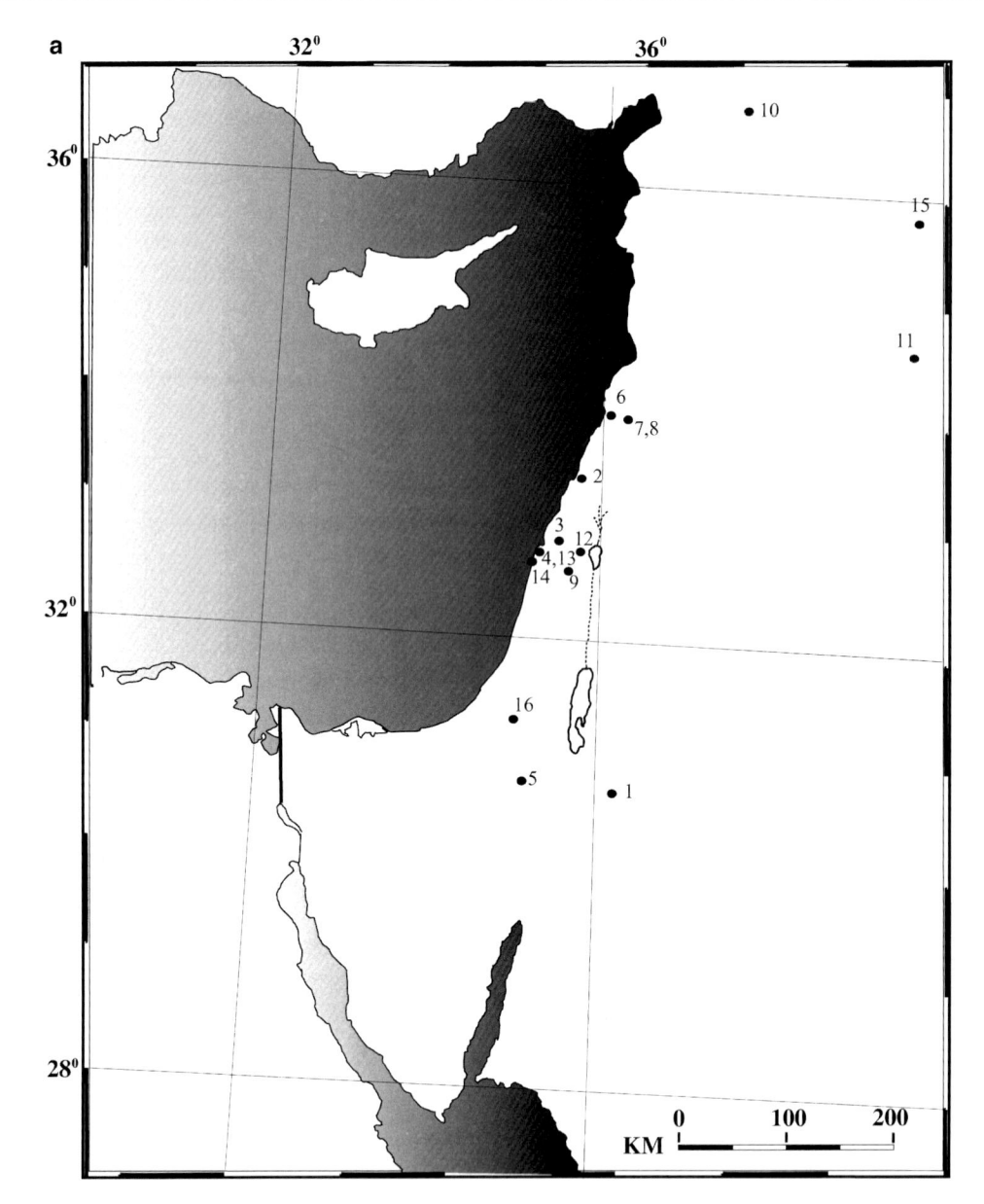

Fig. 2.1a Location of Middle Paleolithic sites mentioned in the text. 1. Ein Difla, 2. Bezez Cave B, 3. Hayonim (E > 405; E < 405), 4. Tabun (D, C, B), 5. Rosh Ein Mor, 6. Ksar Akil (18-10), 7. Naamé, 8. Ras el Kalb A-O, 9. Qafzeh (V–XV; XVI–XXIV), 10. Dedariyeh (11; 3), 11. Douara (III; IV), 12. Amud B, 13. Skhul B, 14. Kebara F, 15. Umm el Tlel (V0Ba; V11a; V13b'1), 16. Far'ah II (L.1; L.0)

(P value < 0.001), while four pair-wise comparison were significant for Axis 2: mesic Mediterranean vs. arid (P value = 0.005), mesic Mediterranean vs. hyper mesic Mediterranean (P value < 0.001), xeric Mediterranean vs. hyper mesic Mediterranean (P value = 0.004) and arid vs. hyper mesic Mediterranean (P value = 0.005).

These results can be visualized in the scatter plots. Figure 2.5 presents the NMMDS PCA when the sites were coded for *environment*. Axis 1 explains the difference between the Mediterranean habitats (mesic and xeric Mediterranean); however, since this does not appear to distinguish between the hyper mesic Mediterranean region and arid region, it does not appear to be related to rainfall. However, this may be related to vegetation cover, with sites from more open habitats, i.e., xeric and arid Mediterranean having more positive values, and sites with more closed habitats, i.e., mesic and hyper mesic Mediterranean, having negative values.

Axis 2 distinguishes a gradient along rainfall. Sites with the most precipitation have the most negative values, sites with intermediate rainfall, both mesic Mediterranean and xeric Mediterranean have neutral values, and arid sites have positive values. Along this axis, the difference in precipitation between the mesic (ca. 780–540 mm) and xeric (540–300 mm) Mediterranean cannot be distinguished.

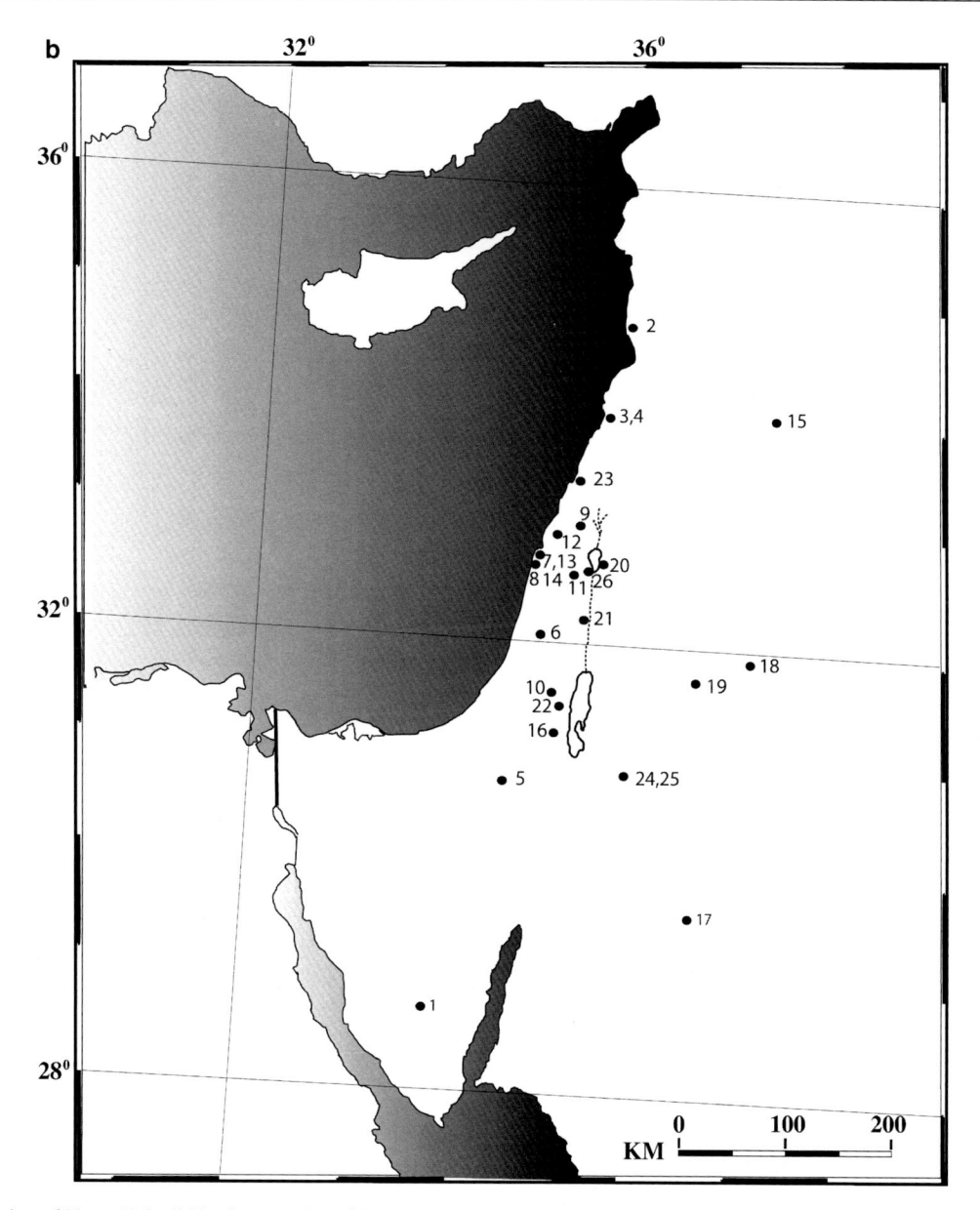

Fig. 2.1b Location of Upper Paleolithic sites mentioned in the text: 1. Abu Noshra (I; II), 2. Abu Halka (IVc; IVd; IVe; IVf), 3. Antelias, 4. Ksar Akil (06-09; 05-19; 9-6), 5. Ain Aqev (D31), 6. Shukbah D, 7. El Wad (D; E; F), 8. Kebara III, 9. Emirah, 10. Erq el Ahmar (B; C; D; E; F), 11. Qafzeh, 12. Hayonim (D3; D1/0), 13. Rakefet, 14. Sefunim, 15. Yabrud (II), 16. El Quesir (C; D), 17. Tor Hamar (F; G), 18. Uwaynid 18, 19. Jilat 9, 20. Nahal Ein Gev I, 21. Fazael (X; XI; IX), 22. Masraq e-Naj, 23. El Bezez A, 24. WHS 784, 25. WHS 618, 26. Ohalo II

In sum, the results from the micromammal database support the hypothesis that the distribution in mosaic habitats, rather than a temporal shift in environment, accounts for the variance in community distribution across sites in the Last Glacial of the Southern Levant. While there is evidence for a temporal shift in the micromammal community, it does not span the MIS 4-3 transition.

Discussion and Conclusions

Shifts in the community structure of macromammals and micromammals in the Southern Levant indicate two different patterns. On the one hand, there is a clear pattern that emerges, which reflects the local mosaic of habitats. The majority of the variance in both large and small mammals

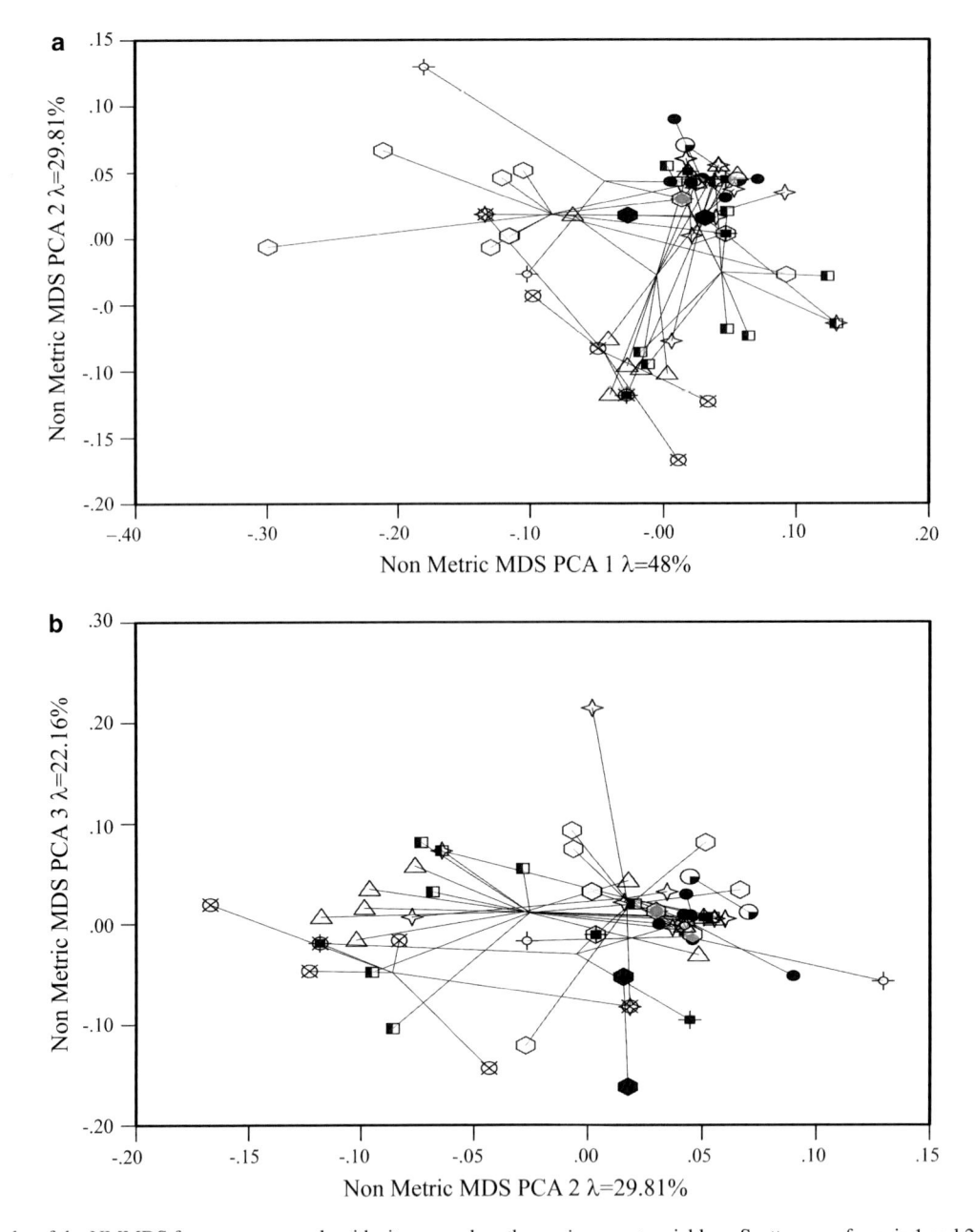

Fig. 2.2 Results of the NMMDS for macromammals with sites scored on the *environment* variable. **a** Scattergram for axis 1 and 2; **b** Scattergram for axis 2 and 3. Legend: ▲ Hyper Mesic Mediterranean, ⊞ Mesic Mediterranean, ◇ Xeric Mediterranean, ✦ Semi arid, ■ Arid

over time can be explained by spatial variation. In addition, there is a smaller percent of the variance that can be explained by temporal variation. However, in both cases, the temporal difference is only observed between the very early sites (MIS 6 and 5) and those from the Late Pleistocene and MIS 2. There is no significant difference between MIS 4 and MIS 3 or between the MP and the MP-UP transition as expected from the hypothesis of climatic forcing. Thus, this difference cannot be used as an indicator for solely a temporal climate change.

The pattern that emerges from this study is a temporal shift in the mammalian community that occurred after the Levantine Aurignacian or towards the UP-Early Epipaleolithic transition, but not prior to that time. There is no unequivocal evidence which supports a climate temporal change and that cannot be related to the environmental ecotonal mosaic structure of the Southern Levant. Furthermore, none that can be assigned specifically to the MIS 4-3 transition and the H5 event and that may be associated with the demise of the Neanderthal population in the region as suggested by Shea (2008).

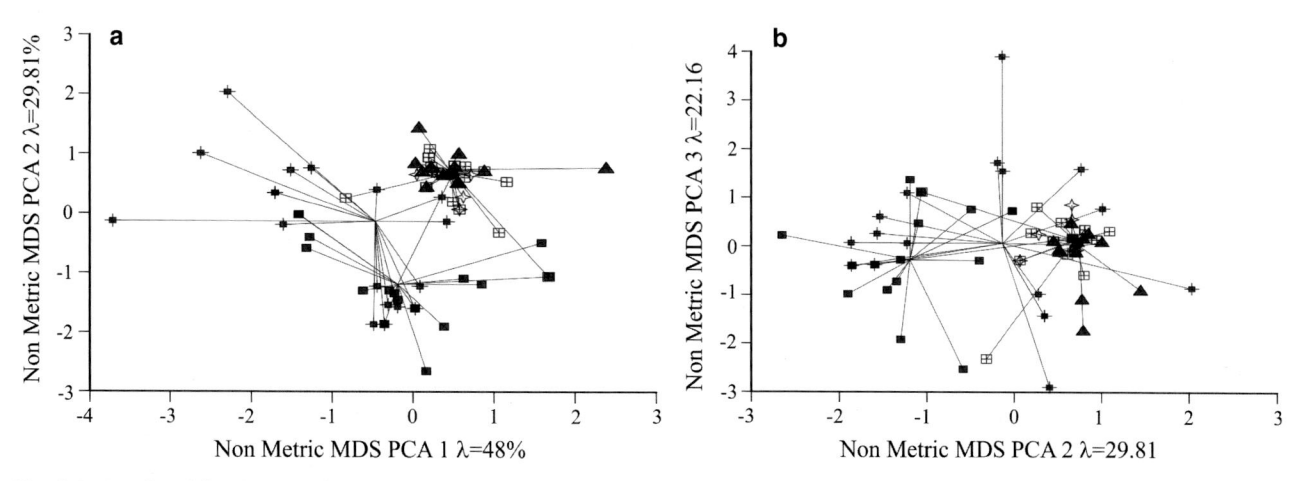

Fig. 2.3 Results of the NMMDS for macromammals with sites scored on the *period* variable. **a** Scattergram for axis 1 and 2; **b** Scattergram for axis 2 and 3. Legend: ✧ Tabun D, ● Tabun C, ◆ Tabun C-B, ○ Tabun B, ✱ MP-UP transition, △ Early Ahmarian, ◇ Levantine Aurignacian, ◐ Atlitian, ⊠ Nebekian, ▢ UP-Early Epipaleolithic

However, the lack of change in community structure among mammals stands in contrast to evidence derived from other paleoclimatic proxies obtained from the region. Stable isotope data derived from speleothems in Levantine caves for the late Middle Paleolithic have suggested a shift in climate throughout the Last Glacial. Climate change recorded in the isotopic record from Peq'iin Cave (Bar-Matthews and Ayalon 2003) is interpreted to represent changes in rainfall that indicate a shift from a wet to a dry habitat as well as a decrease in environmental productivity throughout MIS 4. This pattern is confirmed by other proxies, such as stable isotope data from foraminifera in East Mediterranean sediment cores (Bar-Matthews and Ayalon 2003) and increased pollen from steppe desert taxa in marine sediment cores in the Eastern Mediterranean (Almogi-Labin et al. 2004), This has indicated a period of overall lower ecological productivity during MIS 4 and leading up to the Heinrich event at 50–45 ka. Analysis of pollen core 95–09 in the Eastern Mediterranean, dating from 75.5–56.3 ka, indicates a low proportion (up to 7%) of arboreal pollen along the entire sequence. Deciduous oak is more prevalent in the lower part of the core, which is supportive of a dry Last Glacial in general and of a decline in regional productivity throughout the MIS 4 in particular (Langgut 2007).

How can these seemingly contradictory data sets be reconciled? The relationship between climate, the biotic environment, and humans is a complex one. In order to fully understand how climate affected human population dynamics, we need to understand the relationship between climate–plants–animals and humans. Within this paradigm, I suggest applying a hierarchical model following Rahel (1990), which describes the tiered response of plants and animals to climate change (Fig. 2.6).

This model looks at the relationship between the amplitude (which could comprised of frequency, intensity and/or the variance) of climate change against the level of biological response in a mammalian community. In low amplitude climate change, there is often no change in the biological community. This is called stasis, persistence or stability. In higher amplitudes of climate change, there is only a small level of change. This is manifested in the change in the diets of herbivores from browse to graze as their proportions change in the environment. In even higher amplitudes of climate change, there is a more noticeable change in relative abundance. This is often the result of a change in the distribution of the population across the region. This may be due to a decrease in local resources, which can no longer sustain the population at its former levels. At higher amplitudes of climate change, there is a change in presence-absence of species, i.e., some species become locally extinct and new species appear where they did not occur before; however, the overall community structure does not change. Only at the highest amplitude of climate change do we have an overall shift in biome structure, which corresponds to a total shift in species composition, community structure, and niche composition.

The response to tiered levels of climate change differs depending on the trophic level and the size of the organism. Specifically, herbivores are more sensitive to climate change than carnivores, and smaller taxa are more sensitive to lower amplitudes of climate change than larger taxa are.

If we interpret the results of this paper in light of the hierarchical model, we can observe that this is indeed the case in the Last Glacial in the Southern Levant. We have evidence for changes in precipitation i.e., climate and also for a shift in plant remains for this time period, as is

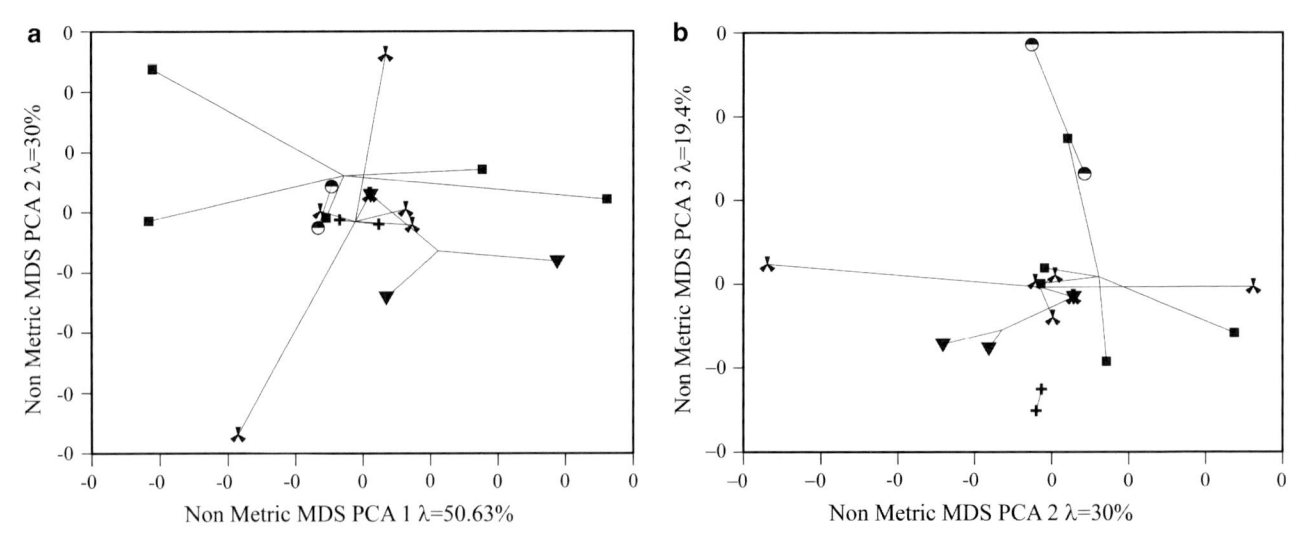

Fig. 2.4 Results of the NMMDS for micromammals with sites scored on the *period* variable. **a**. Scattergram for Axis 1 and 2; **b**. Scattergram for Axis 2 and 3. Legend: ✚ Tabun D, ▼ Tabun C, ⊼ Tabun B, ■ Upper Paleolithic, ◐ Natufian

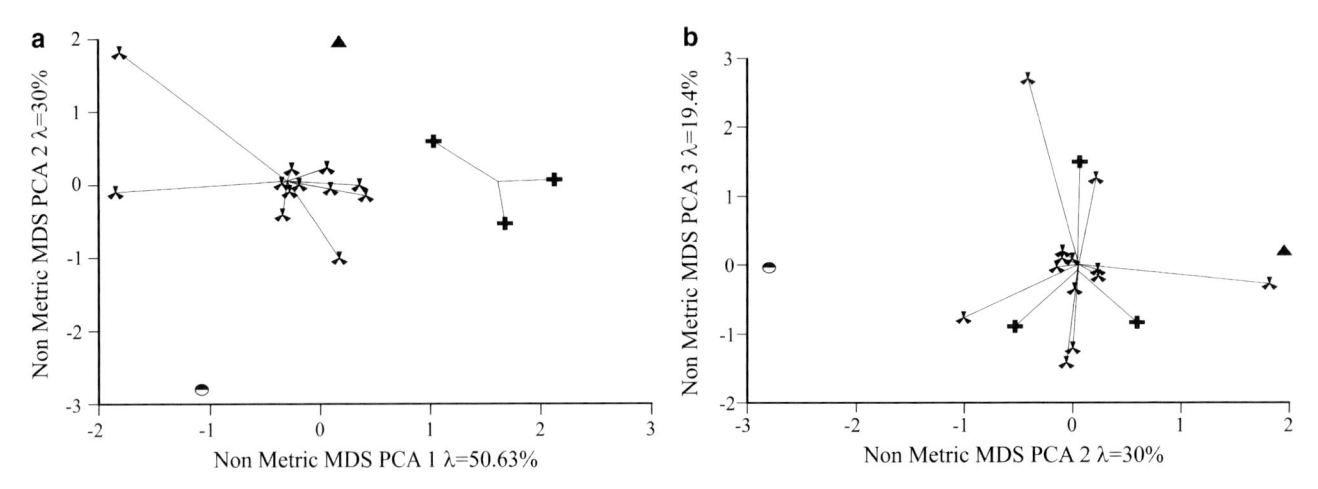

Fig. 2.5 Results of the NMMDS for micromammals with sites scored on the *environment* variable. **a**. Scattergram for axis 1 and 2; **b**. Scattergram for axis 2 and 3. Legend: ◐ Hyper Mesic Mediterranean, ⊼ Mesic Mediterranean, ✚ Xeric Mediterranean, ▲ Arid

demonstrated by the pollen (Bar-Matthews and Ayalon 2003; Bar-Matthews et al. 2003; Langgut 2007).

Other evidence for climate change comes from stable isotope data derived from ungulate teeth. Stable oxygen and carbon analyses of ungulate teeth from the sites of Amud, Qafzeh and Skhul have indicated a difference in climate at the three sites. The ungulates from Amud indicate a cooler and wetter climate than today, which differs from the climate in Qafzeh, which is dryer and warmer. While these two findings can be consistent with either hypothesis presented in this paper, it is interesting to note that the data from Skhul indicate a C_3 environment compared to the C_4 environment in Qafzeh, which is what would be expected from the regional mosaic hypothesis (Hallin 2004).

The question that is relevant for hominin population dynamics is not whether we can detect evidence for climate change, but whether the amplitude of this was large enough to evoke a shift in the large and small mammal communities. Therefore the question is not if we can observe climate change but if we can observe an *effective* climate change i.e., one that has an appreciable effect on hominins.

This paper has shown that there is no shift in community structure in either large or small mammals between MIS 4 and 3. Only at around 30 ka did climate change in a large enough degree to shift the presence-absence of taxa in the region. Can we track changes in mammalian community structure at lower levels of climate change, i.e., relative abundance and/or a shift in diet? Either of these would

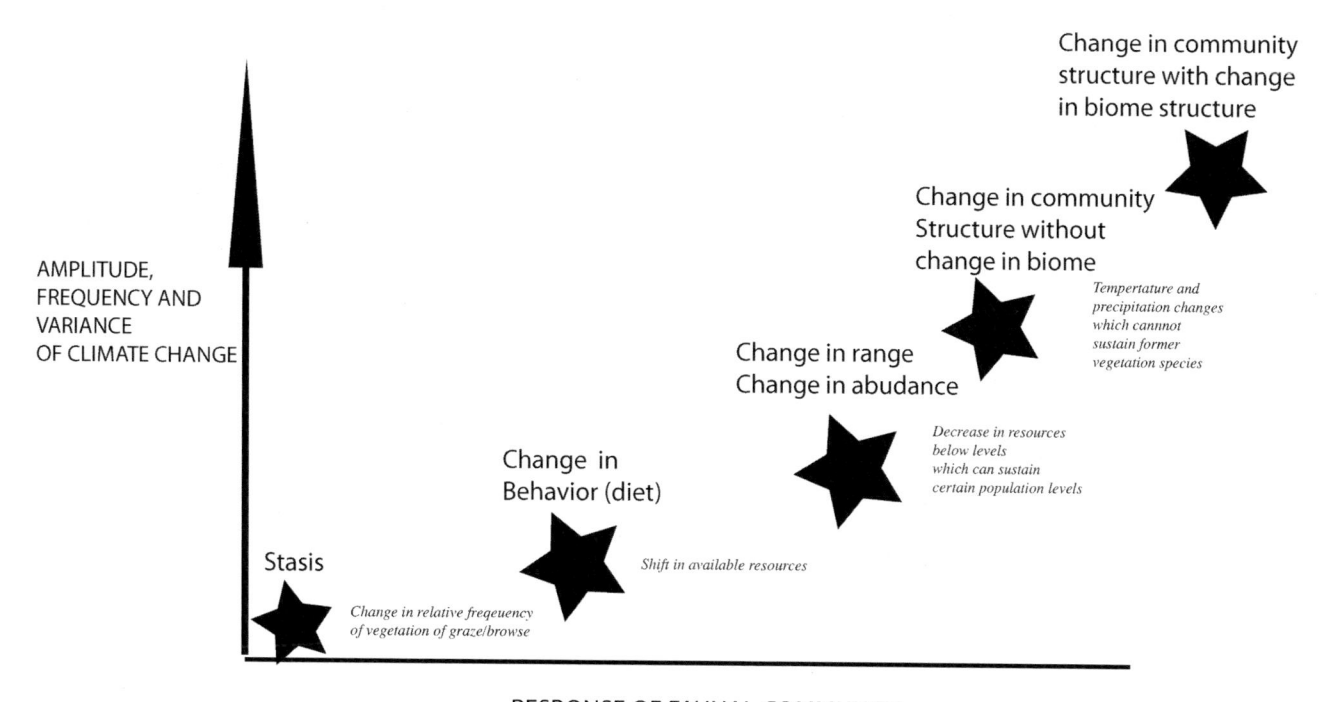

Fig. 2.6 Schematic model describing the relationship between amplitude of climatic change and response of mammalian community

suggest a climate change of lesser amplitude, which may have affected hominins as well.

Mesowear analysis measures the abrasion of upper molars on selenodont ungulates and has been shown to be a reliable proxy for diet in both extant and extinct species, specifically in distinguishing the degree of browse vs. graze in the diet (e.g., Fortelius and Solounias 2000; Franz-Odendaal et al. 2003; Kaiser and Solounias 2003; Kaiser and Franz-Odendaal 2004; Mihlbachler and Solounias 2006; Rivals and Semprebon 2006). Mesowear is a function of dental wear over ca. the last 6 months of an individual's lifetime and is independent of the effects of seasonality, migration patterns or stochastic effects, such as the effects of the individual's last meal, which is common in other paleodietary methods (e.g., Semprebon et al. 2004). Mesowear measures two variables, occlusal relief and occlusal shape, on the paracone of upper M1/M2 of selenodont ungulates. Occlusal relief has two variables: high and low, which are dependent on the height of the cusps above the valley between them. Occlusal shape has three variables: sharp, round and blunt. It has been shown that Southern Levantine ungulates did not shift their diet from browse to graze between MIS 6-3) (Belmaker 2008). That a shift did occur in ungulate diet and could be observed only around 35 ka, which post-dates the disappearance of Neanderthals in the region and the appearance of *Homo sapiens* in the Southern Levant and also is consistent with the pattern of an increase in aridity, which can be observed in the changes in community structure (Fig. 2.7).

Paleodietary studies of small rodents have been difficult to obtain because of the animals' small size, although current analysis currently is being undertaken using various methods, including microwear texture analysis (Belmaker and Ungar 2010). Observing the next levels of changes, those that occur along with a shift in relative abundance of individuals, are highly susceptible to bias induced by taphonomic changes in the fossil record (Kowalewski et al. 2003; Tomasovych and Kidwell 2009; Terry 2010). This is particularly true for Levantine micromammal assemblages, which are dominated by *Microtus guentheri*, at times comprising up to 90% of the assemblages (Belmaker and Hovers 2011).

Given this situation, we may conclude that results observed from relative abundance analyses may be inconclusive due to any known or unknown taphonomic biases. We have shown that the macromammals do not shift their diet over the time period in question (Fig. 2.7). Therefore, based on the proposed hierarchal model, we would not expect any change in relative abundance, as it would be inconceivable to have a climate change of a high enough amplitude to cause a shift in relative abundance, but not a change in diet. While this expectation is not testable, due to the aforementioned taphonomic biases, the relative proportions of the two most common ungulates, Fallow deer

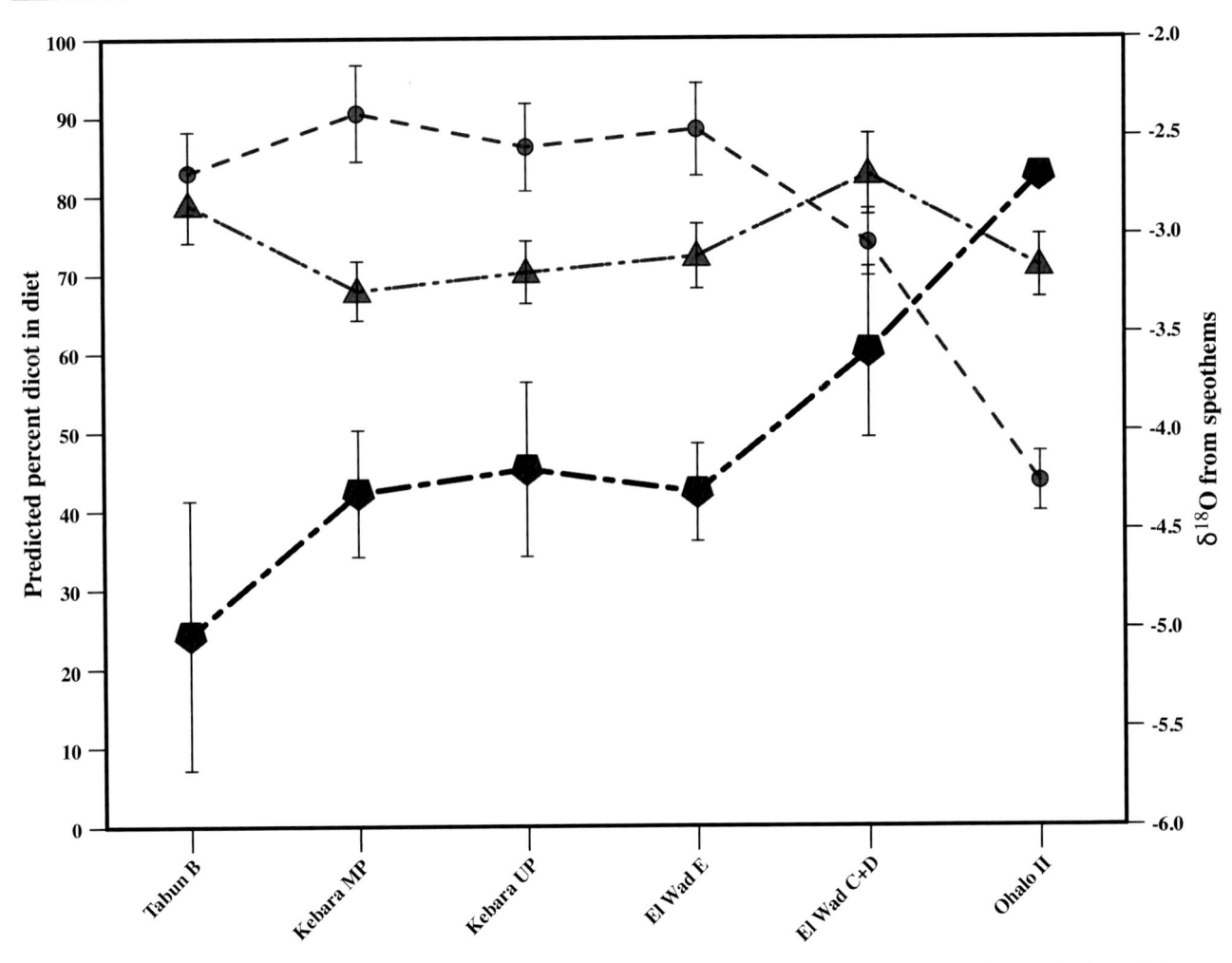

Fig. 2.7 Mesowear result of two ungulate taxa for selected Southern Levantine sites compared to stable isotope from speleothems. Mesowear results presented as percent dicot in diet calculated from prediction regression equations as with 95% confidence intervals (Belmaker 2008). Legend: — ●— Fallow deer (*Dama mesopotamica*, --▲-- Mountain gazelle (*Gazella gazella*), $\delta^{18}O$ —⬢— Data from Bar Matthews et al. (2003)

(*Dama mesopotamica*) and Mountain gazelle (*Gazella gazella*), from Levantine archaeological assemblages from MIS 6 through MIS 2 were compared. The ratio between the two taxa indicates a non-linear change through time and does not indicate a clear change around 45 ka, which would have been expected from the climate forcing hypothesis (Shea 2008). Based on the data observed from this study suggesting a faunal change ca. 35 ka and supported by the mesowear study (Fig. 2.7), the assemblages were divided into two main groups: pre and post 35 ka. Sites that predate 35 ka have a mean of 30% fallow deer with a range of 11–77%, while sites that post-date 35 ka have mean of 16% with a range of 4–54%. This difference supports the results that suggest that a marked climatic shift occurred ca. 35 ka and post-dated the Neanderthal extinction in the region.

Climate also affects morphology of taxa and can be used as a proxy for climate change. Ecogeographic rules such as Bergman and Allen's rule suggest that individual from different latitudes differ in their proportion. The application of the rules has also been used to track climate change through time (Cussans 2017; Davis 1981). However, studies that have focused on the size of Middle Pleistocene ungulates in the Southern Levant (Davis 1981) have not found changes in size of ungulates that may correspond to a large change in climate.

Thus, despite clear evidence for a climatic shift between MIS 4 and 3, there is no evidence for response to climate change from the lowest tier of response from large mammals (diet), no evidence for change in relative abundance of large mammals (medium tier of response to climate change) and no evidence for change from the highest tier of response from both micromammal and macromammals (presence-absence).

What are the implications for hominin population dynamics in the region and specifically for Neanderthal local extinction? Shea (2001, 2003, 2008) posits that the extinction of the Levantine Neanderthals was due to decreased productivity throughout the MIS 4 and the climatic crisis that occurred during the Heinrich 5 (H5) event of 50–45 kyr. The assertion that climate forcing was responsible for Neanderthal extinction in the region is based on the underlying hypotheses that the climatic change was severe enough to cause depletion in either plant or animal resources. More specifically, climate change during the H5, or during the MIS 4 leading up to it, was so severe as to cause a noticeable decrease in resources available for Neanderthals. Neanderthals subsisted on a hunter-gatherer diet (Speth 1987, 2004, 2006; Lev 1993; Albert et al. 2000; Speth and Tchernov 2001; Rabinovich and Hovers 2004; Lev et al. 2005). We can hypothesize what affect an environmental decrease in productivity, resulting from less rainfall in the Southern Levant, would have on the Neanderthal diet. A decrease in productivity would have resulted in a decrease in the productivity of oak and other fruit bearing trees, which would have resulted in less fruits and nuts for gathering. In addition, a shift in the total ratio of trees to open grassland, favoring an increase in grassland, may have resulted in fewer trees used for bedding and fire. Moreover, the decrease in browse would have led to a dietary shift in ungulates from predominantly browse to graze and the decrease in food availability would lead to a decrease in the fecundity of fallow deer and gazelles available for hunting.

However, when climate change is severe enough to evoke an extinction of ecosystem, we often see the extinction of complete ecosystems, trophic levels and/or body sizes. This can be observed in the study presented in this volume, which discusses the extinction in the Quaternary of Northern South America and presents the differential extinction of larger mammals and the elimination of several Quaternary habitats due to climate change (Ferrusquia et al. 2017). The situation in Northern South America described by Ferrusquia et al. (2017) differs from that which we present in this paper. Specifically, in Northern South America, the climatic fluctuation in the Pleistocene led to extensive changes in species composition (i.e., presence-absence) and changes in the geographic distribution of species. In contrast, in the Southern Levant, as presented in this paper, the only species to become locally extinct at the H5 were the Neanderthals. Since Neanderthal were at the top of the tropic level, if climate change contributed to their extinction, we expect to see evidence of change in the population dynamics of taxa in trophic levels below that of Neanderthals. These would include both large and smaller herbivores.

This paper suggests that there is no change in the species composition in both large and small herbivore fauna throughout the time span of Neanderthal presence in the Levant and shortly after Neanderthal extinction in the Levant. While the climate fluctuations in Northern South America were severe enough or of sufficient amplitude to evoke a shift in mammalian species composition, the amplitude of climate shift in the Southern Levant was of not as sever as to led to a total shift in species composition in the ungulate and small mammal communities. Thus, a major climate shift did not occur between MIS 4 and 3 (contra Shea 2003, 2008). As suggested by the hierarchy model (Rahel 1990), lower amplitude climate shifts may affect the diet of herbivores. However, we have shown in the mesowear study that the ungulates did not shift their diet. This further supports the hypothesis that there is little evidence for the presence of climate change during the H5.

Shea (2008) has suggested that the extinction of the Neanderthals was a result of a decrease in productivity resulting from a climate change during the H5 event. It may be argued, that a prolonged decrease in productivity would not have led to a shift in community composition nor shift in diet, but a smaller population in the region, both in species, size of population and size of individuals. An interesting study by Speth and Clark (2006) points to the fact that the hunted herbivores in Kebara Cave decrease between 75 and 45 ka from large *Aurochs*, *Cervus* and *Dama* to the smaller sized *Gazelle* and further more to the smaller sized juvenile gazelle.

This archaeological pattern may be produced by several phenomena. It may be the results in a decrease in prey abundance due to a decrease in local productivity or an increase in human abundance and better hunting efficiency. Given the evidence presented here which do not support a high amplitude climate change in the Levant during the H5, it appear more prudent to attribute this phenomena to local over-hunting by Neanderthals populations, perhaps due to an increase in local Neanderthal population. Furthermore, this pattern of a decrease in size of prey throughout MIS 4/3 has not been repeated in other Neanderthals sites in the Levant (e.g., Amud see Rabinovich and Hovers 2004).

Neanderthals and other Middle Paleolithic hominins are large mammals, and it stands to reason that they were affected primarily by large amplitudes of climate change and not by smaller fluctuations, which may have occurred during this time period. Thus, even if climate changes did occur during the MIS 4 and the H5 in the Levant, they were of such small amplitude as to not have a noticeable effect on the mammalian community. Since hominins were highly adaptive, these small amplitude climate changes would probably not have such an effect on their population as to lead to their complete extinction in the Southern Levant. While there is a clear record of climate change throughout MIS 4 in the Southern Levant present in spleothems, pollen spectra and other climatic proxies there is little evidence of an *effective climate* change during the time period, which may have contributed to the local extinction of Neanderthals in the Southern Levant.

Acknowledgments I thank the editor, Gregory Monks, for inviting me to participate in this volume. Generous funding for this work was received from the Wenner Gren and Irene Levy Sala Foundations and from the Harvard University American School of Prehistoric Research (ASPR) and the Department of Anthropology, College of William and Mary; I thank them all for their support during research for this publication. I am indebted to Ofer Bar-Yosef, Erella Hovers, Anna Belfer Cohen and John Speth for discussions on the subjects of climate change, Neanderthals and modern humans. I also thank two anonymous reviews for their valuable insights during preparation of this manuscript.

References

Albert, R. M., Weiner, S., Bar-Yosef, O., & Meignen, L. (2000). Phytoliths of the Middle Palaeolithic deposits of Kebara Cave, Mt. Carmel, Israel: Study of the plant materials used for fuel and other purposes. *Journal of Archaeological Science, 27*, 931–947.

Almogi-Labin, A., Bar-Matthews, M., & Ayalon, A. (2004). Climate variability in the Levant and Northeast Africa during the Late Quaternary based on marine and land records. In N. Goren-Inbar & J. D. Speth (Eds.), *Human paleoecology in the Levantine corridor* (pp. 117–134). Oxford: Oxbow Books.

Andrews, P. (1990). *Owls, caves and fossils: Predation, preservation and accumulation of small mammal bones in caves, with an analysis of the Pleistocene cave faunas from Westbury-Sub-mendip, Somerset, UK*. Chicago: University of Chicago Press.

Bar-Matthews, M., & Ayalon, A. (2003). Climatic conditions in the Eastern Mediterranean during the Last Glacial (60–10 ky) and their relations to the Upper Palaeolithic in the Levant as inferred from oxygen and carbon isotope systematics of cave deposits. In A. N. Goring-Morris & A. Belfer-Cohen (Eds.), *More than meets the eye: Studies on Upper Palaeolithic diversity in the Near East* (pp. 13–18). Oxford: Oxbow Books.

Bar-Matthews, M., Ayalon, A., Gilmour, M., Matthews, A., & Hawkesworth, C. (2003). Sea-land oxygen isotopic relationships from planktonic foraminifera and speleothems in the Eastern Mediterranean region and their implication for paleorainfall during interglacial intervals. *Geochimica et Cosmochimica Acta, 66*(24), 1–19.

Bar-Yosef, O. (1989). Geochronology of the Levantine Middle Paleolithic. In P. A. Mellars & C. B. Stringer (Eds.), *The human revolution* (pp. 589–610). Edinburgh: Edinburgh University Press.

Bar-Yosef, O. (1992). Middle Paleolithic human adaptations in the Mediterranean Levant. In T. Akazawa, K. Aoki, & T. Kimura (Eds.), *The evolution and dispersal of modern humans in Asia* (pp. 189–216). Tokyo: Hokusen-Sha.

Bar-Yosef, O. (2000). The middle and early Upper Palaeolithic in southwest Asia and neighboring regions. In O. Bar-Yosef & D. Pilbeam (Eds.), *The geography of Neanderthals and modern humans in Europe and the greater Mediterranean* (pp. 107–156). Cambridge, MA: Peabody Museum, Harvard University, Bulletin No. 8.

Bate, D. M. A. (1927). On the animal remains obtained from the Mugharet el-Emireh in 1925. In F. Turville-Petre (Ed.), *Researches in prehistoric Galilee, 1925–1926 and a report on the Galilee skull* (pp. 9–13). Jerusalem: Bulletin of the British School of Archaeology in Jerusalem No. XIV.

Behrensmeyer, A. K., Kidwell, S. M., & Gastaldo, R. A. (2000). Taphonomy and paleobiology. *Paleobiology, 26*(4), 103–147.

Belmaker, M. (2008). Analysis of ungulate diet during the Last Glacial (MIS 5-2) in the Levant: Evidence for long-term stability in a Mediterranean ecosystem. *Journal of Vertebrate Paleontology, 28* (Supplement to Number 3), 50A.

Belmaker, M., & Hovers, E. (2011). Ecological change and the extinction of the Levantine Neanderthals: Implications from a diachronic study of micromammals from Amud Cave, Israel. *Quaternary Science Reviews, 30*(21–22), 3196–3209.

Belmaker, M., & Ungar, P. S. (2010). Micromammal microwear texture analysis – preliminary results and application for paleoecological study. *PaleoAnthropology* A2.

Belmaker, M., Nadel, D., & Tchernov, E. (2001). Micromammal taphonomy in the site of Ohalo II (19 Ky., Jordan Valley). *Archaeofauna, 10*, 125–135.

Cussans, J. E. (2017). Biometry and climate change in Norse Greenland: The effect of climate on the size and shape of domestic mammals. In G. G. Monks (Ed.), *Climate change and human responses: A zooarchaeological perspective* (pp. 197–216). Dordrecht: Springer.

Clark, G. A., Lindly, J., Donaldson, M., Garrard, A. N., Coinman, N. R., Fish, S., & Olszewski, D. (2000). Paleolithic archaeology in the southern Levant: A preliminary report of excavations at Middle, Upper and Epipaleolithic sites in the Wadi al-Hasa, west-central Jordan. In N. R. Coinman (Ed.), *The Archaeology of the Wadi al-Hasa, west-central Jordan, Vol. 2: Excavations at Middle, Upper and Epipaleolithic Sites* (pp. 17–66). Tempe: Anthropological Research Papers No. 52, Arizona State University.

Danin, A., & Orshan, G. (1990). The distribution of Raunkiaer life forms in Israel in relation to the environment. *Journal of Vegetation Science, 1*(1), 41–48.

Davis, S. J. M. (1981). The effects of temperature change and domestication on the body size of Late Pleistocene to Holocene mammals in Israel. *Paleobiology, 7*, 101–114.

Davis, S. J. M. (1982). Climatic change and the advent of domestication of ruminant artiodactyls in the Late Pleistocene-Holocene period in the Israel region. *Paléorient, 8*, 5–16.

Enzel, Y., Amit, R., Dayan, U., Crouvi, O., Kahana, R., Ziv, B., et al. (2008). The climatic and physiographic controls of the eastern Mediterranean over the Late Pleistocene climates in the southern Levant and its neighboring deserts. *Global and Planetary Change, 60*(3–4), 165–192.

Ferrusquía-Villafranca, I., Arroyo-Cabrales, J., Johnson, E., Ruiz-González, J., Martínez-Hernández, E., Gama-Castro, J., et al. (2017). Quaternary mammals, peoples and climate change: A view from Southern North America. In G. G. Monks (Ed.), *Climate change and human responses: A zooarchaeological perspective* (pp. 27–67). Dordrecht: Springer.

Finlayson, C. (2004). *Neanderthals and modern humans. An ecological and evolutionary perspective*. Cambridge UK: Cambridge University Press.

Finlayson, C., & Carrion, J. S. (2007). Rapid ecological turnover and its impact on Neanderthal and other human populations. *Trends in Ecology & Evolution, 22*(4), 213–222.

Finlayson, C., Pacheco, F. G., Rodriguez-Vidal, J., Fa, D. A., Gutierrez Lopez, J. M., Santiago Perez, A., et al. (2006). Late survival of Neanderthals at the southernmost extreme of Europe. *Nature, 443* (7113), 850–853.

Fleisch, H. (1970). Les habitat du Paléolithique moyen à Naamé (Liban). *Bulletin de la Musée de Beyrouth, 23*, 25–93.

Fortelius, M., & Solounias, N. (2000). Functional characterization of ungulate molars using the abrasion – attrition wear gradient: A new method for reconstructing paleodiets. *American Museum Novitates, 3301*, 1–36.

Franz-Odendaal, T. A., Kaiser, T. M., & Bernor, R. L. (2003). Systematics and dietary evaluation of a fossil equid from South Africa. *South African Journal of Science, 99*, 453–459.

Frumkin, A., & Stein, M. (2004). The Sahara-East Mediterranean dust and climate connection revealed by strontium and uranium isotopes in a Jerusalem speleothem. *Earth and Planetary Science Letters, 217*(3–4), 451–464.

Gamble, C., Davies, W., Pettitt, P., & Richards, M. (2004). Climate change and evolving human diversity in Europe during the last glacial. *Philosophical Transactions of the Royal Society of London. Series B: Biological Sciences, 359*(1442), 243–253; discussion 253–254.

Garrard, A. N. (1980). Man-animal-plant relationships during the Upper Pleistocene and Early Holocene of the Levant. Ph.D. Dissertation, University of Cambridge.

Garrard, A. N. (1982). The environmental implications of re-analysis of the large mammal fauna from the Wadi-el-Mughara caves, Palestine. In J. L. Bintliff & W. Van Zeist (Eds.) *Paleoclimates, paleoenvironments and human communities in the eastern Mediterranean region in the later prehistory* (pp. 45–65). Oxford: British Archaeological Reports.

Garrard, A. N., Colledge, S., Hunt, C., & Montague, R. (1988). Environment and subsistence during the Late Pleistocene and Early Holocene in the Azraq Basin. *Paléorient, 14*, 40–49.

Garrod, D. A. E., & Bate, D. M. A. (1942). Excavations at the cave of Shukbah, Palestine. *Proceedings of the Prehistoric Society, 8*, 1–20.

Gilead, I., & Grigson, C. (1984). Far'ah II: A Middle Paleolithic open-air site in the northern Negev, Israel. *Proceedings of the Prehistoric Society, 50*, 71–97.

Griggo, C. (1998). Associations fauniques et activités de subsistance au Paléolithique moyen en Syrie. In M. Otte (Ed.), *Préhistoire d'Anatolie: Genèse de deux mondes/Anatolian prehistory: At the crossword of two worlds* (pp. 749–764). Liège: Études et Recherches Archéologiques de L'Université de Liège.

Guiot, J., de Beulieu, J. L., Cheddadi, R., David, F., Ponel, P., & Reille, M. (1993). The climate in western Europe during the last glacial/interglacial cycle derived from pollen and insect remains. *Palaeogeography, Palaeoclimatology, Palaeocology, 103*, 73–93.

Hallin, K. A. (2004). Paleoclimate during Neanderthal and early modern human occupation in Israel: Tooth enamel stable isotope evidence. Ph.D. dissertation, University of Wisconsin.

Harrison, D. L., & Bates, P. J. J. (1991). *The mammals of Arabia*. Kent: Harrison Zoological Museum.

Hausdorf, B., & Hennig, C. (2003). Biotic element analysis in biogeography. *Systematic Biology, 52*(5), 717–723.

Heller, J. (1970). The small mammals of the Geula Cave. *Israel Journal of Zoology, 19*, 1–49.

Hubbell, S. P. (2001). *The unified neutral theory of biodiversity and biogeography*. Princeton and Oxford: Princeton University Press.

Jelinek, A. J. (1982). The Tabun Cave and Paleolithic man in the Levant. *Science, 216*, 1369–1375.

Kaiser, T. M., & Franz-Odendaal, T. A. (2004). A mixed-feeding *Equus* species from the Middle Pleistocene of South Africa. *Quaternary Research, 62*, 316–323.

Kaiser, T. M., & Solounias, N. (2003). Extending the tooth mesowear method to extinct and extant equids. *Geodiversitas, 25*(2), 321–345.

Kersten, A. M. P. (1992). Rodents and insectivores from the Palaeolithic rock shelter of Ksar 'Akil (Lebanon) and their palaeoloecological implications. *Paléorient, 18*, 27–45.

Kidwell, S. M. (2001). Preservation of species abundance in marine death assemblages. *Science, 294*, 1091–1094.

Kidwell, S. M. (2002). Time-averaged molluscan death assemblages: Palimpsests of richness, snapshots of abundance. *Geology, 30*(9), 803–806.

Kidwell, S. M. (2008). Ecological fidelity of open marine molluscan death assemblages: Effects of post-mortem transportation, shelf health, and taphonomic inertia. *Lethaia, 41*(3), 199–217.

Kidwell, S. M., & Flessa, K. W. (1995). The quality of the fossil record: Population, species, and communities. *Annual Review of Ecological Systems, 26*, 269–299.

Kidwell, S. M., & Flessa, K. W. (1996). The quality of the fossil record: Populations, species and communities. *Annual Review of Earth and Planetary Sciences, 24*, 433–464.

Kidwell, S. M., & Holland, S. M. (2002). The quality of the fossil record: Implications for evolutionary analyses. *Annual Review of Ecology and Systematics, 33*, 561–588.

Kidwell, S. M., Rothfus, T. A., & Best, M. M. R. (2001). Sensitivity of taphonomic signatures to sample size, sieve size, damage scoring system, and target taxa. *Palaios, 16*(1), 26–52.

Klein, R. G. (1995). The Tor Hamar fauna. In D. O. Henry (Ed.), *Prehistoric cultural ecology and evolution: Insights from southern Jordan* (pp. 405–416). New York: Plenum Press.

Kowalewski, M., Carroll, M., Casazza, L., Gupta, N. S., Hannisdal, B., Hendy, A., et al. (2003). Quantitative fidelity of brachiopod-mollusk assemblages from modern subtidal environments of San Juan Islands. USA. *Journal of Taphonomy, 1*(1), 43–65.

Kuhn, S. L., Brantingham, P. J., & Kerry, K. W. (2004). The Early Paleolithic and the origins of modern human behavior. In P. J. Brantingham, S. L. Kuhn, & K. W. Kerry (Eds.), *The early Upper Paleolithic beyond western Europe* (pp. 242–248). Berkeley: University of California Press.

Langgut, D. (2007). Late Quaternary palynological sequences from the eastern Mediterranean Sea. Ph.D. dissertation, Haifa University.

Lehmann, U. (1970). Die Tierreste aus den Hohlen von Jabrud (Syrien). In H. V. K. Gripp, R. Shütrumpf, & H. Schwabedissen (Eds.), *Frühe Menscheit und Umwelt* (pp. 181–188). Köln: Bohlau.

Lev, E. (1993). The vegetal food of the "Neanderthal" man in Kebara Cave, Mt. Carmel in the Middle Paleolithic Period. M.A. dissertation, Bar-Ilan University.

Lev, E., Kislev, M. E., & Bar-Yosef, O. (2005). Mousterian vegetal food in Kebara Cave, Mt. Carmel. *Journal of Archaeological Science, 32*(3), 475–484.

Lindley, C., & Clark, G. A. (1987). A preliminary lithic analysis of the Mousterian site of 'Ain Difla (WHS Site 634) in the Wadi Ali, west central Jordan. *Proceedings of the Prehistoric Society, 53*, 279–292.

Mellars, P. (2004). Neanderthals and modern human colonization of Europe. *Nature, 432*, 461–465.

Mellars, P. (2006). A new radiocarbon revolution and the dispersal of modern humans in Eurasia. *Nature, 439*, 931–935.

Mihlbachler, M., & Solounias, N. (2006). Co-evolution of tooth crown height and diet in Oreodonts (Merycoidodontidae, Artiodactyla) examined with phylogenetically independent contrasts. *Journal of Mammalian Evolution, 13*(1), 11–36.

Millard, A. R. (2008). A critique of the chronometric evidence for hominid fossils: I. Africa and the Near East 500–50 ka. *Journal of Human Evolution, 54*(6), 848–874.

Miracle, P. T., Lenardić, J. M., & Barjković, D. (2009). Last glacial climates, "refugia", and faunal change in southeastern Europe: Mammalian assemblages from Veternica, Velika Pećina, and Vindija caves (Croatia). *Quaternary International, 212*(1), 137–148.

Nadel, D., Lengyel, G., Bocquentin, F., Tsatskin, A., Rosenberg, D., Yeshurun, R., et al. (2008). The Late Natufian at Raqefet Cave: The 2006 excavation season. *Journal of the Israel Prehistoric Society, 38*, 59–131.

Payne, S. (1983). The animal bones from the excavations at Douara Cave. In K. Hanihara & T. Akazawa, (Eds.), *The Paleolithic site of Douara Cave and paleogeography of Palmyra Basin in Syria. Part III* (pp. 1–133). Tokyo: University of Tokyo University Museum Bulletin 21.

Perrot, J. (1955). Le Paléolithique supérieur d'El Quseir et de Masarag an Na'aj (Palestine) Inventaire de la collection René Neuville I et II. *Bulletin de la Société Préhistorique de France, 52*, 493–506.

Phillips, J. L. (1988). The Upper Paleolithic of the Wadi Feiran, southern Sinai. *Paléorient, 14*, 183–200.

Rabinovich, R. (1998). Patterns of animal exploitation and subsistence in Israel during the Upper Palaeolithic and Epi-Palaeolithic (40,000–12,500 BP), based upon selected case studies. Ph.D. dissertation, The Hebrew University of Jerusalem.

Rabinovich, R. (2003). The Levantine Upper Paleolithic faunal record. In A. N. Goring-Morris & A. Belfer-Cohen (Eds.), *More than meets the eye: Studies on Upper Paleolithic diversity in the Near East* (pp. 33–48). Oxford: Oxbow.

Rabinovich, R., & Hovers, E. (2004). Faunal analysis from Amud Cave: Preliminary results and interpretations. *Journal of Osteoarchaeology, 14*(3–4), 287–306.

Rabinovich, R., & Tchernov, E. (1995). Chronological, paleoecological and taphonomical aspects of the Middle Paleolithic site of Qafzeh, Israel. In H. Buitenhuis & H.P Uerpmann (Eds.), *Archaeozoology of the Near East II* (pp. 5–44). Leiden: Backhuys.

Rahel, F. J. (1990). The hierarchical nature of community persistence: A problem of scale. *The American Naturalist, 136*(3), 328–344.

Rink, J. W., Bartoll, J., Goldberg, P., & Ronen, A. (2003). ESR dating of archaeologically relevant authigenic terrestrial apatite veins from Tabun Cave, Israel. *Journal of Archaeological Science, 30*, 1127–1138.

Rivals, F., & Semprebon, G. M. (2006). A comparison of the dietary habits of a large sample of the Pleistocene proghorm *Stockoceros onusrosagris* from the Papago Springs Cave in Arizona to the modern *Antilocapra americana*. *Journal of Vertebrate Paleontology, 26*(2), 495–500.

Rogers, R. R., & Kidwell, S. M. (2000). Association of vertebrate skeletal concentration and discontinuity surfaces in terrestrial and shallow marine records: A test in the Cretaceous of Montana. *The Journal of Geology, 109*, 131–154.

Ronen, A. (Ed.). (1984). *The Sefunim prehistoric sites Mount Carmel, Israel*. Oxford: BAR International Series.

Roy, K., Valentine, J. W., Jablonski, D., & Kidwell, S. M. (1996). Scales of climatic variability and time averaging in Pleistocene biotas: Implications for ecology and evolution. *Trends in Ecology & Evolution, 11*(11), 458–463.

Schwarcz, H. P., Buhay, W. M., Grün, R., Valladas, H., Tchernov, E., Bar-Yosef, O., et al. (1989). ESR dating of the Neanderthal site, Kebara Cave, Israel. *Journal of Archaeological Science, 16*, 653–659.

Semprebon, G. M., Godfrey, L. R., Solounias, N., Sutherland, M. R., & Jungers, W. L. (2004). Can low-magnification steromicroscopy reveal diet? *Journal of Human Evolution, 47*(3), 115–144.

Shea, J. J. (2001). The Middle Paleolithic: Early modern humans and Neanderthals in the Levant. *Near Eastern Archaeology, 64*(1/2), 38–64.

Shea, J. J. (2003). The Middle Paleolithic of the east Mediterranean Levant. *Journal of World Prehistory, 17*, 313–394.

Shea, J. J. (2008). Transitions or turnovers? Climatically-forced extinctions of *Homo sapiens* and Neanderthals in the east Mediterranean Levant. *Quaternary Science Reviews, 27*, 2253–2270.

Solecki, R. S., & Solecki, R. L. (1993). The pointed tools from the Mousterian occupations of Shanidar Cave, Northern Iraq. In D. I. Olszewski & H. L. Dibble (Eds.), *The Paleolithic prehistory of the Zagros-Taurus* (pp. 119–146). Philadelphia: The University Museum, University of Pennsylvania.

Speth, J. D. (1987). Early hominid subsistence strategies in seasonal habitats. *Journal of Archaeological Science, 14*, 13–29.

Speth, J. D. (2004). Hunting pressure, subsistence intensification, and demographic change in the Levantine late Middle Paleolithic. In N. Goren-Inbar & J. D. Speth (Eds.), *Human paleoecology in the Levantine corridor* (pp. 149–166). Oxford: Oxbow Books.

Speth, J. D. (2006). Housekeeping, Neanderthal-style: Hearth placement and midden formation in Kebara Cave (Israel). In E. Hovers & S. L. Kuhn (Eds.), *Transitions before the transition: Evolution and stability in the Middle Paleolithic and Middle Stone Age* (pp. 171–188). New York: Springer.

Speth, J. D., & Clark, J. L. (2006). Hunting and overhunting in the Levantine Late Middle Paleolithic. *Before Farming, 3*, 1–42.

Speth, J. D., & Tchernov, E. (2001). Neanderthal hunting and meat-processing in the Near East: Evidence from Kebara Cave (Israel). In C. B. Stanford & H. T. Bunn (Eds.), *Meat-eating and human evolution* (pp. 52–72). Oxford: Oxford University Press.

Stewart, J. R. (2005). The ecology and adaptation of Neanderthals during the non-analogue environment of Oxygen Isotope Stage 3. *Quaternary International, 137*, 35–46.

Stiner, M. C., & Tchernov, E. (1998). Pleistocene species trends at Hayonim Cave: Changes in climate versus human behavior. In T. Akazawa, K. Aoki, & O. Bar-Yosef (Eds.), *Neandertals and modern humans in western Asia* (pp. 241–262). New York: Plenum Press.

Tchernov, E. (1976). Some Late Quaternary faunal remains from the Avdat/Aqev area. In A. E. Marks (Ed.), *Prehistory and paleoenvironments in the central Negev, Israel, Vol. 1. The Avdat/Aqev area* (pp. 69–73). Dallas: SMU Press.

Tchernov, E. (1984). Faunal turnover and extinction rate in the Levant. In P. S. Martin & R. G. Klein (Eds.), *Quaternary extinctions: A prehistoric revolution* (pp. 528–552). Tucson: The University of Arizona Press.

Tchernov, E. (1986). Rodent fauna, chronostratigraphy, and paleobiography of the southern Levant during the Quaternary. *Acta Zoologica Cracoviensis, 39*, 513–530.

Tchernov, E. (1988). The biogeographical history of the southern Levant. In Y. Yom-Tov & E. Tchernov (Eds.), *The Zoogeography of Israel* (pp. 159–250). Dordrecht: Dr. Junk Publishers.

Tchernov, E. (1992). Biochronology, paleoecology, and dispersal events of hominids in the southern Levant. In T. Akazawa, K. Aoki, & T. Kimura (Eds.), *The evolution and dispersal of modern humans in Asia* (pp. 149–188). Tokyo: Hokusen-sha.

Tchernov, E. (1994). New comments on the biostratigraphy of the Middle and Upper Pleistocene of the southern Levant. In O. Bar-Yosef & R. S. Kra (Eds.), *Late quaternary chronology and paleoclimates of the eastern Mediterranean* (pp. 333–350). Oxford: Radiocarbon.

Tchernov, E. (1998). The faunal sequence of the southwest Asian Middle Paleolithic in relation to hominid dispersal events. In T. Akazawa, K. Aoki & O. Bar-Yosef (Eds.), *Neandertals and modern humans in western Asia* (pp. 77–90). New York: Plenum Press.

Terry, R. C. (2010). On raptors and rodents: Testing ecological fidelity and spatiotemporal resolution of cave death assemblages. *Paleobiology, 36*(1), 137–160.

Thomas, H. (1985). The Early and Middle Miocene land connection of the Afro-Arabian plateau and Asia: A major event for hominid dispersal? In E. Delson (Ed.), *Ancestors: The hard evidence* (pp. 42–50). New York: A. R. Liss.

Tomasovych, A., & Kidwell, S. M. (2009). Fidelity of variation in species composition and diversity partitioning by death assemblages: Time-averaging transfers diversity from beta to alpha levels. *Paleobiology, 35*(1), 94–118.

Tzedakis, P. C., Hughen, K. A., Cacho, I., & Harvati, K. (2007). Placing late Neanderthals in a climatic context. *Nature, 449*(7159), 206–208.

Valladas, H., & Joron, J. L. (1989). Application de la thermoluminescence à la datation des niveaux mousteriens de la grotte de Kebara (Israël): Age préliminaires des unites XII, XI et VI. In O. Bar-Yosef & B. Vandermeersch (Eds.), *Investigations in south Levantine prehistory* (pp. 97–100). Oxford: B.A.R. International Series 497.

Valladas, H., Joron, J.-L., Valladas, G., Arensburg, B., Bar-Yosef, O., Belfer-Cohen, A., et al. (1987). Thermoluminescence dates for the Neanderthal burial site at Kebara in Israel. *Nature, 330*, 159–160.

Valladas, H., Mercier, N., Froget, L., Hovers, E., Joron, J.-L., Kimbel, W. H., et al. (1999). TL dates for the Neanderthal site of the Amud Cave, Israel. *Journal of Archaeological Science, 26*, 259–268.

Vaufrey, R. (1951). Etude Paléontologique, I: Mammifères. In R. Neuville (Ed.), *Le Paléolithique et le Mésolithique du Désert de Judée* (pp. 198–217). Paris: Institut de Paléontologie Humaine.

Weissbrod, L., Dayan, T., Kaufman, D., & Weinstein-Evron, M. (2005). Micromammal taphonomy of el-Wad terrance, Mount Carmel, Israel: Distinguishing cultural from natural depositional agents in the Late Natufian. *Journal of Archaeological Science, 32*, 1–17.

Wallace, I. J., & Shea, J. J. (2006). Mobility patterns and core technologies in the Middle Paleolithic of the Levant. *Journal of Archaeological Science, 33*, 1293–1309.

Zohar, I., Belmaker, M., Nadel, D., Gafny, S., Goren, M., Hershkovitz, I., et al. (2008). The living and the dead: How do taphonomic processes modify relative abundance and skeletal completeness of freshwater fish? *Palaeogeography, Palaeoclimatology, Palaeoecology, 258*, 292–316.

Chapter 3
Quaternary Mammals, People, and Climate Change: A View from Southern North America

Ismael Ferrusquía-Villafranca, Joaquín Arroyo-Cabrales, Eileen Johnson, José Ruiz-González, Enrique Martínez-Hernández, Jorge Gama-Castro, Patricia de Anda-Hurtado, and Oscar J. Polaco

Abstract The Pleistocene and modern mammal faunas of southern North America strongly differ in taxonomic makeup, distribution, and physiognomy. The former faunal complexes are part of the ancient landscape in which early peoples may have interacted. Customarily, differences between the Pleistocene and modern faunas have been attributed to climate change or human-impact driven extinctions. Mexico's Pleistocene mammal record is analyzed in time and space, emphasizing the study of the Rancholabrean Chronofauna, which is the most recent North American Land Mammal Age fauna. Palynological and paleosol records are reviewed as an independent check of the interpretation derived from mammals. The integration of the information provides the basis for a proposal regarding Late Pleistocene climate change trends across the country, and whether people were involved in the mammalian community response to climate change in terms of extinction or biogeographic shifting within and outside the country. This approach supports an explanation of the differences between southern North America's Pleistocene and modern mammal faunas.

Keywords Biogeography • Chronofaunas • Mammals • México • Paleoenvironments • Pleistocene • Rancholabrean

Introduction

Climate change is a world-wide phenomenon that has affected, and is affecting, biodiversity. Currently, that change may be driven largely by human activities (Kellogg 1978; Schneider and Temkin 1978; AIMES 2010; Caballero-Miranda et al. 2010). Changes, however, have taken place episodically during the Quaternary, i.e., the last 2.6 million years (Wright et al. 1993; Barnosky 2008; Gibbard et al. 2010), and have occurred many times in the history of the Earth. Nonetheless, given their recent nature, Pleistocene climatic oscillations are understood better than those of previous geologic epochs because their records are well documented and dated. Their impact on the constitution, structure, and distribution of the biota can be assessed more easily because many species involved are extant (Barnosky et al. 2011).

The North American Quaternary mammal record is very extensive, but that of its southern part (meaning Mexico) is comparatively less well known. Mexico's record is crucial for adequately understanding the impact of climate change that affected continental taxonomic makeup, community structure, and biogeography. Mexico's Quaternary mammal record is enormous, and its modern and Pleistocene components show important differences.

The modern terrestrial mammal fauna consists of 11 orders, 36 families, 168 genera, and 496 species (Ramírez-Pulido et al. 2014). The Pleistocene fauna retrieved from more than 15,000 mammal records consists at present of 13 orders (if the questionable Litoptern record is corroborated), 43 families, 146 genera, and 297 species. Thus, the Pleistocene record is more diverse at ordinal and family levels. It includes both extinct and extant taxa, and some of the latter

I. Ferrusquía-Villafranca · J. Ruiz-González ·
E. Martínez-Hernández · J. Gama-Castro · P. de Anda-Hurtado
Instituto de Geología, Universidad Nacional, Autónoma de México, Ciudad Universitaria, 04510 Coyoacan, Mexico, D. F., Mexico
e-mail: ismaelfv@unam.mx

J. Arroyo-Cabrales (✉) · O.J. Polaco
Laboratorio de Arqueozoología, Instituto Nacional de Antropología e Historia, Moneda 16, Centro Histórico, 06060 México, D. F., Mexico
e-mail: arromatu5@yahoo.com.mx

E. Johnson
Museum of Texas Tech University, Box 43191, Lubbock, TX 79409-3191, USA
e-mail: eileen.johnson@ttu.edu

Present Address:
J. Arroyo-Cabrales
Laboratorio de Arqueozoología, Subdirección de Laboratorios y Apoyo Académico Instituto Nacional de Antropología e Historia, Moneda 16, Col., Centro, 06060 México, D. F., Mexico

were wider-ranging than today's distribution. As in the present, the order with the highest number of fossil species is Rodentia, followed by Chiroptera and Carnivora (Wilson and Reeder 2005). All terrestrial extant orders and families also are recorded as fossils, but only 197 (approximately 2/5) of the 496 modern species are found in the Pleistocene record (Arroyo-Cabrales et al. 2007, 2010; Ferrusquía-Villafranca et al. 2010).

On the other hand, 3 orders, 10 families, 38 genera, and 86 species that existed in the Pleistocene are no longer present in the modern fauna. The Orders Notoungulata and Litopterna are extinct world-wide, and the Proboscidea no longer occur in the Americas. Five families are extinct (Gomphotheridae, Mammutidae, Glyptodontidae, Megatheriidae, and Mylodontidae), whereas the families Herpestidae, Equidae, Hydrochoeridae, Camelidae, and Megalonychidae have been extirpated from North America. Twenty-nine (20%) of the 146 genera recorded in the Pleistocene are now extinct, and nine (6%) have been extirpated. At the species level, 77 are extinct (26%) and nine (3%) are extirpated from Mexico (Ferrusquía-Villafranca et al. 2010).

Although the Laurentide Glacier did not reach Mexico, its advances and retreats in the Pleistocene most likely affected its climate and influenced its landscape (Caballero et al. 2010). A wide variety of environmental conditions existed, such as low lake levels and a downward displacement of ~ 1000 m for the limit of pine forests with alpine vegetation, as well as changes in vegetation composition and distribution. This variety, in turn, allowed co-occurrence of very diverse mammalian taxa, including megaherbivores and megacarnivores that functioned as community-dominant or keystone species. For example, during the Pleistocene, the Central Plateau was covered by grassland, and mammoth coexisted with camel, bison, horses, pronghorns, saber-toothed cat, Pleistocene lion, short-faced bear, dire wolf, skunks, hares, rabbits, capybaras, voles, rats, mice, and bats. At that time, the forested areas on the mountain slopes of the contiguous Sierras Madres Occidental and Oriental, and the Trans-Mexican Volcanic Belt were the roaming grounds of mastodon, gomphotheres, ground sloths, toxodonts, deer, spectacled bear, mountain lion, weasel, otter, raccoon, forest rodents, lagomorphs, and shrews. Early peoples may have coexisted in some of those areas, with the earliest record of human occupation in Mexico dating to around 11.0 ^{14}C kBP (Sánchez 2001; Gonzalez and Huddart 2008; Sánchez et al. 2009).

Geographically, Mexico has an important role in regard to current discussion about the First Americans. It has been considered a large biogeographic corridor (after Simpson 1940; Martin and Harrell 1957) for the first human groups coming from north to south. Few data are available, however, regarding the interactions of these early peoples and large Pleistocene mammals in Mexico (e.g., Arroyo-Cabrales et al. 2006; Johnson et al. 2006). Data equally are limited on the relationship, if any, between the extinction of those large mammals and early peoples in southern North America (Sanchez 2001; González and Huddart 2008).

Changes in the biota's physiognomy have been invoked as instances of extinction/migration processes (ultimately) driven by climate change (Barnosky 2008; Ceballos et al. 2010b). For example, a comparison of Mexico's Pleistocene and modern mammal faunas indicates the latter has very few large species. Extinct taxa include among others, Notoungulata, Mammutidae, Gomphotheriidae, *Smilodon*, *Glossotherium*, and *Neotoma magnodonta*. On the other hand, Elephantidae, Camelidae, Equidae, Hydrochoeridae, and *Neotoma floridana* are extant but extirpated from the country. Finally, the present distribution of some mammal species may be due to their movements during glacial times, with vestigial or relict distributions in those areas that they were able to reach, e.g., *Sorex milleri* and *Lepus flavigularis*. Changes in biodiversity and geographic range seem to be greater in temperate than tropical areas. In the former, a southward shift is apparent, such as with Felidae and Ursidae.

Summing up, extinction, extirpation, and the fauna's physiognomic change related to individual species body size may be linked directly to resource depletion. Resources are distributed differently over the landscape, affected by environmental and geologic conditions, and forming part of the ecosystem (Forman and Godron 1986). In depletion, the type, quality, or quantity of a resource is no longer available or at an adequate level to maintain the population. That depletion could have been induced by shrinking and/or loss of primary plant food-source areas, replacement by less nutritional plants, large herbivore density decrease and/or extinction, as well as possible competition by humans for food. The cause(s) may be different for each species, yet they all seem related primarily to climate and ecosystem changes (Koch and Barnosky 2006). This overview provides a synthesis of the Quaternary Mexican mammalian faunas, their geographic and chronologic distribution, and climatic implications. It creates a foundation from which to explore possible causes for extinctions and extirpations, including human influence.

Methods

The Quaternary mammal record of southern North America has been examined in an attempt to characterize the impact of climate change in the fauna's makeup, structure, and distribution. Comparisons are made between the Pleistocene (Table 3.1) and modern faunas, using the mammal species of the latter as environmental indicators. Their presence/absence in particular times and/or places may disclose differences that may be linked to or explained by climate change. Corroborative evidence from other indicators such as Pleistocene palynological assemblages (regarded as flora surrogates) and paleosol (buried soils) records where available also are used (see Ferrusquía-Villafranca et al. 2010).

For the Pleistocene mammal record analysis, the chronological framework used by Bell et al. (2004) is followed. In Mexico, the Rancholabrean Chronofauna is the best represented. Most local faunas are of Late Rancholabrean age (Quaternary Mexican Mammalian Database, QMMDB; Arroyo-Cabrales et al. 2002). Dating is largely biochronological, i.e., commonly based on a few age-diagnostic taxa. Thus, their position within the Rancholabrean North American Land Mammal Age (NALMA; roughly between 200 to 150 ka and 11 ka; Bell et al. 2004) is approximate. Furthermore, less than 30 local faunas have radiocarbon ages (Arroyo-Cabrales et al. 2002, 2007; Ferrusquía-Villafranca 2010), all Late Rancholabrean. In some instances, the faunas, palynological assemblages, and paleosol records can be placed within the 40–11 ^{14}C kBP interval. In most cases, the environmental comparisons of the fossil and modern mammal faunas largely involve Late Rancholabrean local faunas.

Discussion and comparison of results from segments of a territory as large as Mexico ($\sim$2,000,000 km^2), with a bewildering geomorphic and geologic diversity, calls for reference to an ecologically/geologically meaningful spatial framework. That framework must have equally meaningful spatial units in order to promote precision, clarity, and reproducibility of results. Scientific comparisons based on such vaguely bounded territorial segments as central Mexico, northern Mexico, or the state boundaries would render nearly meaningless such comparisons. Hence, the spatial framework used here (Table 3.2 and Fig. 3.1) is based on the morphotectonic province concept (Ferrusquía-Villafranca 1993; Ferrusquía-Villafranca et al. 2010).

Table 3.1 Pleistocene fossil mammals of Mexico, extant species also included. Main sources: Alvarez (1965), Barrios Rivera (1985), QMMDB (Arroyo-Cabrales et al. 2002), Arroyo-Cabrales et al. (2007); and Ferrusquía-Villafranca et al. (2010)

	Body size	Diet
DIDELPHIMORPHIA		
Didelphidae		
Caluromys derbianus	S	O
Didelphis marsupialis	S	O
Didelphis virginiana	S	O
Marmosa canescens	S	O
Marmosa lorenzoi[a]	S	O
Marmosa mexicana	S	O
Philander opossum	S	O
XENARTHRA sensu lato		
Dasypodidae		
Cabassous centralis[d]	S	O
Dasypus novemcinctus	S	O
Holmesina septentrionalis[a]	L	O
Pampatherium mexicanum[a]	L	O
Glyptodontidae[a]		
Glyptotherium cylindricum[a]	L	H
Glyptotherium floridanum[a]	L	H
Glyptotherium mexicanum[a]	L	H
Megalonychidae[a]		
Megalonyx jeffersoni[a]	L	H
Megalonyx wheatleyi[a]	L	H
Megatheriidae[a]		
Eremotherium laurillardi[a]	L	H
Nothrotheriops mexicanum[a]	L	H
Nothrotheriops shastensis[a]	L	H
Mylodontidae[a]		
Paramylodon harlani[a]	L	H
Myrmecophagidae		
Myrmecophaga tridactyla[d]	M	C
Tamandua mexicana	S	C
SORICOMORPHA		
Soricidae		
Cryptotis mayensis	S	C
Cryptotis mexicana	S	C
Cryptotis parva	S	C
Notiosorex crawfordi	S	C
Sorex cinereus	S	C
Sorex milleri	S	C
Sorex oreopolus	S	C
Sorex saussurei	S	C
CHIROPTERA		
Antrozoidae		
Antrozous pallidus[d]	S	C
Emballonuridae		
Balantiopteryx io[d]	S	C
Peropteryx macrotis	S	C
Saccopteryx bilineata	S	C
Molossidae		
Eumops bonariensis	S	C
Eumops perotis[d]	S	C

(continued)

Table 3.1 (continued)

	Body size	Diet
Eumops underwoodi[d]	S	C
Molossus rufus	S	C
Nyctinomops aurispinosus	S	C
Nyctinomops laticaudatus	S	C
Promops centralis	S	C
Tadarida brasiliensis	S	C
Mormoopidae		
Mormoops megalophylla	S	C
Pteronotus davyi	S	C
Pteronotus parmellii	S	C
Natalidae		
Natalus stramineus	S	C
Phyllostomidae		
Artibeus jamaicensis	S	H
Artibeus lituratus	S	H
Carollia brevicauda	S	H
Carollia perspicillata	S	H
Carollia soweli	S	H
Centurio senex	S	H
Chiroderma villosum	S	H
Choeronycteris mexicana	S	H
Chrotopterus auritus	S	C
Dermanura Phaeotis	S	H
Desmodus cf. *D. draculae*[a]	S	B
Desmodus rotundus	S	B
Desmodus stocki[a]	S	B
Diphylla ecaudata	S	B
Enchisthenes hartii	S	H
Glossophaga soricina	S	H
Leptonycteris curasoae	S	H
Leptonycteris nivalis	S	H
Macrotus californicus[d]	S	C
Micronycteris microtis	S	C
Mimon bennettii	S	C
Sturnira lilium	S	H
Tonatia evotis	S	C
Vespertilionidae		
Corynorhinus townsendii	S	C
Eptesicus brasiliensis	S	C
Eptesicus furinalis	S	C
Eptesicus fuscus	S	C
Lasionycteris noctivagans	S	C
Lasiurus cinereus	S	C
Lasiurus ega	S	C
Lasiurus intermedius	S	C
Myotis californicus	S	C
Myotis keaysi	S	C
Myotis thysanodes	S	C
PRIMATES		
Cebidae		
Alouatta palliata[d]	S	O
Alouatta pigra	S	O
Ateles geoffroyi[d]	S	O

Table 3.1 (continued)

	Body size	Diet
CARNIVORA		
Canidae		
Canis cedazoensis[a]	M	C
Canis dirus[a]	M	C
Canis edwardii[a]	M	C
Canis familiaris	M	C
Canis latrans	M	C
Canis lupus	M	C
Canis rufus[c]	M	C
Cuon alpinus[c]	M	C
Urocyon cinereoargenteus	S	C
Felidae		
Herpailurus yagouaroundi	S	C
Leopardus pardalis	M	C
Leopardus wiedii	S	C
Lynx rufus	S	C
Panthera atrox[a]	L	C
Panthera onca	L	C
Puma concolor	M	C
Smilodon californicus	L	C
Smilodon fatalis[a]	L	C
Smilodon gracilis[a]	L	C
Hyaenidae[b]		
Chasmaporthetes johnstoni[a]	M	C
Mustelidae		
Conepatus leuconotus	S	O
Conepatus mesoleucus	S	O
Lontra longicaudis	S	C
Mephitis macroura	S	O
Mephitis mephitis	S	O
Mustela frenata	S	C
Mustela nigripes[d]	S	C
Spilogale putorius[d]	S	O
Taxidea taxus	S	C
Procyonidae		
Bassariscus astutus	S	O
Bassariscus sumichrasti	S	O
Bassariscus ticuli[a]	S	O
Nasua narica	M	O
Potos flavus	S	H
Procyon lotor	M	O
Procyon pygmaeus	M	O
Ursidae		
Arctodus pristinus[a]	L	C
Arctodus simus[a]	L	C
Tremarctos floridanus[a]	L	O
Ursus americanus[d]	L	O
RODENTIA		
Castoridae		
Castor cf. *C. californicus*	M	H
Cuniculidae		
Cuniculus paca	M	H

(continued)

Table 3.1 (continued)

	Body size	Diet
Dasyproctidae		
Dasyprocta mexicana	M	H
Dasyprocta punctata	M	H
Erethizontidae		
Erethizon dorsatum[d]	M	H
Coendou mexicanus	M	H
Geomyidae		
Cratogeomys bensoni[a]	S	H
Cratogeomys castanops	S	H
Cratogeomys gymnurus	S	H
Cratogeomys merriami	S	H
Cratogeomys tylorrhinus	S	H
Orthogeomys grandis	S	H
Orthogeomys hispidus	S	H
Orthogeomys onerosus[a]	S	H
Thomomys bottae	S	H
Thomomys umbrinus	S	H
Heteromyidae		
Chaetodipus hispidus	S	H
Chaetodipus huastecensis[a]	S	H
Chaetodipus nelsoni	S	H
Chaetodipus penicillatus[d]	S	H
Dipodomys nelsoni	S	H
Dipodomys phillipsii	S	H
Dipodomys spectabilis	S	H
Heteromys desmarestianus	S	H
Heteromys gaumeri	S	H
Liomys irroratus	S	H
Perognathus flavus	S	H
Hydrochaeridae[b]		
Neochoerus aesopi[a]	M	H
Muridae		
Baiomys intermedius[a]	S	H
Baiomys musculus	S	H
Baiomys taylori	S	H
Hodomys alleni[d]	S	H
Hodomys sp. nov.[a]	S	H
Microtus californicus[d]	S	H
Microtus guatemalensis	S	H
Microtus meadensis[a]	S	H
Microtus mexicanus	S	H
Microtus oaxacensis	S	H
Microtus pennsylvanicus[d]	S	H
Microtus quasiater	S	H
Microtus umbrosus	S	H
Neotoma albigula[d]	S	H
Neotoma angustapalata	S	H
Neotoma anomala[a]	S	H
Neotoma cinerea[c]	S	H
Neotoma floridana[c]	S	H
Neotoma lepida	S	H
Neotoma magnodonta[a]	S	H
Neotoma mexicana	S	H
Neotoma micropus	S	H
Neotoma palatina	S	H
Neotoma phenax[d]	S	H
Neotoma tlapacoyana[a]	S	H
Neotomodon alstoni	S	H
Nyctomys sumichrasti	S	H
Oligoryzomys fulvescens	S	H
Ondatra nebracensis[a]	S	H
Onychomys leucogaster	S	H
Oryzomys alfaroi	S	H
Oryzomys couesi	S	H
Oryzomys melanotis	S	H
Otonyctomys hatti	S	H
Ototylomys phyllotis	S	H
Peromyscus boylii	S	H
Peromyscus dificilis	S	H
Peromyscus eremicus	S	H
Peromyscus leucopus	S	H
Peromyscus levipes	S	H
Peromyscus maldonadoi[a]	S	H
Peromyscus maniculatus	S	H
Peromyscus melanophrys	S	H
Peromyscus melanotis[d]	S	H
Peromyscus mexicanus	S	H
Peromyscus ochraventer	S	H
Peromyscus pectoralis	S	H
Peromyscus truei[d]	S	H
Peromyscus yucatanicus	S	H
Reithrodontomys fulvescens	S	H
Reithrodontomys megalotis	S	H
Reithrodontomys mexicanus	S	H
Reithrodontomys montanus	S	H
Sigmodon alleni	S	H
Sigmodon arizonae	S	H
Sigmodon curtisi[a]	S	H
Sigmodon fulviventer	S	H
Sigmodon hispidus	S	H
Sigmodon leucotis	S	O
Sigmodon toltecus	S	H
Synaptomys cooperi[d]	S	H
Tylomys nudicaudus	S	H
Sciuridae		
Ammospermophilus interpres	S	H
Cynomys ludovicianus[d]	S	H
Cynomys mexicanus	S	H
Glaucomys volans	S	H
Marmota flaviventris[d]	S	H
Sciurus alleni	S	H
Sciurus aureogaster	S	H
Sciurus deppei	S	H
Sciurus nayaritensis	S	H

(continued)

Table 3.1 (continued)

	Body size	Diet
Sciurus variegatoides[d]	S	H
Sciurus yuctanensis	S	H
Spermophilus mexicanus	S	H
Spermophilus spilosoma	S	H
Spermophilus variegatus	S	H
LAGOMORPHA		
Leporidae		
Aluralagus sp.[a]	S	H
Aztlanolagus agilis[a]	S	H
Lepus alleni	S	H
Lepus californicus	S	H
Lepus callotis	S	H
Romerolagus diazii	S	H
Sylvilagus audubonii	S	H
Sylvilagus bachmani	S	H
Sylvilagus brasiliensis	S	H
Sylvilagus cunicularius	S	H
Sylvilagus floridanus	S	H
Sylvilagus hibbardi[a]	S	H
Sylvilagus leonensis[a]	S	H
PERISSODACTYLA		
Equidae[b]		
Equus alaskae[a]	L	H
Equus calobatus[a]	L	H
Equus conversidens[a]	L	H
Equus excelsus[a]	L	H
Equus ferus[a]	L	H
Equus cf. *E. francisci*	L	H
Equus giganteus[a]	L	H
Equus mexicanus[a]	L	H
Equus pacificus[a]	L	H
Equus parastylidens[a]	L	H
Equus simplicidens[a]	L	H
Equus tau[a]	L	H
Tapiridae		
Tapirus bairdii[d]	L	H
Tapirus haysii[a]	L	H
ARTIODACTYLA		
Antilocapridae		
Antilocapra americana	L	H
Capromeryx mexicana[a]	L	H
Capromeryx minor[a]	L	H
Stockoceros conklingi[a]	L	H
Tetrameryx mooseri[a]	L	H
Tetrameryx shuleri[a]	L	H
Tetrameryx tacubayensis[a]	L	H
Bovidae		
Bison alaskensis[a]	L	H
Bison antiquus[a]	L	H
Bison bison[d]	L	H
Bison latifrons[a]	L	H
Bison priscus[a]	L	H
Euceratherium collinum[a]	L	H

(continued)

Table 3.1 (continued)

	Body size	Diet
Oreamnos harringtoni[a]	L	H
Ovis canadensis[d]	L	H
Camelidae[b]		
Camelops hesternus[a]	L	H
Camelops mexicanus[a]	L	H
Camelops minidokae[a]	L	H
Camelops traviswhitei[a]	L	H
Eschatius conidens[a]	L	H
Hemiauchenia blancoensis[a]	L	H
Hemiauchenia macrocephala[a]	L	H
Hemiauchenia vera[a]	L	H
Procamelops minimus[a]	L	H
Cervidae		
Cervus elaphus[c]	L	H
Mazama americana	M	H
Navahoceros fricki[a]	L	H
Odocoileus halli[a]	L	H
Odocoileus hemionus	L	H
Odocoileus lucasi[a]	L	H
Odocoileus virginianus	L	H
Tayassuidae		
Platygonus alemanii[a]	M	H
Platygonus compressus[a]	M	H
Platygonus ticuli[a]	M	H
Tayassu tajacu	M	H
Tayassu pecari	M	H
PROBOSCIDEA		
Elephantidae[b]		
Mammuthus columbi[a]	L	H
Mammuthus primigenius[a]	L	H
Gomphotheriidae[a]		
Cuvieronius tropicus[a]	L	H
Stegomastodon mirificus[a]	L	H
Mammutidae[a]		
Mammut americanum[a]	L	H
NOTONGULATA[a]		
Toxodontidae[a]		
Myxotoxodon cf. *M. larensis*[a]	L	H
LITOPTERNA[a]		
Macrauchenidae[a]		
Gen. et sp. indet.[a]	L	H

Summary: 13 Orders, 43 Families, 146 Genera, and 297 Species
Notes and abbreviations: Marine taxa excluded. References to taxa identified only at generic level are not included, but Litopterna
[a]Extinct taxon
[b]Suprageneric taxon extinct in Mexico, but extant outside Mexico
[c]Species extinct in Mexico, but extant outside this country
[d]Species extinct in the morphotectonic province(s) bearing the fossil locality (ies), but extant elsewhere in Mexico. Body mass: L, large. M, medium. S, small. Diet: H, herbivore. B, hematophagous. C, carnivore. O, omnivore. Further information in the text

Table 3.2 Location and basic features of the Morphotectonic Provinces of Mexico. Main sources: Ferrusquía-Villafranca (1993)

Province[a]	Location	Surface in km² and percentage[b]	Altitude ranges (m)[c]	Climate[d]	Chief land form
1	Northwestern Mexico 109°30′–117°00′ WL 23°00′–32°30′ NL	144,000 ~7.34%	0–2,130 0–1,000	BWh, BShs, Csa	Sierras and plains
2	Northwestern Mexico 107°00′–116°00′ WL 23°00′–32°30′ NL	236,800 ~12.02%	0–2,200 200–1,000	BWh, BSh	Sierras and plains
3	Western and northwestern Mexico 102°20′–109°40′ WL 20°30′–31°20′ NL	289,000 ~14.68%	200–3,000 2,000–3,000	Cfb, Aw	Sierras and plateaus
4	Northern Mexico 101°31′–110°31′ WL 26°00′–31°45′ NL	255,900 ~12.52%	200–2,000 800–1,200	BShw, BWh, BSk	Sierras and plateaus
5	Northeastern and northcentral Mexico Transverse sector 100°00′–105°00′ WL 24°30′–26°00′ NL Eastern sector 97°30′–101°20′ WL 19°40′–26°00′ NL	145,500 ~7.54%	200–3,000 1,000–2,000	Transverse sector BWh, BSk Eastern sector Cla, Cwa, BSh	Sierras
6	Eastern Mexico Northern sector 96°30′–100°20′ WL 20°00′–26°00′ NL Southern sector 91°15′–96°46′ WL 17°10′–19°20′ NL	170,600 ~8.66%	0–200	Northern sector Aw′, Cw, Cx′w′ Southern sector Afw′, Amw′	Plains
7	Central Mexico 100°00′–104°00′ WL 21°00′–24°00′ NL	85,300 ~4.33%	1,000–3,300 2,000–3,000	BSh	Plateaus
8	Central Mexico 96°20′–105°20′ WL 17°30′–20°25′ NL Main sector 19°00′–21°00′ NL	175,700 ~9.17%	1,000–5,000 1,000–2,000	Aw′, Cfa, Cwa BSh, Cw, Cflb, Aw	Peaks and plateaus
9	Southern Mexico 94°45′–104°40′ WL 15°40′–19°40′ NL	195,700 ~9.93%	0–3,500 1,200–1,800	Aw′, Aw, BShw, Cwa, Cfa	Sierras and depressions
10	Southeastern Mexico 90°30′–95°00′ WL 14°30′–17°40′ NL	105,400 ~5.35%	0–2,500 200–1,000	Aw, Cw, Cf	Sierras, depressions, and plains
11	Eastern Mexico 87°00′–91°00′ WL 17°50′–21°30′ NL	167,600 ~8.46%	0–200	BShw, Amw	Plains and karst topography

[a]These numbers correspond to those on the map in Figure 1. Baja California morphotectonic province (mp). 2. Northwestern Plains and Sierras mp. 3. Sierra Madre Occidental mp. 4. Chihuahuan-Coahuilan Plateaus and Ranges mp. 5. Sierra Madre Oriental mp. 6. Gulf Coast Plain mp. 7. Central Plateau mp. 8. Trans-Mexican Volcanic Belt mp. 9. Sierra Madre del Sur mp. 10. Sierra Madre de Chiapas mp. 11. Yucatan Platform mp

[b]Percentage = ratio of mp to total surface area of Mexico

[c]First entry = total range, second entry = dominant range

[d]BWh, desert-like, Mat > 18°C; BShs, Csa, temperate with dry winter; BSh, dry Mat > 18°C; Cfb, temperate humid with no dry season; BShw, steppe-like, winter dry season, Mat > 18°C; BSk, steppe-like, Mat > 18°C; Cfa, temperate, no defined dry season; Cwa, temperate with dry winter; Aw′, tropical with dry winter and rainy fall; Cw, temperate with dry winter; Cx′w′, temperate with little rain throughout the year; Afw′, tropical rainy with no defined dry season; Cf, temperate with no defined dry season. The key to the letter symbology is: A, warm humid and subhumid Climate Group (lack of a well-defined dry season); m, rainy season restricted to the summer; w, dry winter and warm season from April to September; w′, less rainy summer with a short dry season. B, warm to cold and very arid to semiarid Climate Group; BS, warm to semicold and arid to semiarid Climate Subgroup; BW, warm to semicold and very arid Climate Subgroup; h, semiwarm with cool winter; k, temperate with a warm summer; s, rainy winter. C, temperate to semicold and humid to semihumid Climate Group; a, warm summer; b, cool and long summer; x′, rainy fall. Source: García (1988)

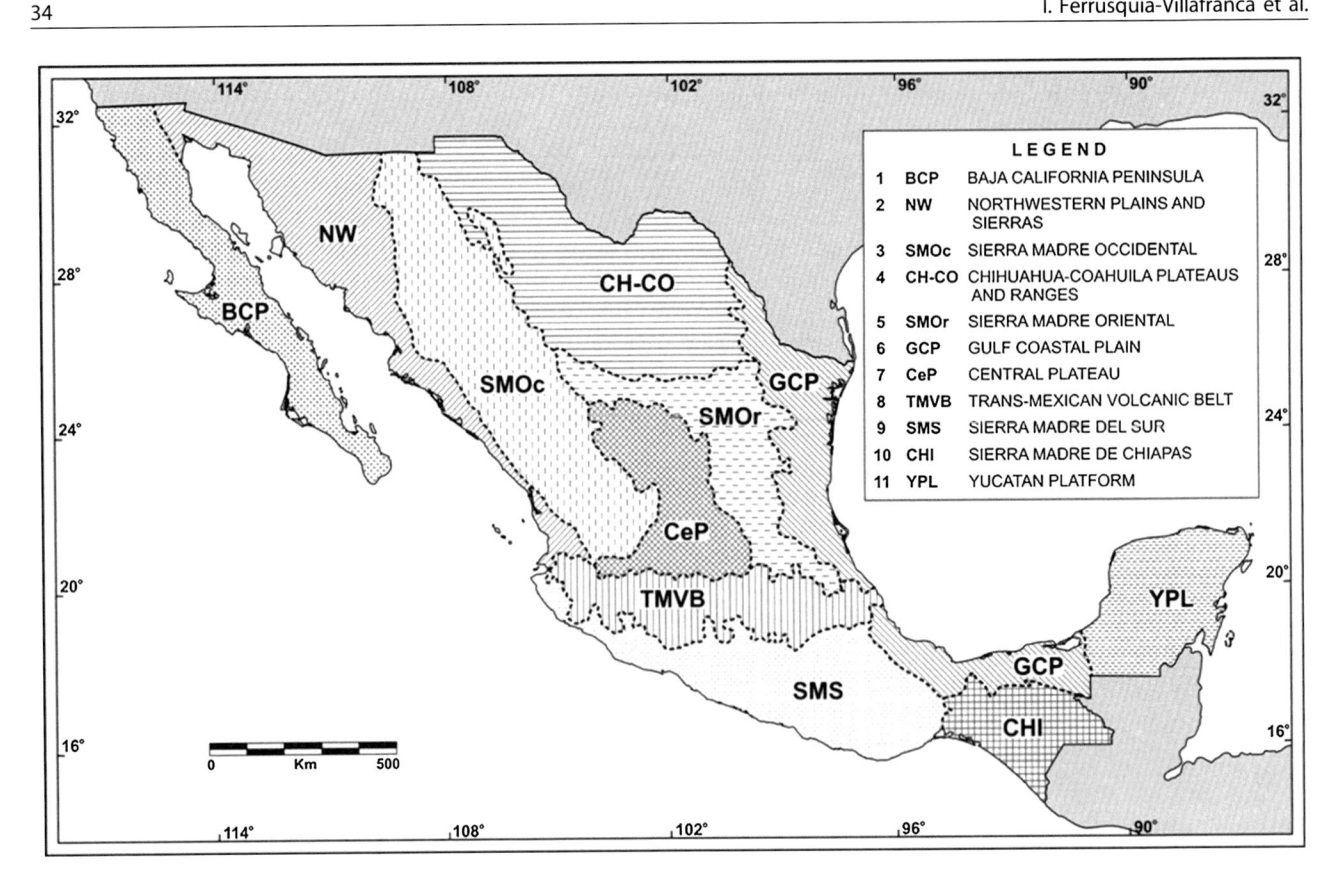

Fig. 3.1 Morphotectonic provinces of Mexico (modified from Ferrusquía-Villafranca 1993, 1998)

Results
The Mammal Record

Mexico's Pleistocene mammal record consists of 297 species, belonging to 146 genera, 43 families, and 13 orders (including Litopterna; Table 3.1), largely drawn from the QMMDB (Arroyo-Cabrales et al. 2002), and supplemented by recent literature (modified from Ferrusquía-Villafranca et al. 2010). This record has been retrieved from ~800 localities unevenly distributed across the country, primarily from the last 100 years (cf. Arroyo-Cabrales et al. 2002). In order to furnish a representative sample of such localities, the 120 most significant ones are plotted on a morphotectonic province template, including those having major local faunas (Fig. 3.2) and listed in Table 3.3; also, those major local faunas are plotted on a modern potential vegetation map (Fig. 3.3) to disclose their ecological congruency (or lack thereof) with their Holocene setting. Due to space limitations, 25 major local faunas (Table 3.4) have been selected and arranged by their morphotectonic province occurrence. This procedure allows assessment of the climate change impact on mammal faunas of particular provinces or groups and detection of major patterns of climate change in space and time across Mexico.

Northern Provinces

The Baja California Peninsula, Northwestern Plains and Sierras, Sierra Madre Occidental, and the Chihuahua-Coahuila plateaus and ranges comprise this morphotectonic province (Fig. 3.1). Only the Northwestern Plains and Sierras and Chihuahua-Coahuila Plateaus and Ranges provinces have yielded major local faunas (Table 3.4 and Fig. 3.2), palynofloras (Fig. 3.4), and paleosol localities (Fig. 3.5). The Northwestern Plains and Sierras have yielded Mexico's only Irvingtonian fauna, El Golfo Local Fauna (l. f.) (Shaw 1981; Croxen et al. 2007), as well as the earliest Rancholabrean fauna, Térapa l.f. (Mead et al. 2006), both in Sonora.

These faunas indicated a more humid climate regime that allowed the presence of a subtropical biota during the Irvingtonian and Early Rancholabrean. The Sierra Madre Occidental paleosol record and the Chihuahua-Coahuila

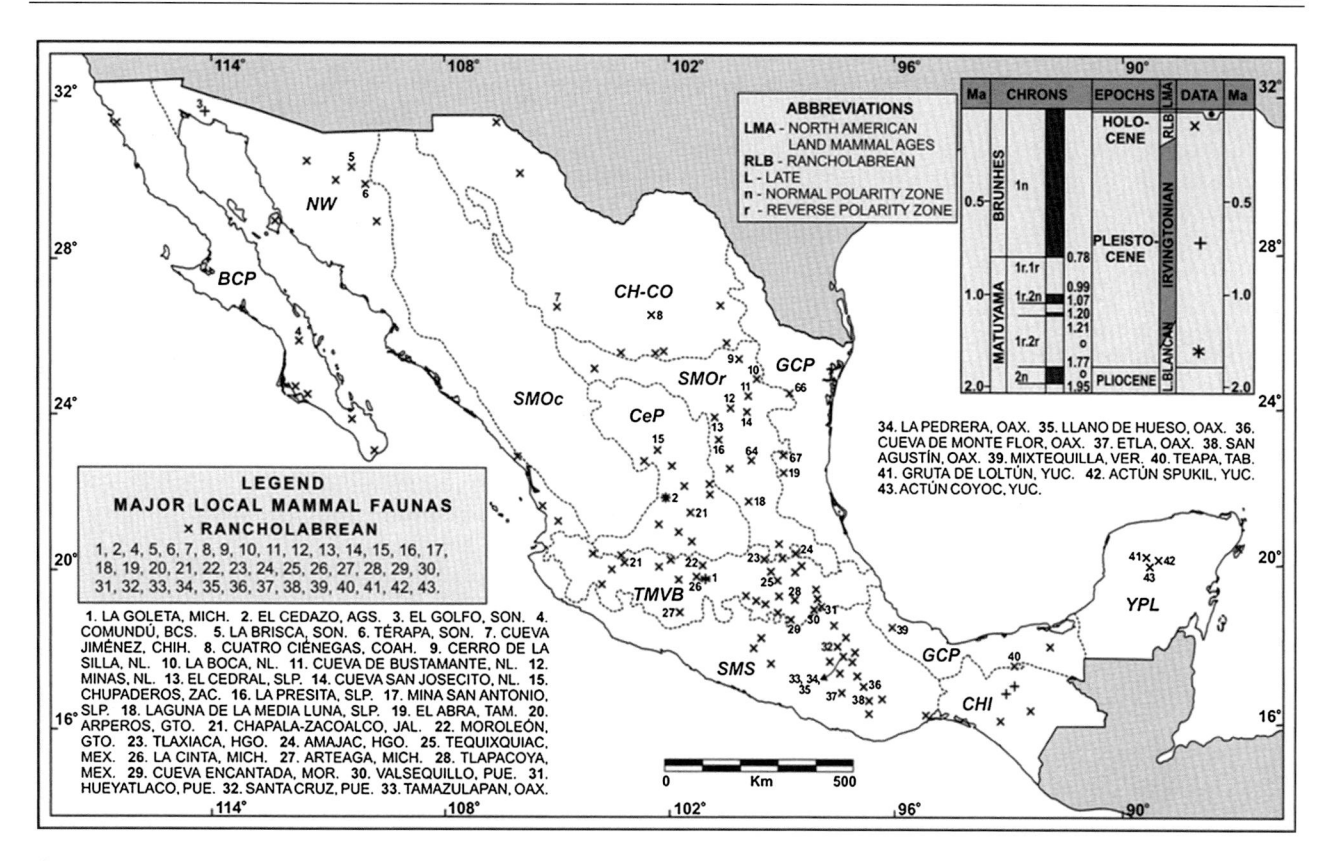

Fig. 3.2 Selected Rancholabrean terrestrial mammal localities of Mexico mapped on the morphotectonic provinces template. Time frame adapted from Bell et al. (2004). The geographic position of the localities is approximated. Main sources: Alvarez (1965), Barrios-Rivera (1985), QMMDB (Arroyo-Cabrales et al. 2002), and Arroyo-Cabrales et al. (2007)

Table 3.3 Selected Pleistocene mammal bearing localities of Mexico. Num. is the locality number shown at Figs. 3.2 and 3.3. The locality names correspond to the nearest town/village or topographic feature. Main sources: Alvarez (1965), Barrios-Rivera (1985), and QMMDB (Arroyo-Cabrales et al. 2002)

Num.	Locality	State	Morphotectonic province	Age
1	La Goleta	Mich	TMBV	Late Blancan/Irvingtonian/Rancholabrean
2	El Cedazo	Ags	CeP	Late Blancan/Irvingtonian/Rancholabrean
3	El Golfo	Son	NW	Irvingtonian
4	Comondú	BCS	BCP	Rancholabrean
5	La Brisca	Son	NW	Rancholabrean
6	Terapa	Son	NW	Early Rancholabrean
7	Cueva Jiménez	Chih	CH-CO	Rancholabrean
8	Cuatro Ciénegas	Coah	CH-CO	Rancholabrean
9	Cerro de la Silla	NL	SMOr	Rancholabrean
10	La Boca	NL	SMOr	Rancholabrean
11	Cueva de Bustamante	NL	SMOr	Rancholabrean
12	Minas	NL	SMOr	Rancholabrean
13	El Cedral	SLP	SMOr	Rancholabrean
14	San Josecito	NL	SMOr	Rancholabrean
15	Chupaderos	Zac	CeP	Rancholabrean
16	La Presita	SLP	SMOr	Rancholabrean
17	Mina San Antonio	SLP	SMOr	Rancholabrean
18	Laguna de la Media Luna	SLP	SMOr	Rancholabrean
19	El Abra	Tam	SMOr	Rancholabrean

(continued)

Table 3.3 (continued)

Num.	Locality	State	Morphotectonic province	Age
20	Arperos	Gto	CeP	Rancholabrean
21	Chapala-Zacoalco	Jal	TMBV	Rancholabrean
22	Moroleón	Gto	TMBV	Rancholabrean
23	Tlaxiaca	Hgo	TMBV	Rancholabrean
24	Amajac	Hgo	TMBV	Pleistocene (?Middle-Late)
25	Tequixquiac	Mex	TMBV	Rancholabrean
26	La Cinta	Mich	TMBV	Rancholabrean
27	Arteaga	Mich	TMBV	Rancholabrean
28	Tlapacoya	Mex	TMBV	Rancholabrean
29	Cueva Encantada	Mor	TMBV	Rancholabrean
30	Valsequillo	Pue	TMBV	Rancholabrean
31	Hueyatlaco	Pue	TMBV	Rancholabrean
32	Santa Cruz	Pue	SMS	Rancholabrean
33	Tamazulapan	Oax	SMS	Rancholabrean
34	La Pedrera	Oax	SMS	Rancholabrean
35	Llano de Hueso	Oax	SMS	Rancholabrean
36	Cueva de Monte Flor	Oax	SMS	Rancholabrean
37	Etla	Oax	SMS	Rancholabrean
38	San Agustín	Oax	SMS	Rancholabrean
39	Mixtequilla	Ver	GCP	Rancholabrean
40	Teapa	Tab	GCP	Rancholabrean
41	Gruta de Loltún	Yuc	YPL	Rancholabrean
42	Actún Spukil	Yuc	YPL	Rancholabrean
43	Actún Coyoc	Yuc	YPL	Rancholabrean
44	Punta San José	BC	BCP	Pleistocene
45	Bahía Magdalena	BCS	BCP	Late Pleistocene
46	Santa Rita	BCS	BCP	Rancholabrean
47	El Carrizal	BCS	BCP	Rancholabrean
48	Santa Anita	BCS	BCP	?Early Pleistocene
49	La Playa	Son	NW	Rancholabrean
50	Arizpe	Son	NW	Pleistocene
51	Isla Tiburón	Son	NW	Pleistocene
52	La Guitarra	Son	NW	Pleistocene
53	El Rosario	Sin	NW	Pleistocene
54	San Blas	Nay	NW	Pleistocene
55	El Pantanal	Nay	NW	Pleistocene
56	Samalayuca	Chih	CH-CO	Pleistocene
57	La Erupción	Chih	CH-CO	Pleistocene
58	La Candela	Coah	CH-CO	Pleistocene
59	Torreón	Coah	CH-CO	Pleistocene
60	Arroyo del Arenal	Coah	CH-CO	Pleistocene
61	Arroyo Ojuelo	Coah	CH-CO	Pleistocene
62	Cuevas del Padre	NL	CH-CO	Pleistocene
63	Cuevas de las Iglesias	Dgo	SMOr	Pleistocene
64	Tula	Tams	SMOr	Pleistocene
65	Mezquital	Hgo	SMOr	Pleistocene
66	San Lázaro	Tams	GCP	Rancholabrean
67	El Salitrillo	Tams	GCP	Pleistocene
68	Zacatecas	Zac	CeP	Pleistocene
69	Laguna de las Tres Cruces	SLP	CeP	Pleistocene
70	Laguna del Salitrillo	Zac	CeP	Pleistocene
71	Brechas Coloradas	SLP	CeP	Pleistocene
72	Rancho Peotillos	SLP	CeP	Pleistocene

(continued)

Table 3.3 (continued)

Num.	Locality	State	Morphotectonic province	Age
73	Rancho la Verdolaga	Jal	CeP	Pleistocene
74	León	Gto	CeP	Pleistocene
75	Guanajuato	Gto	CeP	Pleistocene
76	Ameca	Jal	TMBV	Pleistocene
77	El Salto	Jal	TMBV	Rancholabrean
78	Atotonilco-Zacoalco	Jal	TMBV	Rancholabrean
79	Venustiano Carranza	Jal	TMBV	Pleistocene
80	Ario de Rayón	Mich	TMBV	Pleistocene
81	La Piedad-Santa Ana	Mich-Jal	TMBV	Rancholabrean
82	Zacapú	Mich	TMBV	Rancholabrean
83	Portalitos	Gto	TMBV	Rancholabrean
84	Cuitzeo	Mich	TMBV	Rancholabrean
85	Uruétaro	Mich	TMBV	Rancholabrean
86	Actopan	Hgo	TMBV	Rancholabrean
87	Real del Monte	Hgo	TMBV	Pleistocene
88	Pachuca	Hgo	TMBV	Pleistocene
89	Villa de Tezontepec	Hgo	TMBV	Rancholabrean
90	Huitexcalco	Hgo	TMBV	Rancholabrean
91	Amanalco de Becerra	Mex	TMBV	Pleistocene
92	Tacubaya	DF	TMVB	Rancholabrean
93	Valle de Toluca	Mex	TMBV	Rancholabrean
94	Lerma	Mex	TMBV	Rancholabrean
95	Palpan	Mor	TMBV	Pleistocene
96	Tetela	Mor	TMBV	Rancholabrean
97	Apizaco	Tlax	TMBV	Pleistocene
98	Atlihuetzia	Tlax	TMBV	Pleistocene (?Middle-Late)
99	Nanacatla	Gro	SMS	Pleistocene (?Middle)
100	Huixtac	Gro	SMS	Rancholabrean
101	Tepecoacuilco	Gro	SMS	Rancholabrean
102	Apaxtla	Gro	SMS	Rancholabrean
103	Chichihualco	Gro	SMS	Pleistocene
104	San Pedro Tecamachalco	Pue	SMS	Pleistocene
105	Tehuacan	Pue	SMS	Rancholabrean
106	Tepelmeme	Oax	SMS	Pleistocene
107	Coixtlahuaca	Oax	SMS	Pleistocene
108	Huajuapan de León	Oax	SMS	Pleistocene
109	Yolomecatl	Oax	SMS	Late Pleistocene
110	Nochixtlan	Oax	SMS	Pleistocene
111	La Salina	Oax	SMS	Rancholabrean
112	Santa Marta Ejutla	Oax	SMS	Pleistocene
113	Tehuantepec	Oax	SMS	Pleistocene
114	Villa Corzo	Chis	CHI	Late Pleistocene
115	Aguacatenango	Chis	CHI	Pleistocene
116	Chiapa de Corzo	Chis	CHI	Irvingtonian/Rancholabrean
117	Ixtapa	Chis	CHI	Irvingtonian/Rancholabrean
118	El Ocotlán	Camp	GCP	Pleistocene
119	Actun Lara	Yuc	YPL	Rancholabrean
120	Cozumel	QRoo	YPL	Pleistocene

Abbreviations: **BCP**, Baja California Península. **NW**, Northwestern Plains and Sierras. **CH-CO**, Chihuahuan-Coahuilan Plateaus and Ranges. **SMOr**, Sierra Madre Oriental. **CeP**, Central Plateau. **GCP**, Gulf Coastal Plain. **TMVB**, Trans-Mexican Volcanic Belt. **SMS**, Sierra Madre del Sur. **YPL**, Yucatan Platform

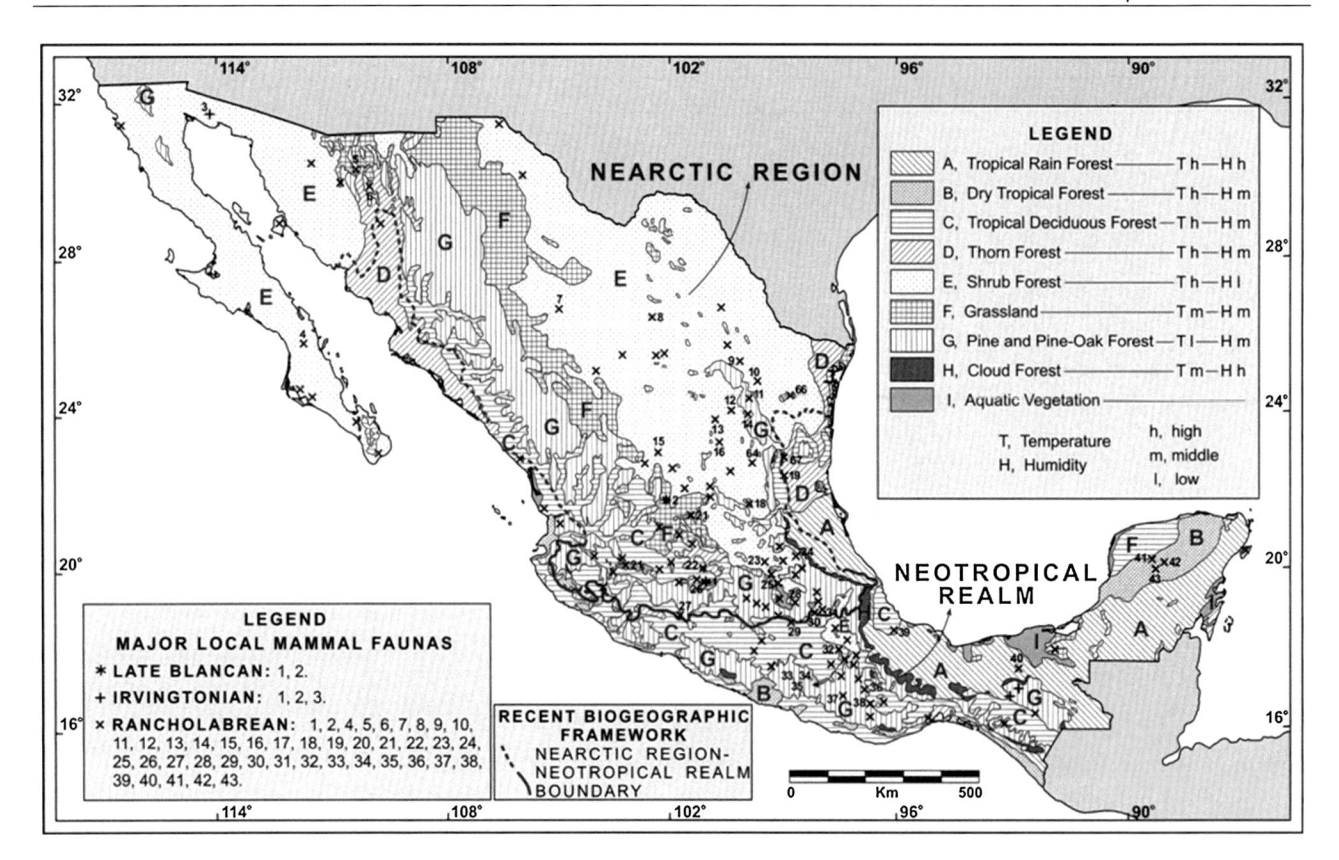

Fig. 3.3 Mexico's chief Pleistocene terrestrial mammal-bearing localities mapped on a modern natural vegetation template (adapted from Rzedowski and Reyna-Trujillo 1990, and Challenger 1998); additionally, the Nearctic Region/Neotropical Realm boundary is plotted). Main sources: Alvarez (1965), Barrios-Rivera (1985), QMMDB (Arroyo-Cabrales et al. 2002), and Arroyo-Cabrales et al. (2007)

Plateaus and Ranges paleo-vegetation record (Betancourt et al. 1990) indicated a cooler and moister climate regime for the latest Rancholabrean (~25–11 kyr). This regime made possible a biotic diversity unparalleled anywhere in Mexico at present. Mammals with diverse ecological requirements coexisted within short distances of each other.

The biota became depleted and modified compositionally as the climate regime changed by the earliest Holocene. Increasing aridity caused xerophilous species largely to replace the former inhabitants. Many mammal taxa, ranging from medium-sized animals weighting between 10 and 100 kg (mesobaric) to large-sized animals weighting over 100 kg (megabaric) (Ferrusquía-Villafranca et al. 2010), became extinct (Table 3.4). Others, depending on their climatic tolerances, experienced major range contractions either northward, e.g., the temperate rodents *Castor* and *Marmota* and the carnivoran *Canis rufus*, or southward, e.g., the tropical capybara *Hydrochoerus*. The extent of this biogeographic shift was such that the extant taxa no longer inhabit Mexico, i.e., they are extirpated. In other instances, some taxa became extirpated from their Rancholabrean morphotectonic province but persist at present elsewhere in the country, e.g., the rodent *Hodomys alleni* and perissodactyl *Tapirus*, currently living in the tropics (Ceballos and Oliva 2005). *Cynomys ludovicianus*, the black-tailed prairie dog, known from the La Playa local fauna (Mead et al. 2010), retreated at least 150 km ENE from its Rancholabrean grounds to easternmost Sonora and adjacent Arizona. The Late Wisconsinan climate regime allowed the existence of these herbivores in northern Sonora, whereas the current climate prevents their presence.

Central and Eastern Provinces

These morphotectonic provinces include the Central Plateau Sierra Madre Oriental and Gulf Coastal Plain. They have yielded eight major local faunas (Table 3.4 and Figs. 3.2 and

Table 3.4 Selected main Pleistocene local mammal faunas of Mexico. Main sources: Alvarez (1965); Barrios-Rivera (1985); QMMDB; Arroyo-Cabrales et al. (2002), and Arroyo-Cabrales et al. (2007). Symbols: [a]Extinct taxon everywhere in the world. [b]Species extinct in Mexico, but still extant outside this country. [c]Species extinct in the morphotectonic province(s) bearing the fossil locality(ies), but extant elsewhere in Mexico

	NW			CH-CO		SMOr					CeP		GCP	TMVB								SMS		YPL	
	A	B	C	D	E	F	G	H	I	J	K	L	M	N	O	P	Q	R	S	T	U	V	W	X	Y
DIDELPHIMORPHIA																									
Didelphidae																									
Didelphis marsupialis																								X	
Didelphis virginiana						X					X			X										X	
Marmosa canescens																								X	
Marmosa lorenzoi[a]																								X	
Marmosa mexicana														X										X	
XENARTHRA sensu lato																									
Dasypodidae																									
Cabassous centralis[c]																									X
Dasypus novemcinctus																									
Holmesina septentrionalis[a]											X			X	X		X	X			sp				
Pampatherium mexicanum[a]	X													X											
Glyptodontidae[a]																									
Glyptotherium cylindricum[a]	X							sp						sp			X					sp			
Glyptotherium floridanum[a]								X							X	X				X					
Glyptotherium mexicanum[a]															X										
Megalonychidae[a]																									
Megalonyx jeffersonii[a]						X											X								
Megalonyx wheatleyi[a]							sp	sp								X									
Megatheriidae[a]																									
Eremotherium laurillardi[a]			X			X								X	X										
Nothrotheriops mexicanum[a]															X										
Nothrotheriops shastensis[a]			X			X	X	X	X					X			X								
Mylodontidae[a]																									
Paramylodon harlani[a]								X			X			X	X										
Paramylodon sp.[a]		sp																			X				
Myrmecophagidae																									
Myrmecophaga tridactyla[c]			X																						

(continued)

Table 3.4 (continued)

	NW			CH-CO		SMOr					CeP		GCP	TMVB								SMS		YPL	
	A	B	C	D	E	F	G	H	I	J	K	L	M	N	O	P	Q	R	S	T	U	V	W	X	Y
SORICOMORPHA																									
Soricidae																									
Cryptotis mayensis																								X	
Cryptotis mexicana						X																			
Cryptotis parva				X						X															
Notiosorex crawfordi					X																				
Sorex cinereus						X																			
Sorex milleri						X																			
Sorex saussurei						X																			
CHIROPTERA																									
Antrozoidae																									
Antrozous pallidus[c]					X																				
Emballonuridae																									
Peropteryx macrotis																								X	
Molossidae																									
Eumops bonariensis																								X	
Eumops perotis[c]				X																					
Eumops underwoodi[c]																								X	
Molossus rufus																								X	
Nyctinomops laticaudatus																								X	
Promops centralis																								X	
Tadarida brasiliensis					X																				
Mormoopidae																									
Mormoops megalophylla						X										X								X	X
Pteronotus davyi																								X	
Pteronotus parnellii																								X	X
Natalidae																									
Natalus stramineus																								X	
Phyllostomidae																									
Artibeus jamaicensis																								X	X
Artibeus lituratus																								X	
Carollia brevicauda																								X	
Centurio senex																								X	

(continued)

Table 3.4 (continued)

	NW			CH-CO		SMOr					CeP		GCP	TMVB								SMS		YPL	
	A	B	C	D	E	F	G	H	I	J	K	L	M	N	O	P	Q	R	S	T	U	V	W	X	Y
Chiroderma villosum																								X	
Choeronycteris mexicana						X																		X	X
Chrotopterus auritus																								X	
Desmodus cf. *D. draculae*[a]																								X	
Desmodus rotundus													X											X	
Desmodus stocki[a]																X									
Diphylla ecaudata																								X	
Glossophaga soricina							X																	X	
Leptonycteris curasoae																									
Leptonycteris nivalis						X																			
Macrotus californicus[c]							X																		
Mimon bennettii																									
Sturnira lilium																								X	
Vespertilionidae																									
Corynorhinus townsendii						X																			X
Eptesicus brasiliensis																									X
Eptesicus furinalis																									
Eptesicus fuscus						X																		X	
Lasiurus cinereus						X																			
Lasiurus ega																								X	
Lasiurus intermedius						X																		X	
Myotis californicus																									
Myotis keaysi					sp		sp	sp																X	
Myotis thysanodes						X																			
CARNIVORA																									
Canidae																									
Canis cedazoensis[a]	X										X														
Canis dirus[a]						X		X	X		X	cf		X	X		X							X	
Canis familiaris													X	X			X							X	
Canis latrans				X	X	X	X	X	X		X			X	X	X	X							X	
Canis lupus						X		X	X	X				X	X									X	
Canis rufus[b]		X													X										
Cuon alpinus[b]																									
Urocyon cinereoargenteus				X		X	X				X													X	

(continued)

Table 3.4 (continued)

	NW			CH-CO		SMOr					CeP		GCP	TMVB							SMS		YPL	
	A	B	C	D	E	F	G	H	I	J	K	L	M	N	O	P	Q	R	S	T	U	V	W	X	Y
Felidae																									
Herpailurus yagouaroundi						X																		X	
Leopardus pardalis																									X
Leopardus wiedii																								X	
Lynx rufus	X			X	X	X				X	X				X	X								X	
Panthera atrox[a]		X		X		X					X			X	X									X	
Panthera onca		X	X	X		X				X	X													X	
Puma concolor				X	X	X							X	X										X	
Smilodon fatalis[a]						X					X			X	X							X		X	
Smilodon gracilis[a]																	X							X	
Smilodon californicus											X														
Hyaenidae *																									
Chasmaporthetes johnstoni[a]			X																						
Mustelidae																									
Conepatus leuconotus													X												
Conepatus mesoleucus				X		X								X		X								X	
Lontra longicaudis														X											
Mephithis macroura							sp																		
Mephitis mephitis				X		X																		X	
Mustela frenata				X		X	sp																		
Mustela nigripes[c]					X		X																		
Spilogale putorius[c]				X		X	X									X	X							X	X
Taxidea taxus				X		X					X													X	
Procyonidae																									
Bassariscus astutus		X		X		X									X									X	
Bassariscus sumichrastri															X									X	
Bassariscus ticuli[a]							X																		
Nasua narica				X																				X	
Procyon lotor	X			X		X								X		X								X	X
Ursidae																									
Arctodus pristinus[a]											X				X		X								
Arctodus simus[a]											X			X	X		X								
Tremarctos floridanus[c]						X		X													cf				
Ursus americanus[c]		X		X		X		X							sp	X									

(continued)

Table 3.4 (continued)

	NW			CH-CO		SMOr					CeP		GCP	TMVB								SMS		YPL	
	A	B	C	D	E	F	G	H	I	J	K	L	M	N	O	P	Q	R	S	T	U	V	W	X	Y
RODENTIA																									
Castoridae																									
Castor cf. *C. californicus*			X																						
Cuniculidae																									
Cuniculus pacca																								X	
Dasyproctidae																									
Dasyprocta mexicana																									
Dasyprocta punctata																									
Erethizontidae																									
Erethizon dorsatum[c]				X		X	X				X														
Coendou mexicanus																								X	
Geomyidae																									
Cratogeomys bensoni[a]																									
Cratogeomys castanops				X	X	X		X		X	X				X	X				X					
Cratogeomys gymnurus						X								X											
Cratogeomys merriami																X									
Cratogeomys tylorrhinus															X										
Geomis sp.	sp																								
Orthogeomys grandis																									
Orthogeomys hispidus																								X	
Orthogeomys onerosus[a]						X																			
Thomomys bottae																									
Thomomys umbrinus				X	X	X					X				X		X			X					
Heteromyidae																									
Chaetodipus hispidus																									
Chaetodipus huastecensis[a]																									
Chaetodipus nelsoni				X																					
Chaetodipus penicillatus[c]				X																					
Dipodomys nelsoni		sp			X		sp																		
Dipodomys phillipsii																									
Dipodomys spectabilis				X	X																				
Heteromys desmarestianus																									
Heteromys gaumeri																								X	

(continued)

Table 3.4 (continued)

	NW			CH-CO		SMOr					CeP		GCP		TMVB								SMS			YPL	
	A	B	C	D	E	F	G	H	I	J	K	L	M	N	O	P	Q	R	S	T	U	V	W	X	Y
Liomys irroratus			sp			X										X				X					
Perognathus flavus					sp		X																		
Hydrochoeridae•																									
Hydrochoerus sp.	X																								
Neochoerus aesopi[a]			sp											sp		X									
Cricetidae																									
Baiomys intermedius[a]															X										
Baiomys musculus										X															
Baiomys taylori					X					sp					X										
Hodomys alleni[c]			sp																						
Hodomys sp. nov[a]															X										
Microtus californicus[c]							X	X								X		X							
Microtus guatemalensis								X																	
Microtus meadensis[a]						X	X	sp		X				X	X	X		X	X						
Microtus mexicanus					X																				
Microtus oaxacensis																									
Microtus pennsylvanicus[c]					X																				
Microtus quasiater					X																				
Microtus umbrosus																									
Neotoma albigula[c]	sp	sp	sp	X	X	X	X	sp		X				X		X		X	X						
Neotoma angustapalata																X									
Neotoma anomala[a]						X																			
Neotoma cinerea[b]					X																				
Neotoma floridana[b]																									
Neotoma lepida					X																				
Neotoma magnodonta[a]															X	X	X								
Neotoma mexicana										X							X			X					
Neotoma micropus					X					X															
Neotoma palatina																			X						
Neotoma phenax[c]														X											
Neotoma tlapacoyana[a]																X									
Neotomodon alstoni															X										

(continued)

Table 3.4 (continued)

	NW			CH-CO		SMOr					CeP	GCP		TMVB								SMS		YPL	
	A	B	C	D	E	F	G	H	I	J	K	L	M	N	O	P	Q	R	S	T	U	V	W	X	Y
Nyctomys sumichrasti																									
Oligoryzomys fulvescens		sp																							
Ondatra nebracensis[a]			sp																						
Onychomys leucogaster					sp																				
Oryzomys alfaroi																									
Oryzomys couesi														X										X	
Oryzomys melanotis																								X	
Otonyctomys hatti																								X	
Ototylomys phyllotis																								X	
Peromyscus boylii					sp	X		sp		X	sp														
Peromyscus dificilis						X	X			X															
Peromyscus eremicus							X																		
Peromyscus leucopus																								X	
Peromyscus levipes						X																			
Peromyscus maldonadoi[a]															X	X									
Peromyscus maniculatus						X	X			X															
Peromyscus melanophrys																									
Peromyscus melanotis[c]																									
Peromyscus mexicanus																									
Peromyscus ochraventer																									
Peromyscus pectoralis							X																		
Peromyscus truei[c]																								X	
Peromyscus yucatanicus																									
Reithrodontomys fulvescens				X				sp		X				X											
Reithrodontomys megalotis				X		X				X				X											
Reithrodontomys mexicanus																									
Reithrodontomys montanus										X															
Sigmodon alleni										X															
Sigmodon arizonae										X															
Sigmodon curtisi[a]	sp	sp	sp		sp																				
Sigmodon fulviventer										X															
Sigmodon hispidus				X										X	sp				X					X	
Sigmodon leucotis						X	X			X															

(continued)

Table 3.4 (continued)

	NW			CH-CO			SMOr					CeP			GCP		TMVB							SMS			YPL	
	A	B	C	D	E	F	G	H	I	J	K	L	M	N	O	P	Q	R	S	T	U	V	W	X	Y
Sigmodon toltecus										X															
Synaptomys cooperi[c]						X																			
Tylomys nudicaudus																									
Sciuridae																									
Ammospermophilus interpres				X																					
Cynomys ludovicianus[c]					sp		sp																		
Cynomys mexicanus																			X						
Glaucomys volans																									
Marmota flaviventris[c]						X																			
Sciurus alleni						X													sp						
Sciurus aureogaster																									
Sciurus deppei																									
Sciurus nayaritensis																									
Sciurus variegatoides[c]						X	X																		
Sciurus yuctanensis																								X	
Spermophilus mexicanus				X														X							
Spermophilus spilosoma				X	X																				
Spermophilus variegatus				X	X																				
LAGOMORPHA																									
Leporidae																									
Aztlanolagus agilis[a]					X																				
Lepus alleni										sp															
Lepus californicus				X	X		X	X	X																
Lepus callotis																									
Sylvilagus audubonii	sp			X			X	X	X	sp				sp											
Sylvilagus brasiliensis																								X	
Sylvilagus cunicularius							X	X								X									
Sylvilagus floridanus						X	X	X								X					U			X	
Sylvilagus hibbardi[a]		X																							
Sylvilagus leonensis[a]						X	X			X															

(continued)

Table 3.4 (continued)

	NW			CH-CO		SMOr					CeP			GCP	TMVB							SMS		YPL	
	A	B	C	D	E	F	G	H	I	J	K	L	M	N	O	P	Q	R	S	T	U	V	W	X	Y
PERISSODACTYLA																									
Equidae[a]																									
Equus alaskae[a]	sp																		sp				sp		
Equus calobatus[a]	sp				sp		sp	sp								X									
Equus complicatus[a]		X	X			X																			
Equus conversidens[a]		sp	X		sp	X	sp	X	X	sp	X	X		X	X		X		X		X			X	
Equus excelsus[a]											X				X		X								
Equus ferus[a]											X														
Equus giganteus[a]															X										
Equus mexicanus[a]								X				X		X	X	X	X								
Equus occidentalis[a]									X						X	X	X								
Equus pacificus[a]								sp			X				X										
Equus parastylidens[a]											X														
Equus simplicidens[a]																					X				
Equus tau[a]		X										X			X										
Tapiridae																									
Tapirus bairdii[c]	sp		sp					sp						sp	sp		X								
Tapirus haysii[a]						X		X	X																
ARTIODACTYLA																									
Antilocapridae																									
Antilocapra americana				X	X				X								sp		sp						
Capromeryx mexicana[a]	sp				sp		sp	X			X		X		X		X		X	sp					
Capromeryx minor[a]		X																							
Stockoceros conklingi[a]	sp		sp			X	X			X	X		X		X		X	X							
Tetrameryx mooseri[a]		sp	sp								X				sp										
Tetrameryx shuleri[a]											X		X				X	X							
Tetrameryx tacubayensis[a]																									
Bovidae																									
Bison alaskensis[a]	sp	sp						sp			X	sp		X	X		X			sp			sp		
Bison antiquus[a]													X	X	X		X								
Bison bison[c]				X																					
Bison latifrons[a]																X								X	
Bison priscus[a]									X					X	X										

(continued)

Table 3.4 (continued)

	NW			CH-CO		SMOr						CeP		GCP	TMVB							SMS			YPL	
	A	B	C	D	E	F	G	H	I	J	K	L	M	N	O	P	Q	R	S	T	U	V	W	X	Y
Euceratherium collinum[a]						X		X							X										
Oreamnos harringtoni[a]						X																			
Ovis canadensis[c]			sp	X																					
Camelidae♦																									
Camelops hesternus[a]	sp	X						X	X		X			X	X	X	X	X					sp		
Camelops mexicanus[a]			sp												X										
Camelops minidokae[a]																	X								
Camelops traviswhitei[a]							X				X														
Eschatius conidens[a]															X										
Hemiauchenia blancoensis[a]			X				sp				sp					sp							sp	sp	
Hemiauchenia macrocephala[a]								X	X						X		X								
Hemiauchenia vera[a]								X																	
Procamelops minimus[a]																	X								
Paleolama sp. ???			sp																						
Titanotylops sp. ???			sp																						
Cervidae																									
Cervus elaphus[b]				X																					
Mazama americana													X												
Navahoceros fricki[a]						X					X											X		X	
Odocoileus halli[a]	sp							X			X			X		X							sp		
Odocoileus hemionus		X		X						X	X														
Odocoileus lucasi						X		X			X														
Odocoileus virginianus				X			X		X				X	M		X								X	X
Tayassuidae																									
Platygonus alemanii[a]	sp										sp				X		X				X				
Platygonus compressus[a]												X													
Platygonus ticuli[a]									X					X											
Tayassu tajacu															sp						sp				
PROBOSCIDEA																									
Elephantidae♦																									
Mammuthus columbi[a]	sp	sp	sp					X	X		X	X		X	X	sp	X				sp		sp	X	X
Mammuthus primigenius[a]														X											

(continued)

Table 3.4 (continued)

	NW			CH-CO		SMOr					CeP			GCP	TMVB						CeP		SMS		YPL
	A	B	C	D	E	F	G	H	I	J	K	L	M	N	O	P	Q	R	S	T	U	V	W	X	Y
Gomphotheriidae[a]																									
Cuvieronius tropicus[a]	sp		sp											sp			sp				sp		sp		
Stegomastodon cf. *S. mirificus*[a]															X										
Mammutidae[a]															X										
Mammut americanum[a]								X	sp		X			sp			X							X	
NOTONGULATA[a]																									
Toxodontidae[a]																									
Myxotoxodon cf. *M. larensis*[a]																									
?LITOPTERNA[a]																									
?Macrauchenidae[a]																		X							
Gen. et sp. indet.[a]																		X							

Morphotectonic Provinces: **NW**, Northwestern Plains and Sierras. **CH-CO**, Chihuahuan-Coahuilan Plateaus and Ranges. **SMOr**, Sierra Madre Oriental. **CeP**, Central Plateau. **GCP**, Gulf Coast Plain. **TMVB**, Trans-Mexican Volcanic Belt. **SMS**, Sierra Madre del Sur. **YPL**, Yucatan Platform

Local Faunas: **A**, Terapa, Son. **B**, La Brisca, Son. **C**, El Golfo, Son. **D**, Cuatro Ciénegas, Coah. **E**, Cueva Jiménez, Chih. **F**, San Josecito, NL. **G**, La Presita, SLP. **H**, El Cedral, SLP. **I**, Minas, NL. **J**, Mina San Antonio, SLP. **K**, El Cedazo, Ags. **L**, Chupaderos, Zac. **M**, La Mixtequilla, **N**, Chapala-Zacoalco, Jal **O**, Tequixquiac, Mex. **P**, Tlapacoya, Mex **Q**, Valsequillo, Pue. **R**, Hueyatlaco, Pue. **S**, La Cinta, Mich., **T**, San Agustín Tlaxiaca, Hgo. **U**, Amajac, Hgo. **V**, San Agustín, Oax. **W**, Planicie de Tamazulapan, Oax. **X**, Gruta de Loltún, Yuc. **Y**, Actún Spukil, Yuc

Sources for the local faunas: **A**: (Paz-Moreno et al. 2003; Bell et al. 2004; Mead et al. 2006, 2007; Carranza-Castañeda and Roldán-Quintana 2007; Hodnett et al. 2009; Núñez et al. 2010; White et al. 2010). **B**: (Van Devender et al. 1990; White et al. 2010). **C**: (Shaw 1981; Shaw and McDonald 1987; Jefferson 1989; Shaw et al. 2005; White et al. 2010). **D**: (Gilmore 1947; Lundelius 1980; Frazier 1981; QMMDB Arroyo-Cabrales et al. 2002). **E**: (Messing 1986; Russell and Harris 1986). **F**: (Furlong 1943; Stock 1943; Cushing 1945; Findley 1953; Jackway 1958; Russell 1960; Kurtén, 1975; Nowak 1979; Kurtén and Anderson 1980; Álvarez and Polaco 1981; Barrios-Rivera 1985; Arroyo-Cabrales and Johnson 1995, 1998, 2008; Arroyo-Cabrales et al. 1996; Polaco and Butron-M 1997; Arroyo-Cabrales and Álvarez 2003; Esteva et al. 2005). **G**: (Polaco and Butron-M. 1997; Arroyo-Cabrales et al. 2004). **H**: (Nowak 1979; Álvarez and Polaco 1981; Polaco 1981; Barrios-Rivera 1985; Lorenzo and Mirabell edits. 1986; Arroyo-Cabrales et al. 1996; Polaco and Butron-M. 1997; Alberdi et al. 2003 Pérez-Crespo et al. 2011). **I**: (Franzen 1994; Arroyo-Cabrales et al. 1996). **J**: (Furlong 1943; Torres-Martinez 1995; Arroyo-Cabrales and Johnson 1998; Ferrusquia-Villafranca and de Anda-Hurtado 2008; de Anda-Hurtado 2009). **K**: (Mooser 1958; Hibbard and Mooser 1963; Dalquest 1974; Mooser and Dalquest 1975a; Mooser and Dalquest 1975b; Frazier 1981; Barrios-Rivera 1985; Montellano-Ballesteros 1992; Reynoso-Rosales and Montellano-Ballesteros 1994; Churcher et al. 1996; QMMDB, Arroyo-Cabrales et al. 2002; Arroyo-Cabrales and Álvarez 2003). **L**: Barrón-Ortiz et al. (2009). **M**: Polaco (1995). **N**: (Furlong 1925; Hibbard and Villa-Ramirez 1950; Hibbard 1955; Downs 1958; Álvarez and Ferrusquia 1967; Guenther 1968; Aviña 1969; Silva-Bárcenas 1969; Álvarez 1971, 1983, 1986; Mones 1973b; Kurtén 1974; Nowak 1979; Berta 1988; Churcher et al. 1996; Edmund 1996; Pichardo 1999; QMMDB Arroyo-Cabrales et al. 2002; McDonald 2002; Alberdi et al. 2004; Lucas 2008; Guzmán Gutiérrez et al. 2009). **O**: (Cuatáparo and Ramírez 1875; Cope 1884; Freudenberg 1921; Furlong 1925; Hibbard and Villa-Ramirez 1950; Hibbard 1955; Downs 1958; Guenther 1968; Aviña 1969; Silva-Bárcenas 1969; Von Thenius 1970; Guenther and Bunde 1973; Kurtén 1974; Nowak 1979; Berta 1988; Churcher et al. 1996; Pichardo 1999; QMMDB Arroyo-Cabrales et al. 2002; Lucas 2008). **P**: (Hibbard and Villa-Ramirez, 1950; Hibbard, 1955; Downs, 1958; Álvarez, 1969, 1986; Kurtén, 1974; Nowak, 1979; Berta, 1988; Álvarez and Hernández-Chávez, 1994; Pichardo 1999; González et al. 2003; Lucas, 2008). **Q**: (Freudenberg 1921; Furlong 1925; Hibbard and Villa-Ramirez 1950; Hibbard 1955; Downs 1958; Kurtén 1967, 1974; Guenther 1968; Aviña 1969; Von Thenius 1970; Guenther and Bunde 1973; Nowak 1979; Berta 1988; Churcher et al. 1996; Pichardo 1997, 1999; Arroyo-Cabrales et al. 2002; McDonald 2002; QMMDB Arroyo-Cabrales et al. 2002; Alberdi and Corona 2005; Lucas 2008; Elizalde-García et al. 2011; Melgarejo-Meraz et al. 2011; Moreno-Fernández et al. 2011; Cruz-Muñoz et al. 2009). **R**: Álvarez (1983). **S**: (García-Zepeda et al. 2008, 2009; Marín-Leyva et al. 2009). **T**: (Bravo Cuevas et al. 2009; Palma-Ramirez et al. 2011). **U**: Kurtén(1975). **W**: (Ferrusquia-Villafranca 1976; Jiménez-Hidalgo et al. 2011). **X**: (Álvarez 1982, 1983; Arroyo-Cabrales and Álvarez 2003; Arroyo-Cabrales and Polaco 2003; Morales-Mejía and Arroyo-Cabrales 2009; Morales-Mejía et al. 2009). **Y**: Arroyo-Cabrales and Álvarez (2003)

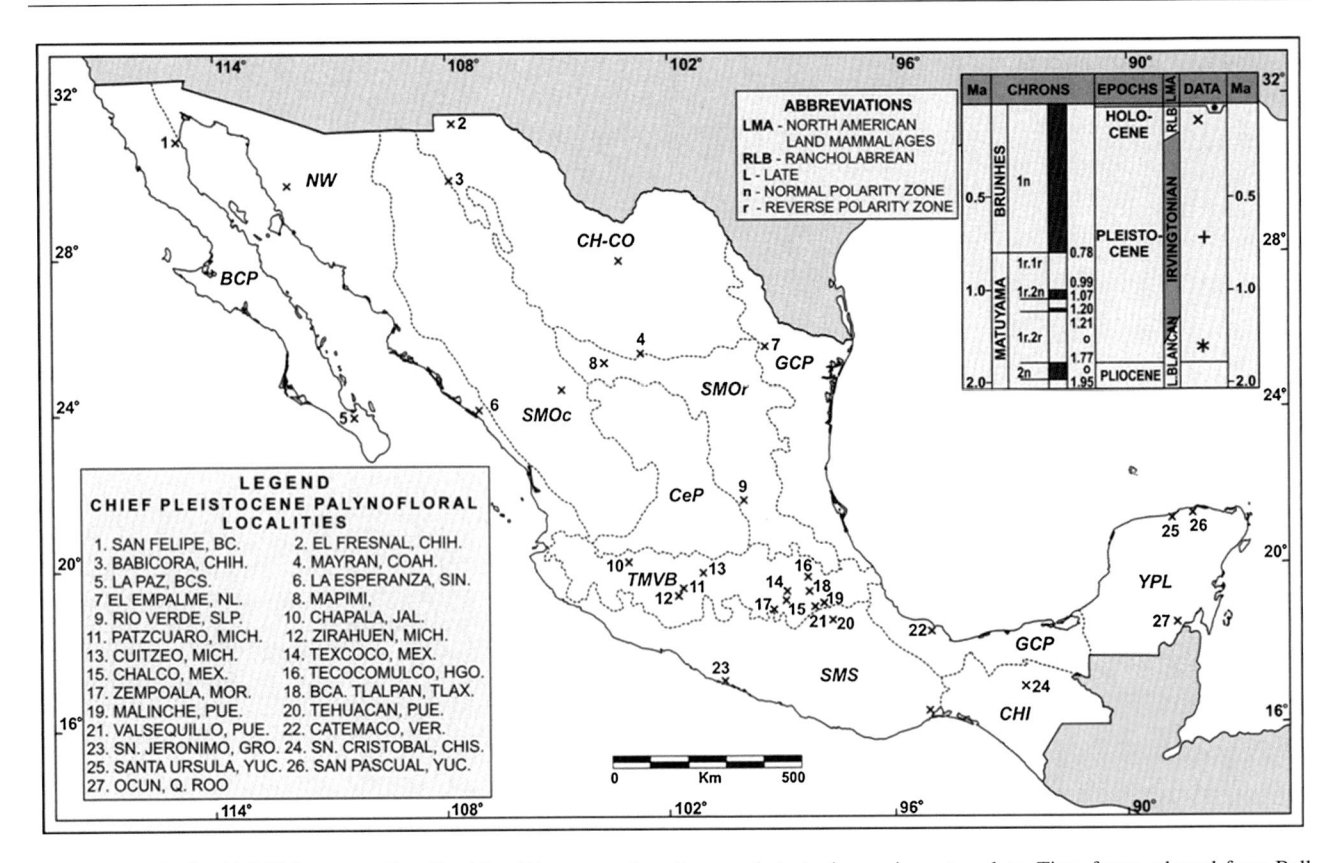

Fig. 3.4 Mexico's chief Pleistocene palynofloral localities mapped on the morphotectonic provinces template. Time frame adapted from Bell et al. (2004). Main sources: (Foreman 1955; Clisby and Sears 1955; Sears and Clisby 1955; Ohngemach 1973, 1977; Bradbury 1971, 1989; Meyer 1973; Brown 1984; González-Quintero 1986; Straka and Ohngemach 1989; Metcalfe 1992; Leyden et al. 1994, 1996; Lozano-García and Ortega-Guerrero 1994, 1997; Anderson and Van Devender 1995; Ortega-Ramírez et al. 1998; Caballero-Miranda et al. 1999; Canul-Montañez 2008; Vázquez et al. 2010; Israde-Alcántara et al. 2010; Caballero-Miranda et al. 2010; Ortega et al. 2010; and Martínez-Hernández, E., unpublished data)

3.3), a few palynofloras (Table 3.5 and Fig. 3.4), and a few paleosol localities, chiefly in the Sierra Madre Oriental province (Fig. 3.5). The Central Plateau province has yielded a biochronologically mixed mammal assemblage including taxa of apparently Late Blancan to Rancholabrean age (Montellano-Ballesteros 1992; Bell et al. 2004) whose time relations are not well understood. Nevertheless, the better part of the assemblage, El Cedazo l.f., is Rancholabrean (Ferrusquía-Villafranca et al. 2010).

By and large, what was inferred from the Northern provinces applies as well to these provinces. The Sierra Madre Oriental and Central Plateau local faunas include tropical/subtropical, temperate, and even xeric species. This combination suggests a variety of ecological settings not extant at present. These provinces would have had a quite different climate during the Rancholabrean, marked by thermal (warm/cold) and humidity/rain (moist/dry) oscillations. These oscillations ultimately were related to advances and retreats of the Laurentide Glacier, as it responded to global climate pattern changes (Broecker 2003). Such climate oscillations also were affected by altitude and latitude, and did not occur in a fixed fashion, i.e., warm with either moist or dry conditions, or cold with moist or dry conditions.

The mammalian and palynological records indicated a latest Rancholabrean (~25–11 ka) cooler and moister climate regime than that of today. The disharmonious character of the mammal fauna, i.e., the non-analogous mammal fauna (Semken 1966), was much more diverse and ecologically varied than today's, involving at least the better part of the Pleistocene. During that longer timeframe, a complex shifting of species distribution, i.e., biogeographic range expansion, contraction, displacement, and/or colonization of new habitats, within relatively short time intervals, as well as extinction, would have taken place. By the end of the Pleistocene, the climate became warmer and drier in general, perhaps to such an extent or intensity that many taxa became extinct.

Medium to large mammals fared the worst, with a major portion of them becoming extinct (Table 3.4). Other species reduced their biogeographic range and currently live outside Mexico, e.g., the carnivoran *Cuon alpinus* and the rodent *Marmota flaviventris* (Hall 2001; Nowak 1991). Others were

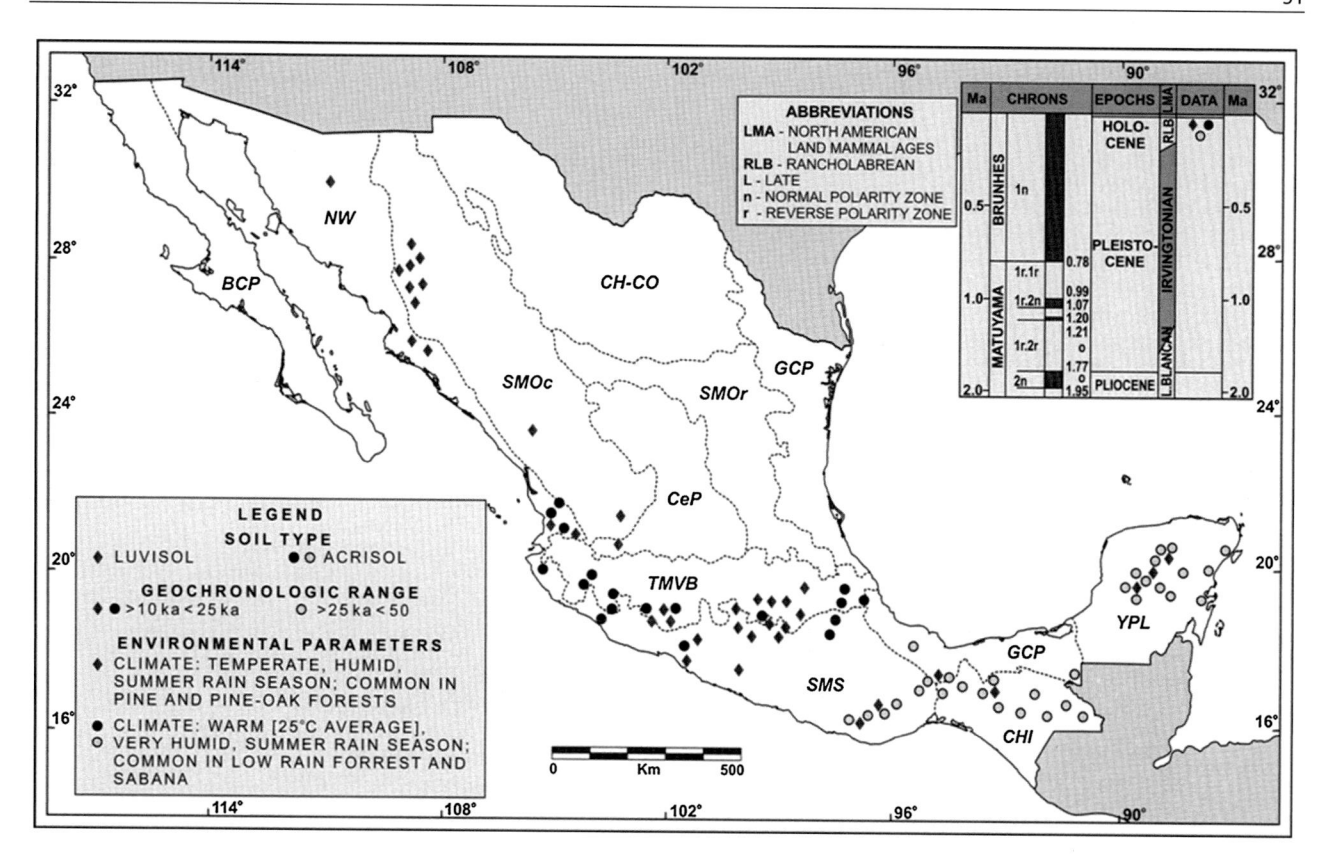

Fig. 3.5 Mexico's chief Pleistocene paleosol localities mapped on the morphotectonic provinces template. Time frame adapted from Bell et al. (2004). Main source: INEGI 2006. Other sources: (Cervantez-Borja et al. 1997; Sedov et al. 2001, 2007, 2008, 2009, 2011; Gama-Castro et al. 2004, 2005; Ortega-Guerrero, et al. 2004; Solleiro-Rebolledo et al. 2003, 2004, 2006; McClung et al. 2005; Jasso-Castañeda et al. 2006, 2007, 2012; Díaz-Ortega et al. 2010; Cruz y Cruz 2011; Gonzáles-Arqueros et al. 2011; Solis et al. 2011; and Tovar and Sedov 2011)

extirpated from their Rancholabrean morphotectonic province, but survived elsewhere within Mexico. For example, the rodent *Hodomys alleni* currently thrives in the tropics. Other rodents such as *Erethizon dorsatum*, *Microtus pennsylvanicus*, *Neotoma albigula*, *Sigmodon arizonae,* and *Synaptomys cooperi* now live in the Northern provinces (Ceballos and Oliva 2005). Finally, rodents such as *Sigmodon alleni* now live in provinces of the southern Northwestern Plains and Sierras, western Trans-Mexican Volcanic Belt, and the Pacific part of Sierra Madre del Sur. While *S. fulviventer* thrives in the Sierra Madre Occidental, small portions of northwestern and southwestern Central Plateau, and a small area in western Trans-Mexican Volcanic Belt (Ceballos and Oliva 2005).

Trans-Mexican Volcanic Belt

This morphotectonic province has a greater number of palynofloral, paleosol, and fossil mammal localities than any of the others (Figs. 3.2, 3.3, 3.4 and 3.5); major local faunas are listed in Table 3.4. Some of the earliest described Pleistocene mammal specials are from the Mexican Basin (Owen 1869). The sampling, however, is biased both geographically (eastern part is more dense) and diachronically (Late Rancholabrean faunas are over-represented, including those few where radiocarbon dating placed them in the ~40–11 ka interval).

The diversity of the Trans-Mexican Volcanic Belt Late Rancholabrean mammal fauna is far greater than that of today. It includes modern taxa of different ecological requirements that are sympatric at the time, e.g., *Neochoerus*, *Platygonus*, *Odocoileus hemionus*, and *Bison*. This association indicates a correspondingly diverse setting, where different vegetation types were either intermixed or separated over short distances. Pleistocene climatic fluctuations acting on a complexly rugged territory such as that of the Trans-Mexican Volcanic Belt could have promoted the large vegetation diversity implied by the diverse mammal record. The pollen record seems to support this contention.

The climate regime changed by the early Holocene, becoming warmer and drier. As in other provinces, most medium to large species became extinct (Table 3.4). Other species became extirpated from Mexico and currently live outside the country, e.g., the carnivoran *Canis rufus* now living in temperate North America (Hall 2001). Others disappeared from their Rancholabrean morphotectonic province but survived elsewhere within Mexico. For example, the

Table 3.5 Late Pleistocene (Rancholabrean) Trans-Mexican Volcanic Belt flora identified from its palynological record. Taxonomical arrangement sensu Cronquist 1981; ranks employed (from higher to lower): Division, Class, Order, Family and Genus. Main sources: (Foreman 1955; Clisby and Sears 1955; Sears and Clisby 1955; Bradbury 1971, 1989, 2000; Ohngemach 1973, 1977; Meyer 1973; Brown 1984; González-Quintero 1986; Straka and Ohngemach 1989; Metcalfe 1992; Lozano García et al. 1993; Lozano-García and Ortega-Guerrero 1994; Caballero-Miranda et al. 1999; Canul-Montañez 2008; Vázquez et al. 2010; Israde-Alcantara et al. 2010; Caballero-Miranda et al. 2010; Ortega et al. 2010; and Martínez-Hernández, E. unpublished data)

BRYOPHYTA
   SPHAGNOPSIDA
      Sphagnales
         Sphagnaceae
           Sphagnum
LYCOPODIOPHYTA
   ISOETOPSIDA
      Isoetales
         Isoetaceae
           Isoetes
   SELAGINELLOPSIDA
      Selaginellales
         Selaginellaceae
           Selaginella
PTERIDOPHYTA
   POLYPODIOPSIDA
      Filicales
         Aspleniaceae
           Asplenium
PINOPHYTA
   CONIFEROPSIDA
      Coniferales
         Cupressaceae
           Juniperus
           Taxodium
           Cupressus
         Pinaceae
           Abies
           Picea
           Pinus
           Tsuga
         Podocarpaceae
           Podocarpus
MAGNOLIOPHYTA
   MAGNOLIOPSIDA
      Apiales
         Araliaceae
           Hydrocotyle
      Asterales
         Asteraceae
      Capparales
         Brassicaceae
           Rorippa
      Caryophyllales
         Amaranthaceae - Chenopodiaceae
         Caryophyllaceae
      Celastrales
         Aquifoliaceae
           Ilex
      Dipsacales
         Adoxaceae
           Viburnum
         Caprifoliaceae

(continued)

Table 3.5 (continued)

         Valerianaceae
           Valeriana
      Ericales
         Ericaceae
      Fabales
         Fabaceae
           Piscidia
           Prosopis
      Fagales
         Betulaceae
           Alnus
           Betula
         Fagaceae
           Fagus
      Gentianales
         Loganiaceae
           Buddleja
      Hamamelidales
         Hamamelidaceae
           Liquidambar
      Juglandales
         Juglandaceae
           Carya
           Engelhardtia
           Juglans
      Lamiales
         Lamiaceae
      Magnoliales
         Magnoliaceae
           Drymys
      Malpighiales
         Hypericaceae
           Hypericum
      Malvales
         Malvaceae
         Tiliaceae
      Myricales
         Myricaceae
           Myrica
      Myrtales
         Melastomataceae
         Myrtaceae
         Onagraceae
           Ludwigia
      Nymphaeales
         Nymphaeaceae
           Nuphar
           Nymphaea
      Piperales
         Chloranthaceae
           Hedyosmun
      Plantaginales
         Plantaginaceae
      Polygonales
         Polygonaceae
           Polygonum
      Ranunculales
         Ranunculaceae
           Ranunculus
           Thalictrum
      Rosales
         Rosaceae
           Holodiscus
      Salicales

(continued)

Table 3.5 (continued)

        Salicaceae
           Populus
           Salix
    Santalales
        Santalaceae
           Arceutholobium
    Sapindales
        Burseraceae
           Bursera
    Saxifragales
        Haloragidaceae
    Scrophulariales

        Lentibulariaceae
           Utricularia
        Oleaceae
           Fraxinus
    Solanales
        Solanaceae
           Datura
    Urticales
        Moraceae
        Ulmaceae
           Celtis
        Urticaceae
LILIOPSIDA
    Alismatales
        Alismataceae
           Sagittaria
        Poramogetonaceae
           Potamogeton
        Ruppiaceae
           Ruppia
    Cyperales
        Cyperaceae
           Eleocharis
           Schoenoplectus
        Poaceae
    Eriocalulales
        Eriocaulaceae
           Eriocaulon
    Juncales
        Juncaceae
           Luzula
    Liliales
        Pontederiaceae
           Heteranthera
    Typhales
        Typhaceae
           Typha

rodent *Cynomys mexicanus* now lives in a narrow and elongated area located in easternmost Chihuahua-Coahuila Plateaus and Ranges and a small portion of northern Sierra Madre Oriental provinces. The rodent *Microtus californicus* now is restricted to the northern Baja California Peninsula province. The rodent *Neotoma albigula* now thrives in the Northwestern Plains and Sierras province, with *N. palatina* restricted at present to a small area in the southeastern Sierra Madre Occidental province and *N. phenax* to the southern Northwestern Plains and Sierras province. The perissodactyl *Tapirus bairdii* now is restricted to the tropics (Ceballos and Oliva 2005).

Southern Provinces

These provinces include the Sierra Madre del Sur Sierra Madre de Chiapas the southern part of the Gulf Coastal Plain, and the Yucatan Platform (Fig. 3.1). They have yielded four major local faunas (Table 3.4, and Figs. 3.3 and 3.6), along with a few important palynofloras (Fig. 3.4), and relatively numerous paleosol localities (Fig. 3.5). Sampling again is strongly biased toward the Late Pleistocene (Rancholabrean NALMA). The paleosol (Acrisol) record points to an equable, largely tropical climate regime for the last 40–50 kyr. The palynological record, particularly that of Tehuacán, Puebla (northeastern Sierra Madre del Sur just south of the Trans-Mexican Volcanic Belt), indicates moister and cooler conditions in the latest Pleistocene (Canul-Montañez 2008). The now prevailing xeric vegetation is a Holocene phenomenon.

The mammal record (Table 3.4; Figs. 3.2 and 3.3) largely includes tropical taxa. Temperate and cosmopolitan taxa are less frequent, but not uncommon. The latter taxa include *Ursus americanus, Eumops perotis,* and *Spilogale putorius*. This composition indicates that during the Late Rancholabrean, the climate regime, although warm and moist in general, probably was punctuated at least locally, by cooler and/or drier episodes that allowed temperate taxa to expand their range southward or to occupy parts of southern habitats in a discontinuous manner. This region's geomorphic complexity, as well as limited sampling both in space and time, does not permit recognition of a particular pattern of climate changes, only delineation of broad climate change trends.

Nevertheless, as in the other provinces, important mammal composition changes took place at the end of the Rancholabrean. Several medium and large taxa became extinct, such as the carnivoran *Canis dirus*, the perissodactyl *Equus*, the artiodactyls *Navahoceros fricki, Odocoileus lucasi,* and *Hemiauchenia*, as well as the proboscideans. Other largely cosmopolitan or temperate taxa survived farther north in temperate habitats, e.g., the chiropterans *Eptesicus fuscus* and *Eumops underwoodi*; carnivoran *Ursus americanus*; and artiodactyl *Bison* (Hall 2001; Ceballos and Oliva 2005), perhaps returning to their primary range.

A Summary of Early Peoples in Mexico

In the last 20 years, studies referring to the early peopling of the Americas have increased, especially those for western North America, e.g., Bonnichsen 1999; Bonnichsen and Turnmire 1999; Parfit 2000; Haynes 2002; Bonnichsen et al. 2005; Jiménez López et al. 2006a, b; Meltzer 2009. In Mexico, interest in early peopling has existed for over a century, e.g., Reyes 1881; Mercer 1896). Research has not yet been able to define when and where the earliest people came into Mexico (see Lorenzo and Mirambell 1999 vs. Dixon 1999). Recent reports on previously known localities (González et al. 2006), however, have enhanced the development of predictive models that assist with the search for new sites.

Much of the controversy about the early peopling of the Americas deals with the value that is given to the indirect evidence of human presence when human skeletal remains are lacking in the sites. Such evidence may include lithics, hearths, and culturally modified bone. Indirect evidence requires further detailed analyses that distinguish between natural processes the materials may have undergone and those processes that are signatures of human intervention. The following brief synthesis of current knowledge regarding early peoples in the Mexican Late Pleistocene underscores the very limited data available.

In northern Mexico, several North American Late Pleistocene Clovis sites now are known. More than a dozen sites have been found in Sonora, some of which have stratigraphically-controlled excavations such as at El Fin del Mundo (Sanchez 2001; Gaines and Sanchez 2009; Sanchez et al. 2014). At this site, possible interaction between people and gomphotheres indicates either hunting or scavenging activities (Sanchez et al. 2014). To the south, evidence for the presence of Clovis peoples greatly diminishes. A few Clovis points have been recovered from Baja California to Costa Rica (Sanchez 2001). Other sites with evidence of Clovis culture are found in the State of Hidalgo in eastern Mexico (Sanchez 2001). These occurrences may be explained by Clovis groups moving along the Gulf Coast from Texas.

Central Mexico is the most explored area in the country. The enormous amount of construction that continues to occur in the Basin of Mexico has resulted in the discovery of numerous paleontological localities and archaeological sites. These occurrences indicate that early peoples were in the area by 11.0 ^{14}C kBP, exemplified by Peñón woman dating to 10,755 ± 75 ^{14}C BP (González et al. 2003; González and Huddart 2008). This early age indicates that Peñón woman is one of the oldest human skeletal remains in the Americas (Dillehay 2000; Meltzer 2009). Sites contain hearths or lithics, e.g., El Cedral San Luis Potosí; Tlapacoya, State of Mexico (Lorenzo and Mirambell 1999), or human-modified bone, e.g., Santa Isabel Ixtapa or Tocuila State of Mexico (Arroyo-Cabrales et al. 2006; Johnson et al. 2012). The Late Pleistocene Basin of Mexico is a highly rich environment that supported a large Columbian mammoth (*Mammuthus columbi*) population. Over 100 mammoth localities are known for the area, yet very few show evidence of human interaction with the carcass (Arroyo-Cabrales et al. 2006).

One of the most controversial sites in the Americas regarding early peopling is the basin of Valsequillo near the capital city of the state of Puebla, east of Mexico City. Several archaeological excavations since the late 1950s and early 1960s have provided inconclusive evidence about the presence of the earliest people in the area. Occupation as early as the Sangamonian interglacial, between 132 and 119 ka, (González et al. 2006) has been proposed, with the latest hypothesis about human footprints having been questioned and recently rejected (Feinberg et al. 2009; Mark et al. 2010). Further research is warranted. Access, however, is prohibited by the pollution in the reservoir dam, where some of the specific sites like Hueyatlaco or Los Hornos are located (Gonzalcz ct al. 2006). Private settlements now extend along and over the edge of the dam means that expectations of being able to conduct excavations in the near future are minimal (Patricia Ochoa-Castillo 2011: personal communication).

Further south, Guila Naquitz is a small shelter near the Valley of Oaxaca, in central Oaxaca. Flannery's (1986) excavation has yielded both seeds and peduncles of squash (*Cucurbita pepo*) with indications of domestication as early as 9.0 ^{14}C kBP (Smith 1997). This date coincides with views about the earliest Naquitz phase being attributed to the early Archaic period (Flannery 1986). Nearby, within the Tlacolula Valley, a few Paleoindian projectile points have been found on the surface (Marcus Winter 2008: personal communication).

In southern Mexico, research seems to indicate the presence of the Americas' two early cultural traditions, North American Clovis from Oaxaca and Chiapas, and Fish-tail fluted points from Central and South America (Santamaría and García-Bárcena 1989). Recent studies from rock-shelters nearby Ocozocuautla, Chiapas, have provided strong evidence of human presence in the state around 11.0–10.0 ^{14}C kBP. These sites have yielded lithics reflecting expedient technology and also milling stones and botanical samples that may indicate incipient horticulture starting at the end of the Pleistocene to early Holocene. Small and medium-sized animals such as deer, peccary, and rabbit are the most hunted prey, while megafaunal remains were not found (Acosta 2010).

Finally, submerged caves near Tulum in the state of Quintana Roo, on the Yucatan Peninsula, contain a diverse megafaunal assemblage of latest Pleistocene age, along with hearths with burned bones, artifacts, and human skeletal remains that date between 11.6 and 8.0 ^{14}C kBP (González-González et al. 2008).

Most recently, a remarkable finding of a human female was reported from Hoyo Negro cenote. The skeleton dates to between 13,000 and 12,000 calendar years ago. She has Paleoamerican craniofacial characteristics and a Beringian-derived mitochondrial DNA (mtDNA) haplogroup (D1), meaning that differences between Paleoamericans and Native Americans probably resulted from in situ evolution rather than separate ancestry (Chatters et al. 2014).

Discussion

The pollen record in Mexico is greatly biased towards the Trans-Mexican Volcanic Belt province in central Mexico and dominated by the Basin of Mexico. Half the records lie in the Trans-Mexican Volcanic Belt and two-thirds of the localities are in or north of the Trans-Mexican Volcanic Belt province. During the latter part of the Rancholabrean, the Basin of Mexico had a cooler, more humid climate regime that allowed the development of numerous freshwater lakes in low-lying areas. Vegetational changes involve altitudinal timberline shifts more than latitudinal displacements (Bradbury 1989; Metcalfe 1992; Caballero-Miranda et al. 2010). This general pattern can be discerned with greater or lesser similarity in other morphotectonic provinces studied.

In the northern part of the Sierra Madre del Sur province, a more humid flora existed at the end of the Pleistocene and was replaced by a xeric flora between 12.0 and 10.0 ^{14}C kBP (Canul-Montañez 2008). To the north in the Chihuahua-Coahuila Plateaus and Ranges province, a more equable, humid climate regime was in place from ~22.0 to 11.0 ^{14}C kBP. A longitudinal east-west expansion of grassland and shrub forest occurred at the expense of pine and pine-oak forest (Van Devender et al. 1987; Betancourt et al. 1990; Van Devender 1990a, b; Van Devender and Bradley 1990).

The Late Pleistocene paleosol record indicates two major contrasting types occurring in different parts of Mexico. Luvisols are found in the Trans-Mexican Volcanic Belt and Sierra Madre Occidental provinces. They have developed within silicic pyroclastic sediments around 20.0–18.0 ^{14}C kBP under humid to subhumid conditions and moderate temperature (Solleiro-Rebolledo et al. 1999, 2003, 2006; Cabadas-Báez 2007; Cabadas-Báez et al. 2010). Acrisols occur in the Sierra Madre del Sur province. They have developed within heterogeneous sediments around 50.0–25.0 ^{14}C kBP under hot and humid conditions within a forested environment (Isphording 1974; Bautista et al. 2003).

The mammal, palynological, and paleosol records (Figs. 3.3, 3.4, 3.5 and 3.6) show time and space biases; the temporal bias favors the Late Pleistocene, and the space bias favors the Trans-Mexican Volcanic Belt province. Both the palynological and paleosol records are sparse north of this province. On the other hand, the mammal record is relatively dense in the Sierra Madre Oriental and Sierra Madre del Sur provinces, yielding some major local faunas (Table 3.4). Interpretation of each data set from the provinces may differ. The environmental sensitivity of components from the individual records varies, indicating different conditions in a given place. For example, the Loltún local fauna includes temperate (ursid) and tropical (dasyproctid and xenarthran) taxa. Not surprisingly, the environmental information obtained from one data set coincides only in general with that of another. In addition, the records disclose important gaps in space and time that must be filled to gain a better understanding of climate change across the country.

Biological communities in Mexico experienced profound changes in species composition (species that are represented) and structure (relationships among those species present) as a consequence of the environmental fluctuations during the Pleistocene. Comparison of Pleistocene and Holocene zoogeographic ranges disclosed different patterns. Many species expanded their distribution to different latitudes or higher/lower altitudes or moved further north/south during the Pleistocene. Also detected were the presence of biogeographic corridors, refugia, and centers of speciation in isolated regions (Caballero-Miranda et al. 2010).

The general situation for mammals is mirrored partially by human populations. Central Mexico is the region where the earliest archaeological sites are located, clearly pointing to their presence at around 11.0 ^{14}C kBP based on dating of human skeletal remains. Evidence for the utilization of faunal resources, however, is very limited (Arroyo-Cabrales et al. 2006; Johnson et al. 2006). For Mexico in general, modern taphonomic studies of Late Pleistocene faunal remains are lacking except for a few cases (Polaco and Heredia-C. 1988; Polaco et al. 1989; Solórzano 1989; Johnson et al. 2012). Although more than 270 mammoth localities are known throughout the country, only six have modified mammoth bone. Of these six, only three have good potential for demonstrating human involvement with mammoth, and those three are located in the Basin of Mexico (Arroyo-Cabrales et al. 2006).

To the northwest and differing from the faunal pattern, evidence is growing for a large presence of early sites that demonstrate a relationship with cultures in the southwestern U.S. Overall for northern Mexico, evidence of early peoples being hunter-gatherers is slowly accumulating (Sanchez 2001), but further discussion on climate change and human response are warranted because the possible questions about such relationships are not yet formulated (see Pilaar Birch and Miracle 2017). To the south, few localities have provided strong evidence for early peoples, most likely due to the poor preservation conditions in tropical soils and the emphasis of current research on advanced cultures. The finding of domesticated squash around 9.0 ^{14}C kBP may indicate that decreasing mobility began very early in the human occupation of Mexico (Sanchez 2001). In general, then, what influence early peoples may have had on medium and large size mammal populations at the end of the Pleistocene cannot be addressed. Using the very limited mammoth data as a potential indication, however, early peoples' hunting activities would not seem to have been the cause of extinctions. A similar situation appears to be the case for South America. Although extinctions may have been more common after early peoples arrived, some medium and large

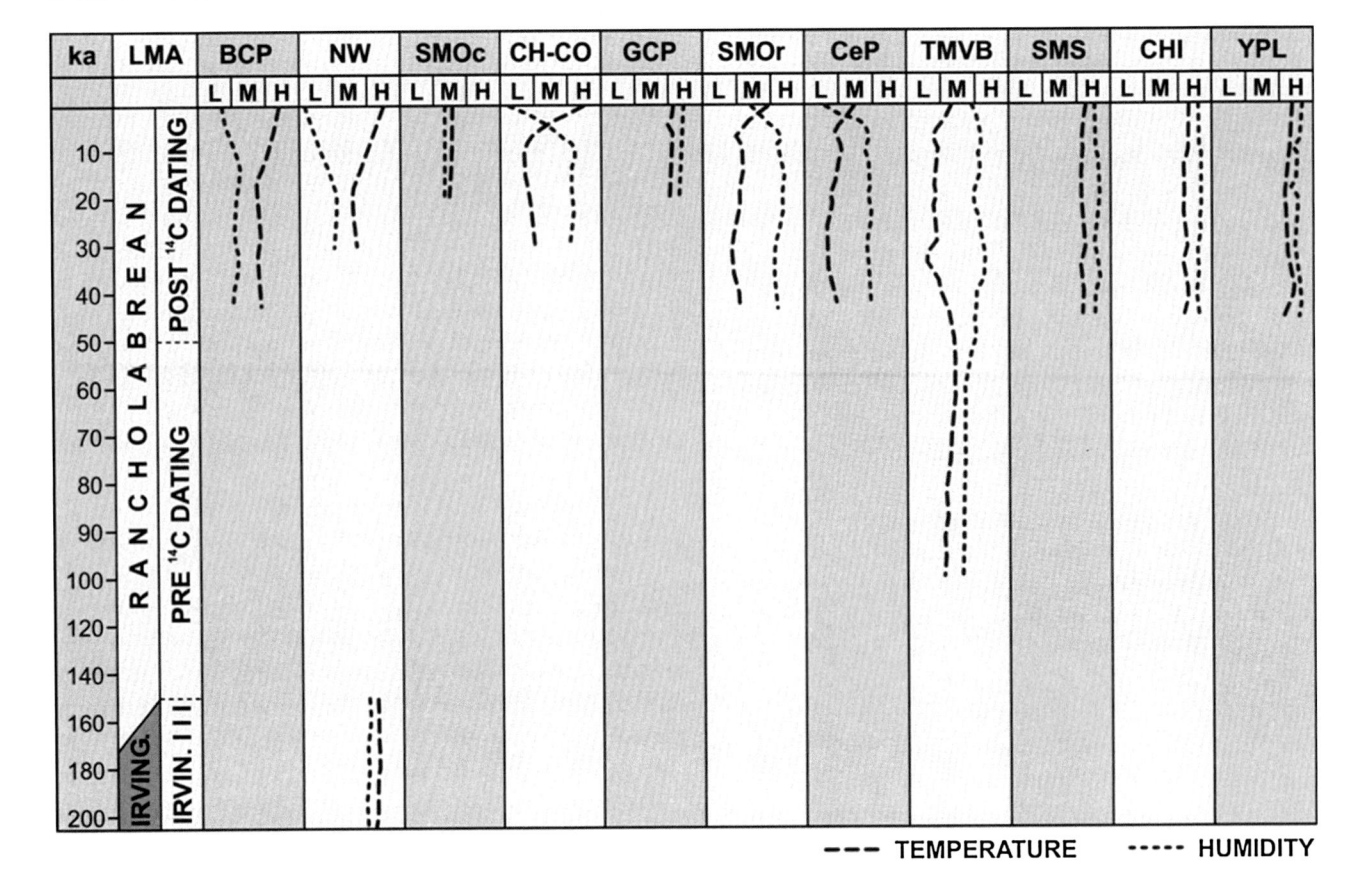

Fig. 3.6 Late Pleistocene probable climate trends in Mexico's morphotectonic provinces (sensu Ferrusquía-Villafranca 1993, 1998). Time frame adapted from Bell et al. (2004). The reviewed palynologic, paleosol, lake sediment, and mammal records were used to assess the trends. Abbreviations (from left to right): **LMA**, North American Land Mammal Ages. **BCP**, Baja California Peninsula. **NW**, Northwestern Plains and Sierras. **SMOc**, Sierra Madre Occidental. **CH-CO**, Chihuahuan-Coahuilan Plateaus and Ranges. **GCP**, Gulf Coastal Plain. **SMOr**, Sierra Madre Oriental. **CeP**, Central Plateau. **TMVB**, Trans-Mexican Volcanic Belt. **SMS**, Sierra Madre del Sur. **CHI**, Sierra Madre de Chiapas. **YPL**, Yucatan Platform

sized taxa persist for up to several thousand years after human arrival (Barnosky and Lindsey 2010). Australia is another case in which humans have always been implicated in megafaunal extinction. Nevertheless, it seems today that the problem was more complex, and the coexistence interval for human and megafauna remains imprecise (Brooks and Bowman 2002; Johnson and Brook 2011). Humans in Mexico coexist with medium and large sized animals rather than being the cause of their extinctions. The lack of genetic signature or any distinctive range dynamics that would distinguish the potential for extinction or survival emphasizes the challenges associated with predicting future responses of extant mammals to climate and human-mediated habitat change (Lorenzen et al. 2011; Ochoa and Piper 2017).

An initial conservative integration of data from the different sets for each province provides the basis for pattern recognition (Fig. 3.6). Modern average climate conditions for each morphotectonic province have been taken from García (1990), Hernández (1990), and Vidal-Zepeda (1990a, b). The general Wisconsinan fluctuating climate pattern and timing have been recognized broadly long ago (Flint 1947), and those of Mexico shortly thereafter (Foreman 1955; Clisby and Sears 1955; Sears and Clisby 1955; Heine 1984). Currently available information, however, does not yet allow the establishment across the country of detailed climatic changes for the whole Wisconsinan, let alone the Pleistocene. For now, only trends can be portrayed.

The coexistence in Mexico of a highly diverse group of mammals during the Late Rancholabrean (and the entire Pleistocene for that matter), as shown by the fossil record, indicated environmental conditions quite different from those of today. Such conditions, among other things, allowed stenotopic species (restricted tolerance to a narrow range of environmental conditions) to extend their range beyond narrow parameters, and eurytopic ones (broad tolerance to a wide range of environmental conditions) to thrive extensively across the country. This situation underscored the fact that the shifting of ecological and climatic zones, at least during the Late Pleistocene, was not a simple matter of displacement or range reduction/extension. New ecological conditions were created as zones overlapped and new and different faunal and floral communities emerged. In other words, Mexico's Pleistocene biome tapestry dynamically adjusted to environmental changes acting on a territory of quite complex relief. The net result was that by Late Rancholabrean, mammal fauna diversity (and by extension biotic diversity as well) was greater than that of today. Community structure, therefore, would have been organized differently to facilitate the complex relationships engendered by shifting and overlapping ecological zones.

The biotic response to these environmental conditions may have involved extinction. Extinctions are common throughout the Pleistocene (Kurtén and Anderson 1980; Bell et al. 2004), with most extinct mammals being medium and large species. The extent of extinctions and the complex of environmental changes involved point to a variety of causes that induced Pleistocene extinctions. The two main causes proposed are climate change in terms of geological-biological impacts or human-driven impacts (Koch and Barnosky 2006).

One such cause could have been disruption of biotic interactions creating a coevolutionary disequilibrium (Graham and Lundelius 1984). Coevolution is the common evolution of multiple taxa (plants and animals) that share close ecological relationships. Through reciprocal selective forces, the evolution of one taxon may be somewhat dependant on the other (Graham and Lundelius 1984:227). Taxa are not isolated on a landscape or in an ecosystem. Coevolution, then, is the interdependent interaction of taxa acting at the evolutionary level. The destabilization of the coevolutionary relationship through various types of disruptions (such as habitat destruction, climatic change, extinctions, extirpations) affects the balance and creates a breakdown in the structure and relationships (disequilibrium). Following that reasoning, the rapid decrease in size and eventual extinction of caballoid horses in Alaska has been linked to climatic shift, changing vegetation, and the collapse of the ecosystem at the end of the Pleistocene (Guthrie 2003). Similarly, the decline in genetic diversity in North American bison appears to be linked to environmental changes with the onset of the last Laurentide Glacial Maximum (Shapiro et al. 2004). The temporal mode of extinction (gradual through a long time span or sudden, nearly instantaneous) has received much attention and is the subject of ongoing debate (Martin and Klein 1989; Bell et al. 2004).

Recently, Faith and Surovell (2009) argued for the possibility of the second alternative, i.e., human driven impacts. They suggest the absence of extinct genera from the fossil record is a result of sampling error. The analysis of Mexico's record, incomplete and biased as the record may be, seems not to bear out this contention. By arguing on negative evidence, the hypothesis lacks evidence for testing its validity.

One other hypothesis that has been contentious over the past 10 years has been the possible meteorite airburst, similar to the famous K-T impact, and how that event could cause the extinction of megafauna and strong cultural changes and population decline in Paleoindian populations (Firestone et al. 2007). Most recently, evidence for such an airburst has been proposed at several Mexican sites where megafauna were found (Gonzalez et al. 2014). For at least one of those sites, a claim has been made that the dating procedures were compromised, and, because of that, the isochrony of the events cannot be confirmed (Meltzer et al. 2014).

Finally, the available information on Mexico's Pleistocene mammals allows only rough discriminations of a few of the many environmental factors involved in this complex environment/biota interplay. Under these circumstances, inferring Pleistocene climate in southern North America from bioevents alone, i.e., the fossil record and actualistic comparisons thereof, allows at best the tracing of broad qualitative patterns for each morphotectonic province. Nonetheless, the environmental factors contribute to a better understanding of the mammalian response, expressed in extinctions, biogeographic shifts, and extirpations that significantly change Pleistocene and Holocene mammal physiognomy. Furthermore, the temporal and spatial gaps of the mammal, palynological, and paleosol records must be filled before a more complete understanding of the Quaternary climate and its changes can be gained in this part of the Americas.

Concluding Remarks

The Holocene and Late Pleistocene Mexican faunas are quite different. This difference is the combined results of individual species extinctions and range modifications that affected and changed the vertebrate biota physiognomy and taxonomic makeup.. The available fossil record, however, does not portray this major biogeographic shifting of species in detail due to the lack of associated chronometric data. The analysis of disjunct (i.e., separated from the main range) and of demonstrably relict species may be an alternative to providing greater detail and understanding about the response of individual species to climate change during the Late Pleistocene. The following summary points are made to illuminate what is known from the Pleistocene record and directions for further research.

The Late Pleistocene mammal record was analyzed by morphotectonic provinces (n = 11) that were grouped into four larger geographic units to examine zoogeographic distribution, any variance in distribution (extinctions, extirpations), and environmental conditions inferred from that distribution and variance.

Northern provinces

- Biotic diversity unparalleled anywhere in Mexico today
- Both extinctions and extirpations (northward and southward) occurred that shaped the modern fauna
- More humid climatic regime during the Irvingtonian and early Rancholabrean than that of today that allowed a subtropical biota
- A cooler and moister regime during the latest Rancholabrean than that of today with a non-analogous fauna and expansion of grassland and shrub forest

Central and Eastern provinces

- Thermal (warm/cold) and humidity/rain (moist/dry) oscillations during the Rancholabrean with a non-analogous fauna
- Both extinctions and extirpations (northward, southward, and westward) occurred that shaped the modern fauna
- A cooler and moister regime during the latest Rancholabrean than that of today with a non-analogous fauna and flora

Trans-Mexican Volcanic Belt

- Mammalian diversity during late Rancholabrean far greater than that of today
- Both extinctions and extirpations (northward, southward, and westward) occurred that shaped the modern fauna
- A cooler and moister regime during the late Rancholabrean than that of today with a non-analogous fauna
- Numerous freshwater lakes developed in low-lying areas and vegetation experienced altitudinal timberline shifts

Southern provinces

- Equable, tropical climate regime during Rancholabrean with a largely tropical fauna and humid flora
- General tropical climate punctuated by cooler and drier episodes during the late Rancholabrean with a non-analogous fauna
- Both extinctions and extirpations (northward) occurred that shaped the modern fauna

Early peoples

- arrived in Mexico by the latest Pleistocene (~ 11 ^{14}C kBP)
- Growing evidence for a strong presence of Clovis culture in the Northern provinces and Gulf Coastal Plain of Eastern provinces
- Most of the known early sites are concentrated in the Basin of Mexico
- Both Clovis (North American tradition) and Fish-tail fluted points (Central and South American tradition) are found in southern Mexico, perhaps representing a cultural transitional zone
- A lack of taphonomic studies coupled with only a very few early sites having a solid association between medium to large animals and humans suggests that human impact on the Late Pleistocene populations appears to have been negligible and was not a cause for extinctions.

Time and space biases exist, with records favoring the Late Pleistocene (Rancholabrean) and the Trans-Mexican Volcanic Belt, particularly the Basin of Mexico. Nevertheless, biological communities throughout the country experience profound changes in species composition and structure. Such a pattern was due not to direct human impact but to the consequences of environmental changes throughout and particularly at the end of the Pleistocene. The shifting of ecological and climatic zones was not a simple matter of displacement and range adjustments. The coexistence of a highly diverse group of mammals indicates a community structure organized differently than that of today in order to facilitate the complex relationships that coexistence would have required. The disruption of these biotic interactions would have created a coevolutionary disequilibrium situation.

The available information on Mexico's Pleistocene mammal, palynological, and paleosol records allows only broad trends to be discerned in the complex environmental-biota interplay and what role, if any, early peoples played in extinction. A critical need exists to fill in the time and space gaps in these records. Solid radiocarbon chronologies need to be developed that can anchor the various records and provide the framework for more in-depth analyses of environmental changes and individual species response. In a focused radiocarbon dating program, the Trans-Mexican Volcanic Belt province and mammoth would be a reasonable target. The most concentrated research has been in this province (particularly the Basin of Mexico) and mammoth is the most ubiquitous Late Pleistocene mammal. Research needs to continue and expand in the other provinces in order to have a representative sample across the country. While research in the Northern provinces has been most fruitful in terms of mammal-human interactions during the latest Pleistocene, the Gulf Coastal Plains (Eastern province) has great potential in illuminating that interaction as well. Mexico's record is critical in understanding the continent-wide affects of Pleistocene climatic changes on plants, animals, and humans. This initial synthesis forms a first-order interpretation and basis for future research directions.

Acknowledgments Several institutions supported parts of the research: Instituto de Geología, Universidad Nacional Autónoma de México, México; Comisión Nacional para Aprovechamiento de la Biodiversidad; Subdirección de Laboratorios y Apoyo Académico, Instituto Nacional de Antropología e Historia, México; and Museum of Texas Tech University. For Johnson, the manuscript represents part of the ongoing Lubbock Lake Landmark regional research program into late Quaternary climate and paleoecological change on the Southern Plains. Jaqueline Torres, student and assistant of the senior author via a *Sistema Nacional de Investigadores* (SNI México) assistantship, helped in the preparation of the manuscript. To all of them, the authors are most grateful.

References

Acosta, O. G. (2010). Late-Pleistocene/Early-Holocene tropical foragers of Chiapas, Mexico: Recent studies. *Current Research in the Pleistocene, 27*, 1–4.

AIMES, Analysis, Integration and Modeling of the Earth System. (2010). Science plan and implementation strategy. *International Geosphere-Biosphere Programme Report, 58*, 1–30.

Alberdi, M. T., Arroyo-Cabrales, J., & Polaco, O. J. (2003). Cuántas especies de caballo hubo en una sola localidad del Pleistoceno Mexicano. *Revista Española de Paleontología, 18*, 205–212.

Alberdi, M. T., Prado, J. L., & Salas, R. (2004). The Pleistocene Gomphotheriidae (Proboscidea) from Perú. *Neues Jarbuch für Geologie und Paläontologie Abhandlungen, 231*, 423–452.

Alberdi, M. T., & Corona, M. E. (2005). Revisión de los gonfoterios en el Cenozoico tardío de México. *Revista Mexicana de Ciencias Geológicas, 22*, 246–260.

Alvarez, T. (1965). Catálogo paleomatozoológico Mexicano. México: Instituto Nacional de Antropología e Historia. *Publicaciones del Departamento de Prehistoria, 17*, 1–70.

Alvarez, T. (1969). Restos fósiles de mamíferos de Tlapacoya, Estado de México (Pleistoceno-Reciente). Kansas: University of Kansas. *Museum of Natural History, Miscellaneous Publications, 51*, 93–112.

Alvarez, T. (1971). Variación de la figura oclusal del premolar inferior en carpinchos fósiles (Rodentia: Hydrochoeridae) de Jalisco, México. México: Instituto Nacional de Antropología e Historia. *Serie Investigaciones del Instituto Nacional de Antropología e Historia, 21*, 1–40.

Alvarez, T. (1982). Restos de mamíferos recientes y pleistocénicos procedentes de la Gruta de Loltún, Yucatán. México: Instituto Nacional de Antropología e Historia. *Departamento de Prehistoria, Cuaderno de Trabajo, 26*, 7–35.

Alvarez, T. (1983). Notas sobre algunos roedores fósiles del Pleistoceno en México. Instituto Politécnico Nacional. *Anales de la Escuela Nacional de Ciencias Biológicas, 27*, 149–163.

Alvarez, T. (1986). Fauna pleistocénica. In J. L. Lorenzo & L. Mirambell (Eds.), *Tlapacoya: 35,000 años de historia del Lago de Chalco* (pp. 173–192). Mexico: Instituto Nacional de Antropología e Historia, Colección Científica, Serie Prehistoria.

Alvarez, T., & Ferrusquía-Villafranca, I. (1967). New records of fossil marsupials from the Pleistocene of Mexico. *The Texas Journal of Science, 19*, 107.

Alvarez, S. T., & Hernández-Chávez, J. J. (1994). Estudio de los restos de *Neotoma* (Rodentia; Cricetidae) del Pleistoceno-Holoceno de Tlapacoya, Estado de México, México con descripción de dos nuevos taxa. *Revista de la Sociedad Mexicana de Paleontología, 7*(2), 1–11.

Alvarez, T., & Polaco, O. J. (1981). Anexo 1. Fauna obtenida de las excavaciones realizadas en el sitio Rancho La Amapola-El Cedral, S.L.P. In J. L. Lorenzo & L. Mirambell (Eds.), *El Cedral, S.L.P., México: Un sitio con presencia humana de más de 30,000 AP* (pp. 123–124). México: X Congreso de la Unión Internacional de Ciencias Prehistóricas y Protohistóricas.

Anderson, R. S., & Van Devender, T. R. (1995). Vegetation history and paleoclimates of the costal lowlands of Sonora Mexico-Pollen records from packrats middens. *Journal of Arid Environments, 30*, 295–306.

Arroyo-Cabrales, J., & Alvarez, T. (2003). A preliminary report of the Late Quaternary mammal fauna from Loltún Cave, Yucatán, Mexico. In B. W. Schubert, J. I. Mead, & R. W. Graham (Eds.), *Ice age cave faunas of North America* (pp. 262–272). Denver, Colorado: University Press and Denver Museum of Nature & Science.

Arroyo-Cabrales, J., & Johnson, E. (1995). A reappraisal of fossil vertebrates from San Josecito Cave, Nuevo Leon. In E. Johnson (Ed.), *Ancient peoples and landscapes* (pp. 217–231). Lubbock, Texas: Museum of Texas Tech University.

Arroyo-Cabrales, J., & Johnson, E. (1998). La Cueva de San Josecito, Nuevo León, México: Una primera interpretación paleoambiental. In C. O. Carranza & M. D. A. Córdoba (Eds.). *Avances en investigación. Paleontología de vertebrados*. México:. Universidad Autónoma del Estado de Hidalgo (*Instituto de Investigaciones en Ciencias de la Tierra, Publicación Especial, 1*, 120–126).

Arroyo-Cabrales, J., & Johnson, E. (2008). Mammalian additions to the faunal assemblages from San Josecito Cave, Nuevo León, Mexico. In L. C. E. Espinosa & J. Ortega (Eds.), *Avances en el estudio de los mamíferos de México*. México, D.F.: Asociación Mexicana de Mastozoología, A.C. (*Publicaciones Especiales, II*, 65–87).

Arroyo-Cabrales, J., & Polaco, O. J. (2003). Caves and the Pleistocene vertebrate paleontology of Mexico. In B. W. Schubert, J. I. Mead, & R. W. Graham (Eds.), *Ice age cave faunas of North America* (pp. 273–291). Bloomington, Indiana: University Press and Denver Museum of Nature & Science (Denver, Colorado).

Arroyo-Cabrales, J., Polaco, O. J., Álvarez, T., & Johnson, E. (1996). New records of fossil tapir from northeastern Mexico. *Current Research in the Pleistocene, 13*, 93–95.

Arroyo-Cabrales, J., Polaco, O. J., & Johnson, E. (2002). La mastofauna del Cuaternario Tardío en México. In M. Montellano-Ballesteros & J. Arroyo-Cabrales (Eds.), *Avances en los estudios paleomastozoológicos en México*. México: Instituto Nacional de Antropología e Historia, Colección Científica 443, 103–123.

Arroyo-Cabrales, J., Polaco, O. J., & Johnson, E. (2004). Quaternary Mammals from Mexico. In L. C. Maul & R.-D. Kahlke (Eds.) (R.A. Meyrick, language editor), *18th International Senckenberg Conference/VI International Paleontological Colloquium in Weimar. Late Neogene and Quaternary biodiversity and evolution: Regional developments and interregional correlations*. Weimar (Germany) (Schriften der Alfred-Wegener-Stiftung), 25th–30th April, 2004. Conference Volume. *Terra Nostra, 2*, 69–70.

Arroyo-Cabrales, J., Polaco, O. J., & Johnson, E. (2006). A preliminary view of the coexistence of mammoth and early peoples in Mexico. *Quaternary International, 142*(143), 79–86.

Arroyo-Cabrales, J., Polaco, O. J., & Johnson, E. (2007). An overview of the Quaternary mammals of Mexico. In R. D. Kahlke, L. C. Maul, & P. A. Mazza (Eds.), *Late Neogene and Quaternary biodiversity and evolution: Regional developments and interregional correlations*, (Vol. II, 259, pp. 191–203). Frankfurt: Courier Forschungsinstitut Senckenberg.

Arroyo-Cabrales, J., Polaco, O. J., Johnson, E., & Ferrusquía-Villafranca, I. (2010). A perspective on mammal biodiversity and zoogeography in the Late Pleistocene of Mexico. *Quaternary International, 212*, 187–197.

Aviña, C. E. (1969). Nota sobre carnívoros fósiles del Pleistoceno de México. *Instituto Nacional de Antropología e Historia, Departamento de Prehistoria, Paleoecología, 5*, 1–20.

Barrios-Rivera, H. (1985). *Estudio analítico del registro paleovertebradológico de México*. B. S. Thesis. Universidad Nacional Autónoma de México.

Barnosky, A. D. (2008). Climatic change, refugia, and biodiversity: Where do we go from here? An editorial comment. *Climatic Change, 86*, 29–32.

Barnosky, A. D., & Lindsey, E. L. (2010). Timing of Quaternary megafaunal extinction in South America in relation to human arrival and climate change. *Quaternary International, 217*, 10–29.

Barnosky, A. D., Matzke, N., Tomiya, S., Wogan, G. O. U., Swartz, B., Quental, T. B., et al. (2011). Has the Earth's sixth mass extinction already arrived? *Nature, 471*, 51–57.

Barrón-Ortiz, C. R., González-Sánchez, V. H., & Guzmán-Gutierréz, J. R. (2009). Mastofauna de Chupaderos (Pleistoceno tardío), Municipio de Villa de Cos, Zacatecas. *XI Congreso Nacional de Paleontología*, 25 al 27 de Febrero de 2009, Centro de Geociencias, Campus-UNAM, Juriquilla, Querétaro, México (pp. 88). México: Sociedad Mexicana de Paleontología A.C.

Bautista, Z. F., Jiménez, O. J., Navarro, A. J., Manu, A., & Lozano, R. (2003). Microrelieve y color del suelo como propiedades de diagnóstico en Leptosoles cársticos. *Terra, 21*(1), 11.

Bell, C. J., Lundelius, E. L., Jr., Barnosky, A. D., Graham, R. W., Lindsay, E. H., Ruez, D. R., Jr., et al. (2004). The Blancan, Irvingtonian, and Rancholabrean mammal ages. In M. O. Woodburne (Ed.), *Late Cretaceous and Cenozoic mammals of North America* (pp. 232–314). New York: Columbia University Press.

Berta, A. (1988). *Quaternary evolution and biogeography of the large South American Canidae (Mammalia: Carnivora)* (p. 32). Publications in Geological Sciences: University of California.

Betancourt, J. L., Van Devender, T. L., & Martin, P. S. (Eds.). (1990). *Packrat middens: The last 40 000 years of biotic change*. Tucson: University of Arizona Press.

Bonilla-Toscano, L., & Castañeda-Posadas, C. (2011). Descripción e identificación de la mastofauna fósil Municipio de Panotla Tlaxcala, México. Libro de Resúmenes, *XII Congreso Nacional de Paleontología*, 22 al 25 de Febrero de 2011, Edificio Carolino, Benemérita Universidad Autónoma de Puebla, Puebla, México (pp. 38). México: Sociedad Mexicana de Paleontología A.C.

Bonnichsen, R. (Ed.). (1999). *Who were the first Americans?* Corvallis, Oregon: Center for the Study of the First Americans, Oregon State University Press, 158 pp.

Bonnichsen, R., & Turnmire, K. L. (Eds.). (1999). *Ice age peoples of North America: Environments, origins, and adaptations of the First Americans*. Corvallis, Oregon: Center for the Study of the First Americans, Oregon State University Press, 536 pp.

Bonnichsen, R., Lepper, B. T., Stanford, D., & Waters, M. R. (Eds.). (2005). *Paleoamerican origins: Beyond Clovis*. College Station, Texas: Center for the Study of the First Americans, Texas A&M University, 367 pp.

Bradbury, J. P. (1971). Paleolimnology of Lake Texcoco, Mexico. Evidence from diatoms. *Limnology and Oceanography, 16*(2), 180–200.

Bradbury, J. P. (1989). Late Quaternary lacustrine paleoenvironments in the Cuenca de México. *Quaternary Science Review, 8*, 75–100.

Bradbury, J. P. (1997). Sources of glacial moisture in Mesoamerica. In J. Urrutia-Fucugauchi, S. E. Metcalfe, & M. Caballero-Miranda (Guest Eds.), Climate change-Mexico, First International Conference on Climate Change in Mexico, Taxco, 1993. (*Quaternary International, 43/44*, 97–110).

Bradbury, J. P. (2000). Limnologic history of Lago de Pátzcuaro, Michoacán, Mexico, for the past 48000 years: Impacts of climate and man. *Paleogeography, Paleoclimatology, Paleocology, 163*, 69–95.

Bravo-Cuevas, V. M., Ortíz-Caballero, E., & Cabral-Pedromo, M. A. (2009). Gliptodontes (Xenarthra, Glyptodontidae) del Pleistoceno Tardío (Rancholabreano) de Hidalgo, Centro de México. *Boletín de la Sociedad Geológica Mexicana, 61*(2), 267–276.

Broecker, W. (2003). Does the trigger for abrupt climate change reside in the ocean or in the atmosphere? *Science, 300*(5625), 1519–1522.

Brook, B. W., & Bowman, D. M. J. S. (2002). Explaining the Pleistocene megafaunal extinctions: Model, chronologies, and assumptions. *Proceedings of the National Academy of Sciences, 99*, 14624–14627.

Brown, R. B. (1984). *The palecology of the northern frontier of Mesoamerica*. Ph.D. Dissertation, University of Arizona.

Cabadas-Báez, H. (2007). *Paleosuelos del Centro de México como indicadores de cambios ambientales ocurridos durante los últimos 30,000 años*. M. S. Thesis. Universidad Nacional Autónoma de México.

Cabadas-Báez, H., Solleiro-Rebolledo, E., Sedov, S., Pi-Puig, T., & Gama-Castro, J. (2010). Pedosediments of karstic sinkholes in the eolianites of NE Yucatán: A record of Late Quaternary soil development, geomorphic processes and landscape stability. *Geomorphology, 122*, 323–337.

Caballero-Miranda, M. (1997). The last glacial maximum in the Basin of Mexico: The diatom record between 34,000 and 15,000 years BP from Lake Chalco. In J. Urrutia-Fucugauchi, S. E. Metcalfe, & M. Caballero-Miranda (Guest Eds.), Climate change-Mexico, First International Conference on Climate Change in Mexico, Taxco, 1993. (*Quaternary International, 43/44*, 125–136).

Caballero-Miranda, M., Lozano, S., Ortega, B., Urrutia-Fucugauchi, J., & Macías, J. L. (1999). Environmental characteristics of Lake Tecocomulco, northern basin of Mexico, for the last 50,000 years. *Journal of Paleolimnology, 22*, 399–411.

Caballero, M. M., Ortega, B., Valadez, F., Metcalfe, S., Macías, J. L., & Sugiura, Y. (2002). Santa Cruz Atizapán: A 22-ka lake level record and climatic implications for the late Holocene human occupation in the upper Lerma basin, central Mexico. *Palaeogeography, Palaeoclimatology, Palaeoecology, 186*, 217–235.

Caballero-Miranda, M., Lozano-García, S., Vázquez-Selem, L., & Ortega, B. (2010). Evidencias de cambio climático y ambiental en registros glaciales y en cuencas lacustres del centro de México durante el último máximo glacial. *Boletín de la Sociedad Geológica Mexicana, 62*(3), 359–377.

Canul-Montañez, M. E. (2008). *Reconstrucción Paleoclimática (Cuaternario Tardío) de la porción occidental del Valle de Tehuacan, Puebla, México: Estudio palinológico*. Ph.D. Dissertation. Universidad Nacional Autónoma de México.

Carranza-Castañeda, O., & Roldán-Quintana, J. (2007). Mastofáunula de la Cuenca Moctezuma, Cenozoico Tardío de Sonora. Mexico. *Revista Mexicana de Geociencias, 24*(1), 81–88.

Ceballos, G., & Oliva, G. (Coords.). (2005). *Los mamíferos silvestres de México*. México: Comisión Nacional para el Conocimiento y Uso de la Biodiveridad-Fondo de Cultura Económica, 988 pp.

Ceballos, G., Arroyo-Cabrales, J., & Ponce, E. (2010a). Effects of Pleistocene environmental changes on the distribution and community structure of the mammalian fauna of Mexico. *Quaternary Research, 73*, 464–473.

Ceballos, G., García, A., & Ehrlich, P. R. (2010b). The sixth extinction crisis loss of animal populations and species. *Journal of Cosmology, 8*, 1821–1831.

Cervantez-Borja, J. F., Meza-Sánchez, M., & Alfaro-Sánchez, G. (1997). Sedimentological characterization of paleosoils in the northern part of the Basin of Mexico. In J. Urrutia-Fucugauchi, S. E. Metcalfe, & M. Caballero-Miranda (Guest Eds.), Climate change-Mexico, First International Conference on Climate Change in Mexico, Taxco, 1993. (*Quaternary International, 43/44*, 75–86).

Challenger, A. (1998). *Utilización y conservación de los ecosistemas terrestres de México: Pasado, presente y futuro*. Comisión Nacional para el Conocimiento y uso de la Diversidad, Universidad Nacional

Autónoma de México-Instituto de Biología, and Agrupación Sierra Madre, México, D.F., 847 pp.

Chatters, J. C., Kennett, D. J., Asmerom, Y., Kemp, B. M., Polyak, V. Nava, Blank, A., et al. (2014). Late Pleistocene human skeleton and mtDNA link paleoamericans and modern native Americans. *Science, 344*, 750–754.

Churcher, C. S., Turnbull, W. D., & Richards, R. L. (1996). Distribution and size variation in North American short-faced bears, *Arctodus simus*. In K. M. Stewart & K. L. Seymour, (Eds.), *Palaeoecology and palaeoenvironments of late Cenozoic mammals: Tributes to the career of C. S. (Rufus) Churcher* (pp. 191–246). Toronto, Ontario: University of Toronto Press.

Clisby, K. H., & Sears, P. B. (1955). Palynology in southern North America, Part 3. *Geological Society of America Bulletin, 66*, 511–520.

Cope, E. D. (1884). The extinct mammalia of the Valley of Mexico. *Proceedings of the American Philosophical Society, 22*(117), 1–21.

Croxen, F. W. III, Shaw, C. A., & Sussman, D. R. (2007). Pleistocene geology and paleontology of the Colorado River delta at Golfo de Santa Clara, Sonora, Mexico. In R. E. Reynolds (Ed.), *Wild, scenic & rapid – a trip down the Colorado River Trough* (pp. 84–89). The 2007 Desert Symposium Field Guide and Abstracts from Proceedings. California State University, Desert Studies Consortium and LSA Associates, Inc., Fullerton.

Cruz-Muñoz, V., Arroyo-Cabrales, J., & Graham, R. W. (2009). Rodents and lagomorphs from the Late Pleistocene deposits at Valsequillo, Puebla, México. *Current Research in the Pleistocene, 26*, 147–149.

Cruz y Cruz, T. (2011). *Paleoambientes del Cuaternario Tardío en Sonora a partir del registro paleopedológico*. M. S. Thesis, Universidad Nacional Autónoma de México.

Cuataparo, N. J., & Ramírez, S. (1875). Descripción de un mamífero fósil de especie desconocida perteneciente al Género *Glyptodon"* encontrado entre las capas post-terciarias en el Distrito de Zumpango. *Sociedad Mexicana de Geografía y Estadística, Boletín Tercera Época, 2*, 354–362.

Cushing, J. E., Jr. (1945). Quaternary rodents and lagomorphs of San Josecito Cave, Nuevo León, Mexico. *Journal of Mammalogy, 26*, 182–185.

Dalquest, W. W. (1974). A new species of four-horned antilocaprid from Mexico. *Journal of Mammalogy, 55*, 96–101.

De Anda-Hurtado, P. (2009). La fauna local Mina de San Antonio, Pleistoceno de San Luis Potosí: Taxonomía, comparación actualística y significación geológico-paleontológica. M. S. Thesis, Universidad Nacional Autónoma de México.

Díaz-Ortega, J., Solleiro-Rebolledo, E., Sedov, S., & Cabadas, H. (2010). Paleosuelos y tepetates del Glacis de Buenavista Morelos (México): Testigos de eventos climáticos de la transición Pleistoceno-Holoceno. *Boletín de la Sociedad Geológica Mexicana, 62*(3), 469–486.

Díaz-Sibaja, R., García-Zepeda, Ma. L., López-García, J. R., Marín-Leyva, A. H., & Gutiérrez-Bedolla, M. (2011). Breve reporte de dos nuevas localidades fosilíferas del Pleistoceno Tardío Rancholabreano en los Valles Centrales de Oaxaca. Libro de Resúmenes, *XII Congreso Nacional de Paleontología*, 22 al 25 de Febrero de 2011, Edificio Carolino, Benemérita Universidad Autónoma de Puebla, Puebla, México (pp. 51). México: Sociedad Mexicana de Paleontología A.C.

Dillehay, T. D. (2000). *The settlement of the Americas: A new prehistory*. New York: Basic Books.

Dixon, E. J. (1999). *Bones, boats and bison: Archeology and the first colonization of western North America*. Albuquerque, New Mexico: The University of New Mexico Press, xiv + 322 pp.

Downs, T. (1958). Fossil vertebrates from Lago Chapala, Jalisco. Congreso Geológico Internacional, XXa. Sesión, México, D. F. *Publicación, 7*, 76–77.

Edmund, A. G. (1996). A review of Pleistocene giant armadillos (Mammalia, Xenarthra, Pampatheriidae). In K. M. Stewart & K. L. Seymour (Eds.), *Palaeoecology and palaeoenvironments of Late Cenozoic mammals: Tributes to the career of C. S. (Rufus) Churcher* (pp. 300–321). Toronto, Ontario: University of Toronto Press.

Elizalde-García, M., Moreno-Fernández, S. M., Melgarejo-Meraz, R., Palma-Ramírez, A., & Castillo-Cerón, J. M. (2011). Restos de un armadillo gigante (Pampatheriidae) en el área de Valsequillo, Puebla. Libro de Resúmenes, *XII Congreso Nacional de Paleontología*, 22 al 25 de Febrero de 2011, Edificio Carolino, Benemérita Universidad Autónoma de Puebla, Puebla, México (pp. 52). México: Sociedad Mexicana de Paleontología A.C.

Esteva, M., Arroyo-Cabrales, J., Flores-Martínez, A., Johnson, E., & Polaco, O. J. (2005). Fossil shrews from San Josecito Cave, Nuevo León, Mexico. *International Society of Shrew Biologists, Special Publication, 1*, 31–47.

Faith, J. T., & Surovell, T. A. (2009). Synchronous extinction of North America's Pleistocene mammals. *Proceedings on the National Academy of Sciences, 106*(49), 20641–20645.

Feinberg, J. M., Renne, P. R., Arroyo-Cabrales, J., Waters, M. R., Ochoa-Castillo, P., & Pérez-Campa, M. (2009). Age constraints on alleged "footprints" preserved in the Xalnene Tuff near Puebla, Mexico. *Geology, 37*, 267–270.

Ferrusquía-Villafranca, I. (1976). Estudios Geológico-Paleontológicos en la Región Mixteca, Parte 1: Geología del Area Tamazulapan-Teposcolula-Yanhutlán, Mixteca Alta, Estado de Oaxaca, México. *Universidad Nacional Autónoma de México, Instituto de Geología, Boletín, 97*, 1–160.

Ferrusquía-Villafranca, I. (1993). Geology of Mexico: A synopsis. In T. P. Ramamoorthy, R. Bye, A. Lot, & J. Fa (Eds.), *Biological diversity of Mexico: Origins and distribution* (pp. 3–107). New York: Oxford University Press.

Ferrusquí'a-Villafranca, I. (1998). La geología de México: Una sinopsis. In T. P. Ramamoorthy, R. Bye, A. Lot, & J. Fa (Eds.), *La diversidad biológica de México, Publicación Especial* (pp. 1–107). Instituto de Biología. Universidad Nacional Autónoma de México.

Ferrusquía Villafranca, I., & De Anda-Hurtado, P. (2008). A new Late Pleistocene fauna from central Mexico and its paleobiological-environmental significance. In *Society of Vertebrate Paleontology, 68th Annual Meeting*, Cleveland, OH (*Journal of Vertebrate Paleontology 28* (Suppl. Nbr. 3), Abstracts of Papers (pp. 77A)).

Ferrusquía-Villafranca, I., Arroyo-Cabrales, J., Martínez-Hernández, E., Gama-Castro, J., Ruiz-González, J., Polaco, O. J., et al. (2010). Pleistocene mammals of Mexico: A critical review of regional chronofaunas, climate change response and biogeographic provinciality. *Quaternary International, 217*, 53–104.

Fiedel, S. (2009). Sudden deaths: The chronology of terminal Pleistocene megafaunal extinction. In G. Haynes (Ed.), *American Megafaunal Extinctions at the End of the Pleistocene* (pp. 21–37). Dordrecht: Springer.

Findley, S. J. (1953). Pleistocene Soricidae from San Josecito Cave, Nuevo León, Mexico. *University of Kansas, Museum of Natural History, Publications, 5*(36), 633–639.

Firestone, R. B., West, A., Kennett, J. P., Becker, L., Bunch, T. E., Revay, Z. S., et al. (2007). Evidence for an extraterrestrial impact 12,900 years ago that contributed to the megafaunal extinctions and the Younger Dryas cooling. *Proceedings of the National Academy of Sciences, 104*, 16016–16021.

Flannery, K. V. (1986). *Guilá Naquitz: Archaic foraging and early agriculture in Oaxaca*. New York: Academic Press.

Flint, R. F. (1947). *Glacial geology and the Pleistocene epoch*. New York: Wiley.

Foreman, S. (1955). Palynology in southern North America, Part 2. *Geological Society of America Bulletin, 66*, 475–509.

Forman, R. T., & Godron, M. (1986). *Landscape ecology*. New York: Wiley.

Franzen, J. L. (1994). Eine Rancho-Labrea Fauna aus Nordost-Mexiko (Bundesstaat Nuevo Leon). *Natur und Museum, 124*, 241–272.

Frazier, M. K. [1981(1982)]. A revision of the fossil Erethizontidae of North America. *Bulletin of the Florida State Museum, Biological Sciences, 27*(1), 1–76.

Freudenberg, W. (1921). *Geologie von Mexiko*. Berlin: Verlag von Gebrüder Borntrager.

Freudenberg, W. (1922). Die Säugetierfauna des Pliocäns und Postpliocäns von Mexiko. II. Tiel: Mastodonden und Elefanten. *Geologische und Palaeontologische Abhandlungen, 14*, 103–176.

Furlong, E. L. (1925). Notes on the occurrence of mammalian remains in the Pleistocene of Mexico, with a description of a new species, *Capromeryx mexicana*. *University of California, Publications in Geological Sciences, 15*(5), 137–152.

Furlong, E. L. (1943). A new Pliocene antilope Stockoceras conklingi from San Josecito Cave, Mexico. *Carnegie Institution of Washington Publication, 551*, 1–8.

Gaines, E. P., & Sánchez, G. (2009). Current Paleoindian research in Sonora, Mexico. *Archaeology Southwest, 23*, 4–5.

Gama-Castro, J. (1996). Los suelos tropicales de México. Génesis, dinámica y degradación. Ph.D. Dissertation, Universidad Nacional Autónoma de México.

Gama-Castro, J., Flores-Román, D., Solleiro-Rebolledo, E., Jasso-Castañeda, C., Rocha-T., A. M., & Villalpando, J. L. (2004). Neosols, relict paleosols and alterites characterization and sptcial distribution in the Trans-Mexican Volcanic Belt, Morelos State: A regional approximation. *Revista Mexicana de Ciencias Geológicas, 21*(1), 160–174.

Gama-Castro, J., Solleiro-Rebolledo, E., McClung, E., Villalpando, J. L., Sedov, S., Jasso-Castañeda, C., et al. (2005). Contribuciones de la Ciencia del Suelo a la investigación arqueológica: El caso de Teotihuacán. *Terra, 23*, 1–11.

García, M. E. (1988). *Modificaciones al Sistema de Clasificación Climática de Köppen para adaptarlo a las condiciones de México*. México: Ed. por M.E. García.

García, E. (1990). IV.4.10. Clima, Mapa Esc. 1:4 000 000, *Atlas Nacional de México, Tomo II, Sección IV Naturaleza, Apartado 4 Clima*. México: Universidad Nacional Autónoma de México, Instituto de Geografía.

García-Zepeda, M. L., Pérez-González, M. S., Godínez-García, V., & Arroyo-Cabrales, J. (2008). Roedores fósiles de la Cinta. Michoacán. *Geos, 28*(2), 186.

García-Zepeda, M. L., Pérez-González, M. S., Godínez-García, V., & Arroyo-Cabrales, J. (2009). Roedores y Herpetofauna Fósil de La Cinta, Michoacán y Portalitos, Guanajuato. *XI Congreso Nacional de Paleontología*, 25 al 27 de Febrero de 2009, Centro de Geociencias, Campus-UNAM, Juriquilla, Querétaro, México (pp. 26). México: Sociedad Mexicana de Paleontología A.C.

Gibbard, P. L., Head, M. J., Walker, M. J. C., & The Subcommission on Quaternary Stratigraphy (2010). Formal ratification of the QuaternarySystem/Period and the Pleistocene Series/Epoch with a base at 2.58 Ma. *Journal of Quaternary Science, 25*, 96–102.

Gilmore, R. M. (1947). Report on a collection of mammal bones from archaeologic cave-sites in Coahuila, Mexico. *Journal of Mammalogy, 28*, 147–165.

Gómez-Peréz, L., & Carbot-Chanona, G. (2011). El registro más sureño de *Glyptotherium florindanum* para Norteamérica. Libro de Resúmenes, *XII Congreso Nacional de Paleontología*, 22 al 25 de Febrero de 2011, Edificio Carolino, Benemérita Universidad Autónoma de Puebla, Puebla, México (pp. 66). México: Sociedad Mexicana de Paleontología A.C.

Gonzáles-Arqueros, L, Vázquez-Selem, L., McClung de Tapia, E., & Gama-Castro, J. E. (2011). Geomorphological mapping and terrain analysis of the Teotihuacan Valley, central Mexico. In *Landscapes and soils trougth time* (pp. 78–79). IUSS Commission on Paleopedology and IUSS Comission on Soil Geography.

González, S., & Huddart, D. (2008). The Late Pleistocene human occupation of Mexico. *FUMDHAMentos, Publicação da Fundação Museu do Homem Americano, 7*, 236–259.

Gonzalez, S., Huddart, D., & Bennett, M. (2006). Valsequillo Pleistocene archaeology and dating: Ongoing controversy in central Mexico. *World Archaeology, 38*, 611–627.

Gonzalez, S., Jiménez-López, J. C., Hedges, R., Huddart, D., Ohman, J. C., Turner, A., et al. (2003). Earliest humans in the Americas: New evidence from Mexico. *Journal of Human Evolution, 44*, 379–387.

Gonzalez, S., Huddart, D., Israde-Alcántara, I., Dominguez-Vazquez, G., & Bischoff, J. (2014). Tocuila mammoths, Basin of Mexico: Late Pleistocene-Early Holocene stratigraphy and the geological context of the bone accumulation. *Quaternary Science Reviews, 96*, 222–239.

González-González, A. H., Rojas Sandoval, C., Terrazas Mata, A., Benavente Sanvicente, M., Stinnesbeck, W., Aviles, O., et al. (2008). The arrival of humans on the Yucatan Peninsula: Evidence from submerged caves in the state of Quintana Roo, Mexico. *Current Research in the Pleistocene, 25*, 1–24.

González-Quintero, L. (1986). Análisis polínicos de los sedimentos. In J. L. Lorenzo & L. Mirambel (Eds.), *Tlapacoya: 35,000 años de historia del Lago de Chalco* (pp. 157–166). Mexico: INAH, Colección Científica, Serie Prehistoria.

Götz, C. M., & De Anda-Alanís, G. G. (2011). Discusión morfométrica y taxonómica de cuatro cráneos de osos cara corta (Tremarctinae) hallados en un contexto subacuático en el Estado de Yucatán, México. Libro de Resúmenes, *XII Congreso Nacional de Paleontología*, 22 al 25 de Febrero de 2011, Edificio Carolino, Benemérita Universidad Autónoma de Puebla, Puebla, México (pp. 69). México: Sociedad Mexicana de Paleontología A.C.

Graham, R. W., & Lundelius, E. L. (1984). Coevolutionary disequilibrium and Pleistocene extinction. In P. S. Martin & R. G. Klein (Eds.), *Quaternary extinctions: A prehistoric revolution* (pp. 223–249). Tucson: The University of Arizona Press.

Guenther, E. W. (1968). Untersuchungen zur jungeiszeitlichen und Nacheiszeitlichen Geologischen und Paläontologischen Geschichte. In F. Tichy (Ed.), *El proyecto México de la Fundación Alemana para la Investigación Científica-Franz Steiner Verlag. GMBH, 1*, 32–36.

Guenther, E. W., & Bunde, H. (1973). Investigaciones geológicas y paleontológicas en México durante los años de 1965 a 1969. *Comunicaciones, 7*, 19–20.

Guthrie, R. D. (2003). Rapid body size decline in Alaskan Pleistocene horses before extinction. *Nature, 426*, 169–171.

Gutiérrez-Bedolla, M., García-Zepeda, Ma. L., López-García, J. R., Marín-Leyva, A. H., & Díaz-Sibaja, R. (2011). Estudio Paleontológico de Uruétaro, Municipio de Álvaro Obregón, Michoacán. Libro de Resúmenes, *XII Congreso Nacional de Paleontología*, 22 al 25 de Febrero de 2011, Edificio Carolino, Benemérita Universidad Autónoma de Puebla, Puebla, México (pp. 75). México: Sociedad Mexicana de Paleontología A.C.

Guzmán-Gutiérrez, J. R., Rodríguez-De la Rosa, R. A., Blanco-Piñón, A., & Hernández-Ávila, J. (2009). Coprolitos de vertebrados del Terciario Tardío de Jalisco, México. *XI Congreso Nacional de Paleontología*, 25 al 27 de Febrero de 2009, Centro de Geociencias, Campus-UNAM, Juriquilla, Querétaro, México (pp. 107). México: Sociedad Mexicana de Paleontología A.C.

Guzmán-Gutiérrez, J. R., Jimenéz-Betts, P., & Carrillo-Rodríguez, C. A. (2011). Nueva localidad de vertebrados del Pleistoceno Tardío en Villa Hidalgo, Zacatecas. Libro de Resúmenes, *XII Congreso Nacional de Paleontología*, 22 al 25 de Febrero de 2011, Edificio Carolino, Benemérita Universidad Autónoma de Puebla, Puebla, México (pp. 77). México: Sociedad Mexicana de Paleontología A. C.

Guzmán-Guzmán, S. (2011). Nuevo registro de un caballo en el centro de Veracruz. Libro de Resúmenes, *XII Congreso Nacional de Paleontología*, 22 al 25 de Febrero de 2011, Edificio Carolino, Benemérita Universidad Autónoma de Puebla, Puebla, México (pp. 78). México: Sociedad Mexicana de Paleontología A.C.

Hall, E. R. (2001). *The mammals of North America*. (2nd ed.) (Vols. 1 and 2). New York: Wiley.

Haynes, G. (2002). *The early settlement of North America: The Clovis era* (p. 360). Cambridge, Inglaterra: Cambridge University Press.

Heine, K. (1984). The classical late Weichselian climatic fluctuations in Mexico. In N. A. Mörner & W. D. Karcén (Eds.), *Climatic changes on a yearly to millenial basis* (pp. 95–115). Dordrecht: Riedel.

Hernández, M. E. (1990). IV.4.9. Medidas de aridez, Mapa Esc. 1:8 000 000, *Atlas Nacional de México, Tomo II, Sección IV Naturaleza, Apartado 4 Clima*. México: Universidad Nacional Autónoma de México, Instituto de Geografía.

Hibbard, C. W. (1955). Pleistocene vertebrates from the Upper Becerra (Becerra Superior) Formation, Valley of Tequixquiac, Mexico, with notes on other Pleistocene forms. *University of Michigan, Contributions from the Museum of Paleontology, 12*(5), 47–96.

Hibbard, C. W., & Mooser, O. (1963). A porcupine from the Pleistocene of Aguascalientes, Mexico. *University of Michigan, Contributions from the Museum of Paleontology, 18*(16), 245–250.

Hibbard, C. W., & Villa-Ramírez, B. (1950). El bisonte gigante de México. *Universidad Nacional Autónoma de México, Instituto de Biología, Anales, 21*(1), 243–251.

Hodnett, J. P. M., Mead, J. I., & Baez, A. (2009). Dire wolf, *Canis dirus* (Mammalia, Carnivora, Canidae), from the Late Pleistocene (Rancholabrean) of East Central Sonora, Mexico. *The Southwestern Naturalist, 54*, 74–81.

Isphording, W. C. (1974). Weathering of Yucatán limestones: The genesis of Terra Rossa (pp. 78–93). *Field Trip 2, 1974, Annual Meeting, Miami*. New Orleans: Geological Society.

Israde-Alcantara, I., Velázquez-Durán, R., Lozano-García, Ma. S., Bischoff, J., Domínguez-Vázquez, G., & Garduño-Monroy, V. H. (2010). Evolución Paleolimnológica del Lago Cuitzeo, Michoacán durante el Pleistoceno-Holoceno. *Boletín de la Sociedad Geológica Mexicana, 62*(3), 345–357.

Jackway, G. E. (1958). Pleistocene Lagomorpha and Rodentia from the San Josecito Cave, Nuevo León, Mexico. *Transactions of the Kansas Academy of Science, 61*, 313–327.

Jasso-Castañeda, C. (2007). *La memória de los paleosuelos del Nevado de Toluca: Un registro de estabilidad geomórfica y cambio ambiental durante el Cuaternario Tardío*. Ph. D. Dissertation, Universidad Nacional Autónoma de México.

Jasso-Castañeda, C., Sedov, S., Gama-Castro, J., & Solleiro-Rebolledo, E. (2006). Los paleosuelos: Un índice del paleoambiente y de la estabilidad del paisaje del Nevado de Toluca. *Terra Latinoamericana, 24*(2), 151–161.

Jasso-Castañeda, C., Gama-Castro, J. E., Solleiro-Rebolledo, E., Sedov, S., & Díaz-Ortega, J. (2012). Morfogénesis, procesos y evolución del horizonte Bw Cámbico en tefra-paleosuelos del Volcán Nevado de Toluca. *Boletín de la Sociedad Geológica Mexicana, 64*, 37–47.

Jefferson, G. T. (1989). Late Cenozoic tapirs (Mammalia: Perissodactyla) of western North America. Natural History Museum of Los Angeles County. *Contributions in Science, 406*, 1–21.

Jezkova, T., Jaeger, J. R., Marshall, Z. L., & Riddle, B. R. (2009). Pleistocene impacts on the phylogeography of the Desert Pocket Mouse (*Chaetodipus penicillatus*). *Journal of Mammalogy, 90*, 306–320.

Jiménez-Hidalgo, E., Guerrero-Arenas, R., McFadden, B. J., & Cabrera-Pérez, L. (2011). The Late Pleistocene (Rancholabrean) *Viko Vijin* local fauna from La Mixteca Alta, Northwestern Oaxaca. Southern Mexicio. *Revista Brasileira de Paleontologia, 14*(1), 15–28.

Jiménez López, J. C., González, S., Pompa y Padilla, J. A., & Ortiz Pedraza, F. (2006a). El hombre temprano en América y sus implicaciones en el poblamiento de la cuenca de México. Primer Simposio Internacional. *Colección Científica, Instituto Nacional de Antropología e Historia, México, 500*, 1–274.

Jiménez López, J. C., Polaco, O. J., Martínez Sosa, G., & Hernández Flores, R. (2006b). *2º Simposio Internacional El hombre temprano en América* (p. 197). México: Instituto Nacional de Antropología e Historia.

Johnson, C. N., & Brook, B. W. (2011). Reconstructing the dynamics of ancient human populations from radiocarbon dates: 10,000 years of population growth in Australia. *Proceedings of the Royal Society, Series B, 278*, 3748–3754.

Johnson, E., Arroyo-Cabrales, J., & Polaco, O. J. (2006). Climate, environment, and game animal resources of the Late Pleistocene Mexican grassland. In J. C. Jiménez López, S. González, J. A. Pompa & F. Ortiz-Pedraza (Eds.), *El Hombre Temprano en América y sus Implicaciones en el Poblamiento de la Cuenca de México*. México: Instituto Nacional de Antropología e Historia (*Colección Científica, 500*:231–245).

Kellogg, W. W. (1978). Global influences of mankind on the climate. In J. Gribbin (Ed.), *Climatic change* (pp. 205–227). Cambridge: Cambridge University Press.

Koch, P. L., & Barnosky, A. D. (2006). Late Quaternary extinctions: State of the debate. *Annual Review of Ecology Evolution and Systematics, 37*, 215–250.

Kurtén, B. (1967). Präriew und Sabelzahntiger aus dem Pleistozän des Valsequillo, Mexiko. *Quärter, 18*, 173–178.

Kurtén, B. (1974). A history of coyote-like dogs (Canidae, Mammalia). *Acta Zoologica Fenica, 140*, 1–38.

Kurtén, B. (1975). A new Pleistocene genus of American mountain deer in North America. *Journal of Paleontology, 56*, 507–508.

Kurtén, B., & Anderson, E. (1980). *Pleistocene mammals of North America* (p. 442). New York: Columbia University Press.

Leyden, B. W., Brenner, M., Hodell, D. A., & Curtis, J. H. (1994). Orbital and internal forcing of climate on the Yucatan Peninsula for the past ca. 36 ka. *Palaeogeography, Palaeoclimatology, Palaeoecology, 109*, 193–210.

Leyden, B. W., Brenner, M., Whitmore, T. J., Curtis, J. H., Piperno, D. R., & Dahlin, B. H. (1996). A record of long- and short-term climate variation from northwest Yucatan: Cenote San Jose Chulchaca. In S. L. Fedick (Ed.), *The managed mosaic: Ancient Maya agriculture and resource use* (pp. 30–50). Salt Lake City: University of Utah Press.

Lorenzen, E. D., Nogués-Bravo, D., Orlando, L., Weinstock, J., Binlanden, J., Marske, K. A., et al. (2011). Species-specific responses of Late Quaternary megafauna to climate and humans. *Nature, 479*, 359–364.

Lorenzo, J. L., & Mirambell, L. (1986). Mamutes excavados en la Cuenca de México (1952–1980). México: Departamento de

Prehistoria. *Instituto Nacional de Antropología e Historia, Cuaderno de Trabajo, 32*, 1–151.

Lorenzo, J. L., & Mirambell, L. (1999). The inhabitants of Mexico during the Upper Pleistocene. In R. Bonnichsen & K. L. Turnmire (Eds.), *Ice age peoples of North America. Environments, origins, and adaptations of the first Americans* (pp. 482–496). Corvallis, Oregon: Center for the Study of the First Americans, Oregon State University Press, 536 pp.

Lozano-García, M. S., & Ortega-Guerrero, B. (1994). Palynological and magnetic susceptibility records of Lake Chalco, central Mexico. *Palaeogeography, Paleoclimatology, Palaecology, 109*, 177–191.

Lozano-García, S., & Ortega-Guerrero, B. (1997). Late Quaternary environmental changes of the central part of the Basin of Mexico: Correlation between Texcoco and Chalco Basins. *Review of Paleobotany and Palynology, 99*, 77–93.

Lozano-García, M. S., & Xelhuantzi-López, M. S. (1997). Some problems in the late Quaternary pollen records of Central Mexico: Basins of Mexico and Zacapu. In J. Urrutia-Fucugauchi, S. E. Metcalfe, & M. Caballero-Miranda (Guest Eds.), *Climate change-Mexico, First International Conference on Climate Change in Mexico, Taxco, 1993*. (*Quaternary International, 43/44*, 117–123).

Lozano-García, M. S., Ortega-Guerrero, M., & Urrutia-Fucugauchi, J. (1993). Late Pleistocene and Holocene paleoenvironments of Chalco Lake, Central Mexico. *Quaternary Research, 40*, 332–342.

Lucas, S. G. (2008). Late Cenozoic mammals from the Chapala Rift Basin, Jalisco, Mexico. In S. G, Lucas, G. S. Morgan, J. A. Spielmann, & D. R. Prothero (Eds.), Neogene mammals. *New Mexico Museum of Natural History and Science, Bulletin, 44*, 39–50.

Lundelius, E. L., Jr. (1980). Late Pleistocene and Holocene mammals from Northern Mexico and their implications for archaeological research. *Bulletin of the Florida State Museum, Biological Sciences, 27*(1), 1–76.

Marín-Leyva, A. H., García-Zepeda, Ma. L., & Arroyo-Cabrales, J. (2009). Caballos Fósiles (*Equus*: Equidae) de La Cinta, Michoacán. *XI Congreso Nacional de Paleontología, 25 al 27 de Febrero de 2009*, Centro de Geociencias, Campus-UNAM, Juriquilla, Querétaro, México (pp. 40). México: Sociedad Mexicana de Paleontología A.C.

Marín-Leyva, A. H., Alberdi, M. T., Arroyo-Cabrales, J., Ponce-Saavedra, J., García-Zepeda, Ma. L., & Tejeda-Alvarado, F. (2011). Caballos fósiles de La Piedad-Santa Ana (Michoacán, Guanajuato) y sus estimaciones de masa corporal. Libro de Resúmenes, *XII Congreso Nacional de Paleontología, 22 al 25 de Febrero de 2011*, Edificio Carolino, Benemérita Universidad Autónoma de Puebla, Puebla, México (pp. 91). México: Sociedad Mexicana de Paleontología A.C.

Mark, D. F., Gonzalez, S., Huddart, D., & Böhnel, H. (2010). Dating of the Valsequillo volcanic deposits: Resolution of an ongoing archaeological controversy in Central Mexico. *Journal of Human Evolution, 58*, 441–445.

Martin, P. S., & Klein, R. G. (Eds.) (1989). *Quaternary extinctions: A prehistoric revolution*. Tucson, Arizona: The University of Arizona Press, 988 pp.

McClung, E., Domínguez, I., Gama-Castro, J., Solleiro-Rebolledo, E., & Sedov, S. (2005). Radiocarbon dates from soil profiles in the Teotihuacá, Valley, Mexico: Geomorphological processes and vegetation change. *Radiocarbon, 47*, 159–175.

McDonald, H. G. (2002). Fossil Xenarthra of Mexico: A review. In M. Montellano-Ballesteros & J. Arroyo-Cabrales (Coords.), *Avances en los estudios paleomastozoológicos en México* (pp. 227–248). México, D.F.: Instituto Nacional de Antropología e Historia, *Colección Científica*.

MacFadden, B. J., & Hulbert, R. C., Jr. (2009). Calibration of mammoth (*Mammuthus*) dispersal into North America using rare earth elements of Plio-Pleistocene mammals from Florida. *Quaternary Research, 71*, 41–48.

Martin, P. S., & Harrell, B. E. (1957). The Pleistocene history of temperate biotas in Mexico and eastern United States. *Ecology, 38*, 468–480.

Mead, J. I., Baez, A., Swift, S. L., Carpenter, M. C., Hollenshead, M., Czaplewski, et al. (2006). Tropical marsh and savanna of the Late Pleistocene in northeastern Sonora, Mexico. *The Southwestern Naturalist, 51*, 226–239.

Mead, J. I., Swift, S. L., White, R. S., McDonald, H. G., & Baez, A. (2007). Late Pleistocene (Rancholabrean) glyptodont and pampathere (Xenarthra, Cingulata) from Sonora, Mexico. *Revista Mexicana de Ciencias Geológicas, 24*, 439–449.

Mead, J. I., White, R. S., Baez, A., Hollenshead, M. G., Swift, S. L., & Carpenter, M. C. (2010). Late Pleistocene (Rancholabrean) Cynomys (Rodentia, Sciuridae: prairie dog) from northwestern Sonora, Mexico. *Quaternary International, 217*, 138–142.

Melgarejo-Meraz, R., Elizalde-García, M., Moreno-Fernández, S. M., Palma-Ramírez, A., & Castillo-Cerón, J. M. (2011). Carnívoros pleistocénicos de Valsequillo, Puebla, México. Libro de Resúmenes, *XII Congreso Nacional de Paleontología, 22 al 25 de Febrero de 2011*, Edificio Carolino, Benemérita Universidad Autónoma de Puebla, Puebla, México (pp. 93). México: Sociedad Mexicana de Paleontología A.C.

Meltzer, D. J. (2009). *First Peoples in a New World: Colonizing Ice Age America* (p. 464). Berkeley, California: The University of California Press.

Meltzer, D. J., Holliday, V. T., Cannon, M. D., & Miller, D. S. (2014). Chronological evidence fails to support claim of an isochronous widespread layer of cosmic impact indicators dated to 12,800 years ago. *Proceedings of the National Academy of Sciences, 111*(21), E2162–E2171.

Mercer, H. C. [1896 (1975)]. *The Hill-Caves of Yucatan: A search for evidence of man's antiquity in the caverns of Central America*. Norman, Oklahoma: University of Oklahoma Press, xliv + 183 pp.

Messing, H. J. (1986). A Late Pleistocene-Holocene fauna from Chihuahua, Mexico. *The Southwestern Naturalist, 31*, 277–288.

Metcalfe, S. E. (1992). *Changing environments of the Zacapu Basin, central Mexico: A diatom-based history spanning the last 30,000 years*. England: University of Oxford, School of Geography, Research Paper 48.

Meyer, E. R. (1973). Late Quaternary paleoecology of the Cuatro Cienegas Basin, Coahuila, Mexico. *Ecology, 54*, 982–985.

Mones, A. (1973). Nueva especie de pecarí fósil del Estado de Jalisco. México, D.F.: Instituto Nacional de Antropología e Historia (*Anales Séptima Época, 3*, 119–128).

Montejano-Esquivias, M., Jardón-Nava, E., & Ladrón de Guevara-Ureña, E. (2009). Recientes hallazgos paleontológicos en El Salto, Jalisco. *XI Congreso Nacional de Paleontología, 25 al 27 de Febrero de 2009*, Centro de Geociencias, Campus-UNAM, Juriquilla, Querétaro, México (pp. 48). México: Sociedad Mexicana de Paleontología A.C.

Montellano-Ballesteros, M. (1992). Una edad del Irvingtoniano al Rancholabreano para la fauna Cedazo del Estado de Aguascalientes. México: Universidad Nacional Autónoma de México (*Instituto de Geología, Revista, 9*, 195–203).

Mooser, O. (1958). La fauna "Cedazo" del Pleistoceno en Aguascalientes. *Universidad Nacional Autónoma de México, Instituto de Biología, Anales, 1–2*, 409–452.

Mooser, O., & Dalquest, W. W. (1975a). Pleistocene mammals from Aguascalientes, Central Mexico. *Journal of Mammalogy, 56*, 781–820.

Mooser, O., & Dalquest, W. W. (1975b). A new species of camel (Genus *Camelops*) from the Pleistocene of Aguascalientes, Mexico. *The Southwestern Naturalist, 19*, 341–345.

Morales-Mejía, F. M., & Arroyo-Cabrales, J. (2009). Los carnívoros (Mammalia, Carnivora) del Cuaternario procedentes de las excavaciones de la Gruta de Loltún, Yucatán, México. *XI Congreso Nacional de Paleontología*, 25 al 27 de Febrero de 2009, Centro de Geociencias, Campus-UNAM, Juriquilla, Querétaro, México (pp. 111). México: Sociedad Mexicana de Paleontología A.C.

Morales-Mejía, F. M., Arroyo-Cabrales, J., & Polaco, O. J. (2009). New records for the Pleistocene fauna from Loltun Cave, Yucatan, Mexico. *Current Research in the Pleistocene, 26*, 166–168.

Moreno-Fernández, S. M., Melgarejo-Meraz, R., Elizalde-García, M., Palma-Ramírez, A., & Castillo-Cerón, J. M. (2011). Pequeños mamíferos pleistocénicos de Valsequillo, Puebla, México. Libro de Resúmenes, *XII Congreso Nacional de Paleontología*, 22 al 25 de Febrero de 2011, Edificio Carolino, Benemérita Universidad Autónoma de Puebla, Puebla, México (pp. 100). México: Sociedad Mexicana de Paleontología A.C.

Nowak, R. M. (1979). North American Quaternary Canis. *University of Kansas, Museum of Natural History, Monograph, 6*, 1–154.

Nowak, R. M. (1991). *Walker's mammals of the World* (5th ed.). Baltimore: The John Hopkins University Press.

Nunez, E. E., Macfadden, B. J., Mead, J. I., & Baez, A. (2010). Ancient forests and grasslands in the desert: Diet and habitat of Late Pleistocene mammals from Northcentral Sonora, Mexico. *Palaeogeography, Palaeoclimatology, Palaeoecology, 297*(2), 391–400.

Ochoa, J., & Piper, P. (2017). Holocene large mammal extinctions in Palawan Island, Philippines. In G. G. Monks (Ed.), *Climate change and human responses: A zooarchaeological perspective* (pp. 69–86). Dordrecht: Springer.

Ohngemach, D. (1973). Análisis polínico del Pleistoceno reciente y del Holoceno en la región Puebla-Tlaxcala. *Proyecto Puebla-Tlaxcala. Comunicaciones, 7*, 47–49.

Ohngemach, D. (1977). Pollen sequence of the Tlaloqua crater (La Malinche volcano, Tlaxcala, Mexico). *Sociedad Botánica de México, Boletín, 36*, 33–40.

Ortega-Guerrero, B., & Urrutia-Fucugauchi, J. (1997). A paleomagnetic secular variation record from Late Pleistocene-Holocene lacustrine sediments from Chalco Lake, Basin of Mexico. In J. Urrutia-Fucugauchi, S. E. Metcalfe, & M. Caballero-Miranda (Eds.), *Climate change-Mexico, First International Conference on Climate Change in Mexico, Taxco 1993* (*Quaternary International, 43/44*, 87–96).

Ortega-Guerrero, B., Sedov, S., Solleiro-Rebolledo, E., & Soler, A. (2004). Magnetic mineralogy in Barranca Tlalpan exposure paleosols, Tlaxcala. Mexico. *Revista Mexicana de Ciencias Geologicas, 21*(1), 120–132.

Ortega, B., Vázquez, G., Caballero, M., Israde, I., Lozano-García, S., Schaaf, P., et al. (2010). Late Pleistocene: Holocene record of environmental changes in Lake Zirahuen, Central Mexico. *Journal of Paleolimnology, 44*, 745–760.

Ortega-Ramírez, J. R., Valiente-Banuet, A., Urrutia-Fucugauchi, J., Mortera-Gutierrez, C. A., & Alvarado-Valdez, G. (1998). Paleoclimatic changes during the Late Pleistocene-Holocene in Laguna Babícora, near the Chihuahua Desert, Mexico. *Canadian Journal of Sciences, 35*, 1168–1179.

Ortega-Ramírez, J., Maillol, J. M., Bandy, W., Valiente-Banuet, A., Urrutia-Fucugauchi, J., Mortera-Gutierrez, C. A., et al. (2004). Late Quaternary evolution of alluvial fans in the Playa, El Fresnal region, northern Chihuahua Desert, Mexico: Paleoclimatic implications. *Geofísica Internacional, 43*(3), 445–466.

Owen, R. (1869). On fossil remains of equines from Central and South America referable to *Equus conversidens* Ow., *Equus tau* Ow., and *Equus arcidens* Ow. *Royal Society of London, Philosophical Transactions, Series B, Biological Sciences, 159*, 559–573.

Palma-Ramírez, A., Martínez-García, A. L., & Castillo-Cerón, J. M. (2009a). Roedores del Pleistoceno-Holoceno de la región centro del Estado de Hidalgo, México. *XI Congreso Nacional de Paleontología*, 25 al 27 de Febrero de 2009, Centro de Geociencias, Campus-UNAM, Juriquilla, Querétaro, México (pp. 55). México: Sociedad Mexicana de Paleontología A.C.

Palma-Ramírez, A., Martínez-García, A. L., Vázquez-Vázquez, C., & Reyes-Corte, M. A. (2009b). Mastofáunula de Huitexcalco de Morelos, Municipio de Chilcuautla, Hidalgo, México. *XI Congreso Nacional de Paleontología*, 25 al 27 de Febrero de 2009, Centro de Geociencias, Campus-UNAM, Juriquilla, Querétaro, México (pp. 54). México: Sociedad Mexicana de Paleontología A.C.

Palma-Ramírez, A., Moreno-Fernández, S. M., Elizalde-García, M., Melgarejo-Meraz, R., & Castillo-Cerón, J. M. (2011). Bioestratigrafía del área de Santa María Amajac, Centro de Hidalgo, México. Libro de Resúmenes, *XII Congreso Nacional de Paleontología*, 22 al 25 de Febrero de 2011, Edificio Carolino, Benemérita Universidad Autónoma de Puebla, Puebla, México (pp. 106). México: Sociedad Mexicana de Paleontología A.C.

Parfit, M. (2000). The dawn of humans. *National Geographic Magazine, 198*, 40–67.

Paz-Moreno, F., Demant, A., Cocheme, J. J., Dostal, J., & Montigny, R. (2003). The Quaternary Moctezuma volcanic field: A tholeiitic to alkali basaltic episode in the central Sonoran Basin and Range Province, Mexico. *Geological Society of America Special Paper, 374*, 1–17.

Peña-Serrano, J., & Miranda-Flores, F. A. (2009). Presencia de restos de mastofauna pleistocénica en la región de las grandes montañas del Estado de Veracruz, México. *XI Congreso Nacional de Paleontología*, 25 al 27 de Febrero de 2009, Centro de Geociencias, Campus-UNAM, Juriquilla, Querétaro, México (pp. 58). México: Sociedad Mexicana de Paleontología A.C.

Pérez-Crespo, V. A., Arroyo-Cabrales, J., Alva-Valdivia, L. M., Morales-Puente, P., Cienfuegos-Alvarado, E., & Otero-Trujano, F. J. (2011). La dieta y el hábitat de los megaherbívoros de El Cedral (Pleistoceno Tardío), San Luis Potosí, México. Libro de Resúmenes, *XII Congreso Nacional de Paleontología*, 22 al 25 de Febrero de 2011, Edificio Carolino, Benemérita Universidad Autónoma de Puebla, Puebla, México (pp. 107). México: Sociedad Mexicana de Paleontología A.C.

Pichardo, M. (1997). Valsequillo biostratigraphy: New evidence for Pre-Clovis date. *Anthropologischer Anzeiger, 55*, 233–246.

Pichardo, M. (1999). Valsequillo Biostratigraphy II: Bison, tools, correlate with Tequixquiac. *Anthropologischer Anzeiger, 57*, 11–24.

Pilaar Birch, S., & Miracle, P. T. (2017). Human response to climate change in the northern Adriatic during the Late Pleistocene and early Holocene. In G. G. Monks (Ed.), *Climate change and human responses: A zooarchaeological perspective* (pp. 87–100). Dordrecht: Springer.

Polaco, O. J. (1981). *Restos fósiles de Glossotherium and Eremotherium (Edentata) en México* (pp. 819–833). Porto Alegre: Brazil, II Congreso Latino-Americano de Paleontología, Anais.

Polaco, O. J. (1995). *Z-471: Análisis de la arqueofauna de la Mixtequilla*. México: Instituto Nacional de Antropología e Historia, Laboratorio de Paleozoología.

Polaco, O. J., & Butron-M. L. (1997). Mamiferos Pleistocénicos de la Cueva la Presita, San Luis Potosí, México. In J. Arroyo-Cabrales & O. J. Polaco (Coords.), *Homenaje al Profesor Ticul Alvarez* (pp. 279–376). México: Instituto Nacional de Antropología e Historia, Colección Científica.

Polaco, O. J., & Heredia-C. H. (1988). Hueso modificado: Un estudio tafonómico contemporáneo. *Trace, 14*, 73–81.

Polaco, O. J., Méndez-B. A., & Heredia-C. H. (1989). Los carnívoros como agentes tafonómicos. *Trace, 15*, 70–73.

Ramírez-Cruz, G. A., & Montellano-Ballesteros, M. (2011). Descripción de dos gliptodontes (Mammalia: Xenarthra) del Pleistoceno Tardío, de los estados de Tamaulipas y Tlaxcala, México. Libro de Resúmenes, *XII Congreso Nacional de Paleontología*, 22 al 25 de Febrero de 2011, Edificio Carolino, Benemérita Universidad Autónoma de Puebla, Puebla, México (pp. 115). México: Sociedad Mexicana de Paleontología A.C.

Ramírez–Pulido, J., González–Ruiz, N., Gardner y, A. L., & Arroyo–Cabrales, J. (2014). List of recent land mammals of Mexico, 2014. Special Publications, The Museum of Texas Tech University, 63:1–69.

Repenning, C. A. (1983). *Pitymys meadensis* Hibbard from the Valley of Mexico and the classification of North American species of *Pitymys* (Rodentia: Cricetidae). *Journal of Vertebrate Paleontology, 2*, 471–482.

Reyes, J. M. (1881). Breve reseña de la emigración de los pueblos en el Continente Americano y especialmente en el territorio de la República Mexicana con la descripción de los monumentos de la Sierra Gorda del Estado de Querétaro, distritos de Cadereyta, San Pedro Toliman y Jalpan, y la extincion de la raza chichimeca. *Boletín de la Sociedad de Geografía y Estadística de la República Mexicana, Tercera época, Tomo V*, 385–490.

Reynoso-Rosales, V. H., & Montellano-Ballesteros, M. (1994). Revisión de los équidos de la Fauna Cedazo del Pleistoceno de Aguascalientes, México. *Revista Mexicana de Ciencias Geológicas, 11*, 87–105.

Ruiz-Martínez, V. C., Osete, M. L., Vegas, R., Nuñez-Aguilar, J. I., & Urrutia-Fucugauchi, J. (2000). Paleomagnetism of late Miocene to Quaternary volcanics from the eastern segment of the Trans-Mexican Volcanic Belt. *Tectonophysics, 318*, 217–233.

Russel, B. D. (1960). Pleistocene pocket from San Josecito Cave, Nuevo Leon. *University of Kansas Publications, Museum of Natural History, 9*, 541–548.

Russell, B. D., & Harris, A. H. (1986). A new leporine (Lagomorpha: Leporidae) from Wisconsinan deposits of the Chihuahuan Desert. *Journal of Mammalogy, 67*, 632–639.

Rzedowski, J., & Reyna-Trujillo, T., 1990. IV.8.2. Vegetación Potencial, Mapa Esc. 1:4 000 000, Atlas Nacional de México, Tomo II, Sección IV Naturaleza, Apartado 8 Biogeografía. Universidad Nacional Autónoma de México, Instituto de Geografía, México, D.F.

Salgado-Rosas, I., Ramirez-Álvarez, S., Beltrán-M., I., Ramírez-G., A., Garibay-Romero, L., Critín-Ponciano, A., et al. (2009). Fauna Pleistocénica correspondiente a la zona norte del Estado de Guerrero, México. *XI Congreso Nacional de Paleontología*, 25 al 27 de Febrero de 2009, Centro de Geociencias, Campus-UNAM, Juriquilla, Querétaro, México (pp. 69). México: Sociedad Mexicana de Paleontología A.C.

Sanchez, M. G. (2001). A synopsis of Paleo-Indian archaeology in Mexico. *The Kiva, 67*, 119–136.

Sánchez, M. G., Gaines, E. P., & Holliday, V. (2009). El Fin del Mundo, Sonora: Cazadores Clovis de Megafauna del Pleistoceno Terminal. *Arqueología Mexicana, 17*, 46–49.

Sanchez, G., Holliday, V. T., Gaines, E. P., Arroyo-Cabrales, J., Martínez-Tagüeña, N., Kowler, A., et al. (2014). Human (Clovis)–gomphothere (*Cuvieronius* sp.) association ~13,390 calibrated yBP in Sonora, Mexico. *Proceedings of the National Academy of Sciences of the United States of America, 111*(30):10972–10977.

Santamaría Estévez, D., & García-Bárcena, J. (1989). Puntas de proyectil, cuchillos y otras herramientas de la Cueva de los Grifos, Chiapas. *Instituto Nacional de Antropología e Historia, Departamento de Prehistoria, Cuadernos de Trabajo, 40*, 1–40.

Schneider, S. H., & Temkin, R. L. (1978). Climatic changes and human affairs. In J. Gribbin (Ed.), *Climatic change* (pp. 228–246). Cambridge: Cambridge University Press.

Sears, P. B., & Clisby, K. H. (1955). Palynology in southern North America, Part 4. *Geological Society of America Bulletin, 66*, 521–530.

Sedov, S., Solleiro-Rebolledo, E., Gama-Castro, J., Vallejo-Gómez, E., & González-Velázquez, A. (2001). Buried paleosols of Nevado de Toluca: An alternative record of Late Quaternary environmental change in Central Mexico. *Journal of Quaternary Science, 16*(4), 375–389.

Sedov, S., Solleiro-Rebolledo, E., & Gama-Castro, J. (2003). Andosol to Luvisol evolution in central Mexico: Timing mechanisms and environmental setting. *Catena, 54*, 495–513.

Sedov, S., Solleiro-Rebolledo, E., Fedick, S. I., Gama-Castro, J., Palacios-Mayorga, S., & Vallejo Gómez, E. (2007). Soil genesis in relation to landscape evolution and ancient sustainable land use in the northeastern Yucatan Peninsula, Mexico. *Atti della Società Toscana di Scienze Naturali, Memorie Serie A, 112*, 115–126.

Sedov, S., Solleiro-Rebolledo, E., Feddick, S. L., Pi-Puig, T., Vallejo-Gómez, E., & Flores-Delgadillo, M. (2008). Micromorphology of soil catena in Yucatan: Pedogenesis and geomorphological processes in a tropical karst landscape. In S. Kapur, A. Mermut, & G. Stoops (Eds.), *New trends in soil micromorphology* (pp. 19–37). Berlin and Heidelberg: Springer.

Sedov, S., Solleiro-Rebolledo, S., Terhorst, B., Solé, J., Flores-Delgadillo, M. L., & Werner, G. (2009). The Tlaxcala basin paleosol sequence: A multiscale proxy of middle to late Quaternary environmental change in central Mexico. *Revista Mexicana de Ciencias Geológicas, 26*(2), 448–465.

Sedov, S., Lozano-García, M. S., Solleiro-Rebolledo, E., McClung de Tapia, E., Ortega-Guerrero, B., & Sosa-Nájera, S. (2011). Tepexpan revisited: A multiple proxy of local environmental changes in relation to human occupation from a paleolake shore section in central Mexico. *Geomorphology, 122*, 309–322.

Semken, H. A. Jr. (1974). Micromammal distribution and migration during the Holocene. American Quaternary Association, 3rd Annual Meeting Abstracts, pp. 25.

Shapiro, B., Drummond, A. J., Rambaut, A., Wilson, M. C., Matheus, P. E., Sher, A. V., et al. (2004). Rise and fall of the Beringian steppe bison. *Science, 306*, 1561–1565.

Shaw, C. A. (1981). *The Middle Pleistocene El Golfo local fauna from northwestern Sonora, Mexico*. M.S. Thesis. California State University, Long Beach.

Shaw, C. A., & McDonald, H. G. (1987). First record of giant anteater (Xenarthra, Myrmecophagidae) in North America. *Science, 236*, 186–188.

Shaw, C. A., Croxen, F. W., & Sussman, D. R. (2005). *El Golfo de Santa Clara, Sonora, Mexico, Fieldtrip*. USA: Society of Vertebrate Paleontology, 65th, Annual Meeting, Fieldguide.

Silva-Bárcenas, A. (1969). Localidades de vertebrados fósiles en la República Mexicana. *Paleontología Mexicana, 28*, 1–34.

Simpson, G. G. (1940). Mammals and land bridges. *Journal of the Washington Academy of Sciences, 30*, 137–163.

Smith, B. D. (1997). The initial domestication of *Cucurbita pepo* in the Americas 10,000 years ago. *Science, 276*, 932–934.

Solis-Castillo, B. Solleiro-Rebolledo, E., Liendo, R., Ortíz, M. A., & Teranishi, K. (2011). *Environmental changes in Holocene alluvial sequences at Maya lowlands: A preview of natural and cultural impacts*. Knoxville, Tennessee: Developing International Geoarchaeology Conference.

Solleiro-Rebolledo, E., Gama-Castro, J. E,, & Palacios-Mayorga, S. (1999). Late Pleistocene Paleosols from Chichinautzin group in the Transmexican Volcanic Belt, Mexico. *Simposio Universitario de*

Edafología (pp. 17–18). México: Universidad Nacional Autónoma de México, Facultad de Ciencias.

Solleiro-Rebolledo, E., Sedov, S., Gama-Castro, J., Flores, R. E., & Escamilla, S. G. (2003). Paleosol-sedimentary sequences of the Glacis de Buenavista, Central Mexico: Interaction of Late Quaternary pedogenesis and volcanic sedimentation. *Quaternary International, 106–107*, 185–201.

Solleiro-Rebolledo, E., Macías, J. L., Gama-Castro, J., & Sedov, S. (2004). Quaternary pedostratigrapgy of the Nevado de Toluca Volcano. *Revista Mexicana de Ciencias Geológicas, 21*(1), 101–109.

Solleiro-Rebolledo, E., Sedov, S., McClung, E., Cabadas-Báez, H., Gama-Castro, J., & Vallejo-Gómez, E. (2006). Spatial variability of environment change in the Teotihuacan valley during late Quaternary: Paleopedological inferences. *Quaternary International, 156–157*, 13–21.

Solórzano, F. A. (1989). Pleistocene artifacts from Jalisco, Mexico: A comparison with some pre-Hispanic artifacts. In R. Bonnichsen & M. H. Sorg (Eds.), *Bone Modification* (pp. 499–514). Orono, Maine: Center for the Study of the First Americans, University of Maine, Orono.

Stock, C. (1943). The cave of San Josecito, Mexico. *California Institute of Technology, Balch Graduate School of Geological Sciences, Contributions, 361*, 1–5.

Straka, H., & Ohngemach, D. (1989). Late Quaternary vegetation history of the Mexican highland. *Plant Systematics and Evolution, 162*, 115–132.

Targulian, V. O., & Sokolova, T. A. (1996). Soil as a bio-abiotic natural system; a reactor, memory and regulator of biospheric interactions. *Eurasian Soil Science, 29*(1), 34–47.

Than-Marchese, B. A., Montellano-Ballesteros, M., & Carbot-Chanona, G. (2009). El perezoso terrestre más grande de México. *XI Congreso Nacional de Paleontología*, 25 al 27 de Febrero de 2009, Centro de Geociencias, Campus-UNAM, Juriquilla, Querétaro, México (pp. 74). México: Sociedad Mexicana de Paleontología A.C.

Torres-Martínez, J. C. (1995). *Fauna local Mina de San Antonio, Pleistoceno tardío de San Luis Potosí, y su significación geológica-paleontológica*. B. S. Thesis, Universidad Autónoma del Estado de Morelos.

Torres-Martínez, A. (2011). Reporte preliminar de la alometría de premolariformes deciduales de *Mammut* (*Mastodon*) *americanum* Kert (Mammalia Proboscidea Mammutidae) del Pleistoceno Tardío de Morelos y Guerrero, México. Libro de Resúmenes, *XII Congreso Nacional de Paleontología*, 22 al 25 de Febrero de 2011, Edificio Carolino, Benemérita Universidad Autónoma de Puebla, Puebla, México (p. 136). México: Sociedad Mexicana de Paleontología A.C.

Tovar, R. E., & Sedov, S. (2011). *The Late Pleistocene environments in the south of Puebla: Paleosols, fossils and other biological proxys from alluvial sequences* (pp. 80–81). IUSS Commission on Paleopedology and IUSS Commission on Soil Geography.

Urrutia-Fucugauchi, J., Lozano-García, M. S., Ortega-Guerrero, B., Caballero-Miranda, M., Hansen, R., Böhnel, H., et al. (1994). Paleomagnetic and paleoenvironmental studies in the southern Basin of Mexico- I. Volcanosedimentary sequence and basin structure of Chalco Lake. *Geofísica Internacional, 33*(3), 421–430.

USDA, United State Department of Agriculture, Soil Conservation Service. (1988). *Soil taxonomy: A basic system of soil classification for making and interpreting soil surveys*. R. E. Krieger (Ed.). Malabar, Florida.

USDA, United State Department of Agriculture. (1996). *Soil survey laboratory methods manual*. Soil Survey Investigations Report No. 42, U.S. Department of Agriculture, National Resources Conservation Services, National Soil Survey Center, Washington.

Van Devender, T. R. (1990a). Late Quaternary vegetation and climate of the Chihuahuan Desert, United States and Mexico. In J. L. Betancourt, T. R. Van Devender, & P. S. Martin (Eds.), *Packrat middens-the last 40,000 years of biotic change* (pp. 105–133). Tucson: University of Arizona Press.

Van Devender, T. R. (1990b). Late Quaternary vegetation and climate of the Sonoran Desert, United States and Mexico. In J. L. Betancourt, T. R. Van Devender, & P. S. Martin (Eds.), *Packrat middens-the last 40,000 years of biotic change* (pp. 134–165). Tucson: University of Arizona Press.

Van Devender, T. R., & Bradley, T. G. (1990). Late Quaternary mammals from the Chihuahuan Desert: Paleoecology and latitudinal gradients. In J. L. Betancourt, T. R. Van Devender, & P. S. Martin (Eds.), *Packrat middens-the last 40,000 years of biotic change* (pp. 350–362). Tucson: University of Arizona Press.

Van Devender, T. R., Thompson, R. S., & Betancourt, J. R. (1987). Vegetation history in the Southwest: The nature and timing of the late Wisconsin-Holocene transition. In W. F. Ruddiman, & H. E. Wright., Jr. (Eds.), *North America and adjacent oceans during the last deglaciations* (pp. 323–352). Geological Society of America.

Vázquez, G., Ortega, B., Davies, S. J., & Aston, B. J. (2010). Registro sedimentario de los últimos *ca.*17000 años del Lago de Ziraguén, Michoacán, México. *Boletín de la Sociedad Geológica Mexicana, 62*(3), 325–343.

Vidal-Zepeda, R. (1990a). IV.4.4. Temperatura Media, Mapa Esc. 1:4 000 000, *Atlas Nacional de México, Tomo II, Sección IV Naturaleza, Apartado 4 Clima*. México: Universidad Nacional Autónoma de México, Instituto de Geografía.

Vidal-Zepeda, R. (1990b). IV.4.6. Precipitación, Mapa Esc. 1:4 000 000, *Atlas Nacional de México, Tomo II, Sección IV Naturaleza, Apartado 4 Clima*. México: Universidad Nacional Autónoma de México, Instituto de Geografía.

Villa-Ramírez, B., & Cervantes, F. A. (2003). *Los mamíferos de México*. Grupo Editorial Iberoamérica, S.A: de C.V. e Instituto de Biología, UNAM, México.

Von Thenius, W. R. (1970). Einige jungplistozäne Sugetiere (*Platygonus, Arctodus* und *Canis dirus*) aus dem Valsequillo, Mexiko. *Quartär, 21*, 57–66.

White, R. S., Mead, J. I., Baez, A., & Swift, S. L. (2010). Localidades de vertebrados fósiles del Neógeno (Mioceno, Plioceno y Pleistoceno): Una evaluación preliminar de la biodiversidad del pasado. In F. E, Molina-Freaner & T. R. Van Devender (Eds.), *Diversidad Biológica de Sonora* (pp. 51–72). México: Universidad Nacional Autónoma de México.

Wilson, D. E. & Reeder, D. A. M. (Eds.). (2005). *Mammal species of the world- a taxonomic and geographic reference*. (3rd ed.). (Vol. 1, pp. 1–743, Vol. 2, pp. 744–2142). Baltimore, Maryland: Johns Hopkins University Press.

Wright, H. E., Jr., Kutzbach, J. E., Webb, T., III, Ruddiman, W. F., Street-Perrott, F. A., & Bartlein, P. J. (1993). *Global climates since the last Glacial Maximum*. Minneapolis: University of Minnesota Press.

Zavaleta-Villareal, V., & Castillo-Cerón, J. M. (2011). Taxonomía y variación intraespecífica de *Equus conversidens* del Pleistoceno de Villa de Tezontepec, Hidalgo. Libro de Resúmenes, *XII Congreso Nacional de Paleontología*, 22 al 25 de Febrero de 2011, Edificio Carolino, Benemérita Universidad Autónoma de Puebla, Puebla, México (pp. 146). México: Sociedad Mexicana de Paleontología A.C.

Chapter 4
Holocene Large Mammal Extinctions in Palawan Island, Philippines

Janine Ochoa and Philip J. Piper

Abstract Zooarchaeological assemblages from northern Palawan, Philippines document the changing composition of the island's mammal fauna during the Late Quaternary. Ille Cave site has a well-dated archaeological sequence dating from the Terminal Pleistocene to the Holocene that includes identifications of tiger, two species of deer and a canid. This faunal record is compared with that of Pasimbahan Cave, which has an assemblage of Middle to Late Holocene age based on artifact associations, biostratigraphic correlation and preliminary radiocarbon dates. At least three large mammals were extirpated in the Holocene. The asynchronous timing of the extinctions signals different trajectories and dynamics of extinction, likely resulting from a combination of climatic, geographic and anthropogenic factors. These records also chronicle human response to these environmental changes. As deer populations on the island diminish by the Middle Holocene, human foragers in the Dewil Valley switch to the Palawan bearded pig as their main large mammal resource.

Keywords Faunal change • Island Southeast Asia • Human impact • Deer (*Rusa, Axis*) • Tiger (*Panthera tigris*) • Canid

J. Ochoa (✉)
Department of Anthropology, University of the Philippines, Diliman, Quezon City, Philippines
e-mail: janine.ochoa@yahoo.com

P.J. Piper
School of Archaeology and Anthropology, Australian National University, Canberra, ACT 2612, Australia
e-mail: phil_piper2003@yahoo.ie

P.J. Piper
Archaeological Studies Program, Palma Hall, University of the Philippines, Diliman, Quezon City, Philippines

Introduction

Zooarchaeological assemblages are often accumulated through a combination of human agency and natural processes, and this permits us to assess the diachronic effects of anthropogenic activities and changing environments on the composition and structure of faunal communities. Islands in particular comprise unique and bounded foci of study as 'laboratories' for evolutionary processes and for human 'experiments' on colonization and occupation. These ecosystems are often composed of isolated vertebrate communities that have not coevolved with humans, making them susceptible to anthropogenic perturbation following colonization. The Philippine archipelago provides exemplary cases. The Philippines, along with the islands of Southeast Asia, have extremely complex geological histories (Hall 2002) that have influenced the evolution of the faunal communities inhabiting them. The changing geological configuration and climate of the region have shaped the dispersal and diversification of terrestrial vertebrates, including its human inhabitants. Insularity entails isolation, but these archipelagos also form chains of connected territories in both an evolutionary and an archaeological sense. Hence during certain periods, such islands have served as isolated regions, as gateways or stepping stones to other territories, as evolutionary dead ends, or as fresh habitat for new colonists. The connectedness of these islands becomes literal during glacial periods when vast tracts of land became exposed due to the decrease in sea levels, joining some islands to the Asian continent or oceanic islands with one another (Fig. 4.1; Heaney 1985; Voris 2000).

Palaeogeographic and climatic changes have also shaped the distribution and contributed to the demise of various mammal species during the Late Quaternary (Harcourt 1999; Meijaard 2003; Louys et al. 2007). Deep archaeological sequences coming from specific sites in the region provide evidence for the changing faunal characteristics of these islands and the possible roles and responses of ancient

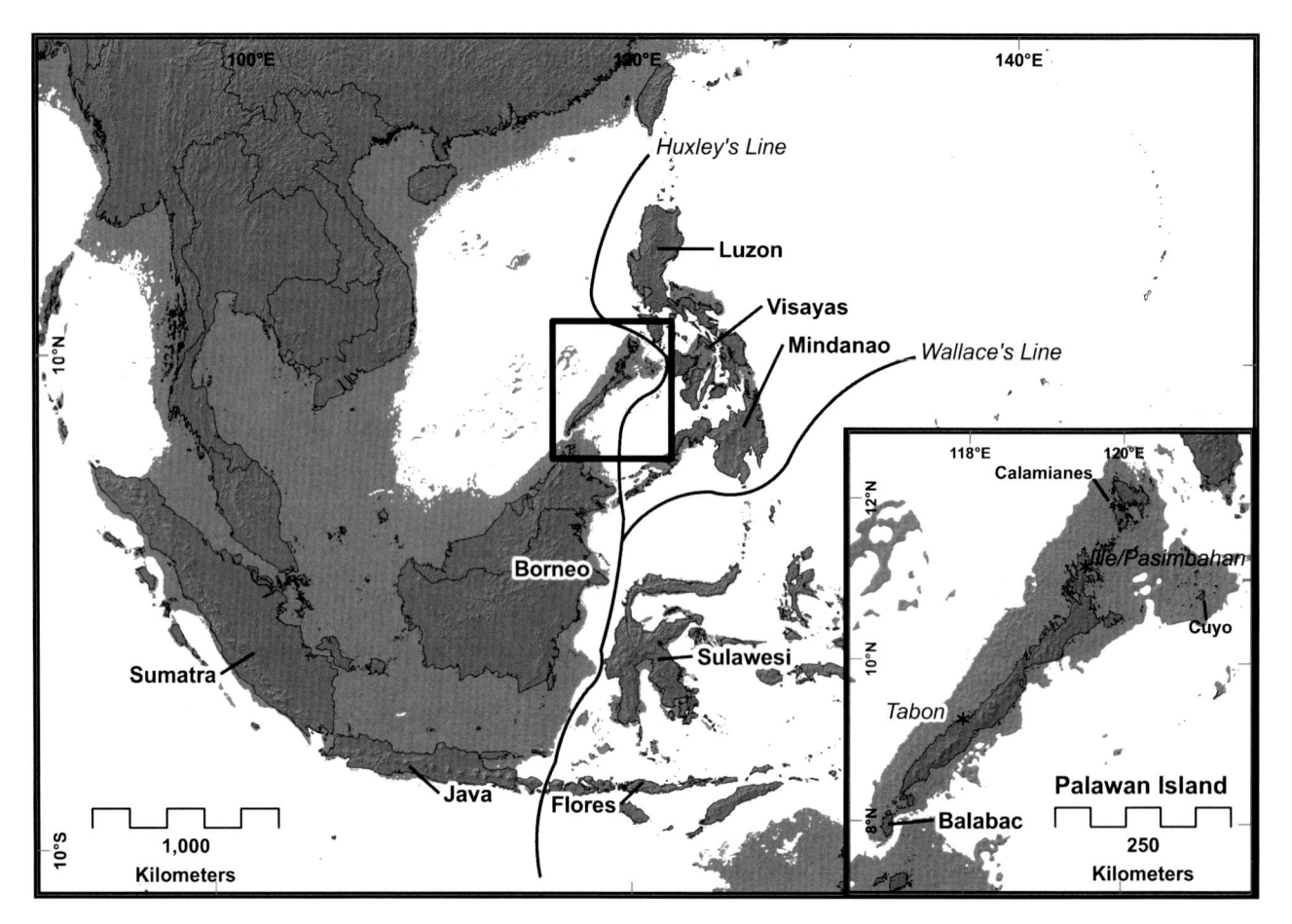

Fig. 4.1 Map of Island Southeast Asia and Palawan showing land distribution in the Last Glacial Maximum (light gray) and at present (dark gray), the location of archaeological sites and biogeographic lines. Inset is the LGM extent (light gray) of the Greater Palawan faunal region (*sensu* Heaney 1985), which includes the Calamianes, Cuyo and Balabac group of islands. (Drawn by Emil Robles)

humans (e.g., Cranbrook 2000; Cranbrook and Piper 2007; Morwood et al. 2008). There is regional evidence for Pleistocene faunal turnovers and large mammal extinctions in Indo-Australia (e.g., Louys et al. 2007; van den Bergh et al. 2001, 2009), but more palaeoenvironmental data and analyses are still needed to understand the dynamics of these events. In the Holocene, the increasing impacts of human activities in island environments have left archaeological traces that lend substance to debates surrounding the modern global biodiversity crisis and the historical roles that human communities have played in it.

Recent excavations on the island of Palawan have produced just such a faunal record that documents the impacts of people on their environments and *vice versa* (Piper et al. 2008, 2011; Ochoa 2009). The island of Palawan is located between Borneo and the main Philippine archipelago on the northeastern edge of the Sunda Shelf in Island Southeast Asia (Fig. 4.1). The focus of this paper is the vertebrate assemblages from two archaeological sites located in the north of the island: Ille and Pasimbahan Caves. These assemblages have been vital in chronicling human and faunal responses to the extensive landscape reconfigurations and climatic changes that occurred at the end of the last glaciation and into the Holocene, and situate the region within the global debate on Late Quaternary extinctions. Dating from the Terminal Pleistocene (*ca.* 14,000 cal BP) to the sub-recent, the faunal assemblages of these sites provide local extirpation records of at least three, and possibly four, large mammals of Palawan: the tiger (*Panthera tigris*), the Calamian hog deer (*Axis calamianensis*), another larger deer species belonging to the genus *Rusa* and perhaps the dhole (*Cuon alpinus*). The varying timing of extirpations and archaeological context of the animal remains indicate different trajectories to extinction brought about by a combination of environmental and anthropogenic factors.

Palawan Biogeography and Southeast Asian Palaeoenvironments

In the Southeast Asian region, climate-forced fluctuation of sea levels was one of the major drivers of palaeoenvironmental changes throughout the Pleistocene. The continental and oceanic islands of Southeast Asia have seen massive reconfigurations during glacial and interglacial periods as demonstrated by bathymetric reconstructions of ancient coastlines and land distribution for the region (Voris 2000; Sathiamurthy and Voris 2006; Robles et al. 2014). Reconfigurations of the landmasses have had predictable cascading effects on the climate, vegetation and fauna of the islands. In the case of terrestrial vertebrates, exposed landmasses and landbridges during periods of lower sea levels potentially presented new habitats and corridors for dispersal when suitable vegetation regimes were present, whereas the inundation of lands during sea level highstands and interglacial periods would have produced barriers for colonization, habitat loss and isolation.

During periods of sea-level lowstands in the Late Quaternary, a vast tract of the Sunda continental shelf was exposed and connected the large islands of Sumatra, Java and Borneo, and numerous smaller islands to Mainland Southeast Asia. This landmass, of which a large proportion is currently submerged beneath the West Philippine ("South China") and Java Seas, would have consisted of an additional 2,300,000 sq. km of landscape during the Last Glacial Maximum (LGM) (Sathiamurthy and Voris 2006). In the last 500,000 years, sea-level lowstand estimates based on oxygen isotope records range from −116 to −139 meters below present sea levels (mbpsl) (Rohling et al. 1998; Hanebuth et al. 2000; Lambeck et al. 2002; Waelbroeck et al. 2002; Bintanja et al. 2005). Sumatra would have been connected to the Malay Peninsula with a minimum sea level reduction of about −25 mbpsl, while Borneo and Java would have been connected by landbridges to the mainland when sea levels were at least about −30 and −35 mbpsl, respectively (Sathiamurthy and Voris 2006). Some authors have previously suggested the presence of a landbridge connection between Palawan and Borneo at some point during the Pleistocene (Fox 1970; Brown and Alcala 1970; Diamond and Gilpin 1983; Heaney 1986) as the former is considered as an important 'stepping stone' for humans and other terrestrial animals migrating into the Philippine archipelago. The timing of this connection however is disputed and the alternative remains that Palawan may never have been connected to the Sundaland in the Pleistocene.

Decrease in sea levels and the exposure of the continental shelf were accompanied by changes in vegetation and precipitation patterns. The surface area of ocean water available for evaporation in the Indo-Pacific was significantly reduced (Verstappen 1975; Bird et al. 2005) which in turn had a series of climatic effects. A prevailing hypothesis regarding Island Southeast Asia's climate and vegetation during the LGM was developed by Heaney (1991) and reanalyzed by Bird et al. (2005). Labelled as the 'savanna corridor hypothesis', it proposes that an expanse of open vegetation extended down from the Malaysian peninsula, into the presently inundated area between Java and Borneo. Bird et al. (2005) suggest that a maximal interpretation of the data would imply the existence of a 150-km wide expanse of open vegetation that would have connected similar floral communities north and south of the equator. In this hypothesized scenario, Palawan and the western flank of the Philippine archipelago may also have been predominantly covered by open vegetation regimes. Carbon isotopic analysis of guano sequences from Makangit and Gangub Caves in northern and southern Palawan respectively indicate forest contraction and the expansion of open savanna on the island during the Last Glacial Period (Bird et al. 2007; Wurster et al. 2010). Rainforest was said to be widespread again in the south of Palawan by around 13.5 ka and in the north by 8 ka.

The warming trend after the Pleistocene led to a suite of palaeoenvironmental changes including sea level rise, the expansion of forested environments and a strengthening of the Asian monsoon (van der Kaars and Dam 1997; van der Kaars et al. 2000; Kershaw et al. 2007; Wang et al. 2008). This shift from cold/dry to very warm/humid conditions is supported by speleothem records from Borneo (Partin et al. 2007) and Flores (Griffiths et al. 2009) showing evidence for an increase in precipitation. The sea continued to rise and inundate the Sunda Shelf: at the beginning of the Holocene, sea level was at around −48 m, and it continued to rise to levels even higher than present at around 7,000 BP (Tjia 1996; Maeda et al. 2004; Horton et al. 2005).

These periodic climatic and geological processes have shaped the modern-day faunas of Southeast Asia. The region is at present broadly split into three major biogeographic divisions: the Indochinese, Sundaic and Wallacean regions. In the Philippine archipelago, the boundaries of present-day faunal regions are demarcated by the extent of land exposed during Pleistocene sea-level lowstands (Fig. 4.1; Heaney 1985; Brown and Diesmos 2001). With the exception of Palawan, the rest of the Philippine islands are traditionally grouped into the highly diverse Wallacean region, following Huxley's modification of Wallace's Line. Recent taxonomic and phylogenetic studies have taken another look into the validity of these biogeographic demarcations and our understanding of the evolution of the region's island faunas, and they demonstrate many exceptions to the 'rule' (Brown and Guttman 2002; Evans et al. 2003; Esselstyn et al. 2010). Palawan itself presents a unique mix of continental and insular affinities, hence providing an interesting case for testing hypothesis about colonization and diversification, as well as the effects of climate and palaeogeographic changes on the terrestrial fauna (Esselstyn et al. 2004).

During Pleistocene glacial maxima, Palawan was much larger than at present and conjoined to many smaller adjacent islands such as Culion, Coron and Busuanga to the north, Cuyo to the west and Balabac to the south, forming what is known as the Greater Palawan faunal region (Heaney 1985). Furthermore, the Balabac Strait between Palawan and Borneo was considerably narrower and at times of sea level lowstands, the two islands might have even been joined by a narrow corridor, facilitating the migration of Sundaic species to and from Palawan (Cranbrook 2000; Tougard 2001). Recent bathymetric and GIS-based reconstructions of Palawan's palaeocoastlines by Robles et al. (2014) have modeled that a minimal connection may have been present in the Middle Pleistocene at *ca*. 420 ka. Along with biogeographic and fossil evidence (Heaney 1985, 1986; Piper et al. 2011; Reis and Garong 2001), the current evidence indicates that this may have been the most recent land connection and that there was no landbridge between Borneo and Palawan in the LGM.

The most recent survey of the extant mammal fauna of Palawan has been conducted by Esselstyn et al. (2004). It is comprised of 58 native species and four non-native species. Of these 58 native mammals, 24 species are non-volant and 13 are endemic. Hutterer (2007) adds to this list a novel shrew species, *Crocidura batakorum*, an endemic insectivore that is also present in the Ille fossil record (Piper et al. 2011). Heaney (1986) observes that over 90% of Palawan's mammalian genera are shared with Borneo and that most species have their closest relatives on the Sunda Shelf. Species richness relative to island size for the non-flying mammals is relatively low compared to the islands of the Sunda shelf. For instance, compared to Borneo's 132 species of land mammals (Cranbrook 2009), the tally from Palawan reflects a relatively depauperate fauna. On the other hand, species richness is still greater in Palawan compared to other Philippine islands (Heaney et al. 2002).

More recently, several authors have also explored Palawan's affinities with the oceanic faunas of the Philippine archipelago. One example is the molecular study of the Palawan bearded pig (*Sus ahoenobarbus*), which was previously categorized as a subspecies of the bearded pig (*S. barbatus*) of Borneo but is now considered as a separate species more closely allied to the Philippine wild pigs (Lucchini et al. 2005). Furthermore, the authors suggest that because of the basal position of *S. ahoenobarbus* in their gene trees and its closer phylogenetic affinities to *S. cebifrons*, the dispersal may have occurred via Palawan to the Philippine archipelago sometime during the Pliocene. A similar pattern is also reflected in the phylogenetic relationships between other Palawan and Philippine vertebrate lineages. Esselstyn and colleagues (2010) have synthesized recent phylogenetic studies and observe that Palawan Island has played different roles in the evolution of Philippine and Sundaic faunal communities. They note that some of these phylogenetic relationships are probably due to the invasion of the oceanic Philippines via Palawan as classically modelled by earlier authors (Brown and Alcala 1970; Diamond and Gilpin 1984). The Palawan colonization route has further led to the diversification of many taxa in the oceanic Philippines in the past several million years, as is well demonstrated for instance by the remarkable murid adaptive radiation and the extraordinary endemism of many vertebrate and invertebrate taxa (Heaney and Mittermeier 1997; Jansa et al. 2006). Many Palawan taxa are indeed shared with the Sunda Shelf only, or with both the Sunda Shelf and the Philippines, but there are also several lineages that are most closely shared with the oceanic Philippines. The latter includes the Palawan spotted stream frog (*Rana moellendorffi*), mountain leaf-warbler (*Phylloscopus trivirgatus*) and the Palawan spadefoot toad (*Leptobrachium tagbanorum*), among others (Brown and Guttman 2002; Jones and Kennedy 2008; Brown et al. 2009, respectively). Furthermore, Esselstyn and colleagues (2010) note that some Palawan populations are nested within clades otherwise restricted to the oceanic Philippines suggesting invasion of Palawan from the oceanic Philippines during much older geological phases, such as in the case of the Palawan flying lizard (*Draco palawanensis*) (McGuire and Kiew 2001).

Gauging Human Impact and Response from the Zooarchaeological Record

The investigation of human impacts on island ecosystems is an enduring theme in island zooarchaeology and more recently has become an important and highly debated body of evidence in light of the modern biodiversity crisis (e.g., Burney 1997; Grayson 2001; Anderson 2002; Giovas and LeFebvre 2006; Pregill and Steadman 2009). The depletion of faunal communities during the Pleistocene-Holocene transition and more recent periods has traditionally been explained by the dichotomy of human-induced or climate-driven extinctions. The overkill hypothesis as originally proposed by Martin (1984) and Martin and Steadman (1999) has been a dominant framework used to explain extinctions of megafauna and slow-breeding animals in North America and other continents. Other authors have favored a 'natural' explanation for such disappearances in light of mounting palaeoenvironmental data and have been skeptical of the demographic reach and technological capabilities of ancient human communities to bring forth such cataclysmic consequences (e.g., Hiscock 2008). Current reviews on the matter of Late Quaternary megafaunal extinctions (Wroe et al. 2004; Koch and Barnosky 2006) all invoke multiple factors and a combination of causes to explain these extinctions and note that there is a paucity of data in support of a global 'blietzkrieg' – or the hunting of megafauna to extinction following first human contact. The case of Australia has often been cited as an example of this overkill phenomenon (e.g., Flannery 1999; Flannery and Roberts 1999), but available direct dates demonstrate that some species were likely extinct before humans arrived, and there were millenia of coexistence between humans and other members of the Australian fauna following the earliest human colonization (Hiscock 2008). Some islands do show concrete evidence for human-driven extinctions after first contact, such as in the case of New Zealand (Towns and Daugherty 1994) and other Pacific Islands (Steadman 1995; Steadman and Martin 2003). In the Japanese archipelago, Norton et al. (2010) propose that most megafaunal extinctions occurred during the MIS 3-2 transition (ca. 30–20 kyr) rather than during the transition to the Holocene and that pre-Jomon foragers may have influenced these events. As this volume demonstrates, the interaction of climatic, environmental and anthropogenic factors vary through time in both continental and island ecosystems (see e.g., Belmaker 2017; Pilaar-Birch and Miracle 2017).

In Island Southeast Asia, various authors have explored explanations that combine both frameworks, positing that Late and Post-Pleistocene environmental changes, combined with direct and indirect human impacts, brought about protracted extinction events (van den Bergh et al. 2009; Earl of Cranbrook 2009). Gauging anthropogenic impact is necessarily dependent on the evidence for human presence in the region. The fossil and artifact records of Island Southeast Asia indicate that several species of *Homo* were present in the region at various times in the Quaternary. The oldest dates for the classic *Homo erectus* fossil sites found in Java range from 1.8 to 1.0 Ma (Swisher et al. 1994; de Vos and Sondaar 1994; Semah and Semah 2012). Stone tools from the site of Mata Menge in the Wallacean island of Flores have fission-track ages between 880 ± 70 ka and 800 ± 70 ka (Morwood et al. 1998). Also in the same island, fossil remains of the enigmatic small hominin *Homo floresiensis* have been recovered from Liang Bua Cave in association with a large faunal assemblage dating between 92,000–13,000 kya (Brown et al. 2004; Morwood et al. 2005; van den Bergh et al. 2009). Some of the earliest and best-known evidence for the remains of *Homo sapiens* in the region come from Niah Cave in Borneo (the "Deep Skull") at *ca.* 42,000 BP (Barker et al. 2007), Tabon Cave in Palawan dating to 47 +10/−11 ka and 31 +8/−7 ka (Detroit et al. 2004) and on the island of Timor more than 42,000 BP (O'Connor and Ono 2012). Mijares et al. (2010) also report a 67,000-year old human (*Homo* sp.) metatarsal from Callao Cave in northern Luzon.

In this region, megafaunal extinctions and faunal turnovers throughout the Pleistocene have been principally influenced by geological and vegetational changes brought about by sea level fluctuations and climate change (Louys et al. 2007). Evidence of the natural turnover of species is demonstrated by the Pleistocene fossil record of Indonesia, where the majority of the Middle Pleistocene megafauna was replaced by a modern tropical rainforest community around the onset of the penultimate interglacial *ca.*125,000 BP (van den Bergh et al. 2001). In the Late Pleistocene and the Holocene though, local disappearances without replacement mimic those seen in other continents and regions, and coincide with the expansion of *Homo sapiens* in Southeast Asia. In the case of oceanic islands with imbalanced and impoverished faunas, animals are often naïve to human predation and are extremely restricted in biogeographic distribution and population size, and hence susceptible to humans and the predatory species that they often introduced (e.g., van den Bergh et al. 2009; van der Geer et al. 2010). The regional archaeological record shows that many mammal, bird and herpetile species have become extinct only in the last few thousand years due to human hunting, habitat modification and translocation of non-native species (Towns and Daugherty 1994; Steadman and Martin 2003; O'Connor and Aplin 2007; Cranbrook 2009; Pregill and Steadman 2009). The zooarcheological record from Ille and Pasimbahan caves in northern Palawan, Philippines also give a good example of how human intervention combined with dramatic changes in landscape and vegetation probably conspired in the local extinction of most of the large mammals that existed on the island at the end of the Pleistocene.

Dewil Valley Archaeology

This zooarchaeological study focuses on the faunal assemblages recovered from two northern Palawan sites: Ille and Pasimbahan Caves in the Dewil Valley of El Nido municipality. The municipality is located 11°20′ N 119°41′ E, and about 240 km northeast of the provincial capital of Puerto Princesa City. The archaeology of the Dewil Valley is anchored on the site of Ille Cave, which has produced tens of thousands of well-stratified and well-preserved cultural remains spanning the last 14,000 years (Lewis et al. 2008).

The Ille karst tower is located in *barangay* New Ibajay of El Nido. It has three mouths at its base, two of which – the East and West mouths – serve as the main access to the tower. The two major excavation trenches are situated in these entrances and are eponymously labeled. They extend out onto a flat platform of silt loam in front of the tower, and several other trenches have been opened around them. During excavations, each stratigraphic layer and feature was allocated a unique *context number*, referred throughout the text as 'Context #'. The data reviewed here for the archaeology and dating of Ille Cave derives from Lewis et al. (2008). The uppermost deposits contain extensive human burial layers that range in date from the Neolithic Period (*ca.* 4,000–2,500 BP) up to historic times based on artifact typology. The burials cut into a similarly extensive shell and bone midden deposit that has been radiocarbon dates from to 5,500–6,500 cal BP (see also Szabo et al. 2004). Beneath this midden, several cremated human burials and a series of contexts with hearth deposits, burnt animal bones, chert and obsidian flakes were uncovered, and dated between 9,000 to 11,000 cal BP by several radiocarbon dates on charcoal and bone. The hearth features are underlain by several steeply sloping sedimentary layers of clay and gravels that are of Terminal Pleistocene age based on a radiocarbon date on charcoal from Context 866 of 13,820–14,116 cal BP (OxA-16666). Immediately below Context 866 is 1306, a sedimentary horizon that produced a sizeable animal bone assemblage dominated by deer and mixed with chert flakes and fragments of charred nuts.

Pasimbahan Cave is located in *sitio* Magsanib of New Ibajay, El Nido, about three kilometers northwest of Ille Cave. It is situated in the southwest face of one of the larger karst formations in the Dewil valley called Istar. Within the thirteen trenches investigated since 2007, one of the oldest culture-bearing layer uncovered so far is an aceramic midden layer of guano-rich dark gray to black silt (Figs. 4.2 and 4.3) (Ochoa et al. 2014; Paz et al. 2008, 2010). This midden deposit has been observed in five trenches (Midden Two trench, S1W5 trench and Trenches C, D, and J) and contains numerous shells, animal bones and angular limestone pebbles that may have been used as heating stones. The upper layers of most of the trenches, particularly Trenches A and B, also have shell middens and other pits and features associated with earthenware, stoneware, Indo-Pacific glass beads and scattered human remains that place them firmly

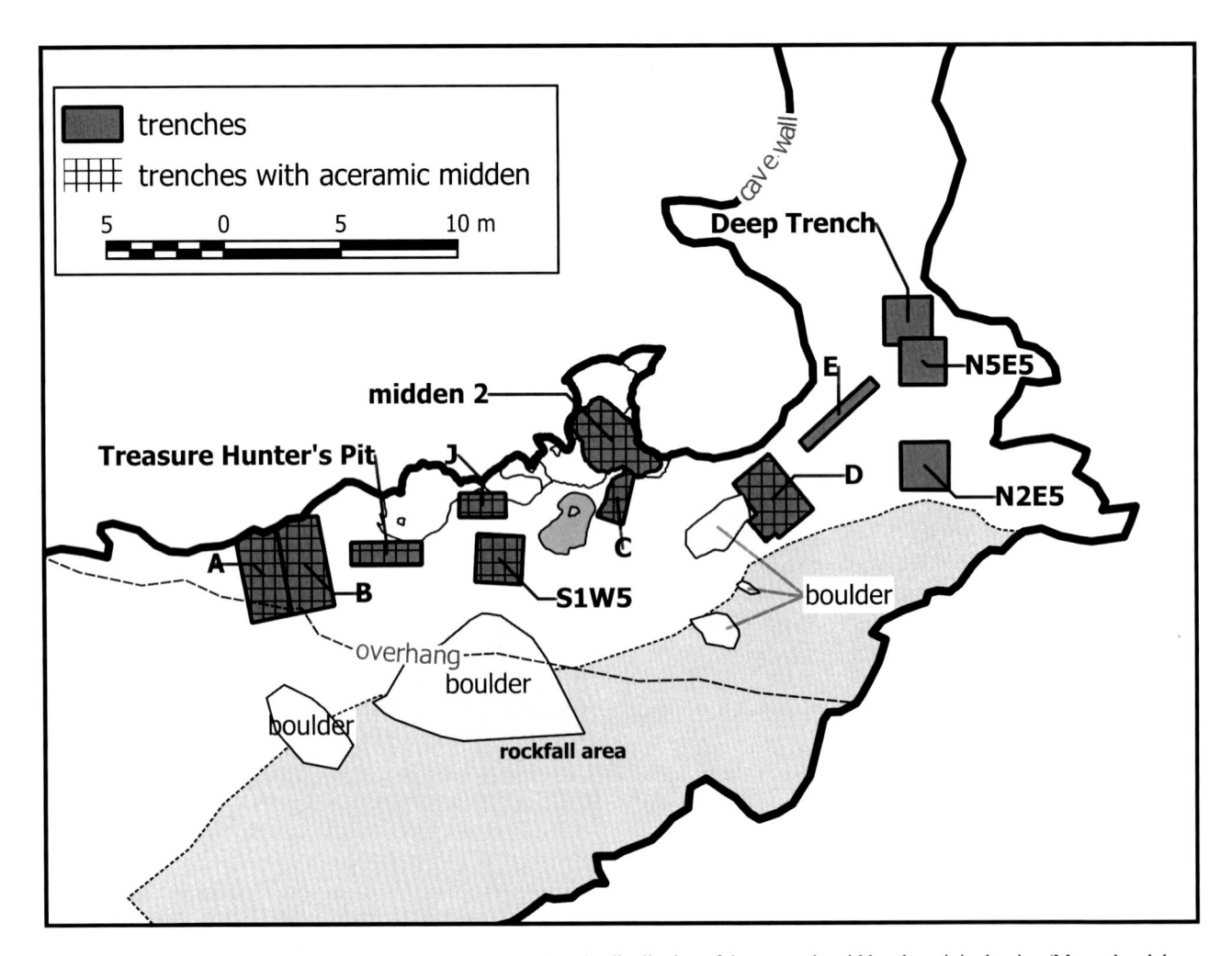

Fig. 4.2 Plan of excavation trenches in Pasimbahan Cave showing the distribution of the aceramic midden deposit in the site. (Mapped and drawn by Emil Robles)

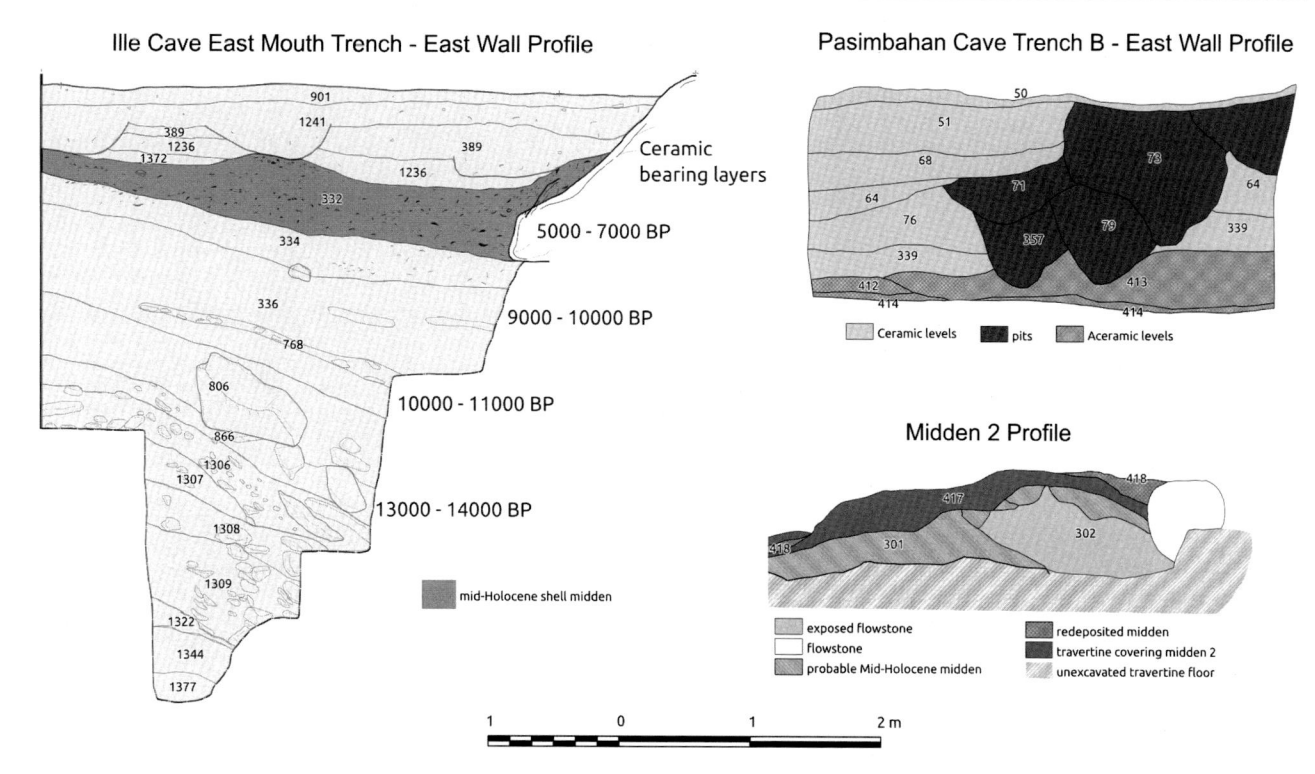

Fig. 4.3 Comparison of stratigraphic profiles from Ille and Pasimbahan Caves showing the radiocarbon-dated sequence of the Ille Cave East Mouth trench (Lewis et al. 2008) and profiles from Pasimbahan Cave Trench B and Midden 2. Numbers on the layers refer to context numbers. (Drawn by Vito Hernandez and Emil Robles)

within the Late Neolithic and Metal Period phases of activity (Fig. 4.3 – East wall profile of Trench A). The deepest Neolithic/Metal Age middens in Trenches A and B stratigraphically overlie layers that are considered to pre-date the introduction of ceramics to the region (i.e., older than *ca.* 4,000 BP).

Several radiocarbon dates have been procured for this site (Ochoa et al. 2014). Direct dating of a deer (*Axis calamianensis*) mandible (sample number IV-2007-Q-2012) from context 309, the dark midden deposit in Trench C, produced a date of 4,333 ± 25 cal BP (Wk-33712). Another radiocarbon date on shell (*Chicoreus capuchinus*, sample number IV-2007-Q-1541) produced a date of 4,697 ± 25 cal BP (Wk-33712). This shell sample also derived from context 309. Additional dates on wood charcoal below these aceramic midden deposits push the age of the site to 10,500 cal BP.

The Ille Cave Bone Assemblage

The Ille vertebrate assemblage has produced new fossil records and taxonomic accounts of several mammal and reptile species (Piper et al. 2011). Included in this assemblage are the remains of deer, wild pig, macaque, monitor lizard, turtles, civet cats, tiger and various small mammals (Fig. 4.4). In this paper, mammalian taxonomy generally follows Esselstyn et al. (2004), except for the Palawan bearded pig (*Sus ahoenobarbus*), which follows Lucchini et al. (2005), and for Southeast Asian deer species, which follows Grubb (2005). Taxonomy of the reptilian fauna follows Diesmos et al. (2005, 2008) for geoemydids and Koch et al. (2010) for varanids. Species identifications, biometric data and specimen counts (e.g., NISP and MNI) for this vertebrate assemblage is reported in Piper et al. (2011) and Ochoa (2009). What follows briefly is a summary of the archaeological context of the assemblage.

There are three major phases represented in the East Mouth trench with high concentrations of bone: the Terminal Pleistocene Context 1306, the Middle Holocene shell midden Context 332 and the Le Holocene layers where later human burials cut into extensive midden remains (Fig. 4.3). In the West Mouth trench, the oldest layers with dense concentrations of bone are Contexts 1626 and 1530 from Early Holocene layers (Fig. 4.6). Above this are midden remains and burial layers homologous to the Middle Holocene and Late Holocene phases of the East Mouth trench.

The Terminal Pleistocene and Early Holocene layers (ca. 14–8 ka) are dominated by deer remains. Two species of deer were identified based on the notable size distinction of teeth

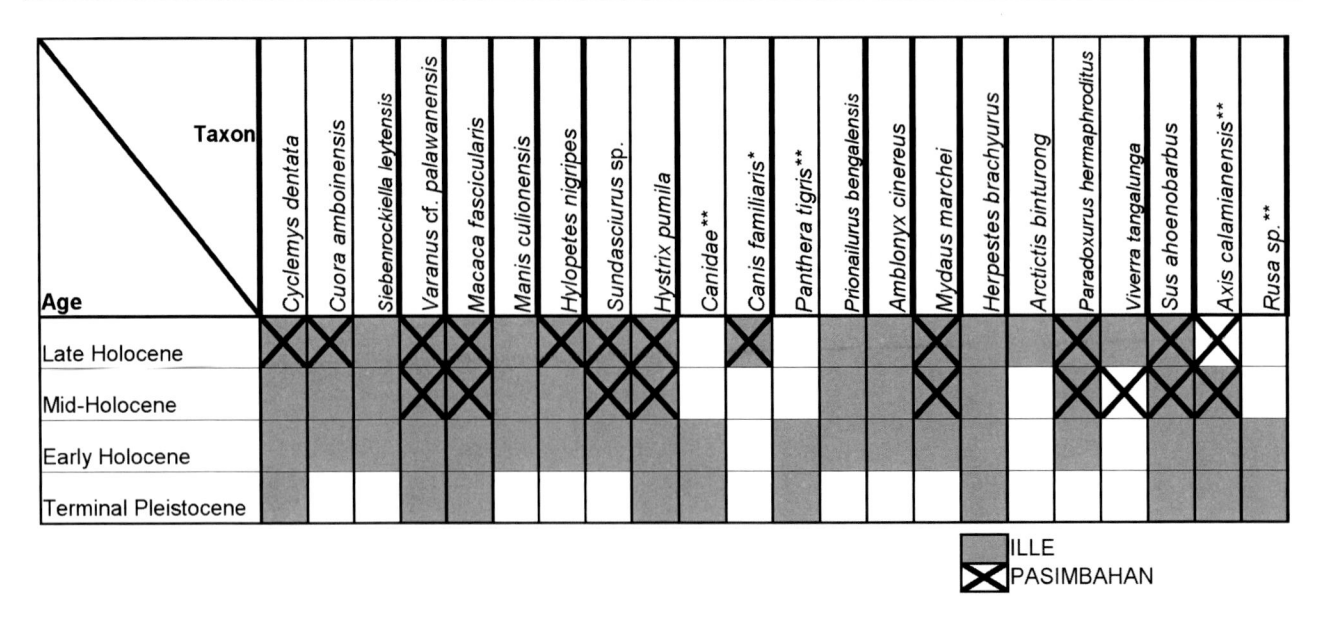

Fig. 4.4 List of *hunted* taxa from Ille and Pasimbahan Caves; *some domestic dog remains from Ille may have been interred (Ochoa 2005); **extirpated/extinct on Palawan Island

and post-cranial remains (see below for further discussion). The larger species is attributed to a member of the genus *Rusa* while the smaller deer species is ascribed to *Axis calamianensis*. Deer are now extinct on Palawan Island itself, and they are not known from old written accounts about the natural history of the island. A few other extant mammals are represented in the Terminal Pleistocene such as the wild pig (*Sus ahoenobarbus*), the long-tailed macaque (*Macaca fascicularis*), the short-tailed mongoose (*Herpestes brachyurus*) and the Palawan porcupine (*Hystrix pumila*). Macaque remains become particularly numerous beginning in the Early Holocene levels and are then fairly common throughout the rest of the sequence. Most of the extant small carnivore species are represented in the Holocene sequence such as the common palm civet (*Paradoxurus hermaphroditus*), binturong (*Arctictis binturong*) and the endemic stink badger (*Mydaus marchei*), and some of them show evidence of butchery in the form of cut marks, indicating they were accumulated as part of human subsistence procurement strategies (Piper and Ochoa 2007; Ochoa 2009). Also represented in the earliest archaeological sequences of the site, but absent from the Middle Holocene onwards (ca. 7,000 ^{14}C BP) are the tiger and a canid. Several arboreal mammals that prefer forest cover such as the arrow-tailed flying squirrel (*Hylopetes nigripes*) and tree squirrels (*Sundasciurus* sp.) are also notably absent from the Terminal Pleistocene, and this could be an indication of the more open environments that prevailed during this period around the Ille karst.

Midden formations are observed in the middle and Late Holocene layers, where thick deposits of shells are mixed with animal bones and various artifacts. At this point in the Ille chronological sequence, pig is now the dominant mammal resource, but the variety of smaller-sized mammals, reptiles and molluscs also indicates expansive exploitation of forest and aquatic resources. Deer remains are extremely rare, and only the smaller deer species (*Axis*) has been observed so far in Late Holocene layers. The reptiles are represented by snakes, the monitor lizard (*Varanus* cf. *palawanensis*) and three species of Geoemydidae, the Asian leaf turtle (*Cyclemys dentata*), box turtle (*Cuora amboinensis*) and the Philippine forest turtle (*Siebenrockiella leytensis*) (Ochoa 2009). Inter-mixed with the hunted taxa are some natural death assemblages that includes some Chiroptera such as the lesser short-nosed fruit bat (*Cynopterus brachyotis*), the diadem roundleaf bat (*Hipposideros diadema*), dusky leaf-nosed bat (*Hipposdieros* cf. *ater*) and Creagh's horseshoe bat (*Rhinolophus creaghi*), as well as the murid rodents Malayan field rat (*Rattus tiomanicus*) and Great Sunda rat (*Sundamys muelleri*).

The Late Holocene bone and shell accumulations of Ille Cave have been largely reworked by numerous human burials. This reflects a common trend in the Southeast Asian region wherein caves are increasingly used as places of burial (Anderson 1997; Barker et al. 2005). This practice has resulted in the mixing of animal remains from various levels of the Late Holocene. Hence only a small sample of archaeological contexts from these levels was analyzed, and they continue to be dominated by the remains of pig. The Late Holocene deposits contain a similar variety of other taxa as the preceding mid- to Early Holocene deposits, demonstrating continuity in techniques of capturing various prey taxa from a range of different habitats.

Pasimbahan Cave Bone Assemblage

Preliminary quantification and analysis of the Pasimbahan Cave assemblage reveal the presence of many of Palawan's extant mammal and reptile species such as those described from Ille Cave (Fig. 4.4). The deepest and oldest accumulations of bone are found in two major aceramic deposits. The first and most widespread across the site is a midden layer consisting of guano-rich dark gray to black silt (Figs. 4.2 and 4.3; Contexts 301, 309, 349, 352, 408). This aceramic midden deposit contains numerous bivalve shell remains and the fragmentary bones of pig, macaques, deer, squirrels, turtles, and monitor lizards. Natural death accumulations of rodents and bats are also present but these still remain to be studied in detail. A few bird remains have also been identified so far, including several heron (*Ardeidae* sp.) bones in the deep midden deposit (Context 408).

The other group of deposits comes from Trenches A and B, where several pits, hearths and accumulations of shell and bone have been uncovered (Figs. 4.2 and 4.3). The presence of chert and obsidian flakes in these features, the absence of pottery, and the radiocarbon dates indicate the antiquity of these layers. Although quantification is still in progress, one familiar pattern already stands out in the Pasimbahan bone assemblage: the pig is the dominant large mammal prey, and deer remains are scarce throughout the archaeological sequence. Pasimbahan is located less than 3 kilometers from Ille within a very similar karst landscape, suggesting that ancient foragers are unlikely to have encountered vastly different environments at the two cave sites. At Ille, the Terminal Pleistocene large mammal fauna is dominated by deer, with an increasing reliance on pig as cervid populations diminished through the Middle Holocene. Thus biostratigraphic correlation with the well-dated Ille assemblage points to the Middle Holocene as the maximum age of the aceramic midden deposits at Pasimbahan described here. Small deer remains are present nonetheless throughout the Pasimbahan stratigraphic sequence in very small quantities, including in the younger layers associated with pottery (Fig. 4.3). Based on artifacts such as metal implements and glass beads, other midden contexts containing deer may be associated with time periods *ca.* 2,000 years and younger, but at present, we take the radiocarbon dating from a midden deposit (context 71) dated to 3,400 cal BP as a marker for the most recent presence of *Axis* deer on Palawan (Ochoa et al. 2014).

Extinct Mammals

The varying distribution of the mammal species recorded in these northern Palawan cave sites signal changes in the local ecology and hunting practices of its human inhabitants (see Ochoa 2009; Piper et al. 2011). Here we focus on the extinct species found in the Late Pleistocene and Holocene record of Palawan and discuss the possible causes that led to their eventual extirpation.

Deer

Biometric analyses of the dentition and selected post-cranial elements indicate that at least two taxa are represented in the archaeological sequence, mainly distinguished by their size (see Piper et al. 2011). Attribution of the 'smaller' cervid archaeological specimens to the Calamian hog deer (*Axis calamianensis*) is based on two lines of evidence: firstly, the measurable dental elements from the archaeological collections are comparable in size and morphology with this extant hog deer, and secondly, *Axis calamianensis* still inhabits the Calamianes group of islands that were, until fairly recently conjoined to Palawan as part of the Greater Palawan biogeographic region. The bigger deer species is attributed to the genus *Rusa*, which includes the two living Philippine deer taxa *Rusa marianna* and *R. alfredi*, as well as *R. unicolor* and *R. timorensis* found in other areas of Southeast Asia (Meijaard and Groves 2004). Biometric comparison of dental remains of the Ille specimens and modern comparatives from across the region suggests an overlap with *R. marianna* from Luzon. However, this deer species shows considerable size variation across its Philippine range (Heaney et al. 1998), and in our small comparative sample of dental measurements, deer from southern Luzon are considerably larger than individuals on Mindanao Island. Additional biometric data of modern Southeast Asian deer taxa was therefore gathered to further analyze the postcranial remains in the assemblages. Limb proportions serve to further distinguish the Ille *Rusa* from many of the extant Southeast Asian deer and generally demonstrate consistent size differences for the members of this genus. This contrasts with the tooth measurements, wherein several overlaps are observed (Piper et al. 2011). Table 4.1 and Fig. 4.5 show selected post-cranial measurements of both fossil material and modern comparatives collected from several natural

Table 4.1 Measurements (in mm) of selected archaeological post-cranial material compared with living deer of Southeast Asia (following von den Driesch 1976). Bt = breadth of trochlea; Bp = medio-lateral breadth of proximal end; Bd = medio-lateral breadth of distal end. The codes ICWM and ICEM refer to specimens recovered from the Ille Cave West and East Mouth trenches, respectively. Comparative material were measured from: FMNH (Field Museum of Natural History), MCZ (Museum of Comparative Zoology, Harvard University), and NMNH (National Museum of Natural History)

Taxon	Locality	Specimen No.	Humerus (Bt)	Radius (Bp)	Radio-ulna (Bd)	Tibia (Bd)	Metacarpal (Bd)	Metatarsal (Bd)
Axis porcinus	India	FMNH-27447	30.72	31.67	–	24.53	–	–
A. porcinus	India	MCZ-37003	–	–	–	–	20.6	22.4
Rusa marianna	Luzon	MCZ-14227	34.4	36.8	33.1	30.9	–	–
R. marianna	Luzon	NMNH-49706	38.16	38.41	34.02	33.66	28.8	28.66
R. marianna	Mindanao	FMNH-61007	29.03	28.85	–	29.94	–	–
R. marianna	Guam	FMNH-186613	34.06	34.6	28.54	28.1	–	–
R. unicolor	Borneo	NMNH-151861	52.79	49.65	42.22	44.23	–	38.25
R. unicolor	Borneo	MCZ-7282	47.8	51.5	47	44	37	36.5
R. unicolor	Borneo	NMNH-151859	55.7	55.03	46.5	48.9	40.4	42.5
R. unicolor	India	MCZ-1381	56.4	63.7	53.9	55.1	45.2	45.2
Rusa sp.	Ille Cave	ICWM-2036	43.28	–	–	–	–	–
Rusa sp.	Ille Cave	ICEM-18425	41.48	–	–	–	–	–
Rusa sp.	Ille Cave	ICWM-2037	–	43.7	–	–	–	–
Rusa sp.	Ille Cave	ICWM-2038	–	43.53	–	–	–	–
Rusa sp.	Ille Cave	ICWM-1172	–	–	39.68	–	–	–
Rusa sp.	Ille Cave	ICWM-2131	–	–	38.19	–	–	–
A. calamianensis	Ille Cave	ICEM-17563	–	–	–	27.22	–	–
Rusa sp.	Ille Cave	ICEM-18427	–	–	–	34.54	–	–
A. calamianensis	Ille Cave	ICEM-17684	–	–	–	–	9.7	–
A. calamianensis	Ille Cave	ICEM-20039	–	–	–	–	13.8	–
A. calamianensis	Ille Cave	ICEM-17513	–	–	–	–	13.84	–
A. calamianensis	Ille Cave	ICEM-18605	–	–	–	–	13.86	–
A. calamianensis	Ille Cave	ICEM-20038	–	–	–	–	–	10.7
A. calamianensis	Ille Cave	ICWM-542	–	–	–	–	–	13.21
Rusa sp.	Ille Cave	ICWM-843	–	–	–	–	–	32.7
Rusa sp.	Ille Cave	ICWM-1356	–	–	–	–	–	32.22
Rusa sp.	Ille Cave	ICWM-812	–	–	–	–	–	30.26

history museums in the United States. Few skeletal elements were complete enough to measure following standard zooarchaeological biometric measurements (von den Driesch 1976) with the exception of the basal and sub-terminal phalanges. All specimens used in the analysis had fused proximal ends and were thus from adult individuals. These measurements show that the larger deer species in the Ille assemblage is intermediate in size between *R. marianna* and *R. unicolor*: they are larger than the Luzon and Mindanao specimens of the Philippine brown deer (*R. marianna*), but considerably smaller than the Bornean sambar (*R. u. brookei*). The Indian subspecies (*R. u. unicolor*) is much larger than all the rest of the members of this genus that were measured for both limb and teeth measurements (see Piper et al. 2011 for teeth metrics). Modern postcranial reference for *Axis calamianensis* and *R. alfredi* was not available, and only *A. porcinus* was measured. This Indochinese hog deer, which is estimated to be about the same size as *A. calamianensis* (Geist 1998), is generally smaller in many limb dimensions than the smallest *R. marianna* individuals measured. Teeth dimensions of *Axis calamianensis* overlap with the lower range of values for *R. marianna* (Piper et al. 2011), but limb proportions of the *A. calamianensis* from Ille (Table 4.1) are consistently much smaller than *R. marianna* reference material that were measured.

A striking pattern is also apparent in the frequencies of large mammal prey in both Ille and Pasimbahan. In the former, deer is only abundant in the Terminal Pleistocene to

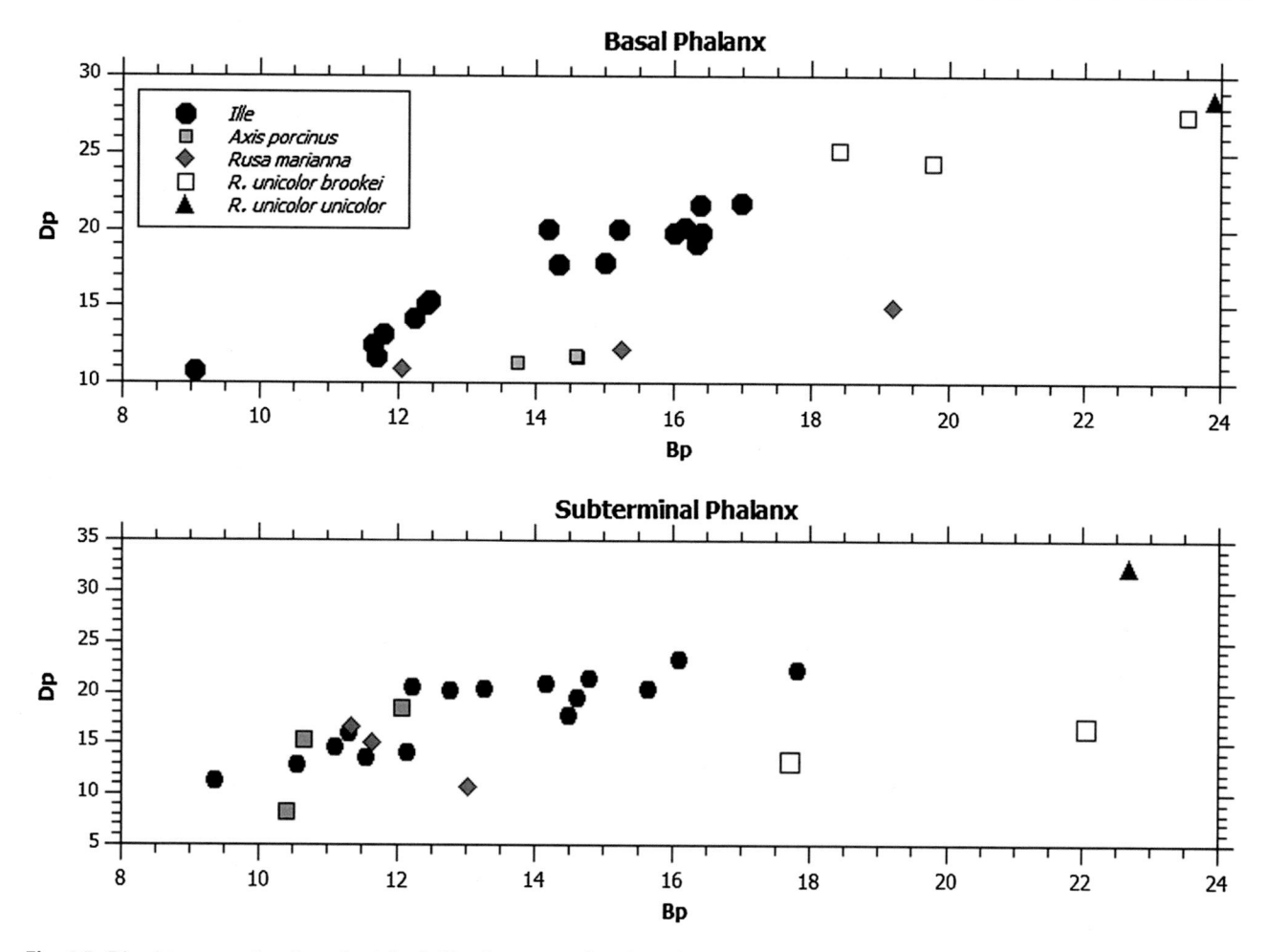

Fig. 4.5 Bivariate scatterplot of proximal depth (Dp, dorso-ventral) and proximal breadth (Bp, medio-lateral) of basal and subterminal phalanges from Ille cervid fossils and modern comparatives of Southeast Asian deer species. All specimens measured (mm) were from adult individuals. The following are the localities from which available reference material derive: *A. porcinus* from Vietnam, *R. marianna* from the Philippines, *R. unicolor brookei* from Borneo, and *R. unicolor unicolor* from India

the Early Holocene, and it becomes increasingly rare by the Middle Holocene, at which time the wild pig becomes the dominant taxon (Fig. 4.6). In Pasimbahan, wild pig is the dominant taxon throughout the sequence and deer are only present in small quantities. The presence of *Axis* in Pasimbahan in well-stratified midden deposits with ceramics and glass beads suggests that this species may have survived into the Metal Period and later. Thus, at Ille it appears that the large-bodied deer (*Rusa* sp.) disappeared during the Early Holocene, whereas the Calamian hog deer perhaps inhabited Palawan until the Late Holocene.

Tiger

One of the rarest and most important finds in the Ille bone assemblage is that of the tiger (*Panthera tigris*). Two bones of this large felid were initially found in the Terminal Pleistocene Context 1306 from the East Mouth trench: a complete basal phalanx and portion of a subterminal phalanx of the left manus. Another fragment of a distal portion and mid-shaft of a basal phalanx was later discovered in an Early Holocene context in the West Mouth trench. The taphonomy and archaeological context of the finds suggest that these remains came from individuals hunted or scavenged in the environments around the cave and that the tiger was a true inhabitant of Palawan (Piper et al. 2008). Contrary to van der Geer et al. (2010) who suggested that these could be traded items partly due to bone surface modifications unlike that of the other fossils in the assemblage, there are no taphonomic differences between the tiger elements and the rest of the macrovertberate material recovered from the same deposits. The tiger has been recorded in Middle Pleistocene deposits at Trinil and Kedung Brubus dating to 900 ka and 800–700 ka respectively on Java (van den Bergh et al. 2001) indicating the antiquity of the species in the islands of Southeast

Asia. Its presence in Terminal Pleistocene and Early Holocene deposits at Madai and Niah Caves on the island of Borneo (Piper et al. 2007) demonstrate that tigers were much more widely distributed throughout the Southeast Asian region than was previously known. The presence of a population of tigers on Borneo also provides evidence for an adjacent source population for colonization to Palawan. There is currently no evidence that the tiger continued to inhabit Palawan beyond the Early Holocene.

Canid

In the Ille assemblage, canid postcranial remains were recorded from Terminal Pleistocene and Early Holocene contexts. Morphometric comparisons indicate that the archaeological material falls within the overlapping limits of variation of the dhole (*Cuon alpinus*) and the earliest known domestic dog introduced to Australasia, the dingo (*Canis familiaris dingo*) (Piper et al. 2011). Recent genetic research has suggested that dogs were already domesticated by 5,400–16,300 BP in the region south of Yangtze River in what is now modern-day China (Pang et al. 2009), but the exact timing is yet to be verified through zooarchaeological evidence. Mitochondrial DNA analysis of dingoes, however, indicates an origin *ca.* 5,000 BP in Mainland Southeast Asia (Savolainen et al. 2004). Furthermore, a recent global review of genetic and archaeological evidence for the origins and translocations of domestic dogs suggests that they were unlikely to have been introduced to regions outside the natural range of grey wolf (*Canis lupus*) populations prior to the movements of human populations associated with agriculture (Larson et al. 2012). Taking both these lines of evidence into consideration, the temporal context of the Ille canid suggests that they are much older than the likely arrival of dingo-type domesticated dogs into the region and thus probably represent the remnants of the dhole. The antiquity of the Ille canid specimens, combined with other taphonomic factors such as the lack of carnivore gnawing on associated mammal bones, also hint that they are perhaps of wild stock. At Ille, dogs are absent during the middle Holocene but re-occur in the later Holocene sequences. The corresponding skeletal elements from the Late Holocene canids and those from the Terminal Pleistocene/Early Holocene demonstrate that the former are considerably smaller than the latter, and correspond in size with other later archaeological dog remains in the region (Clutton-Brock 1959; Medway 1977; Ochoa 2005; Piper et al. 2011 –Table 4a). Hence, we observe a curious gap in the Ille record during the middle Holocene, after which a clearly domesticated form of dog appears in the Late Holocene layers. Early Holocene canid remains have also been recovered in Madai Cave in northern Borneo, and these are also hypothesized to be either the remains of a dhole or an early domestic dog (Cranbrook 1988). As with the tiger, recent zooarchaeological records of the dhole extend its previous biogeographic distributions to the far northeast boundaries of the Sundaic biogeographic region.

Trajectories to Extinction

At least three species of large-bodied mammals of Palawan have become locally extinct in the Holocene. The fragmentary quality of the evidence at present tempers our interpretations, but the data allow us to propose hypotheses about the nature of these extirpations and place them within regional patterns of faunal changes for the time period covered.

The asynchronous demise of the two deer species is notable and likely signifies different causes of their extinction. At present, the larger deer species has not been found in Late Holocene deposits from northern Palawan cave assemblages and the youngest record that we have currently is from the Early Holocene of Ille Cave. A large part of Greater Palawan became inundated during the Pleistocene-Holocene transition leading to a tremendous decrease in land area and separation from the neighboring small island groups of the Calamianes, Cuyo and Balabac, (Fig. 4.1; Robles et al. 2014). The previously drier climate and more open vegetation of the northern part of the island during the LGM was also later on replaced by dense forest cover by *ca.* 8 ka (Bird et al. 2007; Wurster et al. 2010), reducing habitat size for the large mammals. In the case of the tiger, declining deer populations may have also been a factor, reducing available prey resources needed to support viable tiger populations. Still this is unlikely to have been a defining factor in isolation when, like people, the tiger could potentially have switched prey to the wild pig (*Sus ahoenobarbus*). It is possible that on Palawan, the addition of human hunting contributed additional pressure on the large-bodied deer and tiger populations, and it was a combination of these factors that pushed them towards extinction during a period when they were already under ecological stress. Hence, a combination of environmental, ecological and anthropogenic factors may have all influenced the extirpation of these taxa (possibly including the dhole) on Palawan.

The present relict distribution of *Axis calamianensis* and its Late Holocene demise on Palawan Island presents a different scenario. This endemic species is now known to have occupied a large portion of the Palawan faunal region based on the fossil record, but it now only persists in the small islands of the Calamianes. This relict distribution is also mirrored in the condition of the Bawean hog deer (*A.*

kuhlii), which is also in the Holocene fossil record of Java but is no longer extant in the larger island (van den Brink 1982). Their closest extant relative is of Indochinese affinity, *Axis porcinus,* while another Javanese fossil species, *A. lydekkeri*, is proposed to be another close relative; all are included in the subgenus *Hyelaphus* (*sensu* Meijaard and Groves 2004). The Palawan and Javanese fossil records indicate that members of the hog deer taxon were more widely distributed in Southeast Asia and that there was a significant range reduction and extirpations throughout the Pleistocene and Holocene. In the past, other species may have become restricted to patches of refugia when climatic and eustatic changes drove them to retreat into available habitats where they later evolved and became specialists in these habitats. Such is perhaps the case regarding the evolution and diversification of many montane species in the Bornean uplands, some of which have closest relatives that are of Indochinese affinities, as well as of the montane murid specialists of the oceanic Philippines (see Cranbrook 2009; Heaney 2001). These processes of retreat into refugia and the expansion and contraction of habitats influence extinction and speciation events throughout millions of years, as is demonstrated for instance by Abbeg and Thierry (2002) in their model of Southeast Asian macaque evolution.

Among contemporary extinction models, the contagion hypothesis proposed by Lomolino and Channell (1995) states that geographical range collapse is a non-random process, with final populations usually persisting in peripheral areas of the animal's historical range, including montane areas and oceanic islands. In their analysis of historical distributions and geographic analysis of modern-day range collapse, Channell and Lomolino (2000) identify humans and human-related disturbances as the main contagion. In the fossil record, the impacts and roles of ancient humans in extinction events are more difficult to assess, but the archaeological context of fossil remains and the species' biogeographic information can provide further clues regarding anthropogenic impact and human responses.

The Calamian hog deer persisted on Palawan Island well into the Late Holocene, and this indicates that the species survived the extensive palaeogeographic and environmental changes of the Pleistocene-Holocene transition and of previous periods such as the transition from the penultimate glacial to the last interglacial (e.g., van der Kaars and Dam 1997; Hope et al. 2004; Bird et al. 2005). This species only apparently succumbed to ecological pressures, human hunting and other anthropogenic disturbances on the main island of Palawan within the last three millennia, or perhaps even more recently (Ochoa et al. 2014). The same pattern is reflected in the Flores (van den Bergh et al. 2009) and Bornean fossil record (Cranbrook 2009) wherein certain terrestrial vertebrates that survived extensive environmental changes and persisted into the Holocene disappeared only in the last few thousand years because of human-related impacts.

Human Response and Changes in Subsistence Patterns

Deer remains are abundant in the Ille record from the Terminal Pleistocene until the Early Holocene, after which they become progressively rare and eventually disappear altogether from the archaeological record. Deer remains have also been recorded at southern Palawan cave sites in the Late Pleistocene at Tabon (Fox 1970) and Pilanduk (Kress 1977). Fox (1970) also noted the absence of deer in younger assemblages in the Tabon Caves complex. As deer populations diminished in the Holocene, the wild pig became an increasingly important resource (Fig. 4.6). The human response to diminishing deer populations appears to be to switch resources to the Palawan bearded pig, a species adapted to dense forest environments that might have become more common as the rainforests expanded. In addition to the hunting of large game, the human foragers at Ille and Pasimbahan also continued to prey on a variety of locally available smaller mammals and reptiles that would have constituted an important component to the diet. The occasional hunting of the hog deer, on the other hand, continued well into the Late Holocene as demonstrated by the Pasimbahan record. This likely contributed to, or even drove, the extirpation of this species on Palawan Island.

The shift from deer to pig hunting, and the appearance of thick shell midden deposits in the middle Holocene are reflective of the changing environments in the Dewil Valley and of the responses of local foragers to these changes. Robles et al. (2014) hypothesize that the island-wide appearance of dense Middle Holocene shell middens in the archaeology of Palawan is related to changing coastlines and mangrove expansion during the early to Middle Holocene. Previously inland locales became nearer to marine and mangrove resources due to sea level rise and subsequent stabilization of coastlines. The dense Middle Holocene shell middens of Pasimbahan Cave consists predominantly of mangrove and marine species (Ochoa et al. 2014). In the Middle Holocene levels of Ille, a thick shell midden layer is present, also consisting mostly of mangrove and freshwater species (Szabo et al. 2004; Faylona 2003). Prior to this, shell remains were also present in the Early Holocene but in much smaller quantities. This evidence documents the changing accessibility and availability of both vertebrate and invertebrate resources to the foragers of northern Palawan and shows how these ancient humans utilized such resources in the face of environmental changes through time. It also

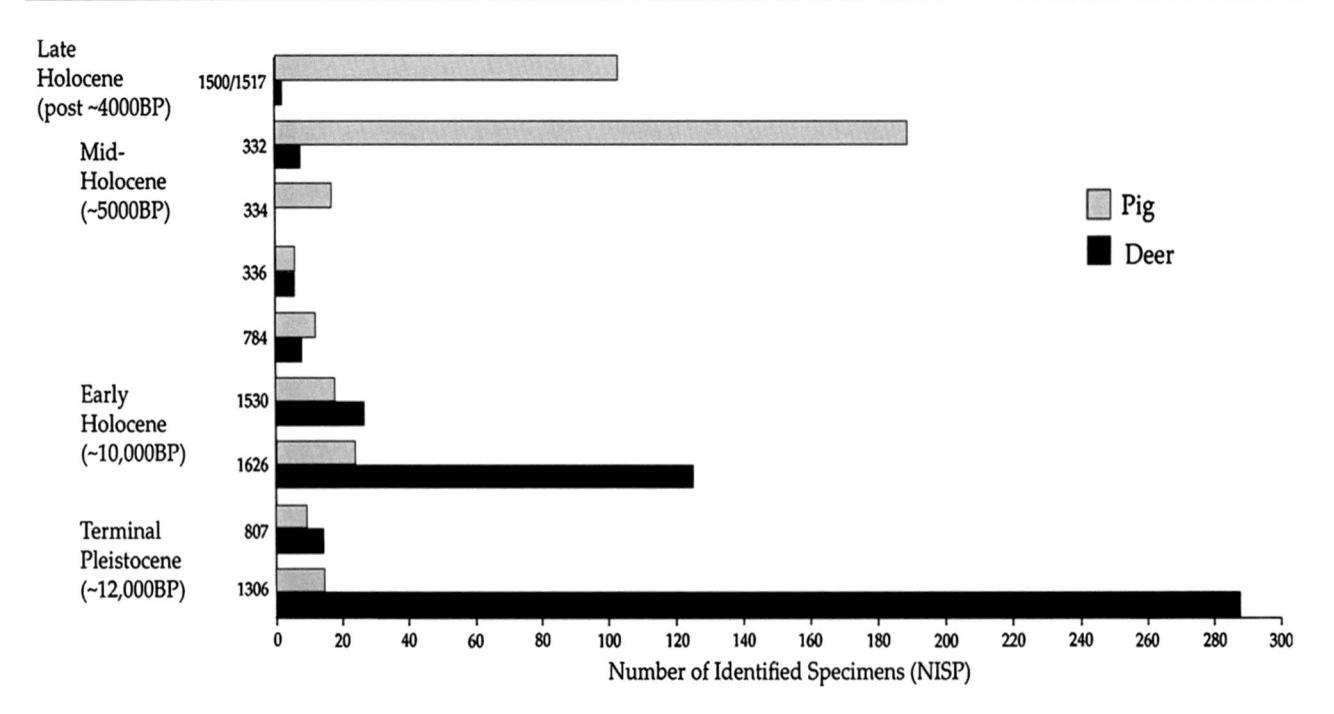

Fig. 4.6 Distribution (NISP) of pig and deer remains in the Ille Cave sequence. Numbers on y-axis refer to context numbers, which are arbitrary numbers allocated to each stratigraphic layer and archaeological feature during excavations. Comparative NISP and MNI data can be found in Piper et al. (2011)

Conclusion

The Palawan zooarchaeological assemblages provide a long record of the interaction between people and environments and their effects on the local faunal communities. The study of these faunal assemblages has demonstrated that Palawan once had a much higher diversity of large mammals, and that these species have been successively lost through a combination of environmental perturbations and human impacts. Combining zooarchaeological with palaeoclimatic and geographic evidence certainly augments our understanding of the historical dynamics of the extinction process and range contractions but there is much more to investigate. For instance, the reasons as to why the Calamian hog deer persists in the small islands to the north are unclear. Little is known of the species' ecology, limiting our ability to test hypotheses regarding its demise on Palawan. Post-Pleistocene environmental changes certainly contributed to the decline of the deer and tiger populations, and this is important to further examine because the reason(s) for a species' final extermination may not be the same as the reason(s) for its protracted decline (Channell and Lomolino 2000). The Late Pleistocene and Early Holocene presence of the tiger in Borneo and Palawan demonstrates that this species was much more widely distributed throughout Island Southeast Asia during the distant past than even historical accounts record. It also indicates that habitat constriction is not a recent phenomenon for these extant fauna, but has been a long-term process going on for millennia. All living Philippine deer currently have a much reduced range and recent extirpations are recorded for *R. marianna* and *R. alfredi* in several islands throughout the archipelago due to a variety of human impacts (Heaney et al. 1998). Understanding the history of climatic and human impacts on the vertebrate faunas of the Philippines is critical to sustaining the unique native faunal communities inhabiting these patches of modern-day human-constrained refugia.

Acknowledgements We thank the Project Directors of the Palawan Island Palaeohistoric Project Prof. Victor Paz (University of the Philippines), Dr. Helen Lewis (University of Dublin), and Mr. Wilfredo Ronquillo (National Museum of the Philippines) for their continuing support in our research endeavours. The authors are also grateful to Dr. Lawrence Heaney (Field Museum of Natural History) for his guidance, and to Mr. Emil Robles for his aid in the environmental reconstructions, maps and profiles. Much gratitude is also dedicated to the curators of the Mammal Division/Section of the following museums for providing access to their reference collections: FMNH, National Museum of Natural History, Harvard University Zooarchaeology Laboratory and Museum of Comparative Zoology, American Museum of Natural History, and the California Academy of Sciences. The research of JO was supported by a Henry Luce Foundation/ACLS East and Southeast Asian Archaeology and Early History Research Fellowship and an outright research grant from the University of the Philippines Office of the Vice Chancellor for Research and Development (UP OVCRD). Radiocarbon dating was funded by the UP OVCRD and conducted by the Waikato Radiocarbon Dating Laboratory (University of Waikato) under the supervision of Dr. Alan Hogg and Dr. Fiona Petchey. The research of PJP was supported by a separate UP OVCRD grant and an Australian Research Council Future Fellowship award FT100100527. Participation in the International Council for Archaeozoology (ICAZ) session "Climate Change, Human Response and Zooarchaeology" (from which this volume derives) was made possible for JO by a UP Research Dissemination Grant and funding from the ICAZ. The authors are also grateful to two anonymous reviewers whose insightful comments and suggestions helped to improve the final version of this contribution.

References

Abegg, C., & Thierry, B. (2002). Macaque evolution and dispersal in insular South-east Asia. *Biological Journal of the Linnaean Society, 75*, 555–576.

Anderson, D. D. (1997). Cave archaeology in Southeast Asia. *Geoarchaeology: An International Journal, 12*(6), 607–638.

Anderson, A. (2002). Faunal collapse, landscape change and settlement history in remote Oceania. *World Archaeology, 33*(3), 375–390.

Barker, G., Reynolds, T., & Gilbertson, D. (2005). The human use of caves in peninsular and island Southeast Asia: Research themes. *Asian Perspectives, 44*(1–15).

Barker, G., Barton, H., Bird, M., Daly, P., Datan, I., Dykes, A., et al. (2007). The 'human revolution' in lowland tropical Southeast Asia: The antiquity and behaviour of anatomically modern humans at Niah Cave (Sarawak, Borneo). *Journal of Human Evolution, 52*, 243–261.

Belmaker, M. (2017). The southern Levant during the last glacial and zooarchaeological evidence for the effects of climatic-forcing on hominin population dynamics. In Monks, G. (Ed.), *Climate change and human responses: A zooarchaeological perspective* (pp. 7–25). Dordrecht: Springer.

Bintanja, R., van den Wal, R. S. W., & Oerlemans, J. (2005). Modelled atmospheric temperatures and global sea levels over the past million years. *Nature, 437*, 125–128.

Bird, M., Taylor, D., & Hunt, C. (2005). Palaeoenvironments of insular Southeast Asia during the last glacial period: A savanna corridor in Sundaland. *Quaternary Science Reviews, 24*, 2228–2242.

Bird, M. I., Boobyer, E., Bryant, C., Lewis, H., Paz, V., & Stephens, W. E. (2007). A long record of environmental change from bat guano deposits in Makangit Cave, Palawan, Philippines. *Earth and Environmental Transactions of the Royal Society of Edinburgh, 98*, 59–69.

Brown, W., & Alcala, A. (1970). The zoogeography of the herpetofauna of the Philippine Islands, a fringing archipelago. *Proceedings of the California Academy of Sciences, 38*(6), 105–130.

Brown, R., & Diesmos, A. (2001). Application of lineage-based species concepts to oceanic island frog populations: The effects of differing taxonomic philosophies on the estimation of Philippine biodiversity. *Silliman Journal, 42*, 133–162.

Brown, R., & Guttman, S. (2002). Phylogenetic systematics of the *Rana signata* complex of Philippine and Bornean stream frogs: Reconsideration of Huxley's modification of Wallace's Line at the Oriental-Australian faunal zone interface. *Biological Journal of the Linnaean Society, 76*, 393–461.

Brown, P., Sutikna, T., Morwood, M., Soejono, R., Jatmiko, Saptomo, E. W., et al. (2004). A new small-bodied hominin from the Late Pleistocene of Flores, Indonesia. *Nature, 431*, 1055–1061.

Brown, R. M., Siler, C. D., Diesmos, A. C., & Alcala, A. C. (2009). Philippine frogs of the Genus *Leptobrachium* (Anura; Megophryidae): Phylogeny-based species delimitation, taxonomic review, and descriptions of three new species. *Herpetological Monographs, 23*(1), 1–44.

Burney, D. (1997). Tropical islands as paleoecological laboratories: Gauging the consequences of human arrival. *Human Ecology, 25*(3), 437–457.

Channell, R., & Lomolino, M. V. (2000). Trajectories to extinction: Spatial dynamics of the contraction of geographical ranges. *Journal of Biogeography, 27*(1), 169–179.

Clutton-Brock, J. (1959). Niah's Neolithic dog. *Sarawak Museum Journal, 9*(13–14), 143–150.

Cranbrook, E. (1988). The contribution of archaeology to the zoogeography of Borneo with the first record of a wild canid of Early Holocene age. *Fieldiana, 1385*(Zoology 12), 1–7.

Cranbrook, E. (2000). Northern Borneo environments of the past 40,000 years: Archaeological evidence. *Sarawak Museum Journal, 76*, 61–110.

Cranbrook, Earl. (2009). Late quaternary turnover of mammals in Borneo: The zooarchaeological record. *Biodiversity and Conservation, 19*(2), 373–391.

de Vos, J., & Sondaar, P. (1994). Dating hominid sites in Indonesia. *Science, 266*, 1726–1727.

Detroit, F., Dizon, E., Falgueres, C., Hameau, S., Ronquillo, W., & Semah, F. (2004). Upper Pleistocene *Homo sapiens* from the Tabon Cave (Palawan, The Philippines): Description and dating of new discoveries. *C. R. Palevol, 3*, 705–712.

Diamond, J., & Gilpin, M. (1983). Biogeographic umbilici and the origin of the Philippine avifauna. *Oikos, 41*, 307–321.

Diesmos, A. C., Parham, J., Stuart, B., & Brown, R. (2005). The phylogenetic position of the recently rediscovered Philippine forest turtle (Bataguridae: *Heosemys leytensis*). *Proceedings of the California Academy of Science, 56*(3), 31–41.

Diesmos, A. C., Brown, R. M., Alcala, A. C., & Sison, R. V. (2008). Status and distribution of nonmarine turtles of the Philippines. *Chelonian Conservation and Biology, 7*(2), 157–177.

Esselstyn, J., Widmann, P., & Heaney, L. (2004). The mammals of Palawan Island, Philippines. *Proceedings of the Biological Society of Washington, 117*(3), 271–302.

Esselstyn, J. A., Oliveros, C. H., Moyle, R. G., Peterson, A. T., McGuire, J. A., & Brown, R. M. (2010). Integrating phylogenetic and taxonomic evidence illuminates complex biogeographic patterns along Huxley's modification of Wallace's Line. *Journal of Biogeography, 37*(11), 2054–2066.

Evans, B., Supriatna, J., Andayani, N., Setiadi, M. I., Cannatella, D., & Melnick, D. (2003). Monkeys and toads define areas of endemism on Sulawesi. *Evolution, 57*(6), 1436–1443.

Faylona, M. G. P. (2003). A preliminary study on shells from Ille Rock Shelter. *Hukay, 3*, 31–49.

Flannery, T. F. (1999). Debating extinction. *Science, 283*, 182–183.

Flannery, T. F., & Roberts, R. G. (1999). Late Quaternary extinctions in Australia: An overview. In R. D. E. McPhee (Ed.), *Extinctions in near time: Causes, contexts and consequences* (pp. 239–256). New York: Plenum Press.

Fox, R. (1970). *The Tabon caves.* Manila: National Museum of the Philippines.

Geist, V. (1998). Three-pronged Old World deer. In V. Geist (Ed.), *Deer of the world: Their evolution, behaviour and ecology* (pp. 55–80). Mechanicsburg, PA: Stackpole Books.

Giovas, C., & LeFebvre, M. (2006). My island, your island, our islands: Considerations for island archaeozoology as a disciplinary community. In *Landscape Zooarchaeology Symposium.* Presented at the ICAZ Conference, Mexico.

Grayson, D. (2001). The archaeological record of human impacts on animal populations. *Journal of World Prehistory, 15*(1), 1–68.

Griffiths, M. L., Drysdale, R. N., Gagan, M. K., Zhao, J. X., Ayliffe, L. K., Hellstrom, J. C., et al. (2009). Increasing Australian-Indonesian monsoon rainfall linked to Early Holocene sea-level rise. *Nature Geoscience, 2*(9), 636–639.

Hall, R. (2002). Cenozoic geological and plate tectonic evolution of SE Asia and the SW Pacific: Computer-based reconstructions, model and animations. *Journal of Asian Earth Sciences, 20*, 353–431.

Hanebuth, T., Stattegger, K., & Grootes, P. M. (2000). Rapid flooding of the Sunda Shelf: A late-glacial sea-level record. *Science, 288* (5468), 1033–1035.

Harcourt, A. H. (1999). Biogeographic relationships of primates on South-East Asian islands. *Global Ecology and Biogeography, 8*, 55–61.

Heaney, L. (1985). Zoogeographic evidence for Middle and Late Pleistocene land bridges to the Philippine islands. *Modern Quaternary Research in SE Asia, 9*, 127–143.

Heaney, L. (1986). Biogeography of mammals in SE Asia: Estimates of rates of colonization, extinction and speciation. *Biological Journal of the Linnean Society, 28*, 127–165.

Heaney, L. (1991). A synopsis of climatic and vegetational change in Southeast Asia. *Climate Change, 19*, 53–61.

Heaney, L. R. (2001). Small mammal diversity along elevational gradients in the Philippines: An assessment of patterns and hypotheses. *Global Ecology and Biogeography, 10*(1), 15–39.

Heaney, L., & Mittermeier, R. A. (1997). The Philippines. In R. A. Mittermeier, P. R. Gil, & C. G. Mittermeier (Eds.), *Megadiversity, earth's biologically wealthiest nations*. CEMEX: Monterrey, Mexico.

Heaney, L., Balete, D., Dolar, M. L., Alcala, A., Dans, A., Gonzales, P., et al. (1998). A synopsis of the mammalian fauna of the Philippine Islands. *Fieldiana Zoology, 88*.

Heaney, L. R., Walker, E., Tabaranza, B., & Ingle, N. (2002). Mammalian diversity in the Philippines: An assessment of the adequate of current data. *Sylvatrop, 10*(1–2), 6–27.

Hiscock, P. (2008). *Archaeology of ancient Australia*. London: Routledge.

Hope, G., Kershaw, A. P., van der Kaars, S., Xiangjun, S., Liew, P.-M., Heusser, L. E., et al. (2004). History of vegetation and habitat change in the Austral-Asian region. *Quaternary International, 118–119*, 103–126.

Horton, B., Gibbard, P., Mile, G., Morley, R., Purintavaragul, C., & Stargardt, J. (2005). Holocene sea levels and palaeoenvironments, Malay-Thai Peninsula, southeast Asia. *The Holocene, 15*(8), 1–15.

Hutterer, R. (2007). Records of shrews from Panay and Palawan, Philippines, with the description of two new species of *Crocidura* (Mammalia: Soricidae). *Lynx (Praha), 38*, 5–20.

Jansa, S. A., Barker, F. K., & Heaney, L. R. (2006). The pattern and timing of diversification of Philippine endemic rodents: Evidence from mitochondrial and nuclear gene sequences. *Systematic Biology, 55*(1), 73–88.

Jones, A., & Kennedy, R. (2008). Evolution in a tropical archipelago: Comparative phylogeography of Philippine fauna and flora reveals complex patterns of colonization and diversification. *Biological Journal of the Linnean Society, 95*(3), 620–639.

Kershaw, A. P., Bretherton, S. C., & van der Kaars, S. (2007). A complete pollen record of the last 230 ka from Lynch's Crater, north-eastern Australia. *Palaeogeography, Palaeoclimatology, Palaeoecology, 251*(1), 23–45.

Koch, P., & Barnosky, A. (2006). Late quaternary extinctions: State of the debate. *Annual Review of Ecology, Evolution, and Systematics, 37*, 215–250.

Koch, A., Gaulke, M., & Bohme, W. (2010). Unravelling the underestimated diversity of Philippine water monitor lizards (Squamata: *Varanus salvator* complex), with the description of two new species and a new subspecies. *Zootaxa, 2446*, 1–54.

Kress, J. (1977). Contemporary and prehistoric subsistence patterns on Palawan. In W. Wood (Ed.), *Cultural-ecological perspectives on Southeast Asia*. Athens, Ohio: Ohio University Press.

Lambeck, K., Esat, T., & Potter, E. K. (2002). Links between climate and sea levels for the past three million years. *Nature, 419*, 199–206.

Larson, G., Karlsson, E. L. K., Perri, A., Webster, M. T., Ho, S. Y. W., Peters, J., et al. (2012). Rethinking dog domestication by integrating genetics, archeology, and biogeography. *Proceedings of the National Academy of Sciences*. doi:10.1073/pnas.1203005109.

Lewis, H., Paz, V., Lara, M., Barton, H., Piper, P., Ochoa, J., et al. (2008). Terminal Pleistocene to Mid-Holocene occupation and an early cremation burial at Ille Cave, Palawan, Philippines. *Antiquity, 82*(316), 318–335.

Lomolino, M. V., & Channell, R. (1995). Splendid isolation: Patterns of geographic range collapse in endangered mammals. *Journal of Mammalogy, 76*(2), 335–347.

Louys, J., Curnoe, D., & Tong, H. (2007). Characteristics of Pleistocene megafauna extinctions in southeast Asia. *Palaeogeography, Palaeoclimatology, Palaeoecology, 243*, 152–173.

Lucchini, V., Meijaard, E., Diong, C., Groves, C., & Randi, E. (2005). New phylogenetic perspectives among species of southeast Asian wild pigs (*Sus* sp.) based on mtDNA sequences and morphometric data. *Journal of the Zoological Society of London, 266*, 25–35.

Maeda, Y., Siringan, F., Omura, A., Berdin, R., Hosono, Y., Atsumi, S., et al. (2004). Higher-than-present Holocene mean sea levels in Ilocos, Palawan and Samar, Philippines. *Quaternary International, 115–116*, 15–26.

Martin, P. S. (1984). Prehistoric overkill: The global model. In P. S. Martin & R. G. Klein (Eds.), *Quaternary extinctions: A prehistoric revolution* (pp. 354–403). Tucson: University of Arizona Press.

Martin, P. S., & Steadman, P. W. (1999). Prehistoric extinctions on islands and continents. In R. D. MacPhee (Ed.), *Extinctions in near time* (pp. 17–52). New York: Plenum Publishers.

McGuire, J., & Kiew, B. H. (2001). Phylogenetic systematics of Southeast Asian flying lizards (Iguania: Agamidae: *Draco*) as inferred from mitochondrial DNA sequence data. *Biological Journal of the Linnean Society, 72*(2), 203–229.

Medway, L. (1977). The ancient domestic dogs of Malaysia. *Journal of the Malaysian Branch Royal Asiatic Society, 50*, 14–27.

Meijaard, E. (2003). Mammals of south-east Asian islands and their Late Pleistocene environments. *Journal of Biogeography, 30*, 1245–1257.

Meijaard, E., & Groves, C. (2004). Morphometric relationships between south-east Asian deer (Cervidae, tribe Cervini): Evolutionary and biogeographic implications. *Journal of Zoology, London, 263*, 179–196.

Mijares, A. S., Détroit, F., Piper, P., Rainer, G., Bellwood, P., Aubert, M., et al. (2010). New evidence for a 67,000-year-old human presence at Callao Cave, Luzon, Philippines. *Journal of Human Evolution., 59*, 123–132.

Morwood, M. J., O'Sullivan, P. B., Aziz, F., & Raza, A. (1998). Fission-track ages of stone tools and fossils on the east Indonesian island of Flores. *Nature, 392*, 173–176.

Morwood, M. J., Brown, P., Jatmiko, Sutikna T., Wahyu Saptomo, E., Westaway, K. E., et al. (2005). Further evidence for small-bodied hominins from the Late Pleistocene of Flores, Indonesia. *Nature, 437*, 1012–1017.

Morwood, M. J., Sutikna, T., Saptomo, E. W., Westaway, K. E., Jatmiko, Awe Due, R., et al. (2008). Climate, people and faunal succession on Java, Indonesia: Evidence from Song Gupuh. *Journal of Archaeological Science, 35*(7), 1776–1789.

Norton, C. J., Kondo, Y., Ono, A., Zhang, Y., & Diab, M. C. (2010). The nature of megafaunal extinctions during the MIS 3-2 transition in Japan. *Quaternary International, 211*(1–2), 113–122.

O'Connor, S., & Aplin, K. P. (2007). A matter of balance: An overview of Pleistocene occupation history and the impact of the last glacial phase in East Timor and the Aru Islands, eastern Indonesia. *Archaeology in Oceania, 42*, 82–90.

O'Connor, S., & Ono, R. (2012). Pelagic fishing at 42,000 years before present and the maritime skills of modern humans. *Science, 334*, 1117–1121.

Ochoa, J. (2005). In dogged pursuit: A reassessment of the dog's domestication and social incorporation. *Hukay, 8*, 35–66.

Ochoa, J. (2009). Terrestrial vertebrates of Ille Cave, Northern Palawan, Philippines: Subsistence and palaeoecology in the Terminal Pleistocene to the Holocene. Master's thesis, University of the Philippines.

Ochoa, J., Paz, V., Lewis, H., Carlos, J., Robles, E., Amano, N., et al. (2014). The archaeology and paleobiological record of Pasimbahan-Magsanib Site, northern Palawan, Philippines. *Philippine Science Letters, 7*, 22–36.

Pang, J.-F., Kluetsch, C., Zou, X.-J., Zhang, A.-B., Luo, L.-Y., Angleby, H., et al. (2009). MtDNA data indicate a single origin for dogs south of Yangtze River, less than 16,300 years ago, from numerous wolves. *Molecular Biology and Evolution, 26*(12), 2849–2864.

Partin, J. W., Cobb, K. M., Adkins, J. F., Clark, B., & Fernandez, D. P. (2007). Millennial-scale trends in west Pacific warm pool hydrology since the last glacial maximum. *Nature, 449*(7161), 452–455.

Paz, V., Ronquillo, W., Lewis, H., Piper, P., Carlos, A. J., Robles, E., et al. (2008). *Palawan Island palaeohistoric research project: Report on the 2008 Dewil Valley field season*. Unpublished manuscript.

Paz, V., Ronquillo, W., Lewis, H., Lape, P. V., Robles, E., Hernandez, V., et al. (2010). *Palawan Island palaeohistoric project: Report on the 2010 field season*. Unpublished manuscript.

Pilaar-Birch, S., & Miracle, P. (2017). Human response to climate change in the northern Adriatic during the Late Pleistocene and Early Holocene. In Monks, G. (Ed.), *Climate change and human responses: A zooarchaeological perspective* (pp. 87–100). Dordrecht: Springer.

Piper, P. J., & Ochoa, J. (2007). The first zooarchaeological evidence for the endemic Palawan stink badger (*Mydaus marchei* Huet 1887). *Hukay, 11*, 85–92.

Piper, P. J., Cranbrook, E., & Rabett, R. (2007). Confirmation of the presence of the tiger *Panthera tigris* (L.) in Late Pleistocene and Holocene Borneo. *Malayan Nature Journal, 59*(3), 259–267.

Piper, P. J., Ochoa, J., Lewis, H., Paz, V., & Ronquillo, W. P. (2008). The first evidence for the past presence of the tiger *Panthera tigris* (L.) on the island of Palawan, Philippines: Extinction in an island population. *Palaeogeography, Palaeoclimatology, Palaeoecology, 264*(1–2), 123–127.

Piper, P. J., Ochoa, J., Robles, E., & Paz, V. (2011). Palaeozoology of Palawan Island, Philippines. *Quaternary International, 233*, 142–158.

Pregill, G. K., & Steadman, D. W. (2009). The prehistory and biogeography of terrestrial vertebrates on Guam, Mariana Islands. *Diversity and Distributions, 15*(6), 983–996.

Reis, K. R., & Garong, A. M. (2001). Late Quaternary terrestrial vertebrates from Palawan Island, Philippines. *Palaeogeography, Palaeoclimatology, Palaeoecology, 171*, 409–421.

Robles, E., Piper, P. J., Lewis, H., Paz, V. J., & Ochoa, J. (2014). Late quaternary sea level changes and the palaeohistory of Palawan Island, Philippines. *Journal of Island and Coastal Archaeology., 10*(1), 1–21. doi:10.1080/15564894.2014.880758.

Rohling, E. J., Fenton, M., Jorissen, F. J., Bertrand, P., Ganssen, G., & Caulet, J. P. (1998). Magnitudes of sea-level lowstands of the past 500,000 years. *Nature, 394*(6689), 162–165.

Sathiamurthy, E., & Voris, H. K. (2006). Maps of Holocene sea level transgression and submerged lakes on the Sunda Shelf. *The Natural History Journal of Chulalongkorn University, Supplement, 2*, 1–44.

Savolainen, P., Leitner, T., Wilton, A. N., Matisoo-Smith, E., Lundeberg, J., & Renfrew, C. (2004). A detailed picture of the origin of the Australian dingo, obtained from the study of mitochondrial DNA. *Proceedings of the National Academy of Sciences of the United States of America, 101*(33), 12387–12390.

Sémah, A. M., & Sémah, F. (2012). The rain forest in Java through the quaternary and its relationships with humans (adaptation, exploitation and impact on the forest). *Quaternary International, 249*, 120–128.

Steadman, D. W. (1995). Prehistoric extinctions of Pacific island birds: Biodiversity meets zooarchaeology. *Science, 267*, 1123–1131.

Steadman, D. W., & Martin, P. S. (2003). The late Quaternary extinction and future resurrection of birds on Pacific islands. *Earth-Science Reviews, 61*(1–2), 133–147.

Swisher, C. C., Curtis, G., Jacob, T., Getty, A. G., Suprijo, A., & Widiasmoro. (1994). Age of the earliest known hominids in Java. *Science, 263*(5150), 1118–1121.

Szabo, K., Swete Kelly, M. C., & Penalosa, A. (2004). Preliminary results from excavations in the eastern mouth of Ille cave, northern Palawan. In Paz, V. (Ed.), *Southeast Asian archaeology: Wilhelm G. Solheim II festschrift* (pp. 209–224). Quezon City: University of the Philippines Press.

Tjia, H. D. (1996). Sea-level changes in the tectonically stable Malay-Thai Peninsula. *Quaternary International, 31*, 95–101.

Tougard, C. (2001). Biogeography and migration routes of large mammal faunas in South-East Asia during the Late Middle Pleistocene: Focus on the fossil and extant faunas from Thailand. *Palaeogeography, Palaeoclimatology, Palaeoecology, 168*(3–4), 337–358.

Towns, D. R., & Daugherty, C. H. (1994). Patterns of range contractions and extinctions in the New Zealand herpetofauna following human colonisation. *New Zealand Journal of Zoology, 21*(4), 325–339.

van den Bergh, Gert D., de Vos, J., & Sondaar, P. Y. (2001). The late Quaternary palaeogeography of mammal evolution in the Indonesian archipelago. *Palaeogeography, Palaeoclimatology, Palaeoecology, 171*(3–4), 385–408.

van den Bergh, G. D., Meijer, H. J. M., Due Awe, R., Morwood, M. J., Szabó, K., van den Hoek Ostende, L. W., et al. (2009). The Liang Bua faunal remains: A 95-kyr sequence from Flores, east Indonesia. *Journal of Human Evolution, 57*(5), 527–537.

van den Brink, L. M. (1982). On the mammal fauna of the Wajak cave, Java (Indonesia). *Modern Quaternary Research in Southeast Asia, 7*, 177–194.

van der Geer, A., Lyras, G., de Vos, J., & Dermitzakis, M. (2010). *Evolution of island mammals: Adaptation and extinction of placental mammals on islands*. Oxford: Wiley-Blackwell.

van der Kaars, S., & Dam, R. (1997). Vegetation and climate change in West-Java, Indonesia during the last 135,000 years. *Quaternary International, 37*, 67–71.

van der Kaars, S., Wang, X., Kershaw, P., Guichard, F., & Arifin Setiabudi, D. (2000). A Late Quaternary palaeoecological record from the Banda Sea, Indonesia: Patterns of vegetation, climate and biomass burning in Indonesia and northern Australia. *Palaeogeography, Palaeoclimatology, Palaeoecology, 155*(1–2), 135–153.

Verstappen, H. (1975). On palaeoclimates and landform development in Malesia. *Modern quaternary research in Southeast Asia*. Rotterdam: A.A. Balkema.

von den Driesch, A. (1976). *A guide to the measurement of animal bones from archaeological sites*. Cambridge, MA: Peabody Museum of Archaeology and Ethnology, Harvard University.

Voris, H. (2000). Maps of Pleistocene sea levels in southeast Asia: Shorelines, river systems and time durations. *Journal of Biogeography, 27*(5), 1153–1167.

Waelbroeck, C., Labeyrie, L., Michel, E., Duplessy, J. C., McManus, J. F., Lambeck, K., et al. (2002). Sea-level and deep water temperature changes derived from benthic foraminifera isotopic records. *Quaternary Science Reviews, 21*(1–3), 295–305.

Wang, Y., Cheng, H., Edwards, R. L., Kong, X., Shao, X., Chen, S., et al. (2008). Millennial- and orbital-scale changes in the East Asian monsoon over the past 224,000 years. *Nature, 451*(7182), 1090–1093.

Wroe, S., Field, J., Fullagar, R., & Jermin, L. S. (2004). Megafaunal extinction in the late Quaternary and the global overkill hypothesis. *Alcheringa: An Australasian Journal of Palaeontology, 28*(1), 291–331.

Wurster, C. M., Bird, M. I., Bull, I. D., Creed, F., Bryant, C., Dungait, J. A. J., et al. (2010). Forest contraction in north equatorial southeast Asia during the last glacial period. *Proceedings of the National Academy of Sciences, 107*(35), 15508–15511.

Chapter 5
Human Response to Climate Change in the Northern Adriatic During the Late Pleistocene and Early Holocene

Suzanne E. Pilaar Birch and Preston T. Miracle

Abstract Climate and sea level constrain the abundance of primary producers (plants) and habitat size. These directly affect the seasonal density and distribution of animal species, which inevitably have implications for human decisions regarding what to eat, where to live, how long to stay there, and when to move. Are diversification strategies such as the inclusion of low-ranked terrestrial resources and marine species in the diet effective coping mechanisms for climate-driven environmental change and habitat loss due to sea level rise? Is intensification of resource exploitation indicative of dietary stress? How might these adaptations affect the seasonal round? Our paper discusses these questions, spanning the transition from post-glacial foraging lifestyles at the Pleistocene/Holocene boundary (12,000 BP) to the introduction of pastoralism during the early Neolithic (7,000 BP), using zooarchaeological material from the upland cave site of Vela Špilja on the island of Lošinj in the Kvarner Gulf of Croatia.

Keywords Paleolithic-Mesolithic subsistence • Residential mobility • Optimal foraging theory • Broad spectrum diet • Seasonality • Carcass processing

S.E. Pilaar Birch (✉)
Department of Anthropology and Department of Geography, University of Georgia, Athens, GA, USA
e-mail: sepbirch@uga.edu

P.T. Miracle
Division of Archaeology, Department of Archaeology and Anthropology, University of Cambridge, Cambridge, UK

Introduction

Human response to climate change is currently a major research focus in archaeology, as the number of diverse contributions to this volume suggests. Exactly how humans have adapted to changing local environments, modified landscapes, and accompanying ecological shifts throughout prehistory is a question of particular interest. The Mediterranean and Adriatic basins are critical regions for addressing questions of hunter-gatherer adaptive strategies as they come to bear on the eventual adoption of sedentary lifestyles, use of domestic plants and animals, and increasing societal complexity as the Neolithic spread from Southwest Asia into Europe during the Early Holocene.

Globally, the Terminal Pleistocene is characterized by great climatic instability as environmental amelioration took place following deglaciation. This was accompanied by rapidly rising sea level until the onset of the Younger Dryas cold period 12,900 years ago. Sea level continued to rise during the Preboreal and Boreal, periods of favorable climatic conditions beginning 11,700 years ago (Gibbard 2010; Burroughs 2005, 2007). The Great Adriatic Plain bridged the Balkan and Italian peninsulas; a flat, low-lying area, it is now covered by the northern half of the Adriatic Sea in southern Europe. It has been argued both that the Great Adriatic Plain would have been rich in game, water, and lithic resources (van Andel 1989; Bailey and Gamble 1990; Miracle 1995) and that it was a barren, saline wasteland (Mussi 2001) during the Pleistocene. However, the abundance of upland cave sites on what would have been the edges of the Plain attest to the viability of the landscape. In the Holocene, rising sea levels led to the eradication of the Plain and would have had significant implications for Mesolithic forager populations.

How did climate change, the inundation of the Great Adriatic Plain, and therefore the transformation and loss of lowland habitats in the Late Pleistocene and Early Holocene affect human subsistence and mobility? Did the decrease in availability of inland terrestrial resources and emergence of a coastal landscape cause a shift from a specialist to a broad spectrum diet? If the zooarchaeological assemblage reflects increasing amounts of low-ranked terrestrial and marine resources, then this may suggest a broad spectrum adaptation to the local environmental changes. Does a broad spectrum diet occur as a consequence of dietary stress or does it represent the exploitation of abundant, newly available resources? If there is evidence of intensification in the form of increased carcass processing, then this may suggest dietary changes are a result of resource stress. The site of Vela Špilja on the island of Lošinj in the Kvarner Gulf of Croatia serves as a central case study, and we provide a brief overview of its archaeological and environmental setting below. We then discuss how changes in subsistence at Vela Špilja Lošinj might fit a general model for the Eastern Adriatic (Miracle 1995) and compare our results from this site with evidence from other sites in the Adriatic and wider Mediterranean region.

Vela Špilja Lošinj

The cave of Vela Špilja is located on the island of Lošinj in the Adriatic Sea off the coast of Croatia and is presented here as a case study (Fig. 5.1). It was excavated in 2004 as part of the Paleolithic and Mesolithic Settlement of the Northern Adriatic Project (Miracle 2006). Thus far Pleistocene and Holocene contexts with archaeological evidence of human occupation dating to the Late Upper Paleolithic, Mesolithic, and Early Neolithic have been excavated. Five major cultural horizons have been identified based on ceramics and lithics.

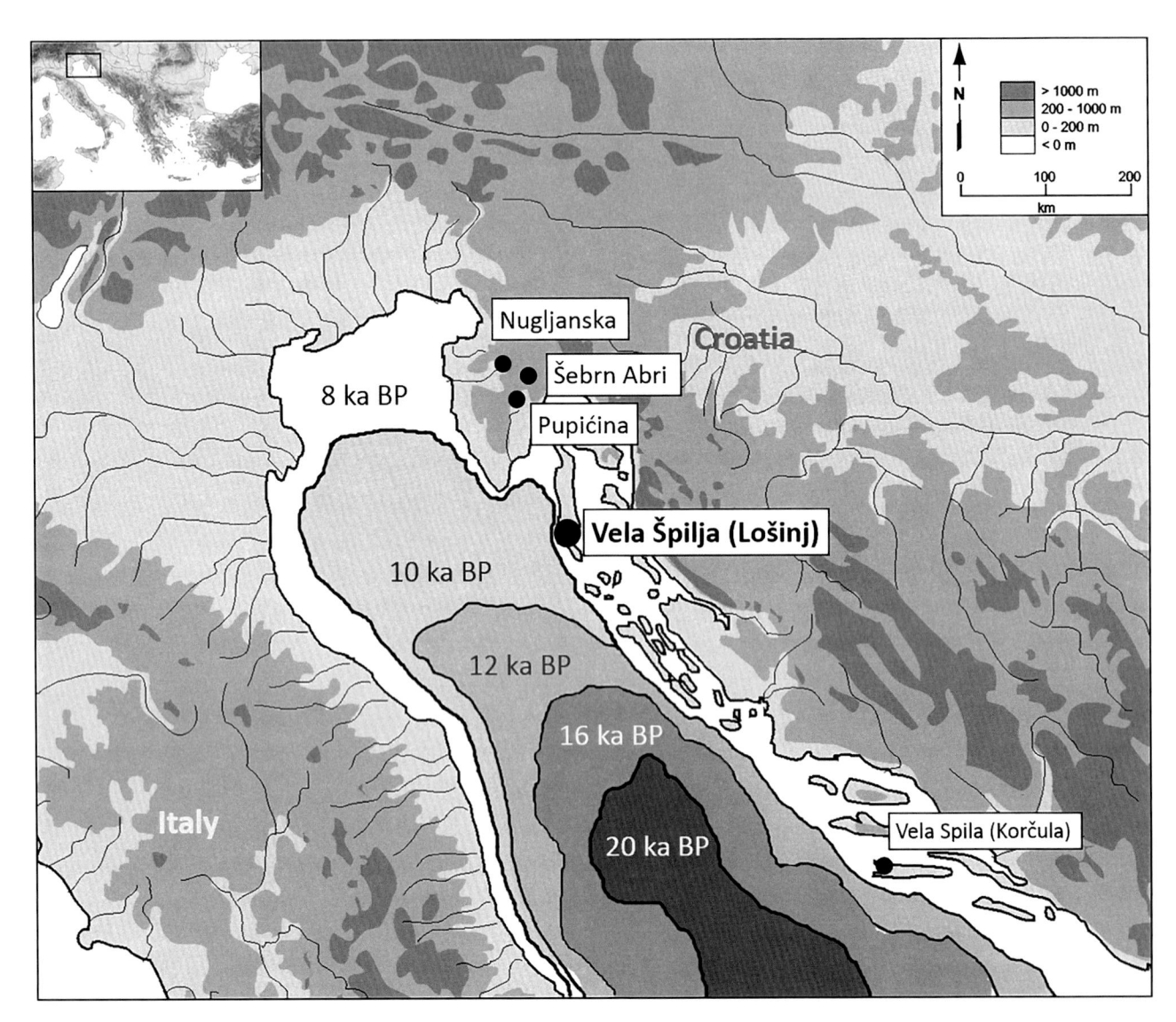

Fig. 5.1 Vela Špilja and the regional settlement system were affected by sea level rise during the Pleistocene-Holocene transition

Table 5.1 Climate and local environment and their correlation to archaeology at VSL

Date (ka)	Global Climate Event and Conditions (Burroughs 2005)	Local Sea Level (Asioli et al. 2001; Lambeck et al. 2004; Surić et al. 2005; Burroughs 2005)	Local Environment and Vegetation (Rossignol-Strick et al. 1992; Huntley and Prentice 1993; Willis 1994; Rossignol-Strick 1999; Asioli et al. 2001; Burroughs 2005)	VSL Layer and Horizon
6.0–8.0	Atlantic Warm and wet	−23 m at 7920 BP	Continuation of warm and wet conditions. Decrease in deciduous trees; increase in *Olea* (cultivation?)	Level 40 Horizon 3
8.0–8.2	Cooling Event Cold snap	Lake Agassiz melt. Possible rapid substantial sea level rise	Increase in coniferous species	–
8.2–9.0	Atlantic Warm and wet	−34 m at 9,160 BP	Increase in deciduous oak pollen and maximum extension of montane deciduous forest. Absence of frost and increased precipitation in winter; possible summer droughts	Level 60 Horizon 4
9.0–11.0	Boreal Warm and dry	−36 m/sl at 10,185 BP; warming of surface waters	Expansion of mixed deciduous woodland; diversification of plant populations in higher elevations. Increased seasonality (warmer summers and colder winters)	Level 70/75 Horizon 4
11.0–11.3	"Preboreal Oscillation" Short-lived cooling			–
11.3–11.6	Preboreal Rapid warming	Warming of surface waters	Expansion of forest; decrease of grass steppe	Level 80 Horizon 5
11.6–12.9	Younger Dryas Very cold and dry	Plateau in rise; decrease of freshwater input to Adriatic. Cold surface waters	Replacement of woodland cover by dry steppe; expansion of lowland semi-desert and upland tundra; intermediary grass steppe and deciduous steppe-forest decrease. Summer drought and cold winters	–
12.9–14.5	Bølling-Allerød Sudden warming	Rapid sea level rise from −125 m at LGM. Warm surface waters	Expansion and increase of temperate tree species; expansion of mid-altitude grass steppe	Level 120 Horizon 5

A double-backed flint piece from the lowest excavated horizon (Horizon 5) attests to use in the Late Upper Paleolithic, while a chert flake and burin from Horizon 4 are typical of Mesolithic stone tool types. This is followed by early Neolithic cultural material including impressed ware ceramics in Horizons 1–3. The geographical setting of the cave and long history of human settlement provide the opportunity to investigate continuity and disruption in patterns of consumption and mobility. Absolute dates for the Pleistocene contexts are as early as 11,315 years cal BP (using OxCal IntCal09; 9805 ± 50) and span well into the Holocene (Forenbaher et al. 2013). Additional radiocarbon dates for the early Holocene have recently been published (Pilaar Birch et al. 2016). These can be correlated with known climatic events, estimated sea levels, and cultural periods (Table 5.1). Vela Špilja Lošinj (hereafter VSL) presently sits at 258 m above sea level. As a high point overlooking the Great Adriatic Plain in the Pleistocene, it offered an ideal location for observing game such as horse and red deer below as well as for targeting chamois and ibex in its mountainous surroundings.

What makes Vela Špilja on Lošinj particularly interesting is its progression from an inland cave to an island cave over the course of a few thousand years. Noticeable change in topography, landscape, and surrounding ecosystems took place within a single human lifetime. This would have influenced both the density and species of game available. For example, red deer, abundant in the Paleolithic assemblage, disappears much earlier on Lošinj than on similar inland sites. While water may have created a barrier for some wildlife, the distance from Lošinj to the mainland and other islands is not prohibitive to short distance sea travel. This ultimately has implications for human decisions about use of limited island resources and tradeoffs in mobility.

Overview of the Study Region

The current distribution of Late Pleistocene (Paleolithic) and Early Holocene (Mesolithic/Neolithic) sites in the northern Adriatic region is most likely due to geomorphology rather than the lack of research. Only a very small number of Upper Paleolithic sites are found in valleys and terraces within the mountains of the northern Trieste Karst, due to the presence of alluvial megafans and meltwater during the late glacial which would have made the area uninhabitable (Boschian and Fusco 2007). Land surface erosion due to these geological processes means that most sites older than 15,000 years have been lost. The lower Great Adriatic Plain would have been much more hospitable and upland cave sites located on the Istrian Peninsula have well-dated layers with cultural material spanning from the Late Upper Paleolithic to the Bronze Age. These sites were important within the regional settlement system and are reviewed in further detail below.

Pupićina

Pupićina cave is located in the canyon of Vela Draga at an elevation of 220 m above present-day sea level and 15 m above the valley floor. It is approximately 10 km from the current coastline, but Mount Učka, at 1,400 m high, stands as an obstacle to direct access. The uppermost Paleolithic occupation layers have been dated to 13,150–11,750 cal BP and contained lithic tools, cores, flakes, and debris from regionally sourced flint (Miracle 2001). There are also some tools made from local and exogenous sources suggesting a large territory within which people were moving and obtaining raw lithic material (Komšo and Pellegatti 2007).

The Mesolithic for this site is securely dated between 11,650–9,750 cal BP, during which a permanent stream flowed on the valley floor (Miracle 2001; Komšo 2006). Perforated shells of the marine mollusc *Columbella* as well as freshwater species have been found in these layers. The numerous faunal remains recovered and their significance will be discussed further below. The site was likely a regional base camp for a settlement system of which VSL, Nugljanska, and Šebrn Abri were a part (Miracle 1997, 2001; Komšo 2006).

The Neolithic at Pupićina has also been extensively documented and the earliest levels date to 7,485 cal BP (Miracle and Forenbaher 2006). Finds included Danilo-Vlaška pottery and ground stone tools, as well as evidence of use of the cave for penning domestic sheep and goat.

Nugljanska

Nugljanska is a large cave situated above an extensive plain, 550 m in elevation. There are currently seven radiocarbon dates from this site, spanning from approximately 15,100–8,700 years before present, encompassing the Late Upper Paleolithic and the Mesolithic at the site (Miracle and Forenbaher 2000; Pilaar Birch and Miracle 2015). The site has been interpreted as the earliest example of human settlement in the uplands of the Ćićarija Mountains (Komšo and Pellegatti 2007). Hunting tools and burins made of regional material comprise the terminal Late Upper Paleolithic toolkit (Komšo and Pellegatti 2007). Lithics from the Mesolithic levels include endscrapers and linear tools (Komšo 2006). A number of hearths were excavated and over 11,000 faunal remains were recovered and analyzed (Miracle and Forenbaher 2000; Pilaar Birch 2012; Pilaar Birch and Miracle 2015). It has been suggested that hunter gatherers used Nugljanska as an outpost for monitoring resources during the Paleolithic, and as a residential base camp in the Mesolithic (Komšo 2006; Miracle 2007).

Šebrn Abri

Šebrn Abri is a rock shelter that is located further inland and higher than Pupićina, Nugljanska, and VSL, about 750 m above sea level at the foot of a 30 m high limestone cliff. A steep valley cuts in front of the rock shelter, affording a view of both the mountain plateau on which it is located and the landscape below. There is no evidence for occupation during the Late Upper Paleolithic. Radiocarbon dating has bracketed use of the cave for a period of up to 800 years in the Boreal period of the Early Holocene, beginning with the Mesolithic at approximately 10,460 cal BP (Miracle et al. 2000). Cultural finds include lithics and perforated *Columbella* shells as at Pupićina. Miracle and colleagues (2000) suggest that the evidence for multiple stages of production within the lithic assemblage, composed mostly of hunting tools, indicates the site was not merely a hunting station but also a place where tools were manufactured and modified. This site is also notable for its early occupation date at a high elevation, which may have to do with specialized seasonal use (Miracle et al. 2000; Komšo 2006).

Other Istrian Sites

A number of smaller sites have been excavated in the region and are briefly mentioned here. Klanjčeva cave is 800 m above sea level and faces southwest, with evidence for Mesolithic use and occupation. There is no evidence for post-Mesolithic use of the cave, and lithic material is consistent with regional Mesolithic stone tool types. AMS dates from uppermost layer date to 11,220 cal BP. However, the integrity of the stratigraphy has been questioned as the underlying level has been dated to 9,510 cal BP (Miracle and Forenbaher 2000). The site of Vela is a small cave about 250 m above sea level and is located directly across the canyon from Pupićina. It has yielded some evidence of human use during the Mesolithic (Komšo 2006) as well as Neolithic and later periods (Forenbaher et al. 2008, 2010).

Vela Spila

Further south on the Dalmatian island of Korčula, Vela Spila (literally "big cave") offers a useful comparison for the Istrian sites. The cave is 130 m above sea level and overlooks the village and bay of Vela Luka. The island is located at the narrowest portion of the Adriatic Sea, and on a clear day it is possible to see the Italian coast from the vantage point of the cave. In the Upper Paleolithic, this would have meant an all-encompassing view of the large ungulate herds moving across the Great Adriatic Plain below. Paleolithic dates range from the LGM and late glacial (14,120 cal BP) during which time the cave would have been 230 m above sea level and the coast 20 km away (Čečuk and Radić 2005; Radić et al. 2008). By the Mesolithic, the sea would have been much closer and the cave located at approximately 150 m above sea level, a change which is reflected in the increase in marine species in the faunal assemblage and modifications to the toolkit. Charcoal dates the early Mesolithic to 9,200 cal BP and uppermost Mesolithic to 8,000 cal BP (Čečuk and Radić 2005; Komšo 2006). As at Pupićina and Šebrn further north, some decorated and perforated shells of *Columbella* were found. There is possible evidence of use of flint sources from the opposite (Italian) shore, though local flint sources have also recently been discovered (Perhoč 2009). Three child burials occur deep within the cave. The earliest Neolithic layers, classified as such by the presence of ceramics and domestic sheep and goat, overlap with the Mesolithic, suggesting continuation of cave use during the Mesolithic-Neolithic transition (Čečuk and Radić 2005). It is of interest to note that both Vela Spila on the island of Korčula and Vela Špilja on the island of Lošinj have very early dates for the Neolithic, suggesting elements of the "Neolithic package" may have first travelled north via a coastal route (cf. Forenbaher et al. 2013).

Climate and Environment

Global climate change at the Pleistocene-Holocene transition can be translated into regional environmental change using multiple proxy records. Large-scale data from Northern Atlantic ice cores is useful for broadly classifying climate change in the eastern Adriatic, as the sensitivity of the Mediterranean region to the North Atlantic Ocean has been established (Allen et al. 1999; Hughes et al. 2006). Davis et al. (2003) provide reconstructions of the temperature across Europe for the last 12,000 years and suggest that differences in climate over the continent varied widely both seasonally and spatially, but that mean annual temperature increased almost linearly up until 7,800 years ago.

Pleistocene

A regional geoarchaeological study by Boschian and Fusco (2007) used macro analysis, micromorphology, and examination of pollen data from cave sites and sediment cores in the northern Adriatic to address the Late Upper Paleolithic regional environmental context. They posit that the Trieste karst was a harsh, unwooded landscape experiencing cold and dry environmental conditions. These would have persisted into the early Mesolithic, perhaps creating a barrier to the movement of human populations during a time of rising sea level further south. The scenario in what is now Istria, and what was then the upland periphery of the Plain, would have been very different. Open woodland and wet conditions persisted throughout the late glacial, perhaps due to the permanence of water basins amongst the hills that later became islands in the Kvarner Gulf. At Pupićina, the sedimentology reflects high energy water flow, suggesting short periods of flood. The faunal assemblage at Pupićina also suggests a slightly wooded environment (wild boar and red deer), and this is complemented by pollen records that suggest an expansion of deciduous forest in this location at the Terminal Pleistocene (Willis 1994; Miracle 2001). In and around the Great Adriatic Plain, multiple proxies (including plant pollen and macrofossils, sedimentology, and molluscs) from Lake Vrana and Valun Bay on the isle of Cres less than 40 km from VSL suggest open grassland vegetation in the late glacial (Schmidt et al. 2000). Across the sea, a core near the present day Italian shoreline at Milano Marittima also contained pollen evidence indicative of an open landscape with herbaceous shrub and pine communities during the Late Pleistocene, with an expansion of deciduous species in the Terminal Pleistocene and Early Holocene (Boschian and Fusco 2007). In addition, Pleistocene faunal remains from VSL suggest mixed open woodland and grassland (red deer and horse) as well as broken upland terrain (chamois/ibex).

Holocene

Sediments from the Polje Čepić core in the Istrian peninsula were deposited during "a period of rapidly changing climate at the late glacial-Early Holocene transition" (Balbo et al. 2006). This is evident in numerous climate and environmental proxies throughout the region. Pollen evidence from Edera cave north of Istria attests to an increase in forest taxa at this time (Gallizia Vuerich and Princivalle 1994). Sediment properties of a core from Lake Vrana on the nearby island of Cres, accompanied by changes in pollen and mollusc species, indicate increased seasonality and increased overall temperature in the study region during the Preboreal (Schmidt et al. 2000, 2004). Climate proxy data from stable isotope analysis ($\delta^{18}O$) of ungulate teeth at Pupićina support this warming trend in the Early Holocene (Pilaar Birch et al. 2016). The Boreal is marked by an increase in beech (*Fagus*) pollen (Schmidt et al. 2000) indicating the emergence of the present day vegetation in the Istrian mountains at around the time Lošinj became island 8,000 years ago. Towards the end of the Boreal there was an expansion in Mediterranean climate species, and Neolithic and Bronze age proxy data reflect the introduction of domestic plants and forest clearing (Schmidt et al. 2000).

Sea Level Rise

In the Early Holocene, rising sea levels in the Mediterranean caused inland flooding and the modern Adriatic Sea was formed. The karstic nature of the landscape and presence of numerous small mountains led to the formation of the multitude of Dalmatian islands and an indented coastline. Many small islands were and continue to be within view of each other and easy swimming or rowing distance, potentially facilitating prehistoric inter-island trade networks and mobility. Speleothem evidence from a submerged cave in Tihovac Bay approximately 40 km southwest of VSL suggests the area was first submerged by freshwater at approximately 8,500 years ago, when sea level was approximately −23 m below present levels. Marine conditions started at approximately 7,920 cal BP (Surić et al. 2002, 2005). This provides an estimated date for the insulation of VSL as well as an indication of the rapid pace of inundation. Lambeck et al. (2004) provide modeled curves for the Eastern Adriatic and suggest the sea took its modern form at approximate 8,000 years ago, but that modern coastlines were not established until around 6,000 years ago, when sea level was an estimated 2.5 m below current levels. It is important to keep in mind not only the large-scale impact of sea formation as the Great Adriatic Plain disappeared, but also smaller scale ecosystem and microhabitat changes as the land surrounding VSL transitioned from an open grassland ecosystem, to lowland estuary and wetland, and finally a coastal ecozone during the Mesolithic.

A Model for Changing Subsistence Patterns at Vela Špilja Lošinj
Optimal Foraging Theory

This study aims to determine whether broadening the resource base and intensifying carcass use were viable survival strategies in the context of rapid environmental change. The current zooarchaeological data best lend themselves to discussion within the framework of optimal foraging theory (MacArthur and Pianka 1966; Smith and Winterhalder 1981, 1992). This diet-breadth model makes a number of assumptions; notably that the frequency with which prey species are encountered is equal to their abundance in the environment. Based on the forager's own knowledge of food quality and the handling time required in obtaining the prey, the forager decides whether or not to pursue the prey animal (Kelly 1995). "Handling time" includes the time it takes to hunt, kill, process and consume the animal. Simply put, "The decision entails an immediate opportunity cost comparison: (a) pursue the encountered resource, or (b) continue searching with the expectation of locating more valuable resources to pursue. If the net return to (b) is greater than (a), even after allowing for additional search time, then the optimizing forager will elect to pass by the encountered resource, and will continue to do so no matter how frequently this type of resource is encountered" (Winterhalder and Kennett 2006:14). This allows us to use archaeologically visible changes in diet breadth through time as a proxy for human response to environmental change through optimization of resource procurement and exploitation strategies. If the resource base is high and the risk of starvation is low, low energy resources will not be exploited; energy expensive but high-energy yielding foods will. Translated into the archaeological record, prey specialization should represent a stable and resource-rich environment, notably demonstrated in Stiner (2000), Miracle (2007). There will be a focus on a small number of species with a high handling time and high energy return such as red deer and large- to medium-sized ungulates, thought to have been abundant on the Pleistocene Adriatic Plain. Environmental change may have caused reduced availability and predictability of preferred game species, though this does not necessarily equate to resource stress. If this is true, we expect to see the broadening of the resource base and use of low-ranked prey species consistent with the expectations of optimal foraging theory. We argue that this scenario should also be accompanied by a significant increase in the intensity of carcass processing if dietary stress is occurring. Finally, a shift in the seasonal mobility patterns is expected and will be discussed briefly, but a detailed consideration of the data regarding this transition is beyond the scope of this paper.

The Model

As Miracle (1995:31) discusses in his model for the late glacial eastern Adriatic, the loss of the Great Adriatic Plain due to rising sea levels, coupled with fluctuations in temperature and rainfall, would have led to a decrease in the local resource base for forager groups. In the case of VSL and other sites in the region, an initial increase in prey density would have occurred at the onset of habitat restriction at the end of the Pleistocene and Early Holocene beginning around 12,000 years BP, but by 10,000 BP and the approximate start of the Mesolithic this resource base would have been affected by the insulation of VSL and the disappearance of the plain, which would have led to a local extirpation of larger prey species and ungulates. This would have had serious implications for changing resource exploitation at VSL from the Pleistocene into the Early Holocene. Miracle (1995) proposes an increased dietary breadth and more complete processing of the carcass as general coping strategies for surviving in the post-glacial Adriatic, and the assumptions of this model are tested below.

Zooarchaeological Data

Data is parsed between the Neolithic, Mesolithic, and Palaeolithic horizons, rather than by levels within these horizons. This is because this paper is both looking at change through three major time periods from four main contexts and because doing so generates more robust sample sizes (Fig. 5.2). The faunal assemblage from VSL totaled 3,103 bones, 1,021 (33%) of which were identifiable to element and species. The remaining 2,082 (67%) were classified as unidentifiable.

Diversification

Overall, the assemblage follows the expected pattern of dietary specialization in the Pleistocene layers, though the significance of these patterns may be somewhat limited by sample size in earlier levels. Red deer (*Cervus elaphus*) make up the majority of the Paleolithic (Horizon 5) assemblage, along with chamois (*Rupicapra rupicapra*) and ibex (*Capra ibex*), hare (*Lepus europaeus*), boar (*Sus scrofa*), and fox (*Vulpes vulpes*). There is also a small number of horse (*Equus* spp.) in this horizon. If this is taken as a proxy for consumption, it would appear people during this time are specializing in red deer and supplementing this high-investment prey with some arguably "mid-range" resources such as chamois/ibex and boar.

In the Mesolithic (Horizon 4), although the species richness in the assemblages is the same (seven), their frequency is very differently distributed. The percentage of red

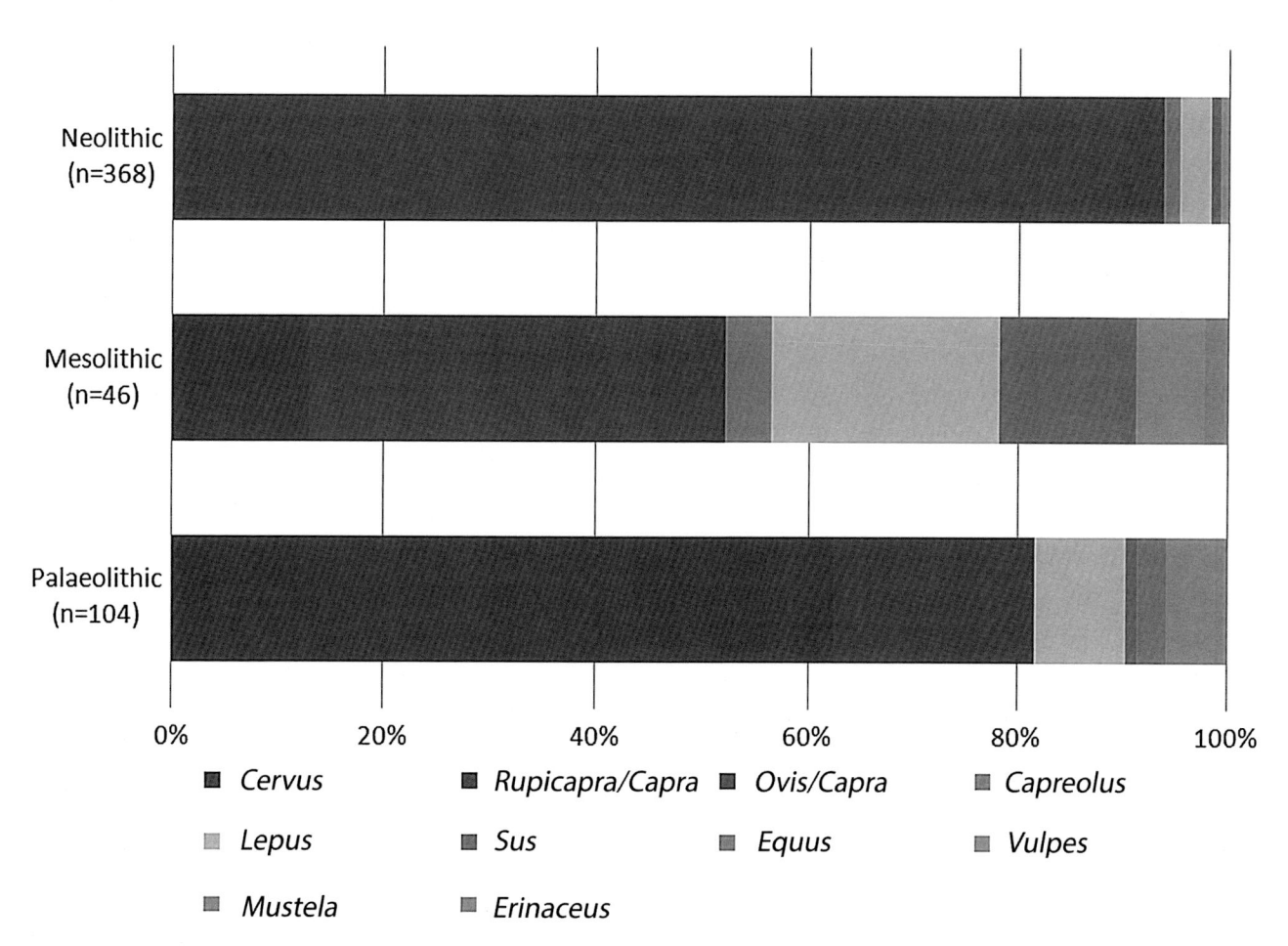

Fig. 5.2 Change in relative species abundance at VSL

deer is reduced. There are increased numbers of chamois and ibex as well as hare. There is a higher relative abundance of boar, fox, and roe deer. The distribution reflects a "broad spectrum" diet. It would seem that movement of red deer further inland as sea level rose affected both prey availability and choice, as additional smaller resources were exploited.

The Neolithic (Horizon 3) is extremely different from the Mesolithic, with 95% of the assemblage made up of (presumably imported) domestic sheep and goat. A very small percentage of fox, hare, and *Sus* (pig or wild boar) are probably incidental and suggest that during this time the cave was most likely used by pastoralists no longer foraging for food.

The presence of microfauna in caves can be indicative of frequency of occupation, but it can also add to the discussion of increased dietary breadth. In general, there are large numbers of microfaunal remains in the Pleistocene, with a marked decrease in the uppermost Paleolithic levels. We suggest that this is potentially related to environmental conditions. In the Mesolithic, we attribute the appearance of large percentages of bird and fish in the microfaunal assemblage (20%) to the possible exploitation of these resources. There is also the appearance of marine mollusc species such as limpets (*Patella* spp.), mussels (*Mytilus galloprovincialis*) and topshells (*Gibulla* and *Phorcus* spp.) for the first time in the Mesolithic layers, confirming use of marine resources.

Intensification

There is evidence for human modification of animal bone in each stratigraphic level at VSL. This includes cut marks resulting from disarticulation, defleshing, and chopping, as well as burning and smashing. Types of fragmentation and body part frequency are used to discuss the possibility of intensified processing. About 87% of bone was fragmented; only 7% of that number was recent breaks.

Spiral fractures occur when fresh or "green" bone breaks, likely near the time of death (Lyman 1994). They are most likely the product of human or carnivore action as a result of obtaining the calorie-rich marrow inside the bone. Intensity of fracturing can be indicative of grease production and storage (Outram 2001). Angular breaks occur after the bone is no longer "green" but has become "dry". This is most likely around the time of burial as pressure is placed on the bone and it cracks. When angular and spiral fractures occur together it is likely that the bone was processed in some form and then broke again during internment. All levels have a fairly well distributed amount of breaks that occurred when the bone was fresh and dry. Spiral breaks alone increase from 13% in the Paleolithic to 24% in the Mesolithic, which would seem to indicate more intense processing for marrow.

However, if the category of combined spiral and angular fractures is included in this percentage, around 50% of all bones for all levels have spiral fractures. Furthermore, spiral fractures can be caused by a number of variables including both human and carnivore activities, so there is not sufficient evidence from this type of analysis to suggest intensified processing of carcasses in the Mesolithic as compared to the Paleolithic; rather, carcass processing appears rather moderate throughout both periods.

Body part representation may offer further insight into carcass processing, although it can sometimes be difficult to interpret due to the number of factors influencing the preservation of bone and teeth. Differences in density and mineral structure contribute to these problems; the desirability of a cut of meat due to the resources it provides such as marrow and fat is another compounding factor influencing which bones are prominent in the assemblage (Lyman 1994, 2008). Contexts were considered separately in order to identify patterning and change through time using caprids (chamois and ibex). Assuming that the abundance of each respective element is correlated to its abundance in reality, it is possible to compare the frequency of certain elements in the Paleolithic with the Mesolithic. In all levels, axial skeleton components appear to make up almost 25% of the NISP. This difference can be attributed to the higher frequency of axial elements over appendicular elements in the ungulate skeleton.

Due to small sample size, only the Late Upper Paleolithic (Horizon 5) and Mesolithic (Horizon 4) caprids are considered in this analysis (Fig. 5.3). The LUP is comprised of a majority of axial skeletal elements and hind limbs, whereas Mesolithic caprids are better represented by anterior elements such as mandible and skull fragments as well as the thoracic vertebrae, ulnae, and humeri. Posterior elements consist of the calcaneus, metatarsal, and 1st and 2nd phalanges. The high percentage of teeth present in the assemblage may suggest that diagenetic and preservational biases have heavily influenced body part representation, and the small sample size of this level makes these results difficult to interpret. However, there is also preservation of less dense skull and vertebral fragments in the Mesolithic, which suggest there may be a real pattern in the data. Of note is the abundance of hind limb elements in the LUP, which are considered high-utility elements, yielding desirable amounts of flesh and marrow. The Mesolithic assemblage is more evenly divided amongst body part areas, suggesting whole bodies may have been brought back to the site more frequently, and were potentially more completely processed. There is a lower frequency of long bones, which could be due to extensive processing for marrow and grease rendering, in which many of the fragments are broken into pieces so tiny they are not identifiable. Metapodials are less desirable elements as they yield some marrow but little amounts

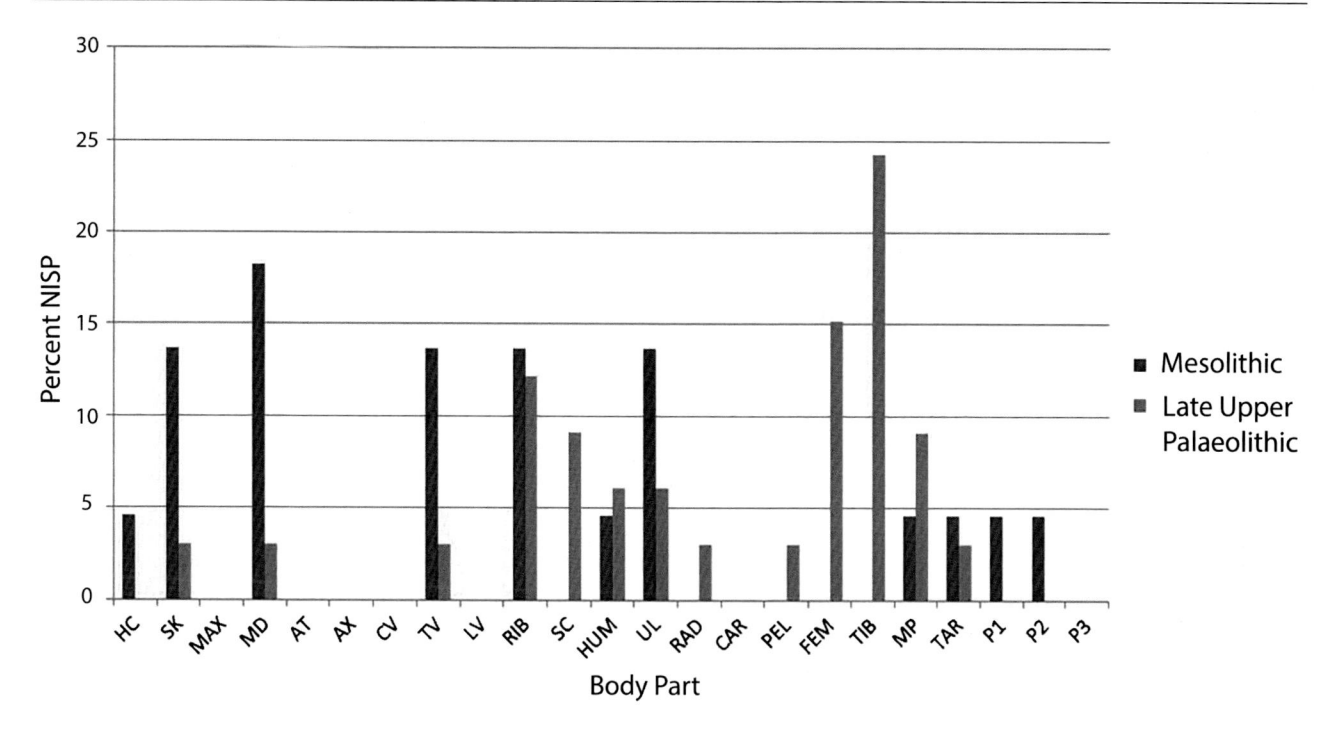

Fig. 5.3 Body part representation of caprids between the Late Upper Paleolithic and Mesolithic

of meat (Lam and Pearson 2005). Many of the metapodials showed signs of splitting and spiral breaks, which could be a sign of more complete carcass use in the form of marrow and fat exploitation in these lower-utility skeletal elements. Fat is much higher in calories than protein and carbohydrates, and body parts high in fat are likely to be recognized by human groups who consume large amounts of animal protein and who are under dietary stress (Binford 1978; Outram 2001; Morin 2007). Evidence for intensification during the Mesolithic is somewhat inconclusive due to the small sample size and number of variables affecting the assemblage, but this may potentially be resolved in future analyses with a larger dataset.

Seasonal Mobility

Changes in relative species abundance and carcass processing at VSL suggest that human groups more extensively exploited locally available resources in the face of environmental change. We suggest that this behavior would result in multi-season use of sites to more efficiently exploit resources within the landscape. The results described here are generated using age curves from zooarchaeological material only, and are considered preliminary in nature. Stable isotope analysis of ungulate teeth and marine shell are more likely to resolve questions regarding seasonal mobility and use of VSL but are outside the scope of this discussion (Pilaar Birch 2012). In the Late Upper Paleolithic, there is a large number of adult specimens and some juveniles; less than 10% are infants. Preliminary cervid data suggests an abundance of juveniles in the Late Upper Paleolithic (75%, n = 13). This may suggest a late summer summer/early autumn use. In the Mesolithic, there is a noticeable shift to a majority of infant and fetal animals. Again, the preliminary and circumstantial nature of the data must be stressed, but springtime use can be inferred from the large abundance of very young animals, although this does not preclude use in additional seasons. The Neolithic data also suggest a spring use of the site. In periods where infant or juvenile remains are rare, spring occupation cannot be discounted; however, where they are present, it is arguable that the site was used in at least this season and potentially multiple seasons.

Discussion
Summary of the Data and its Application to the Model

In the Paleolithic at VSL, there was an emphasis on red deer, as hunters were able to observe and stalk their prey on the Great Adriatic Plain below. There would be some use, perhaps more opportunistic or occasional, of smaller ungulates such as chamois and ibex which lived in the mountainous area immediately adjacent to the cave. With the rise of sea level and isolation of the site, decrease in habitat for the red deer, and the subsequent shrinking of the red deer population, those using Vela Špilja Lošinj would have intensified their predation on the chamois and ibex population, while also diversifying their diet to include more species, such as hare and seafood resources. But they do not conclusively appear to have also engaged in more complete carcass processing, and available environmental proxies suggest that the overall climatic conditions during the Mesolithic were favorable despite habitat loss caused by rising sea level. Though VSL does appear to fit the expectations of optimal foraging theory, the admittedly limited butchery data do not fully support an intensification of carcass processing indicating dietary stress. As Miracle (1995) suggests, long distance residential mobility was probably not a viable option for alleviating dietary stress during the late glacial-Early Holocene transition. Seasonal re-use of sites is more likely as people settled into their local regions, especially in attempt to cope with the increasing environmental strain and increased seasonal variability. In the case of VSL, further investigation of seasonal use of the site through stable isotope analysis may offer the opportunity for a more in-depth interpretation of the role of mobility as an adaptation to environmental change.

Diversification

How does the trend of dietary diversification at VSL compare with other Mesolithic sites in the region and beyond? Miracle (2001) discusses data from Pupićina which suggest that resource diversification occurred at the site throughout the Mesolithic and that remains of large and medium sized ungulates could be refuse from episodic feasting during this time. In contrast to the LUP specialization in red deer at VSL, in the LUP at Pupićina there is an almost even distribution of red deer, boar, and roe deer. There are two lines of evidence which suggest further dietary diversification occurred from the LUP into the Mesolithic. First, species richness increases from 3 species to 5 species in the early Mesolithic, then to 6 in the middle/late Mesolithic. These additional species are "lower ranked" resources (badger and hare). In addition, there is an increase in edible land snails and marine molluscs in association with the increase of hare and badger. Human consumption of land snails is supported by a correlation between the increase of *Helix* with the geometric density of ungulates and marine molluscs (Miracle 2002).

The faunal record at the site of Šebrn Abri stands in contrast to VSL and Pupićina. The data suggest a generalized Mesolithic diet that became more specialized in red deer procurement through time, constituting up to a maximum of 34% of the assemblage. Over 90% of the Mesolithic assemblage is comprised of red deer, roe deer (13%) and wild boar (11%), along with remains identified as small- to medium-sized ungulates. In contrast, red deer make up just fewer than 10% of the assemblage at Mesolithic VSL, while chamois and ibex contribute 36% of the assemblage and hare is the second most abundant species (25% of the assemblage). Wild boar is similar in relative abundance in both assemblages (13% at VSL). At Šebrn, we suggest the main cause for the apparent increase in focus on larger-bodied species, particularly red deer as compared to other sites in the region, could be due to the increasing scarcity of this and other larger prey species in the surrounding lower elevations, including areas undergoing insularization such as Lošinj. In addition, VSL may have been used in different seasons multiple times each year in order to maximize access to both terrestrial and aquatic resources, in contrast to the use of Šebrn as a hunting outpost during a single season when migratory upland forest species were targeted.

The site of Vela Spila on the island of Korčula further south became isolated in the Early Holocene much as Vela Špilja became isolated on Lošinj. There are high proportions of red deer and onager in the Late Upper Paleolithic faunal assemblage at this site (Spry-Marqués 2012). In the Mesolithic, dietary diversification includes the exploitation of fish and shellfish; fine bone tools found at the site have been interpreted as implements used for their consumption (Čečuk and Radić 2005). Large quantities of sea snails and limpets (*Phorcus* and *Patella* spp.) have been recovered in addition to terrestrial animals such as boar, fallow deer, red deer, and small game such as hare. Tuna and swordfish indicate fishing of deepwater species rather than just on-shore gathering of marine and intertidal molluscs, suggesting marine resources were more important further south than they were at the head of the Adriatic.

Regional stable isotope data derived from archaeological humans and animals is also useful to incorporate here. Boar, red and roe deer, chamois, ibex, and hare remains from Pupićina, Šebrn, and Klanjčeva caves and human remains from Pupićina (n = 4) were relatively undifferentiated, exhibiting low variation in $\delta^{13}C$ and $\delta^{15}N$ values; this suggests prey species were procured from a similar geographic area and that human diets consisted of primarily terrestrial resources, with limited consumption of aquatic resources (Robinson 2006; Paine 2009). That interpretation is supported by a larger regional dataset that included samples throughout the eastern Adriatic dated to the Mesolithic-Neolithic transition (Lightfoot et al. 2011). Coastal and island Mesolithic populations showed evidence of elevated marine resource consumption compared to sites of similar date further inland, as well as Neolithic-period sites regardless of location, but are never as enriched in $\delta^{15}N$ as comparable Western European Mesolithic populations inferred to have depended heavily on marine resources.

There is a long history of the Mesolithic characterized as a period where coastal elements are incorporated into a broad spectrum diet, based on both zooarchaeological and stable isotope evidence throughout Europe (e.g., Richards and Hedges 1999; Bailey and Spikins 2008). In Istria, the lack of marine resources prior to the Mesolithic is most likely in part due to distance of the shoreline before Terminal Pleistocene sea level rise. If marine resources are considered a low-ranked resource from a nutritional and energetic point of view, then following the expectations of optimal foraging theory the incorporation of these foods into the diet may be a result of dietary stress. However, their role in the diet can also be viewed as a response to the newfound proximity, accessibility, and abundance of marine foodstuffs. Technology and gender or age roles can make them more efficient to procure and process (Bird and Bliege Bird 2000, 2004; Jochim 1998). We feel that evidence for resource stress as a driver of the dietary shifts taking place at VSL is equivocal. This point of view has implications for viewing broad spectrum diet as a pre-adaptation to agriculture in general, as foragers adjusted food choice and seasonal mobility patterns within a changing landscape.

Intensification

At Pupićina, there are strong positive correlations between food utility and carcass unit frequency in red deer and medium sized ungulates, especially in the middle Mesolithic (Miracle 2001). There is no significant correlation between food utility and carcass unit frequency for roe deer and smaller ungulates in the LUP, but this changes for the entirety of the Mesolithic. There would seem to be convincing evidence for selective transport of higher utility elements to the site during the late Mesolithic as opposed to the LUP and early Mesolithic. Butchery data seem to suggest an increase in range as well as intensity from the LUP to the Mesolithic, perhaps correlated with dismembering of the meatier parts prior to transport to the site. There is also evidence for intense processing of phalanges and long bones during the early Mesolithic relative to the middle and end of Mesolithic (in this case, fragmentation shifts from articular ends to long bone shaft).

At Šebrn, a pattern of differential treatment emerges for medium sized ungulate carcasses (including red deer, but not boar) versus small sized ungulates (including roe deer, chamois, and ibex). The body part distribution of medium sized ungulates reflects the practice of bringing whole carcasses into the cave and then butchering and processing them on site. There is a statistically significant correlation between element representation and marrow availability, suggesting some processing of limbs for marrow. In contrast, for small sized ungulates there is a high frequency of hind limbs, which have a high meat and marrow utility. However, there is a dearth of other notably marrow-rich elements such as forelimbs and other axial skeletal parts in general, suggesting parts of small ungulates were selectively transported to the site for consumption while more intensively processing larger ungulate carcasses. In sum, there is some evidence for change in how carcasses are processed in the region at this time, but it appears to depend on site location and type of game rather than being an across-the-board transition or adaptation to dietary stress.

Seasonal Mobility

Several studies have sought to identify patterns that make clear the relationship between seasonality, land availability, resource use, and human mobility. In attempt to characterize the time of year and duration of site occupation, red deer ecology and season of antler shed was applied at the well-known British Mesolithic site of Star Carr (Clark 1954; Legge and Rowley-Conwy 1988; Milner 2003). The debate over the nature of reindeer herd management and plausibility of incipient husbandry in Paleolithic France has included antler measurements and sexing (Sturdy 1975), dental cementum thin section analysis (Gordon 1988), osteometric analysis (Weinstock 2000, 2002) and more recently, stable isotope analysis (Stevens et al. in prep.). In South Africa, stable isotope analysis of human remains and occupational evidence from shell middens has been used to look at the link between diet and mobility in the Middle Holocene (Sealy 2006; Parkington 2007).

At Pupićina, late summer and early winter use during the LUP has been suggested based on long bone fusion and tooth eruption data from wild boar, red, and roe deer remains. In the Mesolithic, there is a slight change to an autumnal focus. Miracle (2001:182) notes that the trend of diversification and increased exploitation of small game follows the modeled predictions of subsistence intensification due to "local factors of duration of occupation and/or regional changes in ecological abundance and variability." Although the sample size of elements which could be aged was limited at Šebrn (n = 191), there is evidence for spring and autumn use of the site throughout its duration of use (Miracle et al. 2000). In layer 3, which was dated to 9,860 cal BP, 26 fetal/infant specimens were identified, suggesting the springtime procurement of pregnant hinds and newborn animals. A single red deer mandible was estimated to have died during the autumn. While this evidence does not necessarily reflect multi-season use within one year, it does show the site was used in different seasons, and perhaps in different ways in these seasons.

The articulation of stable isotope and faunal data to further investigate the nature of seasonality and seasonal site use in the study region is ongoing (Pilaar Birch et al. 2016, Pilaar Birch in preparation). Discussion of a few sites further afield may be useful for building an interpretive framework for that data. Recent work by Mannino et al. (2007) on the site of Grotta dell'Uzzo in northwestern Sicily has particularly relevant results for discussion here. At a present day elevation of 65 m above sea level, it contains Mesolithic, Neolithic, and so-called "transition" layers and is comparable in temporal distribution to VSL. Season of collection (and therefore death) was determined through edge-sampling the archaeological shell of the species *M. turbinata* (now *Phorcus*). Results showed winter and spring use in the early Mesolithic (9,000–8,500 cal BC), with a shift to *year-round* use and possibly occupation at approximately 8,750 years BP (7,500–7,100 cal BC). In what is considered to be a "transition" phase at the site, between 6,300–6,000 cal BC, the isotope results implicate late autumn and winter use only. Zooarchaeological data suggests the cave was used as a base for fishing, wildfowling and collection of molluscs on a seasonal basis during this time. The successful combination of isotopic and zooarchaeological data to reconstruct the nature of site use as well as season of occupation is encouraging. Changes in intensity of site use are suggestive that there may be a larger pattern of shifting mobility during the Early Holocene.

Franchthi Cave in Greece is another site that offers an extremely useful comparison and perhaps some speculation for the results of this study. The mouth of the cave is currently located 12.5 m above present sea level. Molluscs are not present in the archaeological record until approximately 11,000 years ago, at the end of the Paleolithic, indicating a shift in dietary preferences at this time that coincides with rising sea levels (Rose 1995; Shackleton 1988; Stiner and Munro 2011). The use of mollusc species in the diet continues for the duration of 6,000 years. The Late Upper Paleolithic levels at Franchthi show a strong autumnal bias in *Patella* and *Phorcus* species. There is a transition, as at Grotta dell'Uzzo, to an even seasonal distribution indicating year-round use in the early Mesolithic (Deith and Shackleton 1988). The late Mesolithic returns to a seasonal bias, with most shells being collected in the summer and autumn. There is sporadic year-round use of the site in the Neolithic, thought to be due to a sedentary farming village archaeologically documented as appearing in the area during this time (Deith and Shackleton 1988).

Conclusion
Vela Špilja Lošinj

This study has raised some interesting questions for further research. There does appear to be a broad spectrum diet in the Mesolithic, coinciding with regional environmental and local ecological transitions. This would seem to suggest the change in diet at VSL is a response to changes in local prey availability and predictability. Evidence of intensification, which would support dietary stress as a potential factor in human foraging behavior, is unfortunately inconclusive. Apparent changes in the regional settlement system leads us to question the role of the seasonal round and site use in adaptations to environmental change. Could multi-season site use be a way of mitigating local changes in micro-habitats? What about the subtleties in decadal, yearly and seasonal regional resource scarcity?

Though estimates for the date of the current Adriatic coastline vary, making it difficult to neatly pin changing mobility to changing sea level and related environmental changes, the Mesolithic seems to live up to its reputation as a transitional period where people had to adapt to a changing landscape by modifying their dietary strategies and most likely, seasonal mobility patterns. One can imagine a seasonal round emerging in the Mesolithic, which encompassed an area of 200–500 km^2 and used Vela Špilja Lošinj as a place for obtaining lowland coastal resources and upland terrestrial ones. The exploration of new ecological niches created by the changing coastline affected not only the landscape, but also the role of the human within it.

Implications for Future Research

The Mediterranean and Adriatic basins are important regions for addressing these questions as they come to bear on cultural change and social networks in the form of the eventual adoption of sedentary lifestyles, use of domestic species of plants and animals, and increasing societal complexity (Robb and Miracle 2007). Both large and small scale environmental fluctuations have governed human movements and survival for thousands of years and would have had a significant influence on the nature and rate of spread of the Neolithic throughout Southwest Asia and Eastern Europe. Documenting hunter-gatherer adaptive strategies prior to the transition from highly mobile hunting and gathering to semi-nomadic or largely sedentary herding lifestyles is important for understanding patterns of human response to environmental change through time.

The Future from the Past

What of the relevance of this study and others like it included in this volume, geared towards understanding general patterns of human response to climate change through time? By reconstructing the fluctuations in regional faunal communities, this study directly contributes to our knowledge of past ecosystems and early peoples' reactions to climate change. Testing ecological models based on theories such as optimal foraging has value for our knowledge of present ecosystem functioning and human modification of those ecosystems. Furthermore, by establishing the greater significance of archaeological sites for understanding our common environmental and cultural heritage, we facilitate their stewardship and conservation for future generations.

Acknowledgements This paper presents results from SEPB's MPhil research at the University of Cambridge in 2009, presented at the 2010 International Council for Archaeozoology meeting in Paris, France. The authors would like to gratefully acknowledge the Gates Cambridge Trust and the Department of Archaeology and Anthropology at the University of Cambridge for research funding and support. Many thanks are also due to the volume editor, Greg Monks, and three reviewers, whose thoughtful comments and edits helped improve the final version. SEPB would also like to thank English Heritage and the ICAZ organizational committee for partially funding her participation in the session from which this volume originated.

References

Allen, J. R. M., Brandt, U., Brauer, A. Hubberten, H.-W., Huntley, B., Keller, J., et al. (1999). Rapid environmental changes in southern Europe during the last glacial period. *Nature 400*, 740–743.

Asioli, A., Trincardi, F., Lowe, J. J., Ariztegu, D., Langone, L., & Oldfield, F. (2001). Sub millennial scale climatic oscillations in the central Adriatic during the Lateglacial: Palaeoceanographic implications. *Quaternary Science Reviews, 20*, 1201–1221.

Bailey, G., & Gamble, C. (1990). The Balkans at 18,000 BP: The view from Epirus. In C. S. Gamble & O. Soffer (Eds.), *The world at 18,000 BP* (pp. 148–167). London: Unwin Hyman.

Bailey, G., & Spikins, P. (Eds.). (2008). *Mesolithic Europe*. Cambridge: Cambridge University Press.

Balbo, A. L., Andrič, M., Rubinič, J., Moscariello, A., & Miracle, P. T. (2006). Palaeoenvironmental and archaeological implications of a sediment core from Polje Čepic, Istria, Croatia. *Geologia Croatica, 59*, 107–122.

Binford, L. R. (1978). *Nunamuit ethnoarchaeology*. Ann Arbor: University of Michican Academic Press.

Bird, D. W., & Bliege Bird, R. (2000). The ethnoarchaeology of juvenile foragers: Shellfishing strategies among Meriam children. *Journal of Anthropological Archaeology, 19*, 461–476.

Boschian, G., & Fusco, F. (2007). Figuring out no-one's land: Why was the karst deserted in the Late Glacial? In R. Whallon (Ed.), *Late Paleolithic environments and cultural relations around the Adriatic* (pp. 15–16). BAR International Series 1716.

Burroughs, W. J. (2005). *Climate change in prehistory: The end of the reign of chaos*. Cambridge: Cambridge University Press.

Burroughs, W. J. (2007). *Climate change: A multidisciplinary approach*. Cambridge: Cambridge University Press.

Clark, J. G. (1954). *Excavations at Star Carr: An early Mesolithic site at Seamer near Scarborough, Yorkshire*. Cambridge: Cambridge University Press.

Čečuk, B. & Radić, D. (2005). *Vela Spila*. Vela Luka: Centar za kulturu 'Vela Luka'.

Davis, B. A. S., Brewer, S., Stevenson, A. C., & Guoit, J. (2003). The temperature of Europe during the Holocene reconstructed from pollen data. *Quaternary Science Reviews, 22*, 1701–1716.

Deith, M. R., & Shackleton, N. J. (1988). *Marine molluscan remains from Franchthi Cave*. Bloomington: Indiana University Press.

Forenbaher, S. (2008). Archaeological record of the Adriatic offshore islands as an indicator of long-distance interaction in prehistory. *European Journal of Archaeology, 11*, 223–244.

Forenbaher, S., Kaiser, T., & Miracle, P. T. (2013). Dating the east Adriatic Neolithic. *European Journal of Archaeology, 16*, 589–609.

Gallizia Vuerich, L., & Princivalle, F. (1994). Studio mineralogico e palinologico su alcuni sediment della grotto dell'Edera (Carso Triestino): Un tentative di ricostruzione paleoclimatica. *Il Quaternario- Italian Journal of Quaternary Sciences, 7*, 569–576.

Gibbard, P. L. (2010). The newly-ratified definition of the Quaternary System/Period and redefinition of the Pleistocene Series/Epoch, and comparison of proposals advanced prior to formal ratification. *Episodes, 33*, 152–158.

Gordon, B. C. (1988). Of men and reindeer herds in French Magdalenian prehistory. *BAR International Series, 390*, 197–199.

Hughes, P. D., Woodward, J. C., & Gibbard, P. L. (2006). Late Pleistocene glaciers and climate in the Mediterranean. *Global and Planetary Change, 50*, 83–98.

Huntley, B., & Prentice, I. C. (1993). Holocene vegetation and climates of Europe. In H. E. Wright, J. E. Kutzbach, T. Webb, W. F. Ruddiman, F. A. Street-Perrott, & P. J. Bartlein (Eds.), *Global climates since the Last Glacial Maximum* (pp. 136–168). Minneapolis: University of Minnesota Press.

Jochim, M. A. (1998). *A hunter-gatherer landscape: Southwest Germany in the Late Paleolithic and Mesolithic*. New York: Plenum.

Kelly, R. L. (1995). *The foraging spectrum: Diversity in hunter-gatherer lifeways*. Washington: Smithsonian Institution Press.

Komšo, D. (2006). The Mesolithic in Croatia. *Opuscula Archaeologica, 30*, 55–92.

Komšo, D., & Pellegatti, P. (2007). The late Epigravettian in Istria: Late Paleolithic colonization and lithic technology in the northern Adriatic area. In R. Whallon (Ed.), *Late Paleolithic environments and cultural relations around the Adriatic* (pp. 27–39). BAR International Series 1716.

Lightfoot, E., Boneval, B., Miracle, P. T., Šlaus, M., & O'Connell, T. C. (2011). Exploring the Mesolithic and Neolithic transition in Croatia through isotopic investigations. *Antiquity, 85*, 73–86.

Lam, Y. M., & Pearson, O. M. (2005). Bone density studies and the interpretation of the faunal record. *Evolutionary Anthropology, 14*, 99–108.

Lambeck, K., Antonioli, F., Purcell, A., & Silenzi, S. (2004). Sea-level change along the Italian coast for the past 10,000 yr. *Quaternary Science Reviews, 23*, 1567–1598.

Legge, A., & Rowley-Conwy, P. (1988). *Star Carr revisited: A re-analysis of the large mammals*. London: Centre for Extra-Mural Studies.

Lyman, R. L. (1994). *Vertebrate taphonomy*. Cambridge: Cambridge University Press.

Lyman, R. L. (2008). *Quantitative paleozoology*. Cambridge: Cambridge University Press.

Macarthur, R. H., & Pianka, E. (1966). On optimal use of a patchy environment. *The American Naturalist, 100*, 603–609.

Mannino, M. A., Thomas, K. D., Leng, M. J., Piperno, M., Tusa, S., & Tagliacozzo, A. (2007). Marine resources in the Mesolithic and Neolithic at the Grotta dell'Uzzo (Sicily): Evidence from isotope analyses of marine shells. *Archaeometry, 49*, 117–133.

Milner, N. (2003). Pitfalls and problems in analysing and interpreting faunal remains with regards to seasonality. *Archaeological Review from Cambridge, 16*, 50–65.

Miracle, P. (1995). Broad spectrum adaptations re-examined: Hunter-gatherer responses to late-glacial environmental changes in the eastern Adriatic. Ph.D. Dissertation, University of Michigan.

Miracle, P. T. (1997). Early Holocene foragers in the karst of northern Istria. *Porocilo o raziskovanju paleolitika, neolitika in eneolitika v Sloveniji, 24*, 43–61.

Miracle, P. (2001). Feast or famine? Epi-Paleolithic subsistence in the northern Adriatic basin. *Documenta Praehistorica, 26*, 177–197.

Miracle, P. T. (2002). Mesolithic meals from Mesolithic middens. In P. T. Miracle & N. Milner (Eds.), *Consuming passions and patterns of consumption* (pp. 65–88). Cambridge: McDonald Institute for Archaeological Research.

Miracle, P. T. (2006). Neolithic shepherds and their herds in the northern Adriatic Basin. In D. Sarjeantson & D. Field (Eds.), *Animals in the Neolithic of Britain and Europe* (pp. 63–94). Oxford: Oxbow Books.

Miracle, P. T. (2007). The late glacial 'Great Adriatic Plain': 'Garden of Eden' or 'No Man's Land' during the Epipaleolithic? A view from Istria (Croatia). In R. Whallon (Ed.), *Late paleolithic environments and cultural relations around the Adriatic* (pp. 41–51). BAR International Series 1716.

Miracle, P. T., & Forenbaher, S. (2000). Pupićina Cave Project: Brief summary of the 1998 season. *Histria Archaeologica, 29*, 27–48.

Miracle, P. T. & Forenbaher, S. (2006). *Prehistoric herders of northern Istria. The archaeology of Pupićina Cave* (Vol. 1). Pula: Archaeological Museum of Istria.

Miracle, P., Galanidou, N., & Forenbaher, S. (2000). Pioneers in the hills: Early Mesolithic foragers at Šebrn Abri (Istria, Croatia). *European Journal of Archaeology, 3*, 293–329.

Morin, E. (2007). Fat composition and Nunamiut decision-making: A new look at the marrow and bone grease indices. *Journal of Archaeological Science, 34*, 69–82.

Mussi, M. (2001). *Earliest Italy: An overview of the Italian Paleolithic and Mesolithic*. New York: Springer.

Outram, A. K. (2001). A new approach to identifying bone marrow and grease exploitation: Why the "indeterminate" fragments should not be ignored. *Journal of Archaeological Science, 28*, 401–410.

Paine, C., O'Connell, T., & Miracle, P. T. (2009). Stable isotopic reconstruction of diet at Pupićina Cave. In S. McCartan, R. Schulting, G. Warren, & P. Woodman (Eds.), *Mesolithic horizons* (pp. 210–216). Papers presented at the 7th International Conference on the Mesolithic in Europe, Belfast 2005.

Parkington, J. (2007). On diet and settlement in Holocene South Africa. *Current Anthropology, 48*, 581–582.

Pilaar Birch, S. E. (2012). *Human adaptations to climate change and sea level rise at the Pleistocene-Holocene transition in the northeastern Adriatic*. Ph.D. Dissertation, University of Cambridge.

Pilaar Birch, S. E., & Miracle, P. T. (2015). Subsistence continuity, change, and environmental adaptation at the site of Nugljanska, Istria, Croatia. *Environmental Archaeology, 20*(1), 30–40. doi:10.1179/1749631414Y.0000000051.

Pilaar Birch, S. E., Miracle, P. T., Stevens, R. E., & O'Connell, T. C. (2016). Reconstructing late Pleistocene/early Holocene migratory behavior of ungulates using stable isotopes and its effects on forager mobility. *PLOS ONE, 11*(6): e0155714. doi:10.1371/journal.pone.0155714.

Radić, D., Lugović, B., & Marjanac, L. (2008). Neapolitan Yellow Tuff (NYT) from the Pleistocene sediments in Vela Spila on the island of Korčula: A valuable chronostratigraphic marker of the transition from the Paleolithic to the Mesolithic. *Opuscula Archaeologica, 31*, 7–26.

Richards, M. P., & Hedges, R. E. M. (1999). A Neolithic revolution? New evidence of diet in the British Neolithic. *Antiquity, 73*, 891–897.

Robb, J., & Miracle, P. (2007). Beyond migration versus acculturation: New models for the spread of agriculture. *Proceedings of the British Academy, 144*, 99–115.

Robinson, S. (2006). *Using carbon and nitrogen stable isotopic analysis to reconstruct the food web of the peninsula of Istria, northern Croatia*. MPhil. Thesis, University of Cambridge.

Rose, M. (1995). Fishing at Franchthi Cave, Greece: Changing environments and patterns of exploitation. *Old World Archaeology Newsletter, 18*, 21–26.

Rossignol-Strick, M., Planchais, N., Paterne, M., & Duzer, D. (1992). Vegetation dynamics and climate during the deglaciation in the south Adriatic basin from a marine record. *Quaternary Science Reviews, 11*, 414–423.

Rossignol-Strick, M. (1999). The Holocene climatic optimum and pollen records of sapropel 1 in the eastern Mediterranean, 9000–6000 BP. *Quaternary Science Reviews, 18*(515), 530.

Schmidt, R., Müller, J., Drescher-Schneider, R., Krisai, R., Szeroczynska, K., & Barić, A. (2000). Changes in lake level and trophy at Lake Vrana, a large karstic lake on the Island of Cres (Croatia), with respect to palaeoclimate and anthropogenic impacts during the last approx. 16,000 years. *Journal of Limnology, 59*, 113–130.

Sealy, J. (2006). Diet, mobility, and settlement pattern among Holocene hunter-gatherers in southernmost Africa. *Current Anthropology, 47*, 569–595.

Shackleton, J. C. (1988). *Marine molluscan remains from Franchthi Cave. Excavations at Franchthi Cave, Greece, fascicle 4*. Bloomington: Indiana University Press.

Smith, E. A., & Winterhalder, B. (1981). New perspectives on hunter-gatherer socio-ecology. In B. Winterhalder & E. A. Smith (Eds.), *Hunter-gatherer foraging strategies: Ethnographic and archaeological analyses* (pp. 1–12). University of Chicago Press.

Smith, E. A., & Winterhalder, B. (1992). *Evolutionary ecology and human behavior*. New York: Aldine de Gruyter.

Spry-Marqués, V. P. (2012). *The Adriatic Plain: A glacial refugium? Epigravettian subsistence strategies at Vela Spila (Croatia)*. Ph.D. Dissertation, University of Cambridge.

Stiner, M. C., & Munro, N. D. (2011). On the evolution of diet and landscape during the Upper Palaeolithic through Mesolithic at Franchthi Cave (Peloponnese, Greece). *Journal of Human Evolution, 60*, 618–636.

Stiner, M., Munro, N., & Surovell, T. (2000). The tortoise and the hare. Small game use, the broad-spectrum revolution, and Paleolithic demography. *Current Anthropology, 41*, 39–73.

Sturdy, D. A. (1975). Some reindeer economies in prehistoric Europe. In E. S. Higgs (Ed.), *Palaeoeconomy* (pp. 55–95). Cambridge: Cambridge University Press.

Surić, M., Juračić, M. & Horvatinčić, N. (2002). Late Pleistocene—Holocene changes of the eastern Adriatic coast line. *Littoral 2002, The Changing Coast*, 259–263.

Surić, M., Juracic, M., Horvatincic, N., & Krajcarbronic, I. (2005). Late Pleistocene-Holocene sea-level rise and the pattern of coastal karst inundation: Records from submerged speleothems along the Eastern Adriatic Coast (Croatia). *Marine Geology, 214*, 163–175.

van Andel, T. H. (1989). Late Quaternary sea-level changes and archaeology. *Antiquity, 63*(733), 745.

Weinstock, J. (2000). Late Pleistocene reindeer populations in middle and western Europe: An osteometrical study of *Rangifer tarandus*. *BioArchaeologica 3*.

Weinstock, J. (2002). Reindeer hunting in the Upper Palaeolithic: Sex ratios as a reflection of different procurement strategies. *Journal of Archaeological Science, 29*, 365–377.

Willis, K. J. (1994). The vegetational history of the Balkans. *Quaternary Science Reviews, 13*, 769–788.

Winterhalder, B., & Kennett, D. J. (2006). Behavioral ecology and the transition from hunting and gathering to agriculture. In B. Winterhalder & D. J. Kennett (Eds.), *Behavioral ecology and the transition to agriculture* (pp. 1–21). Berkeley: University of California Press.

Part II
The Early – Mid-Holocene

Chapter 6
Early to Middle Holocene Climatic Change and the Use of Animal Resources by Highland Hunter-Gatherers of the South-Central Andes

Hugo D. Yacobaccio, Marcelo Morales, and Celeste Samec

Abstract The goal of this paper is to study the relationship between the use of animal resources by hunter-gatherers and the environmental modification that occurred with the onset of Middle Holocene aridity in the southern Altiplano of the Andes. This objective will be pursued taking into account both regional and local spatial scales. In the regional analysis, our purpose is to study relevant changes in animal species' patterns represented in the archaeological record from both sides of the Andean range. For these reasons some theoretical issues regarding habitat fragmentation and loss are addressed. In order to reach our objective in the local scale we discuss isotopic values ($\delta^{13}C$ and $\delta^{15}N$) obtained from archaeofaunas recovered in a rock-shelter, located in the Argentinean slope which presents Early and Middle Holocene human occupations.

Keywords Stable isotopes • Environmental change • Puna de Atacama

Introduction

A paradox regarding the relationship between environmental change and the use of animal resources by hunter-gatherers has become commonplace in archaeological and anthropological literature. On the one hand, environmental carrying capacity and its evolution through time has emerged as a key feature in most theoretical models of behavioral ecology and other ecological perspectives in hunter-gatherers studies. On the other hand, the relationship between environmental changes and the actual impact on animals and other resources important for human subsistence remain almost unexplored in empirical terms, with a few exceptions (Grayson 2000; Broughton et al. 2008; Wolverton 2008).

With this in mind, the goal of this paper is to study the relationship between the use of animal resources by hunter-gatherers and the environmental modification that occurred with the onset of Middle Holocene aridity in the southern Altiplano of the Andes. This objective will be pursued taking into account both regional and local spatial scales. In the regional analysis, our purpose is to study relevant changes in animal species' patterns represented in the archaeological record from both sides of the Andean range. The local focus is relevant for discussions of the resilience and internal flexibility of certain habitats containing key resources that supported multiple human occupations throughout the Early to Middle Holocene time-span. For these reasons some theoretical issues regarding habitat fragmentation and loss are addressed. In order to reach our objective in the local scale we discuss isotopic values ($\delta^{13}C$ and $\delta^{15}N$) obtained from archaeofaunas recovered in a rock-shelter, located in the Argentinean slope, containing Early and Middle Holocene human occupations. It is worth noting that there are not many localities that have occupations in both periods, and, for this reason, the study of both environmental conditions and resource use patterns in these localities shows how hunter-gatherer behaviors changed with the onset of arid conditions.

Environment is regarded as the main limiting factor in hunter-gatherers adaptations (Kelly 1995; Binford 2001). Furthermore, high mountain environments have been considered as a set of constraints within which human populations must operate (Aldenderfer 1998). The high operational costs in these conditions must be minimized by hunter-gatherers developing a series of behavioral responses which involved a certain degree of residential mobility, an emphasis on logistical mobility, camps placed in "optimal" locations, and the development of balanced reciprocity, among other features (Aldenderfer 1998). The environment is not a fixed entity; rather, it has internal flexibility. As dynamism is a major feature of the environment, humans do not only culturally respond to habitat modifications, but also transform

H.D. Yacobaccio (✉) · M. Morales · C. Samec
CONICET–Instituto de Arqueología, Universidad de Buenos Aires, Buenos Aires, Argentina
e-mail: hdyacobaccio@gmail.com

their environment by altering landscape for several purposes, such as food acquisition. As Schutkowski (2006:26) states:

Humans do not simply adapt to the conditions and constraints of nature, but actively change the ecological framework through active management and strategies.

This statement highlights the interdependence of environmental conditions and human actions. Thus the study of the history of resource use by hunter-gatherers of the Altiplano during the mid-Holocene is the main aim of this chapter.

This chapter is based on a detailed description of modern-day environmental characteristics and animal resources available to humans, an exhaustive paleoenvironmental data analysis and synthesis, and the use of concepts of the model of fragmented landscapes (Thompson Hobbs et al. 2008) as a framework for explaining the patterns of archaeological faunal record of the Altiplano.

The Altiplano or Puna Region

The Puna region of Argentina and Chile comprises the arid highlands placed between 19° and 27°S latitude and between 3,000 and 4,500 m above sea level (m asl) (Fig. 6.1). This area is defined as a Highland Desert Biome dissected by several mountain chains oriented NE-SW. It is characterized by high solar radiation due to its altitude, wide daily thermal amplitude, marked seasonality in rainfall, and low atmospheric pressure. Primary productivity is mainly concentrated

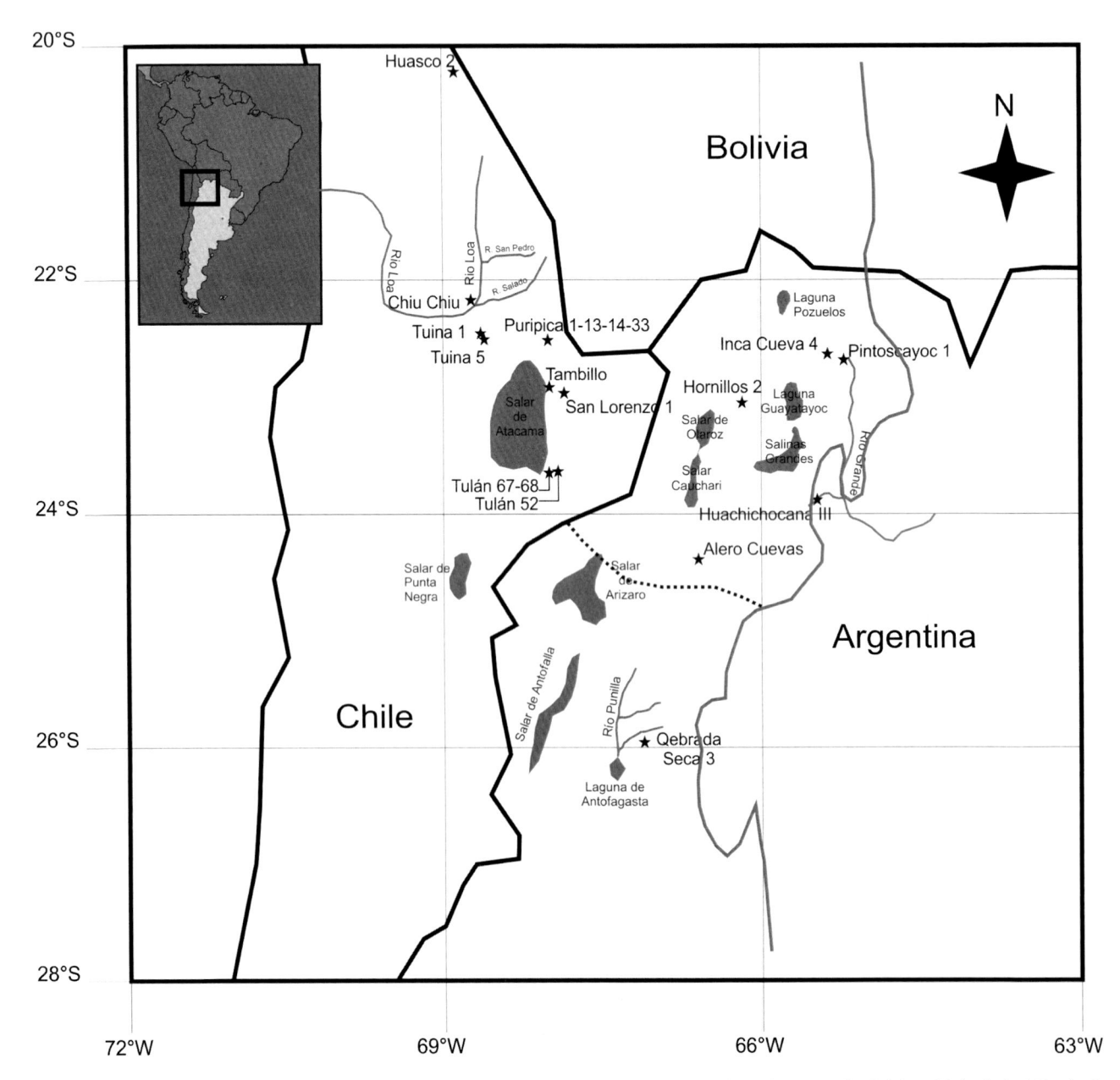

Fig. 6.1 Map with the location of the archaeological sites mentioned in this chapter, and main drainages and salt lakes. Light dashed line is the limit between Dry and Salt Puna, and dark dashed line show the Puna limit at 3000 m asl

on stable hydrological systems like primary basins, high gorges, and wetlands. Few permanent freshwater basins, salt lakes, pans and playas constitute the general hydrological net. A few rivers and several springs scattered in the landscape are the only sources of freshwater, which is a critical resource for human populations. The summertime rainfall in the region is largely ruled by the so-called South American Monsoon-like System (Zhou and Lau 1998). This system produces about 80% of annual precipitation occurring in the Andean highlands between December and February (Vuille and Keiming 2004). In turn, these conditions set a heterogeneous distribution of vegetal and animal resources. Some localized areas, defined as Nutrients Concentration Zones (NCZ, see Yacobaccio 1994), contain most of the available regional biomass. Dominant climatic conditions are somewhat different at both sides of the Andean range.

Within this region, the Chilean Altiplano placed above 4,000 m asl is characterized by the presence of scattered large salars (salt basins) and a few small lakes (Laguna Lejía and Miscanti) (Latorre et al. 2003). Many authors have attributed the dryness of the Chilean highlands to the rain shadow effect of the Andes which block the moisture coming to the interior of the continent (Houston and Hartley 2003), emphasizing the key role of this mountain range in shaping the rainfall distribution over the continent, especially during austral summer (Lenters and Cook 1995; Garreaud 2009). The Atacama Desert (3,000–3,200 m asl) also receives nearly all of the moisture from the east (except for Pacific fogs and occasional winter rains in all coastal areas below 1,000 m) although it is extremely arid because it lies beyond the normal reach of the austral summer precipitation and also lies north of the winter cyclonic precipitation that usually occurs south of 30°S (Messerli et al. 1993). The potential evapotranspiration in the Chilean north exceeds precipitation, at least by a factor of five, even in the southern III Region (27°S) located to the south of the study area, where evapotranspiration can reach 1,000 mm/yr (Earle et al. 2003). Because of these extreme conditions, the wetlands in the Atacama Desert region are only found in deeply incised canyons and near the base of the Andes where ground-water flow paths intersect basin aquifers (Rech et al. 2003). The Puna of Chile is divided in three areas: (1) the moister, located near Arica, (2) the Loa Basin that shares the general characteristics of the Argentinean Dry Puna, and finally (3) the extremely dry southern sector near lake Miscanti that corresponds to the highlands of the Salar de Atacama (above 4,000 m asl).

On the other hand, the Puna of Argentina exhibits a latitudinal gradient in aridity that determines two sub-regions: (1) north of 24°S, the Dry Puna, with a mean annual precipitation of 340 mm/yr (reaching even 400–500 mm/yr in some localities of the north-western corner), and (2) south of 24°S, the Salt Puna in which precipitation barely reaches the average of 100 mm/yr, and salt-lakes and saline soils are dominant features of the landscape (Morales 2011).

These overall conditions set a patchy distribution of vegetal and animal resources. The wide altitudinal range produces variations in plant assemblages from 'tolar' (shrub steppe) to 'pajonal' communities (herbaceous grasslands), with patched 'vegas' (wetlands) placed in both of these main vegetal communities. Four main plant communities can be identified in both sectors of the the Puna, but with higher primary productivity in the Dry Puna (Cabrera 1976; Arzamendia et al. 2006; Borgnia et al. 2006):

1. Shrub steppe (*tolar*) dominated by *Parastrephia lepidophylla* and *Fabiana densa* with a low proportion of herbs (5%) is the most extensive pasture area of the zone and is located between 3,500 and 3,900 m asl.
2. Herbaceous steppe (*pajonal*) dominated by *Festuca* spp. and other grasses, like *Poa* spp. and *Stipa* spp., can be found between 4,100 and 4,700 m asl. It is important to highlight that above 4,300 m asl, shrub presence diminishes almost completely.
3. Wetlands (*vegas*) represent restricted swamp areas composed of short grasses. These dense grasslands of *Deyeuxia* spp. and *Mulembergia* spp. are distributed in both altitudinal settings from 3,500 to 4,700 m asl (Ruthsatz and Movia 1975).
4. Mixed steppes of Gramineae and Compositae (only eight genera are present) can be found between 3,900 and 4,100 m asl.

The quality of forage in these vegetal communities deserves further consideration because it is directly related to the distribution of ungulates, especially wild camelids. The foraging potential varies between different plant communities, and, furthermore, the vegas have distinct properties at diverse altitudes. Recently, Benítez (2005) studied pasture quality in the southern sector of the Salt Puna, analyzing the nutritional components of plants eaten by wild and domesticated camelids. This study included the percentage of proteins, neutral and acid fibres, and lignin of the foraged species. Her results showed that the majority of the shrubs have high yields of proteins, greater than 5%, but only three grasses reach this value (*Deyeuxia brevifolia*, *Stipa frigida*, and *Distichlis spicola*). Thus, it is worth noting that the percentage of proteins is greater in the shrub steppe than in the wetlands. On the other hand, grasses have higher values of fibres, also important in diet because they improve digestibility. Consequently, proteins and fibers, which are key factors in camelid diet, are most consistently available in the shrub steppe.

Holocene Environments (10,000–3,500 BP)

In the past 30 years, a broad set of palaeoenvironmental studies with diverse spatial and temporal scopes have been carried out in the South Central Andes. Several areas like northern Chile or southern Peru have been particularly studied, while the work realized in others, like northwestern Argentina, have been increasing in recent years.

The information obtained from many paleoenvironmental records has shown a complex and heterogeneous picture regarding the different climatic effects on the environment at different spatial scales (i.e., regional, sub-regional, local). This complex picture led to a heated debate about the nature of the regional climate during the Middle Holocene in the Puna de Atacama around a few proxies (Grosjean 2001; Grosjean et al. 2003; Latorre 2002, 2003, 2006). For this reason we present a general multi-scale synthesis of the different environmental scenarios during the 10,000–3,500 BP time-span and the climate modifications and changes behind them. We do not present a detailed review and analysis of each paleoenvironmental record; instead, we summarize the main tendencies and patterns detected in ~2,000 year time blocks.

The climate during most of the Early Holocene (10,000–8,000 BP) was stable and regionally homogeneous (Thompson et al. 1995, 1998, 2000; Bradbury et al. 2001; Ramírez et al. 2003). Atmospheric circulation patterns were less influenced by the South Atlantic Anticyclone (Villagrán 1993) and the virtual absence or reduced intensity and frequency of short term climatic oscillations like cold and wet conditions resulting from ENSO (Villagrán 1993; Bradley 2000) or PDO (Pacific Decadal Oscilation, Mantua et al. 1997). This long-term stability is also evidenced by the presence of typical herbaceous steppe vegetation in lower altitudes than today (below 3,800 m asl), which replaced shrub steppe as noted in many pollen records. In the same way, a regional positive hydrological balance could be inferred from the higher levels of the most important lake systems in the region (Geyh et al. 1999; Bradbury et al. 2001; Abbott et al. 2003). Finally, the increase in the

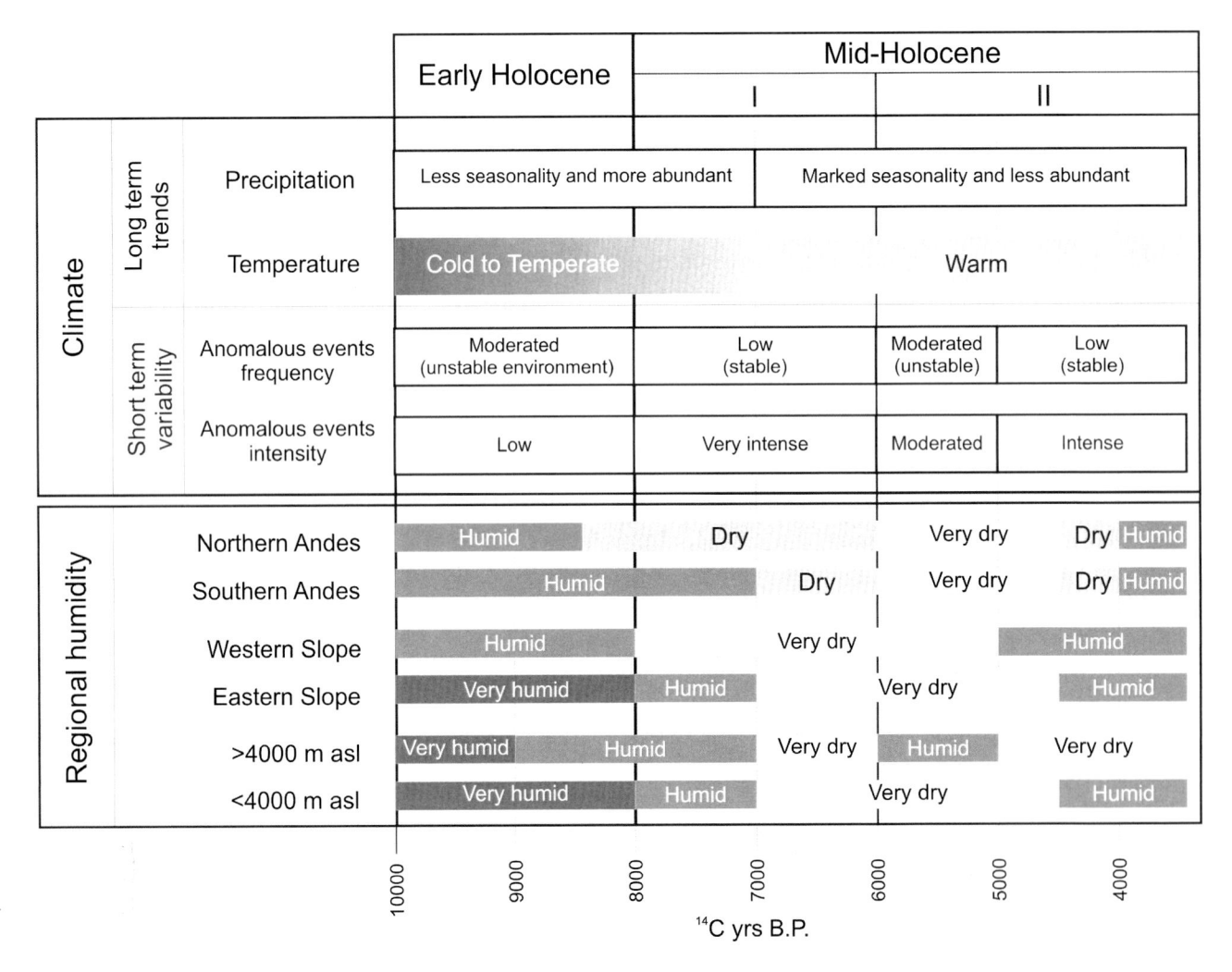

Fig. 6.2 Regional Climatic Trends for the Early and Middle Holocene

regional abundance of pack rat middens (Latorre 2002, 2003, 2006) and paleosols (Morales 2011) also points to more stable and moister conditions during the 10,000–8,000 BP time-span (Fig. 6.2).

The 8,000–6,000 BP time-span represents a transitional environment between the previous moister and colder climate, and a more arid and warmer one, having less stable conditions than the previous period. Modifications in the atmospheric circulation system due to the intensification of South Atlantic Anticyclone (Villagrán 1993), and other changes, like the orbital cycle which modified the insolation levels in the area (Rowe et al. 2002), are responsible for this transformation. The end of the moister Early Holocene conditions has non-synchronous dates. It ends around 8,000 BP, or even a few centuries later, in Northern Chile and NW Argentina, possibly coinciding with the global 8.2K event recorded in the Huascaran Ice Cores by Thompson et al. (1995). This later chronology is noticed in the water bodies with broad moisture catchment areas or located at higher altitudes, above 4,000 m asl (Morales 2011). The higher groundwater levels developed vast wetlands in both slopes of the Andes until at least 7,000 BP, as shown by the deposition of alluvial sediment in the Bolivian altiplano (Servant and Servant-Vildary 2003), the peat and palustrine deposits of Quebrada Puripica (Grosjean 2001) and Quebrada Lapao (Yacobaccio and Morales 2005), and the delay in the transition to dryer conditions in wetlands sediments at Pastos Chicos and Laguna Colorada in NW Argentina (Tchilinguirian 2009; Morales 2011). Moreover, in particular locations moister conditions remained for the most part of the Middle Holocene. For example, water systems with their catchment placed over 5,000 m (Abbott et al. 2003) and in some wetlands located over 4,000 m in NW Argentina, such as Alto Tocomar, remained as late as 5,500 BP (Morales 2011).

The second part of the Middle Holocene, between 6,000 and 3,500 BP, could be characterized as a moment of extreme regional aridity with more humid conditions in the western and lower parts (below ~3,000 m) of the Andes. This seems to be particularly evident in the vegetal composition of the pack rat middens in Chile; in a highstand event in the Salar de Atacama (2,300 m) (Bobst et al. 2001), and in high levels of the Lago Aricota (2,800 m) in Peru. Meanwhile, places located over 3,000 m evidenced more or less synchronic interruption in moisture input at about 6,200 BP and a strong desiccation event near 5,000 BP. This pattern is clearly compatible with the onset of the first El Niño event, recorded around 7,000–6,000 BP (Villagrán 1993; Veit 1996; Riedinger et al. 2002), both in the increase of moisture in the lowlands and also in the extreme aridity of the highlands and eastern slope of the Andes.

Fragmented Landscape

As evidenced by palaeoenvironmental data, the Middle Holocene is characterized by a fragmented environment in comparison with the more stable and regular Early Holocene. The long-term environmental changes that occurred starting at the end of the latter period increased the spatial heterogeneity of the environment, generating a series of behavioral and/or strategic modifications in human groups, in order to average out sudden local shortages in resources (Dincauze 2000). As evidenced in the archaeological record, the Puna hunter – gatherers modified their patterns of mobility by moving and concentrating where crucial resources were available, amplifying their range, introducing technological innovations (i.e., mortars and pestles, more diversified types of projectile points associated with diverse hunting techniques), and, in the later part of the period, changing economic strategies (camelid husbandry, and introduction of cultivars from the lowlands). These changes reveal the complex relationship between people and environment during the long Middle Holocene period.

The transition between Early and Middle Holocene could be characterized as the step from a stable and regular environment to a heterogeneous one, and we explore here the characteristics of this environmental breakup. Fragmentation can be produced by either human action or natural phenomena and can be defined as the process of dissection of natural systems into isolated parts. Fragmentation effects on natural systems could be amplified by habitat loss, which occurs when the area of a habitat suitable for a species or community is diminished (Thompson Hobbs et al. 2008) and the number of biotic interactions is reduced. Fragmentation also generates spatial heterogeneity and therefore high spatial dependence, which means that the number of resources per unit area substantially varies with spatial location (Thompson Hobbs et al. 2008). Also, in arid and semi-arid environments, the ecological processes are shaped by temporal variation in primary production and in the availability of surface water (see Thomas 1997). These environmental pulses of resource availability are interrupted by periods of resource scarcity in decadal, annual or seasonal scales. In these cases, when heterogeneity is high (that is, the number of resources strongly varies in space and time), consumers, both wildlife and human hunter-gatherers, must overcome the possibility of insufficient resources for sustaining population in certain habitat areas or patches. For this reason, patch connectivity exerts significant effects on population dynamics, and mobility operates as the mechanism which offers connectivity between sectors of a fragmented landscape. The degree of connectivity depends not only on the scale of mobility but also in landscape configuration.

Some analyses suggest that climatic seasonality played a fundamental role in driving long-term late Quaternary variation in artiodactyl densities, causing a depression in them (Broughton et al. 2008). A declining index of ungulate abundance may result from decreasing hunting success of larger prey which results in a lower overall return rate (Codding et al. 2010). In the South Andean Puna, on the contrary, the zooarchaeological record shows an increase in artiodactyl index from sites of the Middle Holocene caused by an increase in the amount of camelids. In this case, an alternative scenario can be suggested: climatic change did not generate local extinction and/or population diminution of wild camelids, as these species are well adapted to desert environments. Perhaps Middle Holocene climate change only caused displacements of camelids populations to well watered areas with higher pasture availability. A similar scenario seems to have occurred in southern Patagonia (Rindel et al. 2017).

Highland Animal Resources

The Puna is an environment with low ungulate diversity. Only three species are found: two wild camelids (guanaco and vicuña) and taruca deer (*Hippocamelus antisensis*). Guanacos (*Lama guanicoe*) are spread all over the Andean range from Peru to Tierra del Fuego from 8°S to 55°S, inhabiting a whole variety of open habitats (arid, semi-arid, hilly, mountain, steppe), and temperate forest environments (subpolar *Nothofagus* forest in Patagonia). Although guanaco populations are important in Patagonia, in other areas like Peru, the northernmost mountain regions of Argentina and Chile, and the Bolivian and Paraguayan Chaco, it represents an endangered species (González et al. 2006; Censo 2010). The guanaco social structure in the breeding season comprises three basic social units: territorial family groups; male groups (non-territorial), and solitary males (Franklin 1982). Family groups' territoriality is directly correlated with stable food supply. When a severe drop in food availability occurs, usually in winter, guanaco populations displace, loosing territoriality and forming mixed herds breaking apart family groups (Cajal 1985). On the other hand, out of the breeding season, group composition varies according to environmental conditions. Regarding forage selectivity, González et al. (2006) classified guanacos as an intermediate herbivore or opportunistic (mixed) feeder, foraging on a highly diverse range of food sources, although this feeding behavior can exhibit some variation in correspondence to habitat differences (Puig et al. 2008). According to the archaeological record, guanacos inhabited the Puna in montane grasslands and the meso- and also micro-thermic valleys of lower altitude in the past. Today this species distribution is restricted to small populations (only hundreds of individuals) in high altitude environments (Censo 2010).

The vicuña (*Vicugna vicugna*) lives only in high altitude Puna environments above 3,400 m, from Peru to Argentina and Chile, between 9° and 29°S. This wild camelid species almost became extinct in the middle twentieth century; today the population has recovered in most parts of the Andes, but they require strong conservation policies in order to overcome the pressure for exploiting their valuable fiber (Vilá 2000). Vicuña are adapted to open grasslands and steppes, but they concentrate in especially high numbers in wetlands. The species can be classified as a grazer. It feeds preferably on herbs and grasses, exhibiting great efficiency for processing C_4 plants, although in steppe habitats they feed on variable proportions of shrubs (Borgnia et al. 2006; Arzamendia 2008). The vicuña lives in family groups consisting of one male, three to four females and two offspring. These groups are stable and territorial all year round. The mating system has mixed components of polygynous, resource defense, and harem (females + calves) defense because the alpha male limits and defends an area, but he also conducts the females to the territory when they move far away (Vilá 1999). However, there is an overlap of territories between family groups and tolerance to neighbors when the range of one harem is used by another when the "owners" are absent (Arzamendia 2008). On the other hand, bachelor groups can vary in number and location several times a day (Vilá 1991). Arzamendia and Vilá (2006) have found a differential habitat use: family groups are associated with major vegetation cover, found mainly on piedmont and flat areas, whereas bachelor groups are more related to more open vegetation cover, found mainly on mountain slopes. Contrary to guanacos, the vicuñas are obligate drinkers, thus they must drink water every day; therefore location of water is a limiting factor concerning the distribution of vicuña populations (Vilá 2000).

There are only a few observations about vicuña and guanaco sympatric behavior. In the San Guillermo Reserve, Cajal (1979) recorded a partial overlapping between guanaco and vicuña populations, the former having a wider distribution, more mobility, and less defined territoriality. It is important to mention that the mean size for guanaco territories is 260 hectares and 138 hectares for vicuñas and that there is also some segregation by altitude, guanacos being found in lower habitats than vicuñas.

The taruca deer (*Hippocamelus antisensis*), the third ungulate of the region, is distributed from the Ecuadorian Andes to Argentina and Chile, although today the species is extinct in the Puna areas of these two countries. Small populations remain in the highland grasslands (1,800 to 5,500 m) of the Mountain Forest (Yungas) and also in the high mountain ranges in the southern section of

Table 6.1 Archaeological faunal record from the Puna of Argentina

Argentine Puna		Sites	Artiodactyla	Camelidae	Cervidae	Chinchillidae	*Lagidium sp*	*Chinchilla sp*	Caviidae	*Ctenomys sp*	*Cavia sp*	Rodentia indet	Avis	Carnivora	Dasipodidae	Total	N Artiodactyla	N Small Mammals and Avis	Total	% Artiodactyla	% Small mammals and Avis	% Non fused Artiodactyla
Early Holocene		Pintoscayoc	998	–	25	39	234	70	1382	602	–	81	64	5	–	3500	1023	2472	3495	41.4	58.6	32.9
		CHIIIE3	5	751	2	98	–	–	–	–	–	–	–	3	–	859	758	98	856	88.5	11.5	67
		Inca Cueva 4	127	107	15	741	–	–	–	4	–	1	8	1	10	1014	249	747	996	25	75	50
		QS3 LL	–	687	–	74	–	–	–	1	–	38	16	1	–	817	687	129	816	84.2	15.8	41
		Hornillos 2	480	113	18	1314	–	–	–	–	–	53	–	–	–	1978	611	1314	1925	31.7	68.3	–
		Acuevas F4	532	575	1	55	–	–	–	–	–	50	–	–	–	1213	–	–	1318	91.3	8.6	41
Mid-Holocene	I (8,000–6,000 BP)	Hornillos 2	890	271	8	603	–	–	–	–	–	110	–	–	713	2595	1169	603	1772	66	34	–
		QS3 ML	–	537	–	44	–	–	–	2	–	5	–	–	51	639	537	2	539	99.6	0.4	19
		Pintoscayoc 5	–	62	2	24	–	–	12	49	–	24	1	–	–	174	64	86	150	42.6	57.4	30.6
	II (6,000–3,500 BP)	Acuevas F3	128	243	–	14	–	–	–	–	–	39	1	1	–	426	371	15	386	96.1	3.8	38
		Acuevas F2	371	327	–	18	–	–	–	–	–	41	–	–	–	757	698	18	716	97.5	2.5	60.6
		QS3 UL	–	1708	–	63	–	–	–	15	–	41	18	–	–	1845	1708	96	1804	94.7	5.3	13.5
		Inca Cueva 7	–	20	–	–	–	–	–	–	–	–	–	–	–	20	20	–	40	50	50	–
		Unquillar	–	45	–	5	–	–	–	–	–	–	–	–	–	50	45	5	50	92	8	–
		CHIIIE2	–	57	–	–	–	–	–	–	–	–	–	–	–	57	57	–	57	100	–	–
		Tomayoc	–	365	–	–	–	–	–	–	–	–	–	–	–	365	365	–	365	99.7	0.3	–

Table 6.2 Archaeological Faunal Record from the Puna of Chile and Salar de Atacama

Northern Chile	Sites	Artiodactyla	Camelidae	Cervidae	Chinchillidae	Lagidium sp	Chinchilla sp	Caviidae	Cavia sp	Ctenomys sp	Rodentia indet	Avis	Carnivora	Dasipodidae	Total	N Artiodactyla	N Small Mammals and Avis	Total	% Artiodactyla	% Small Mammals and Avis	% Non Fused Artiodactyla
Early Holocene	San Lorenzo 1	–	5	–	43	–	–	–	–	–	16	4	–	–	68	5	47	52	9.6	90.4	–
	Tuina 1	–	25	–	7	–	–	–	–	–	9	–	–	–	41	25	7	32	78.1	21.9	–
	Tuina 5	–	393	12	85	–	–	–	–	–	226	38	–	–	754	405	123	528	76.7	23.3	66
	Tulán 68	–	165	–	24	–	–	–	–	–	25	926	11	–	1151	165	950	1115	14.7	85.2	12
	Tambillo	–	1047	–	33	–	–	–	–	969	57	93	–	–	2199	1047	1119	2166	48.3	51.6	36
Mid-Holocene I (8,000–6,000 BP)	Tulán 67	–	283	–	76	–	–	–	–	–	17	38	–	131	545	283	76	359	78.8	21.1	18
	Huasco 2	–	64	–	12	–	–	–	–	1	–	2	–	15	84	64	15	79	81	19	48
	Puripica 13-14	–	238	–	–	–	–	–	–	–	5	9	–	14	266	238	5	243	97.9	2	6
Mid-Holocene II (6,000–3,500 BP)	Puripica 33	–	932	–	–	–	–	–	–	–	1	4	–	5	942	932	5	937	99.4	0.5	10
	Tulán 52	–	12096	–	1558	–	–	–	–	277	14	68	–	–	14013	12096	1903	13999	86.4	13.6	32
	Puripica 1	–	3426	–	711	–	–	–	–	21	93	1	143	–	4395	3426	896	4322	81.4	18.5	58
	ChiuChiuCem	–	5861	–	–	–	–	–	–	–	48	42	–	–	5951	5861	42	5903	99.2	0.7	21

northwestern Argentina (Regidor and Rosati 2001). Taruca groups, as a cryptic species, are highly variable throughout the year. "Group size and structure [...] changed seasonally and these changes were highly correlated with the reproductive cycle of the deer. [...] The breeding cycle is strongly seasonal and is determined by precipitation and temperature patterns" (Merkt 1987: 397). Mixed groups of adult males and females with yearlings and/or fawns exhibit an average of 9.5 individuals (three to 40 individuals) for most of the year. During the breeding season (winter), taruca form groups of one male and four to five females, male groups, and mixed groups of young males. Fawns appear between February and April with a sharp peak in March (Merkt 1987; Regidor and Rosati 2001). Taruca presence in the archaeological record is highly variable but mostly low, and it is only present in a few sites (Tables 6.1 and 6.2).

Regarding the use of animal resources in the past, it is important to mention that hunter-gatherers also consumed an important range of small mammals (rodents, Xenarthra), and birds (*Metriopelia* sp, *Anas puna*, *Anas cyanoptera*) that lived in a wide range of local environments. Their importance in the archaeological record decreases through time, but their variation is highly local.

Regional Patterning of the Faunal Record

The faunal samples were obtained from 28 archaeological sites and levels (Fig. 6.1; Tables 6.1 and 6.2). It is worth noting that in the Chilean Puna of Atacama there are no sites occupied for more than one time period (i.e., Early Holocene, Middle Holocene I or Middle Holocene II), whereas in the Argentinian Puna the scenario is very different. In the latter area there are two sites in which the occupations comprise all three periods, and four cases comprising two of them (being contiguous in time or not). As we have already seen, the environmental changes had different impacts on both sides of the Andean range. For this reason, we will compare the archaeozoological samples of these two areas using NISP (Number of Identified Specimens per Taxon) (Lyman 1994). Between the Early Holocene and the Middle Holocene I, the proportion of camelids rises sharply from 45.5 to 85.9%, in the Chilean Puna, whereas in the Argentine Puna they increase from 60 to 76%. The Middle Holocene II frequencies also show an increase in comparison with previous values, reaching 91.6% in Chile and 88.9% in Argentina.

These trends suggest that a marked and abrupt specialization process occurs in the Chilean range before a more gradual change is seen in the highlands of Argentina. This process could represent a response to an earlier installation of arid conditions in the west ($\sim 8,000$ BP), but it could also result from the availability of a greater diversity of habitats for hunter-gatherers to the east of the Andean mountain range. This difference is well illustrated in the box plots (Fig. 6.3a, b) showing greater faunal variation between the localities of the eastern slope throughout the period. In Chile, camelids regularly dominate the faunas of the archaeological sites and, except for the higher percentage of camelids, there are no other differences between the Middle Holocene I and Middle Holocene II. In contrast, in Argentina there are substantial differences in mean values and in frequencies between these two chronological periods of the Middle Holocene.

On the other hand, the use of small mammals and birds is very different on both sides of the cordillera (Fig. 6.3c, d). In the Puna of Argentina, the chinchillas and vizcachas frequencies decrease through time, while these mammals increase in the Puna of Chile towards the Middle Holocene I then later diminish, although they are in greater proportion in the Middle Holocene II than in the Early Holocene. Aside from the vizcachas and chinchillas, other small mammals and birds are concentrated in a few archaeological sites. All these taxa reach to 51.67% in the Early Holocene in the Puna of Chile, whereas in the Argentinean Puna they have an average of 39.6% in the same period. An abrupt fall in small mammal frequencies starts with the onset of Middle Holocene in Chile as indicated by the 14% for the Middle Holocene I and the scarce 8.36% in the next period. In Argentina, this process is more gradual and irregular due to the high variation observed between localities, but the proportions of these species also fall to 23.9% during the Middle Holocene I, and to 11% in the Middle Holocene II.

The proportion of non-fused bone remains of camelids shows interesting differences when comparing both sides of the Andean range, while maintaining the same basic trend (Fig. 6.4). The percentage of non-fused bones changes geographically and across periods, but this difference is not statistically significant ($F = 1.112$, $p = 0.3512$). However, we will analyze them because they are of importance regarding modifications in animal use by hunter-gatherers.

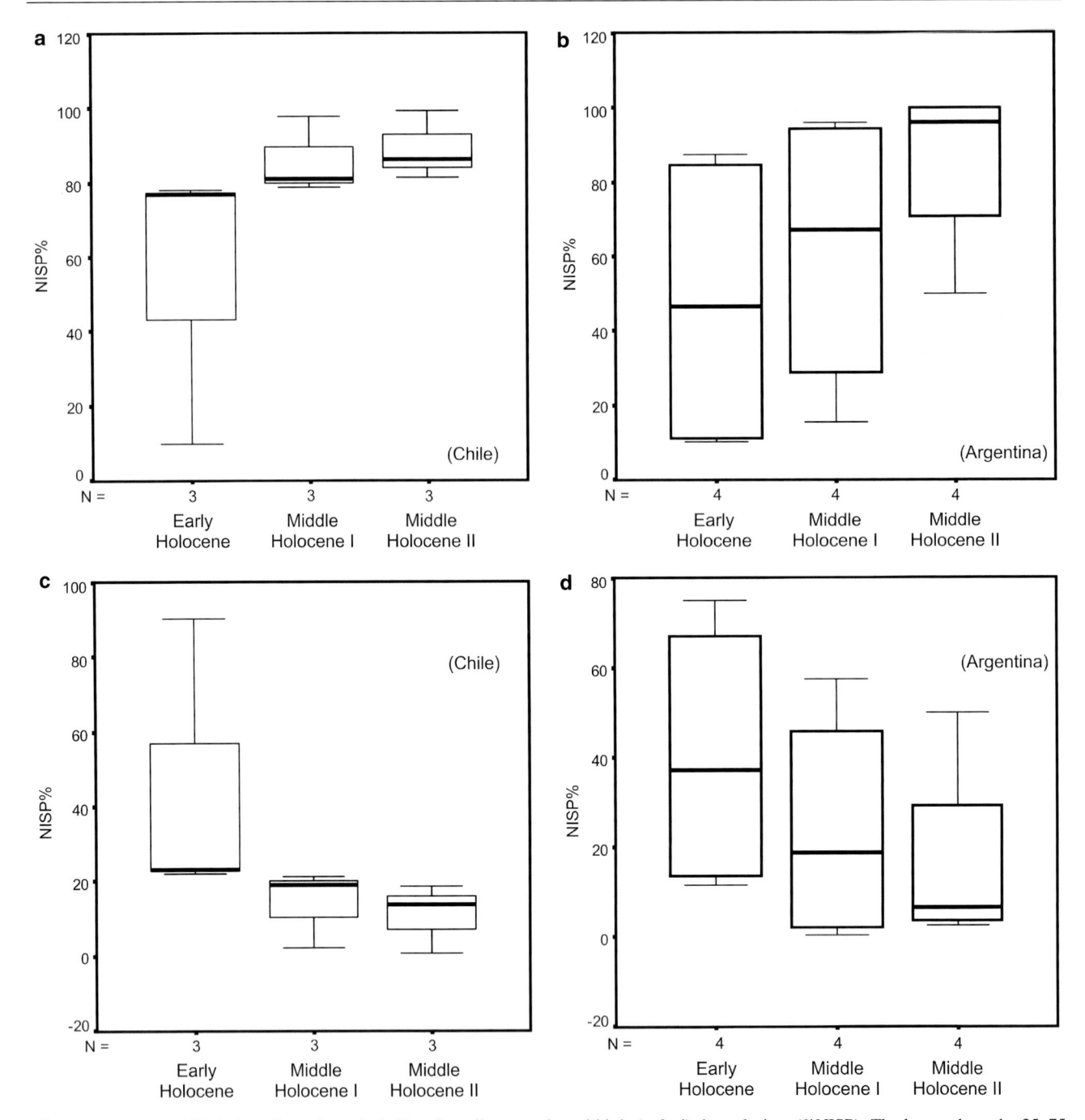

Fig. 6.3 Trends and Variation of ungulates (a & b) and small mammals and birds (c & d) through time (%NISP). The boxes show the 25–75 percent quartiles, the median (horizontal line inside the box), and the minimal and maximal (short horizontal lines)

In the Puna of Chile, the proportion non-fused bones (% NISP) (newborns, young and sub-adults up to 36 months) decreases from 38% (N = 1635) to 24% (N = 585), from the Early Holocene to the following period, but amounted to 30.2% (N = 22,315) in the Middle Holocene II. Besides, in this sector of the Puna there is a great variation in the presence of non-fused bones between sites; this variability does not appear to be caused by taphonomic issues, although we need more research on this topic. In Argentina during the Early Holocene there is a high proportion (46.37%, N = 4436) of non-fused individuals. This percentage drops to 29.2% (N = 1113) in the Middle Holocene I, but then rises to 36.9% (N = 2522) in the Middle Holocene II. The increase of non-fused bones during the later part of the Middle Holocene could be attributed to changing patterns in camelid management, in which protective herding could have been playing an important role (Yacobaccio 2004; Cartajena et al. 2007). The appearance of big size camelids

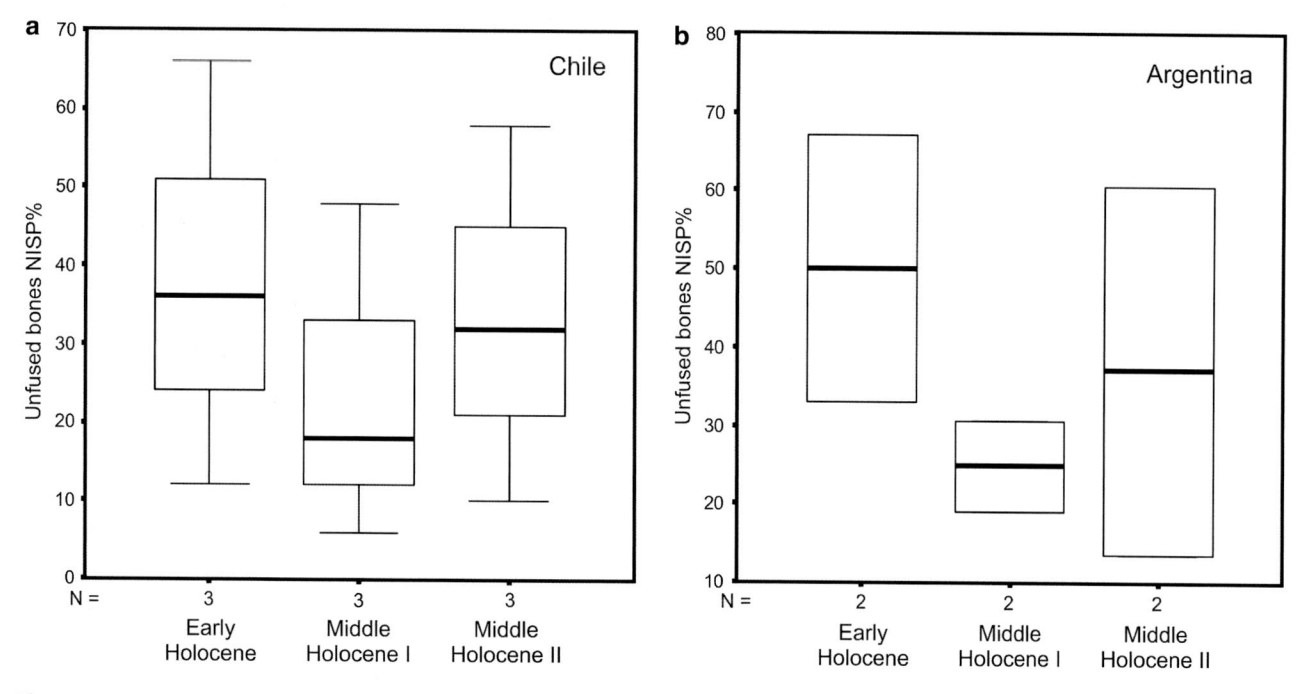

Fig. 6.4 Trend and Variation of Non-fused bones of Camelids, (**a**) Puna of Chile; (**b**) Puna of Argentina

(like llamas), pens, and other contextual evidences (i.e., rock art) points toward the implementation of management techniques of camelids as the initial steps toward their domestication.

Resiliency of Local Habitats

As we have already said, the local scale is useful for discussing the flexibility of certain habitats in the face of environmental changes, which make them persistent places for hunter-gatherer occupation. There are only three localities in the whole Puna region with human occupation in all three time periods discussed here, and we will examine one of them, the Hornillos 2 site. This locality is a rockshelter of 42 m² located at 23° 13′ 47″ S, 66° 27′ 22″ W and at 4,020 m asl, in the Jujuy province of Argentina (Fig. 6.2). It exhibits 9 occupational levels, dated between 9,710 and 6,130 BP. The earliest seven levels correspond to the Early Holocene, while the later two are dated between 7,760 and 6,130 BP and therefore pertain to the Middle Holocene. Several changes have been detected between the occupations of the two periods, such as the extent of the occupied surface, intensity of activities, and modification in the use of faunal and floral resources. Differences have also been observed in the isotopic signal of the camelid bones. Middle Holocene I and II have been clumped improve the quantitative analysis of isotopic samples.

Animal skeletal tissues can provide paleoenvironmental information in the form of isotopic ratios that relate to changes in consumed food and water (Bocherens and Drucker 2007). Thus, we measured stable isotope ratios on camelid bones in order to obtain information about the stability of the floral communities over time and changes in the water balance of the environment. The sample for this analysis was selected from a total of 15,000 bone fragments according to three criteria: (1) the bones must have good preservation (not weathered); (2) identifiable to a species or group-size (large [guanaco], small [vicuña]), and to different individuals, and (3) stratigraphic location, i.e., equivalent representation of diverse strata. Thirty-three stable isotope values ($\delta^{13}C$ and $\delta^{15}N$) measured on bone and dentin collagen obtained from camelid bone remains are discussed in this section (Table 6.3). We consider that these values could be interpreted as a random sampling of the camelid populations that inhabited the Susques area between 9,710 and 6,130 years BP and that were exploited by the site occupants. Initially, thirty eight remains were sampled to obtain isotopic values, but five of them must be excluded due to anomalies in their C:N ratios (>3.6, above the normal value for collagen), and due to highly depleted $\delta^{13}C$ values, indicating low collagen yield (DeNiro 1985; Ambrose 1993).

In analyzing these data, we will consider the modern reference values for this area, generated within the framework of a Dry Puna camelid ecological study (Samec 2011). At the same time, the Hornillos 2 faunal data will be evaluated in a taxonomic dimension, taking into account the

Table 6.3 Carbon and Nitrogen isotopic values of camelid samples from Hornillos 2

Sample	Layer	Skeletal part	Camelid size	$\delta^{13}C$	$\delta^{15}N$	C:N
H2 02	2	Proximal Metacarpal	Large camelid	−18.79	10.16	3.68
H2 03	2	First Phalange diaphysis	Large camelid	−16.79	10.25	3.01
H2 04	2	Phalange 1 proximal end	Large camelid	−16.90	8.69	2.98
H2 05	2	Metapodial diaphysis	Large camelid	−12.38	10.74	2.98
H2 06	2	Metapodial diaphysis	Large camelid	−17.20	9.25	2.91
H2 07	2	Metapodial diaphysis	Vicuña	−15.80	10.02	3.04
H2 08	2	Metapodial diaphysis	Large camelid	−18.70	6.22	3.03
H2 09	2	Phalange 1 proximal end	Large camelid	−18.80	6.03	3.04
H2 10	2	Phalange 1 proximal end	Large camelid	−19.02	6.18	3.06
H2 11	2	Phalange 1 proximal end	Vicuña	−18.23	7.83	3.70
H2 12	2	Scapula	Large camelid	−17.45	7.26	2.95
H2 16	2	Mandible	Vicuña	−16.73	7.05	2.93
H2 18	2	Teeth incisive 3	Large camelid	−19.00	9.01	3.63
H2 20	2	Teeth incisive 1	Large camelid	−18.10	9.18	2.90
H2 25	2	Long bone fragment	Vicuña	−16.38	9.91	3.04
H2 14	3	Long bone fragment	Vicuña	−17.03	7.25	3.03
H2 21	4	Metapodial distal end	Guanaco	−16.68	10.54	3.01
H2 27	4	Proximal Phalange	Guanaco	−17.49	8.18	2.89
H2 28	4	Long bone fragment	Vicuña	−19.32	5.58	3.08
H2 29	4	Proximal Phalange	Vicuña	−17.39	5.59	2.89
H2 43	5	Long bone fragment	Guanaco	−18.34	6.86	3.22
H2 45	5	Long bone fragment	Vicuña	−18.40	7.08	3.22
H2 47	6	Femur	Guanaco	−19.09	6.09	2.93
H2 50	6	Long bone fragment	Vicuña	−17.65	5.02	2.92
H2 51	6	Teeth incisive	Guanaco	−17.56	10.02	2.99
H2 52	6A	Radius ulna	Guanaco	−16.67	7.64	2.84
H2 53	6A	Long bone fragment	Vicuña	−17.18	4.91	2.86
H2 24	6B	Phalange indeterminate	Indeterminate	−17.83	5.94	2.97
H2 31	6B	Long bone fragment	Guanaco	−18.44	4.46	2.84
H2 33	6B	Teeth incisive	Guanaco	−19.23	3.67	2.93
H2 35	6C	Metapodial shaft	Vicuña	−16.68	5.87	2.87
H2 37	6C	Rib proximal	Guanaco	−18.63	5.19	2.99
H2 41	6D	Metapodial	Indeterminate	−16.01	8.11	2.81

differences between the represented camelid species. In this sense, osteometric data points to the presence of vicuñas and guanacos. It is important to mention the presence of faunal remains pertaining to a camelid group of large size, similar to modern llamas, recovered from layer 2, dated between 6,130 and 6,340 years BP. This could be interpreted as big specimens of wild camelids, perhaps the first ones subject to human management, i.e., protective herding (Yacobaccio 2004).

$\delta^{13}C$ Values

The carbon isotopic data comprise a range of values from −19.32 to −12.38‰. In relation to this, we can affirm that these wild camelids were mixed feeders that consumed both C_3 and C_4 plants, although the mean $\delta^{13}C$ value of −17,57‰ seems to point in favour of a high consumption of C_3 vegetation in opposition to a relative low amount of C_4 plants in the diet.

At the same time, the $\delta^{13}C$ values presented in Table 6.3 are in agreement with those measured on modern camelid specimens (*Lama glama*, *Lama guanicoe* and *Vicugna vicugna*, N = 26) (Yacobaccio et al. 2009, 2010; Samec 2011) from the same elevation as the site, i.e., between 3,900 and 4,200 m. As can be seen in Fig. 6.5, there are no statistically significant differences between both data groups (F = 0.5762, p > 0.01), even taking into account the presence of an outlier in the Hornillos 2 sample distribution, which is a value obtained from a "large camelid" specimen, that also presents a high $\delta^{15}N$ value (Table 6.3).

The clear overlap between modern and archaeological carbon isotopic data from the same altitudinal range is significant, although the values measured on the archaeological materials show a certain enrichment that seems to point to a significant presence of C_4 plants in the past diet. Nevertheless, this difference is not statistically significant, thus it

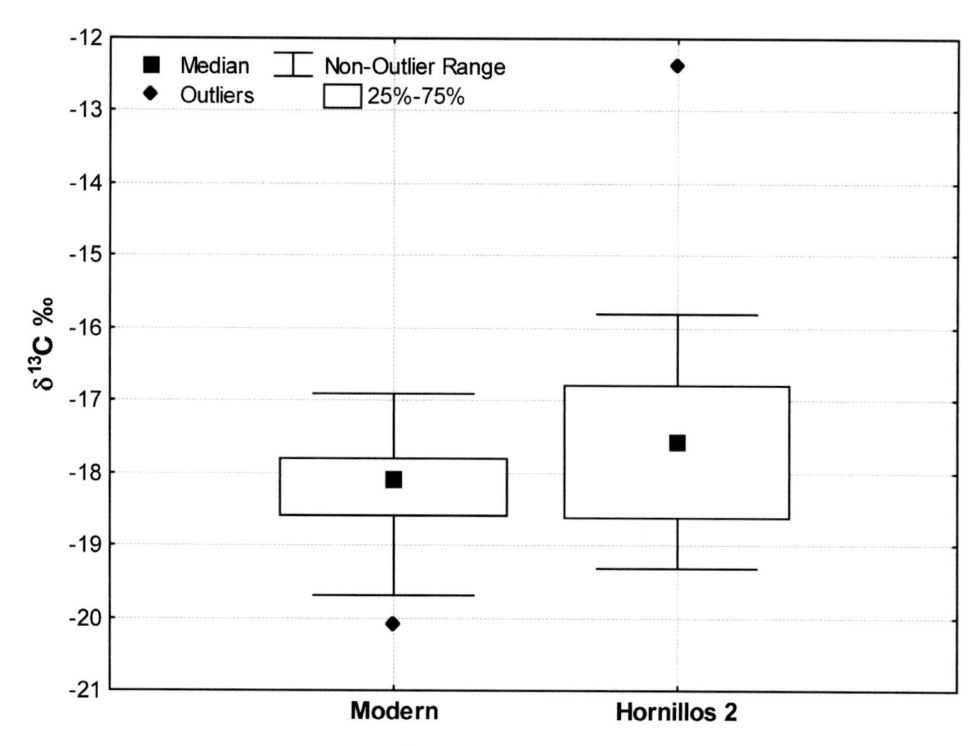

Fig. 6.5 Comparison between modern and archaeological $\delta^{13}C$ isotopic data of camelids

allows us to postulate that Early and Middle Holocene camelid diet did not exhibit great differences from the present day feeding behavior of the same species, at least concerning C_3 and C_4 proportions. Hence, the camelids consumed by the Hornillos 2 inhabitants fed in a mixed steppe area, consuming herbaceous (C_3 and C_4) and shrub (C_3) vegetation. This mixed steppe can be observed today in the vicinity of the site. Even considering the Middle Holocene climatic change towards more arid conditions, the camelid isotopic data presented here, and their correlation with altitude, seem to suggest that the mixed steppe would have changed neither their C_3 and C_4 vegetation community proportions, nor their localization in this altitudinal gradient in comparison with earlier and later times.

As can be seen in the Fig. 6.6a (in which two undetermined specimens were not included) the three groups: vicuñas and guanacos (from layers 3, 4, 5, 6, 6a, 6b, 6c and 6d) and "large camelids" (from layer 2) exhibit no differences, showing similar diets in terms of the C_3 and C_4 plants proportions. Although vicuñas have enriched values compared to the other two groups, the statistical tests do not reveal significant differences (vicuñas and "large camelids", $F = 0.105$, $p > 0.01$; vicuña and guanaco, $F = 2.278$, $p > 0.01$). The preference for herbaceous C_4 plants exhibited by the vicuñas' $\delta^{13}C$ values can be explained in relation to the grazing habits of this species.

The samples corresponding to the Early Holocene (layers 4, 5, 6, 6a, 6b, 6c and 6d) and those to the Middle Holocene (layers 2 and 3) are compared in Fig. 6.6b. Leaving the already mentioned outlier aside, the distributions of $\delta^{13}C$ values clearly overlap, showing no statistically significant differences between both data groups ($F = 0.979$, $p > 0.01$). These results allow us to suggest that the camelid diets did not exhibit great differences in the C_3 and C_4 proportions between both periods. If we assume that these animals were feeding on the local vegetation and were hunted near the site, the composition of the vegetation does not seem to have been affected by the Middle Holocene aridity in a significant manner.

$\delta^{15}N$ Values

The archaeological data regarding nitrogen isotopic values is scarce for the Andes, and there are no modern reference values for this region in general or for the Puna area in particular. The $\delta^{15}N$ values presented here (Table 6.3) show a wide distribution, covering a range between 3.67 and 10.74‰, in agreement with the few samples obtained from camelid remains of the Late Holocene from archaeological sites of the Quebrada de Humahuaca (Jujuy), the Yocavil Valley (Salta) (Mengoni Goñalons 2007), and from modern samples of llamas from Peru (Turner et al. 2010). The nitrogen values do not show significant differences by taxonomic group (Fig. 6.7a). Although Fig. 6.7a shows a slight enrichment in the data belonging to the "large

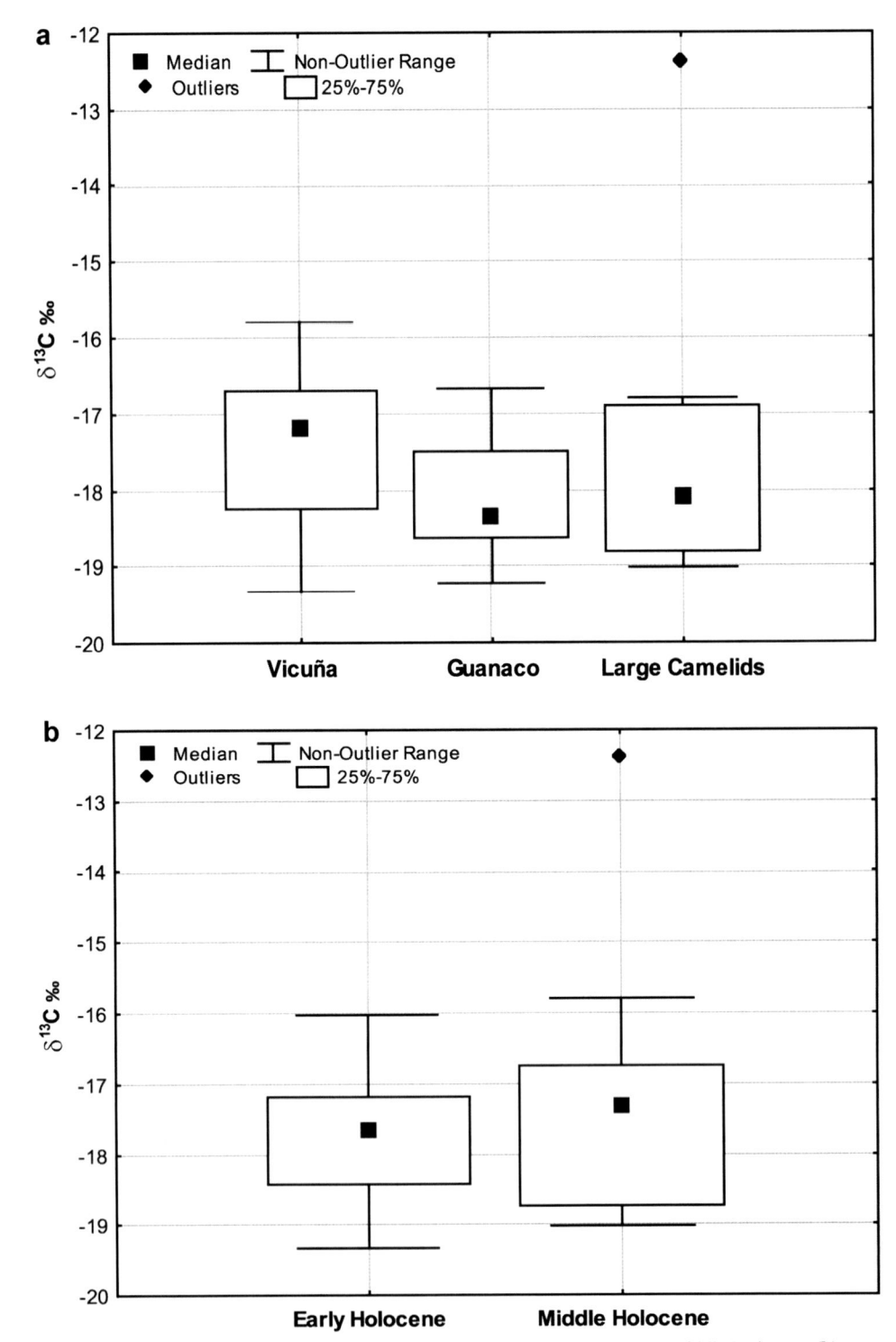

Fig. 6.6 Carbon isotopic values for the different species of camelids (**a**), and between Early and Middle Holocene (**b**)

camelids" group, the difference is not statistically significant ("large camelids" and vicuñas, $F = 4.161$, $p > 0.01$; "large camelids" and guanacos, $F = 2.627$, $p > 0.01$). This enrichment can result from two possible causes and is in line with a long controversy in the literature of the last twenty years. The first possible explanation regards this group as integrated by specimens (guanacos and or llama-size in layer 2) which are drought tolerant herbivores that do not require drinking water on a daily basis in contrast to the vicuñas, which are obligate drinkers. For some authors, this characteristic could have determined a differential excretion of nitrogen by urea, resulting in enriched

$\delta^{15}N$ tissue values (Ambrose and DeNiro 1986; Ambrose 1991; Adams and Sterner 2000).

On the contrary, according to some other investigations, this enrichment can be related to the climatic change towards more arid conditions that occurred during the layer 2 time span, corresponding to the second half of the Middle Holocene. This environmental change, as we have already seen, involved a drop in amount and frequency of precipitation and a decrease in water availability for the whole Puna area which could have changed the nitrogen balance in the soils and the entire ecosystem (Amundson et al. 2003; Hartman 2011). Our archaeological data show that those

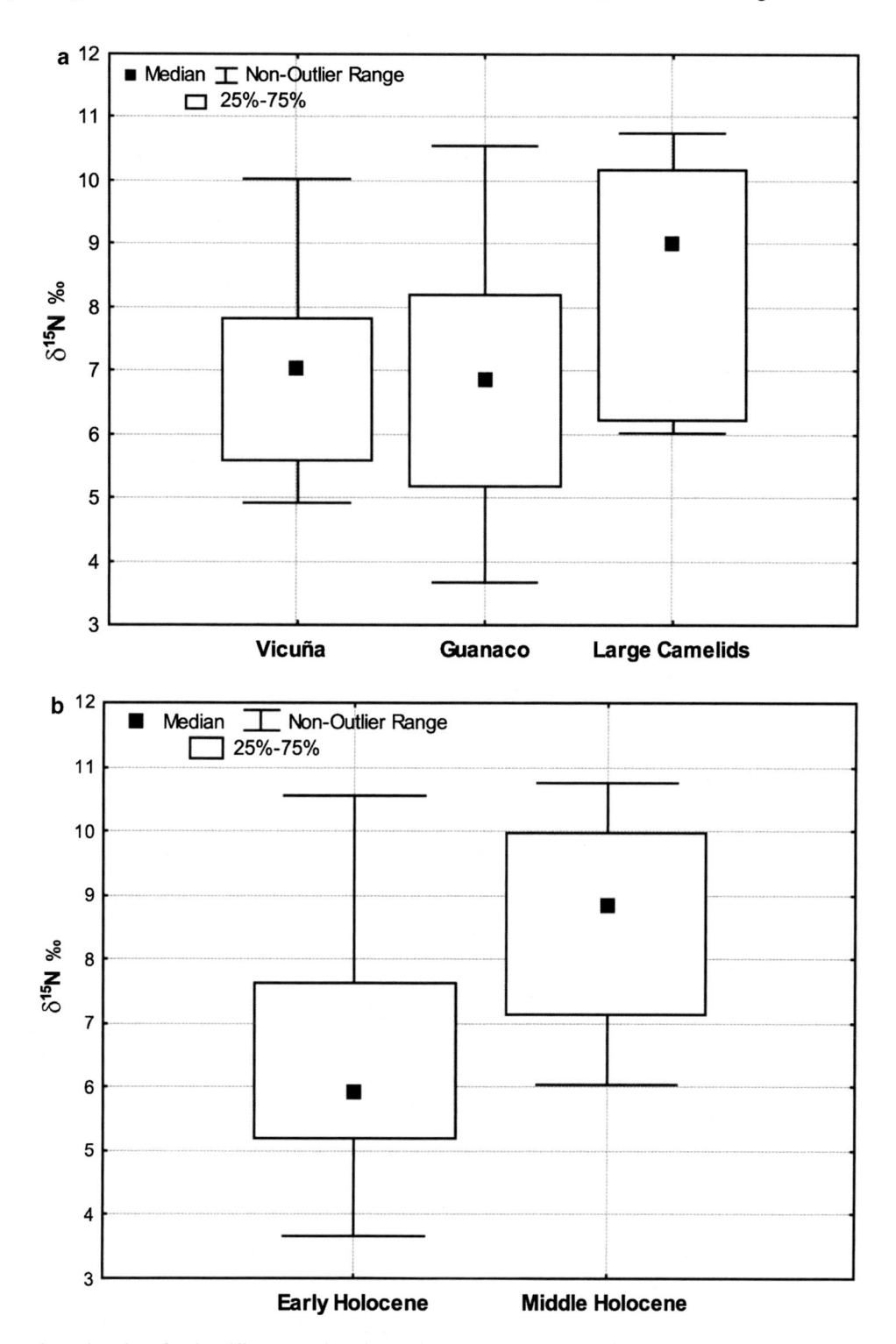

Fig. 6.7 Nitrogen isotopic values for the different species of camelids (**a**), and between Early and Middle Holocene (**b**)

specimens undoubtedly assigned to *Lama guanicoe* or to *Vicugna vicugna* in layers 6D through 3 exhibit similar nitrogen values and are quite different to those from the "large" specimens. This seems to exclude the first hypothesis in favor of an explanation that considers that these "large camelids" (llama-size) would have been exposed to the more arid conditions during the Middle Holocene (Sealy et al. 1987).

Further, when comparing the $\delta^{15}N$ values corresponding to the Early Holocene with those dated in the Middle Holocene, the difference in the distributions and the medians can easily be seen in Fig. 6.7b (F = 9.883, p < 0.01). This plot shows that Middle Holocene $\delta^{15}N$ values are highly enriched and exhibit a median value 3‰ more positive than the Early Holocene one. If we take into account the results of ecological research that emphasize changes in the nitrogen balance of the ecosystems as a consequence of aridity, this enrichment could be related to the more arid conditions established with the onset of the Middle Holocene (Amundson et al. 2003; Murphy and Bowman 2006; Hartman 2011). This assertion is supported by the presence of a strong negative correlation between the $\delta^{15}N$ values of soils and plants and precipitation in diverse parts of the globe, which is related to the nitrogen cycle and the organic or inorganic character of the soil (Heaton 1987; Austin and Vitousek 1998; Amudson et al. 2003; Hartman 2011). It is our position that the prevalence of inorganic soils in the Middle Holocene (Tchilinguirián 2009) would cause enrichment in the vegetal $\delta^{15}N$ values and in all the trophic chain afterwards. For that reason, enriched $\delta^{15}N$ values in Hornillos 2 camelids could be explained by the more arid environmental conditions of this period.

Concluding Remarks

As evidenced by the results presented in the previous sections, several relationships between environmental changes related to the onset of Middle Holocene aridity and the animal resources structure – and therefore their use – can be suggested.

On a regional scale, it is evident that a change in the importance of camelids, becoming the primary resource staple for the inhabitants of the Puna, was triggered by the fragmentation of its habitat induced by Middle Holocene aridity. Notwithstanding, this specialization process seems to have occurred in a marked and abrupt manner, and slightly earlier, in the Chilean slope in comparison to the highlands of Argentina were this process shows a more gradual trend. We think that this specialization is clearly related to at least two causes. First, we think it is due to the increased predictability of the spatial arrangements of camelid groups due to the scarcity of highly productive habitats during Middle Holocene. In these terms, the search costs were markedly reduced during this period in comparison with the more homogeneous Early Holocene in which camelids could have been dispersed throughout the landscape. Second, we think the specialization in camelids also could be generated by the increased cost of acquiring alternative low-rank resources. As suggested by Grosjean et al. (2003), rodents are strongly dependent on moisture; consequently, their populations are usually well correlated to regional moisture. If a marked drop in moisture availability occurs, as in several events during Middle Holocene, these populations will be strongly impacted, affecting their regional abundance and, as a result, their costs as a food source. In terms of other low-ranking resources, the information is too fragmentary to isolate any trend. As well, it appears that Middle Holocene aridity impacted wetlands and shallow ponds and lakes that are the usual habitat for a huge proportion of the regional birds. In consequence, the regional distribution of these kinds of resources was affected, and this was a particularly important change in the Chilean slope as evidenced from several sites, e.g., Tulán 67, Huasco 2, Puripica 13–14, and Tulán 52.

The proportion of non-fused bones of camelids shows variability on both sides of the mountain range. Whereas in Chile there is a great variation in the proportions of non-fused bones between localities, in Argentina the newborns and young camelids are consistently distributed among them during the Middle Holocene. As has been already mentioned, this evidence suggests at least two possibilities that are mutually complementary: (1) greater regularity in the acquisition of adult prey due to the employment of collective hunting techniques, and (2) greater accessibility of camelids in the eastern side of the Andean mountain chain because of the greater availability of suitable habitats. As the reduced abundance of water holes and productive habitats became a common feature of the landscape, animals started to aggregate in the desirable patches of the fragmented landscape, and human groups did the same to optimize their access to high-ranked prey and to minimize travel and handling cost. One consequence was an increase in group size that allowed different hunting techniques, supported by other evidence, such as the diversity projectile point types, and the presence of facilities for hunting (stone made wind-breaks).

Another fact to be mentioned regarding the age-classes of the regional bone sample is the increase of non-fused bones proportion during the later part of the Middle Holocene. As suggested by some authors (Yacobaccio 2004; Cartajena et al. 2007) this pattern could be the consequence of changes in camelid management, in which protective herding was practiced by some hunter-gatherer local groups. The evidence that supports this hypothesis is the appearance of llama-size camelids, pens, and other contextual evidence, i.e., rock art (Gallardo and Yacobaccio 2005), that points toward the implementation of protective herding of camelids as the initial steps toward their domestication. At the end of the period, around 3,700 BP, domesticated camelids are quite common in the archaeological record of the region (Yacobaccio 2004; Cartajena et al. 2007).

On a local spatial scale, isotopic information coming from Hornillos 2 points toward two important issues: (1) no major vegetation composition changes or spatial rearrangements have been detected during the first half of the Holocene regarding camelid diet, and (2) a statistically significant enrichment of N values of camelids reflects more arid conditions in the region during the Mid-Holocene, coinciding with mean regional paleoenvironmental trends. Regarding the first issue, the results concerning the camelid diet from Hornillos 2 point to animals that fed in a mixed steppe area similar to that observed today in the area surrounding the site. The data presented here seem to suggest that the mixed steppe would have changed neither their C_3–C_4 vegetation community proportions nor their localization in this altitudinal gradient in comparison with earlier and later times. A decline in the total biomass of these communities because of the arid conditions cannot, however, be discarded yet. In relation to the second issue, Middle Holocene $\delta^{15}N$ values in Hornillos 2 are highly enriched, especially in those specimens determined as "large camelids". If we take into account the results of ecological research that emphasize changes in the nitrogen balance of the ecosystems as a consequence of aridity, this enrichment could be related to the more arid conditions established with the onset of the Middle Holocene. We think that the proliferation of inorganic soils during the Middle Holocene (Tchilinguirián 2009) is the main cause of the enrichment in the vegetal $\delta^{15}N$ values and in all the trophic chain afterwards.

The spatio-temporal distribution of the zooarchaeological evidence is consistent with the characteristics of a fragmented environment as predicted by the model here reviewed. In places with greater environmental resilience, occupation continuity between the Early and Middle Holocene can be observed. In other locations, however, the favourable habitats disappeared around 7,000 BP, and human groups following wildlife moved away, searching for wetlands or other places, generally located above 4000 m, with concentrated resources.

When Andean hunter-gatherers were faced with a dryer, patchier environment, they appear to have changed their social structure forming larger groups. Also, they changed their resource extraction methods by employing communal hunting/resource extraction, and varied their technology (variety of artifact types). Indeed, they found an advantage in more arid Middle Holocene conditions by reducing search and handling times and increasing productivity by emphasizing highly ranked (large body size) animal prey.

Hopefully, such lessons can be valuable in the context of the currently changing environmental conditions. Interdisciplinary research provides valuable lessons in understanding the ways that plants, animals and people of the Argentinian Puna coped with increasing aridity during the Early to Middle Holocene transition.

Acknowledgments This research has been funded by grants of CONICET (PIP 6322), and the University of Buenos Aires (UBACYT F052). We thank the three anonymous reviewers and Gregory Monks for their insightful comments which helped to improve the final version of this chapter.

References

Abbott, M., Wolfe, B., Wolfe, A., Seltzer, G., Aravena, R., Mark, B., et al. (2003). Holocene palehydrology and glacial history of the central Andes using multiproxy lake sediments studies. *Palaeogeography, Palaeoclimatology, Palaeoecology, 194*, 123–138.

Adams, T. S., & Sterner, R. W (2000). The effect of dietary nitrogen content on trophic level $\delta^{15}N$ enrichment. *Limnology and Oceanography, 45*, 601–607

Aldenderfer, M. S. (1998). *Montane foragers. Asana and the South-Central Andean Archaic.* Iowa City: University of Iowa Press.

Ambrose, S. H. (1991). Effects of diet, climate and physiology on nitrogen isotope ratios in terrestrial foodwebs. *Journal of Archaeological Science, 18*(3), 293–317.

Ambrose, S. H. (1993). Isotopic analysis of paleodiets: Methodological and interpretive considerations. In M. K. Sandford (Ed.) *Investigations of ancient human tissue. Chemical analyses in anthropology* (pp. 59–130). Langhorne: Gordon and Breach Science Publishers.

Ambrose, S. H., & DeNiro, M. J. (1986). The isotopic ecology of east African mammals. *Oecologia, 69*, 395–406.

Amundson, R., Austin, A. T., Schuur, E. A. G., Yoo, K., Matzek, V., Kendall, C., et al. (2003). Global patterns of the isotopic composition of soil and plant nitrogen. *Global Biogeochemical Cycles, 17* (1), 1031.

Austin, A. T., & Vitousek, P. M. (1998). Nutrient dynamics on a precipitation gradient in Hawaii. *Oecologia, 113*, 519–529.

Arzamendia Y. (2008). *Estudios etoecológicos de vicuñas (Vicugna vicugna) en relación a su manejo sostenido en silvestría, en la Reserva de la Biosfera Laguna de Pozuelos (Jujuy, Argentina).* Ph. D. Dissertation, Universidad Nacional de Córdoba.

Arzamendia, Y., Cassini, M. H., & Vilá, B. L. (2006). Habitat use by Vicugna vicugna in Laguna Pozuelos Reserve, Jujuy, Argentina. *Oryx, 40*, 198–203.

Benítez, V. V. (2005). *Calidad de dieta de la vicuña (Vicugna vicugna vicugna) en Reserva de Laguna Blanca, Catamarca, Argentina.* Grade thesis, Universidad Nacional de Luján.

Binford, L. R. (2001). *Constructing frames of reference. An analytical method for archaeological theory building using ethnographic and environmental data sets.* California: University of California Press.

Bobst, A., Lowenstein, T., Jordan, T., Godfrey, L., Ku, T.-L., & Luo, S. (2001). A 106 ka paleoclimate record from drill core of the Salar de Atacama, northern Chile. *Palaeogeography, Palaeoclimatology, Palaeoecology, 173*, 21–42.

Bocherens, H., & Drucker D. G. (2007). Carbonate stable isotopes/terrestrial teeth and bones. In S. Elias (Ed.) *Encyclopedia of Quaternary Sciences* (pp. 309–316). Elsevier.

Borgnia, M., Maggi, A., Arriaga, M., Aued, B., Vilá, B. L., & Cassini, M. H. (2006). Caracterización de la vegetación en la Reserva de Biosfera Laguna Blanca (Catamarca, Argentina). *Ecología Austral, 16*, 29–45.

Bradbury, P. J., Grosjean, M., Stine, S., & Sylvestre, F. (2001). Full and late glacial lake records along the PEP 1 transect: Their role in developing interhemispheric paleoclimate interactions. In V. Markgraf (Ed.), *Interhemispheric climate linkages* (pp. 265–291). San Diego: Academic Press.

Bradley, R. S. (2000). Past global changes and their significance for the future. *Quaternary Science Reviews, 19*, 391–402.

Broughton, J. M., Byers, D. A., Bryson, R. A., Eckerle, W., & Madsen, D. B. (2008). Did climatic seasonality control late Quaternary artiodactyl densities in western North America? *Quaternary Science Reviews, 27*, 1916–1937.

Cabrera, A. L. (1976). *Regiones fitogeográficas Argentinas*. Buenos Aires: Editorial Acme.

Cajal, J. (1979). Estructura social y área de acción del guanaco (Lama guanicoe) en la Reserva de San Guillermo, Pcia. de San Juan. In *3rd Conveción Internacional sobre Camélidos Sudamericanos* (pp. 23–33). Viedma.

Cajal, J. L. (1985). Comportamiento. In J. L Cajal & J. N. Amaya (Eds.), *Estado actual de las investigaciones sobre camélidos en la República Argentina* (pp. 87–100). Buenos Aires: SECYT.

Cartajena, I., Núñez, L., & Grosjean, M. (2007). Camelid domestication on the western slope of the Puna de Atacama, northern Chile. *Anthropozoologica, 42*, 155–173.

Censo (2010). *Manejo de Fauna Silvestre en Argentina. Primer Censo Nacional de Camélidos Silvestres al Norte del Río Colorado*. Buenos Aires: Secretaría de Ambiente y Desarrollo Sustentable de la Nación.

Codding, B. F., Bird, D. W., & Bliege Bird, R. (2010). Interpreting abundance indices: Some zooarchaeological implications of Martu foraging. *Journal of Archaeological Science, 37*, 3200–3210.

DeNiro, M. J. (1985). Postmortem preservation and alteration of in vivo bone collagen isotope ratios in relation to palaeodietary reconstruction. *Nature, 317*, 806–809.

Dincauze, D. F. (2000). *Environmental archaeology*. Cambridge: Cambridge University Press.

Earle, L. R., Werner, B. G., & Aravena, R. (2003). Rapid development of an unusual peat-accumulating ecosystem in the Chilean Altiplano. *Quaternary Research, 59*, 2–11.

Franklin, W. L. (1982). Biology, ecology and relationship to man of the South American Camelids. In M. A. Mares & H. H. Genoways (Eds.) *Mammalian biology in South America*. Pymatuning Laboratory of Ecology. Special Publication, *6*, 457–489.

Gallardo, F., & Yacobaccio, H. (2005). Wild or domesticated? Camelids in Early Formative rock art of the Atacama Desert (Northern Chile). *Latin American Antiquity, 16*(2), 115–130.

Geyh, M., Grosjean, M., Núñez, L., & Schotterer, U. (1999). Radiocarbon reservoir effect and the timing of the Late-Glacial/Early Holocene humid phase in the Atacama Desert (Northern Chile). *Quaternary Research, 52*, 143–153.

González, B. A., Palma, R. E., Zapata, B., & Marín, J. C. (2006). Taxonomic and biogeographical status of guanaco *Lama guanicoe* (Artiodactyla, Camelidae). *Mammal Review, 36*, 157–178.

Grayson, D. K. (2000). Mammalian responses to Middle Holocene climatic change in the Great Basin of the western United States. *Journal of Biogeography, 27*, 181–192.

Grosjean, M. (2001). Mid-Holocene climate in the South-Central Andes: Humid or dry? *Science, 292*, 2391a.

Grosjean, M., Cartagena, I., Geyh, M. A., & Núñez, L. (2003). From proxy data to paleoclimate interpretation: The mid-Holocene paradox of the Atacama Desert, northern Chile. *Palaeogeography, Palaeoclimatology, Palaeoecology, 194*, 247–258.

Garreaud, R. D. (2009). The Andes climate and weather. *Advances in Geosciences, 7*, 1–9.

Hartman, G. (2011). Are elevated $\delta^{15}N$ values in herbivores in hot and arid environments caused by diet or animal physiology? *Functional Ecology, 25*, 122–131.

Heaton, T. H. E. (1987). The 15 N/14 N ratios of plants in South Africa and Namibia: Relationship to climate and coastal/saline environments. *Oecologia, 74*, 236–246.

Houston, J., & Hartley, A. (2003). The central Andean west-slope rainshadow and its potential contribution to the origin of hyperaridity in the Atacama Desert. *International Journal of Climatology, 23*(12), 1453–1464.

Kelly, R. (1995). *The foraging spectrum*. Washington D.C.: Smithsonian Institution Press

Latorre, C., Betancourt, J., & Arroyo, M. (2006). Late Quaternary vegetation and climate history of a perennial river canyon in the Río Salado basin (22°S) of Northern Chile. *Quaternary Research, 65*, 450–466.

Latorre, C., Betancourt, J., Rylander, K., & Quade, J. (2002). Vegetation invasions into absolute desert: A 45,000 yr rodent midden record from the Calama-Salar de Atacama basins, northern Chile (lat 22°–24°S). *Geological Society of America Bulletin, 114*(3), 349–366.

Latorre, C., Betancourt, J. L., Rylander, K. A., Quade, J., & Matthei, O. (2003). A vegetation history from the arid prepuna of northern Chile (22°–23°S) over the last 13,500 years. *Palaeogeography, Palaeoclimatology, Palaeoecology, 194*, 223–246.

Lenters, J., & Cook, K. (1995). Simulation and diagnosis of the regional summertime precipitation climatology of South America. *Journal of Climate, 8*, 2988–3005.

Lyman, R. L. (1994). *Vertebrate taphonomy*. Cambridge: Cambridge University Press.

Mantua, N. J., Hare, S., Zhang, Y., Wallace, J. M., & Francis, R. C. (1997). A Pacific interdecadal climate oscillation with impacts on salmon production. *Bulletin of the American Meteorological Society, 78*, 1069–1079.

Mengoni Goñalons, G. L. (2007). Camelid management during Inca times in N.W. Argentina: Models and archaeozoological indicators. *Anthropozoologica, 42*(2), 129–141.

Merkt, J. R. (1987). Reproductive seasonality and grouping patterns of the north Andean deer or Taruca (*Hippocamelus antisensis*) in southern Peru. In C. Wemmer (Ed.), *Biology and management of the Cervidae* (pp. 388–400). Washington: Smithsonian Institution Press.

Messerli, B., Grojean, M., Bonani, G., BüRgi, A., Geyh, M., Graf, K., et al. (1993). Climate change and naturla resource dynamics of the Atacama Altiplano during the last 18,000 years: Apreliminary synthesis. *Mountain Research and Development, 13*, 117–127.

Morales, M. R. (2011). *Arqueología ambiental del Holoceno Temprano y Medio en la Puna Seca argentina. Modelos paleoambientales multi-escalas y sus implicancias para la Arqueología de Cazadores-Recolectores*. British Archaeological Reports (BAR) S2295, South American Archaeology Series 15. Oxford: Archaeopress.

Murphy, B. P., & Bowman, D. M. J. S. (2006). Kangaroo metabolism does not cause the relationship between bone collagen $\delta^{15}N$ and water availability. *Functional Ecology, 20*, 1062–1069.

Puig, S., Videla, F., Cona, M. I., & Roig, V. G. (2008). Habitat use by guanacos (*Lama guanicoe*, Camelidae) in northern Patagonia (Mendoza, Argentina). *Studies on Neotropical Fauna and Environment, 43*, 1–9.

Ramirez, E., Hoffmann, G., Taupin, J. D., Francou, B., Ribstein, P., Caillon, N., et al. (2003). A new Andean deep ice core from Nevado Illimani (6350 m), Bolivia. *Earth and Planetary Science Letters, 212*, 337–350.

Rech, J., Pigati, J. S., Quade, J., & Betancourt, J. L. (2003). Re-evaluation of Mid-Holocene wetland deposits at Quebrada Puripica, Northern Chile. *Palaeogeography, Palaeoclimatology, Palaeoecology, 194*, 207–222.

Regidor, H. A., & Rosati, V. R. (2001). Taruca. In C. M. Dellafiore & Maceira, N. (Eds.), *Los ciervos autcótonos de la Argentina y la acción del hombre* (pp. 75–82). Buenos Aires: Secretaría de Ambiente y Desarrollo Sustentable de la Nación.

Riedinger, M. A., Steinitz-Kannan, M., Last, W. M., & Brenner, M. (2002). A 6100 14C yr record of El Niño activity from the Galápagos Islands. *Journal of Paleolimnology, 27*, 1–7.

Rindel, D., Goñi, R., Belardi J. B., & Bourlot, T. (2017). Climatic changes and hunter-gatherer populations: Archaeozoological trends in southern Patagonia. In G. G. Monks (Ed.), *Climate change and human responses: A zooarchaeological perspective* (pp. 153–172). Dordrecht: Springer.

Rowe, H., Dunbar, R., Mucciarone, D., Seltzer, G., Baker, P., & Fritz, S. (2002). Insolation, moisture balance and climate change on the South American Altiplano since the last glacial maximum. *Climatic Change, 52*(1–2), 175–199.

Ruthsatz, B., & Movia, C. (1975). *Relevamiento de las estepas andinas del noreste de la provincia de Jujuy, República Argentina*. Buenos Aires: Fundación para la Educación, la Ciencia y la Cultura.

Samec, C. T. (2011). *Perspectiva isotópica sobre la alimentación de camélidos domésticos y silvestres de la Puna Jujeña: Construyendo un marco de referencia para estudios arqueológicos*. Grade thesis, Universidad de Buenos Aires.

Schutkowski, H. (2006). *Human ecology. Biocultural adaptations in human communities*. Berlin-Heidelberg-New York: Springer.

Sealy, J. C., Van Der Merwe, N. J., Lee Thorp, J. A., & Lanham, J. (1987). Nitrogen isotope ecology in southern Africa: Implications for environmental and dietary tracing. *Geochimica et Cosmochimica Acta, 51*, 2707–2717.

Servant, M., & Servant-Vildary, S. (2003). Holocene precipitation and atmospheric changes inferred from river paleowetlands in the Bolivian Andes. *Palaeogeography, Palaeoclimatology, Palaeoecology, 194*, 187–206.

Tchilinguirian, P. (2009). *Paleoambientes Holocenos en la Puna Austral (27°S): Implicaciones geoarqueológicas*. Ph.D. dissertation, Universidad de Buenos Aires.

Thomas, D. S. G. (1997). *Arid zone geomorphology. Process, form and change in drylands*. Chichester: Wiley.

Thompson Hobbs, N., Reid, R. S., Galvin, K. A., & Ellis, J. E. (2008). Fragmentation of arid and semi-arid ecosystems: Implications for people and animals. In K. A. Galvin, R. S. Reid, R. H. Behnke Jr., & N. Thompson Hobbs (Eds.), *Fragmentation in semi-arid and arid landscapes* (pp. 25–44). Dordrecht: Springer.

Turner, B. L., Kingston, J. D., & Armelagos, G. J. (2010). Variation in dietary histories among the inmigrants of Machu Picchu: Carbon and nitrogen isotopic evidence. *Chungara, 42*, 515–534.

Thompson, L., Mosley-Thompson, E., & Henderson, K. (2000). Ice-core palaeoclimate records in tropical South America since the Last Glacial Maximum. *Journal of Quaternary Science, 15*(4), 377–394.

Thompson, L., Davis, M., Mosley–Thompson, E., Sowers, T. A., Henderson, K., Zagorodnov, V. S., et al. (1998). A 25,000-year tropical climate history from Bolivian ice cores. *Science, 282*, 1858–1864.

Thompson, L., Mosley-Thompson, E., Davis, M., Lin, P.-N., Henderson, K., Cole-Dai, J., et al. (1995). Late glacial stage and Holocene tropical ice core records from Huscarán, Peru. *Science, 269*, 46–50.

Veit, H. (1996). Southern westerlies during the Holocene deduced from geomorphological and pedoglacial studies in the Norte Chico, Northern Chile (27–33 S). *Palaeogeography, Palaeoclimatology, Palaeoecology, 123*, 107–119.

Vilá, B. L. (1991). Vicuñas (*Vicugna vicugna*) agonistic behavior during the reproductive season. In F. Spitz, G. Janeau, G. Gonzalez, & S. Aulagnier (Eds.), *Ongulés/Ungulates 9.1 Proceedings of the International Symposium* (pp. 475–482). Toulose: S.F.E.P.M. & I.R.G.M..

Vilá, B. L. (1999). La importancia de la etología en la conservación y manejo de las vicuñas. *Etología, 7*, 63–68.

Vilá, B. L. (2000). Comportamiento y Organización Social de la Vicuña. In B. González, F. Bas, C. Tala, & A. Iriarte (Eds.), *Manejo Sustentable de la Vicuña y el Guanaco* (pp. 175–191). Santiago.

Villagrán, C. (1993). Una interpretación climática del registro palinológico del último glacial-postglacial en Sudamérica. *Bulletin de l'Institute Francaise des Etudes Andines, 22*(1), 243–258.

Vuille, M., & Keimig, F. (2004). Interannual variability of summertime convective cloudiness and precipitation in the central Andes derived from ISCCP-B3 data. *Journal of Climate, 17*, 3334–3348.

Wolverton, S. (2008). Harvest pressure and environmental carrying capacity: An ordinal-scale model of effects on ungulate prey. *American Antiquity, 7*(3), 179–199.

Yacobaccio, H. D. (1994). Biomasa animal y consumo en el Pleistoceno–Holoceno Surandino. *Arqueología, 4*, 43–71.

Yacobaccio, H. D. (2004). Social dimensions of camelid domestication in the southern Andes. *Anthropozoologica, 39*(1), 237–247.

Yacobaccio, H., & Morales, M. (2005). Mid-Holocene environment and human occupation of the Puna (Susques, Argentina). *Quaternary International, 132*, 5–14.

Yacobaccio, H. D., Morales, M. R., & Samec, C. T. (2009). Towards an isotopic ecology of herbivory in the Puna ecosystem: New results and patterns in *Lama glama*. *International Journal of Osteoarchaeology, 19*, 144–155.

Yacobaccio, H. D., Samec, C T., & Catá, M. P. (2010). Isótopos estables y zooarqueología de camélidos en contextos pastoriles de la puna (Jujuy, Argentina). In M. A. Gutiérrez, M. De Nigris, P. M. Fernández, M. Giardona, A. Gil, A. Izeta, G. Neme, & H. Yacobaccio (Eds.), *Zooarqueologia a principios del siglo XXI. Aportes teóricos, metodológicos y casos de estudio* (pp. 77–86). Buenos Aires: Editorial del Espinillo.

Zhou, J., & Lau, K. M. (1998). Does a monsoon climate exist over South America? *Journal of Climate, 11*, 1020–1040.

Chapter 7
Climate Change at the Holocene Thermal Maximum and Its Impact on Wild Game Populations in South Scandinavia

Ola Magnell

Abstract The impact of climate change on wild game populations in South Scandinavia is evaluated based on analysis of faunal remains of red deer, roe deer, moose, aurochs and wild boar to trace variations in the abundance and body size before and during the Holocene thermal maximum (HTM). The abundance of aurochs and moose decreases before the HTM. In Scania the red deer, roe deer and wild boar populations are stable before and during HTM, but on Zealand an increase in red deer is noticed while the abundance of wild boar decreases. A decrease in body size of red deer correlates with the HTM, while wild boar seems to increase in size. No change in size of roe deer could be observed. The change of the wild game populations during the HTM and its relevance to wild game biology are also discussed.

Keywords Aurochs • Mesolithic • Moose (*Alces alces*) • Osteometry • Red deer • Roe deer • Wild boar

Introduction

Climate change due to global warming will, in the near future, affect the conditions for ecosystems and animals (Levinsky et al. 2007; Parry et al. 2007). This means challenges for nature conservation and wildlife management on a regional as well as a global scale. In order to maintain a long-term conservation of viable populations of wild game it is of significance to consider the effects climate change will have on wild game populations. Examples of expected changes caused by climate are spread of diseases and parasites, along with changes of vegetation and habitats that will further affect the wild game populations. Climate change will probably alter the geographical distribution and the abundance of different species and the dynamics of wild game populations. A way to understand how climate changes may affect wild game is to study the past. By analysing palaeoclimatic data and faunal assemblages from archaeological sites it is possible to reconstruct how prehistoric wild game populations were affected by climate changes. Understanding changes in the past can help to model impacts of current and future climate changes on wild game populations.

Various palaeoclimatic records, such as pollen, tree-rings, ice-cores, glacier fluctuations and marine sediments, show, for Northern Europe during the Early Holocene, a long-term trend of an increasingly warmer climate punctuated by rapid climatic shifts between 11,500–7,500 cal BP. This was followed by a period of uninterrupted warmth during the Holocene thermal maximum (HTM), ca. 7,500–5,500 cal BP with a peak around 6,000 cal BP, when annual temperatures were about 2°C warmer than present (Johnsen et al. 2001; Davis et al. 2003; Snowball et al. 2004). This corresponds approximately to the predicted increase of annual temperature caused by global warming during this century (Christensen and Hewitson 2007). The climate during the HTM differed from that of the Early Holocene not only by higher annual temperature, but also by the great discrepancy between winter and summer temperature in each period. During the HTM, the summer temperature seems to have been somewhat higher (about 0.5°C) than during the previous period, but the largest difference is noticed in the winter temperature (about 1°C). However, in comparison with the present climate, the largest difference compared to the HTM is in the summer temperatures (Davis et al. 2003).

The faunal record for the Mesolithic in South Scandinavia is rich with well-preserved bone assemblages from several sites dating to the period before and during the HTM. Since the subsistence strategy was based on gathering plants, fishing and hunting, the faunal assemblages are abundant in remains of wild game, especially ungulates. Southern Scandinavia is also a region where paleoenvironmental development is well studied (Snowball et al. 2004; Björkman 2007). This makes the archaeozoological record from Mesolithic

O. Magnell (✉)
Statens Historiska Museer, Odlarevägen 5, 226 60 Lund, Sweden
e-mail: ola.magnell@arkeologerna.com

South Scandinavia a suitable material for studying the relationship between wild game populations and climate change at the HTM. This study is based on analysis of the abundance of taxa and body size of ungulates. Body size and bone measurements in ungulates correlate with climate, but also other ecological factors such as habitat quality and nutrition (Clutton-Brock and Albon 1983; Langvatn and Albon 1986; Bertouille and de Crombrugghe 1995; Cussans 2017). The purpose of this study is to compare the ungulate fauna in South Scandinavia during the HTM and Early Holocene. Does the abundance of different ungulate taxa and body size of animals change between these periods? If so, is it possible to explain these changes as an effect of climate change? Do different species of ungulates respond differently to climatic change? Is it possible from the changes of the ungulate fauna during the HTM to understand what will happen with wild game populations in the near future considering the effects of global warming? A concern about the effect of climate change on the wild game populations has been raised by the nature conservation authorities. This study is an attempt, through archaezoological analysis, to contribute to an understanding of the dynamics of wild game populations and climate change in a long-term perspective.

The relevance of archaeozoology to nature conservation and wildlife management has been stated by certain scholars for almost a century (Wintemberg 1919). However it is only in the last several decades that archaeozoology has begun to have an impact on issues concerning nature conservation (Lyman 1996, 2006; Lauwerier and Plug 2004; Lyman and Cannon 2004; Frazier 2007). The benefits of archaeozoology, such as the long-time perspective and as an ecological benchmark to "pristine" conditions, are obvious to most people working with animal remains, but archaeozoology is often considered to be of less relevance within nature

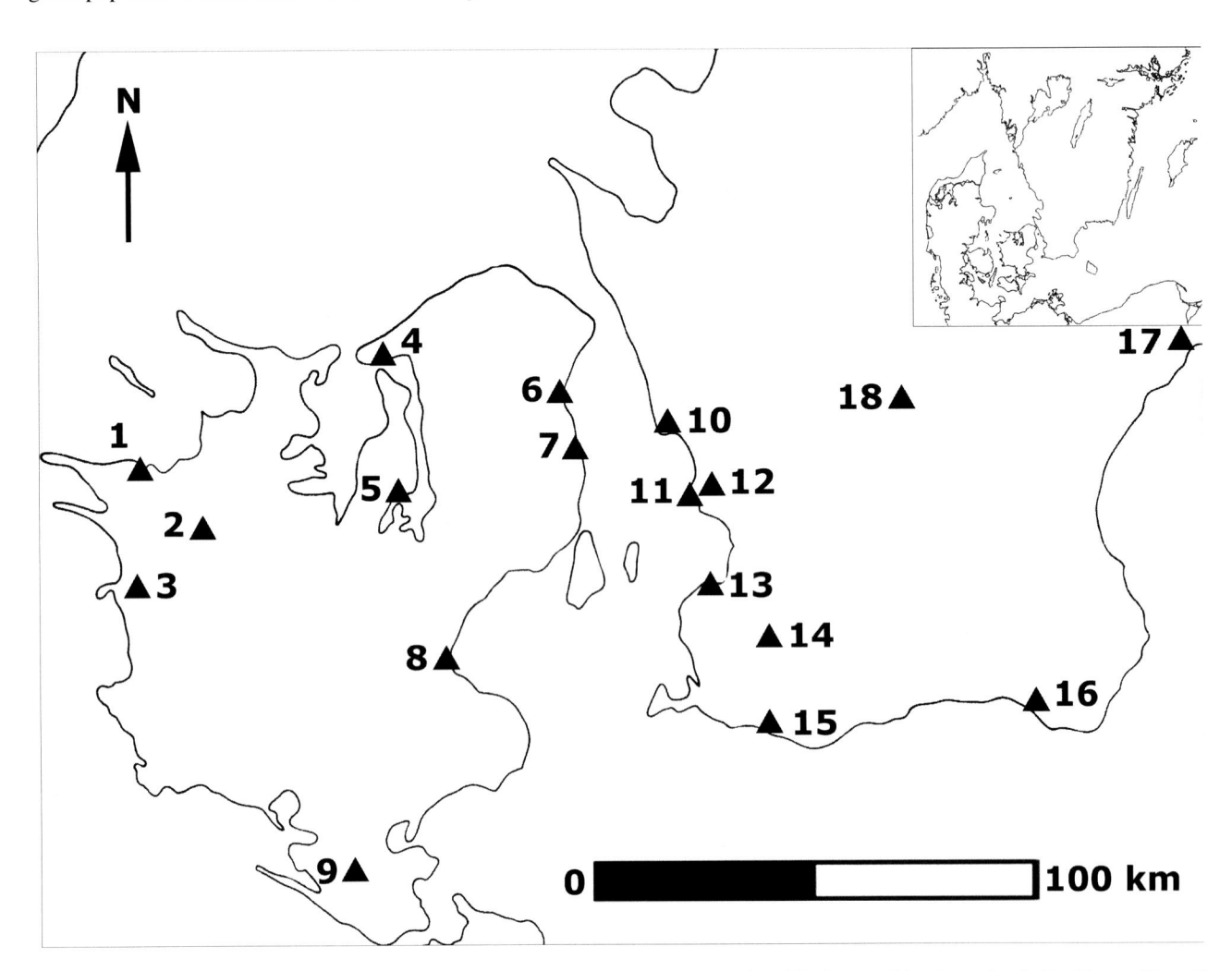

Fig. 7.1 South Scandinavia, Scania (southern Sweden) and Zealand (eastern Denmark) with sites used in the study. 1. Smakkerup Huse, 2. Kongemose, Mosegården III, Muldbjerg I, Nøddekonge, Præstelyngen, Ulkestrup Lyng, Vejkonge and Åkonge, 3. Mullerup south and Mullerup north, 4. Sølager (Ertebølle layer), 5. Lollikhuse, 6. Nivå 10 and Nivågård, 7. Maglemosegaard and Stationsvej, 8. Ølby Lyng, 9. Lundby I, Lundby II, Sværdborg I and Sværdborg II, 10. Tågerup I, II and III, 11. Löddesborg, 12. Hög, 13. Arlöv I and Segebro, 14. Bökeberg III and Södra Lindved, 15. Skateholm I and Skateholm II, 16. Bredasten, 17. Nymölla III, 18. Ageröd I:B, Ageröd I:D, Ageröd I:HC, Ageröd III, Ageröd V, Ringsjöholm and Sjöholmen

conservation and wildlife management (Lyman 2006; Frazier 2007). Archaeozoology is often considered only an interesting curiosity. If wildlife management states that it aims at maintaining a long-term conservation of viable populations of wild game, it is necessary not only to rely on studies of wild game population dynamics over a few years or a decade, but also to become aware of the potential in archaeozoology. However, archaeozoologists also need to deal with issues that are of relevance to wildlife management and to better communicate the results. More co-operation and dialogue is needed between archeozoologists and wild game management. This study is in many ways rooted in archaeozoology, but hopefully it will show that prehistoric animal bones have relevance to wildlife management and help to develop an archaeozoology of more relevance to nature conservation.

Material

The study is based on faunal remains of ungulates; red deer (*Cervus elaphus*), roe deer (*Capreolus capreolus*), moose (*Alces alces*), aurochs (*Bos primigenius*) and wild boar (*Sus scrofa*) from 42 Mesolithic settlements in Scania (southern Sweden) and Zealand (eastern Denmark) dated to about 9,500–6,000 cal BP (Fig. 7.1; Tables 7.1 and 7.2). It is important to consider that the material originates from settlements and hunted animals. This means that changes of frequency of taxa in settlements may not only reflect changes in the fauna but also changes in hunting strategies.

The study is based on a large set of sites excavated between 1900 and 1999, which means that different excavation strategies and use of sieving could have biased the samples. Since the study is based on remains of larger mammals, the effect of different excavation techniques is assumed to have minor effect on the quantification.

This study has not included the subfossil finds from non-archaeological contexts, since this record is heavily biased towards larger taxa, such as aurochs. The fact that remains from male aurochs are clearly more frequent than remains from female aurochs in the collections of non-archaeological subfossils shows that this record is biased towards large and conspicuous bones (Liljegren and Lagerås 1993). Further, relatively few radiocarbon dating of subfossils from non-archaeological finds of roe deer and wild boar are available.

Table 7.1 Faunal remains (NISP) of ungulates from Mesolithic sites on Zealand, Denmark, ordered by date. [a]Reference for dating. [b]Reference for archaeozoological data

	Aurochs	Moose	Roe deer	Red deer	Wild boar	References
Lundby II	103	159	88	97	314	Henriksen (1980)[a]; Rosenlund (1980)[b]
Mullerup-south	317	501	389	403	882	Winge (1904)[b]; Tauber (1971)[a]; quantification by the author, Previous unpublished data
Mullerup north	50	63	19	20	76	Winge (1904)[b]; Blankholm (1992)[a]; quantification by the author, Previous unpublished data
Sværdborg I (1918)	432	403	676	443	925	Winge (1919)[b]; quantification by the author, Previous unpublished data[a]
Sværdborg I (1943)	43	127	1111	196	783	Aaris-Sørensen (1976)[b]; Henriksen (1976)[a]
Sværdborg II	2	0	38	50	25	Rosenlund (1971)[b]; Henriksen (1976)[a]
Ulkestrup Lyng	183	32	157	417	158	Noe-Nygaard (1995)[a, b]
Mosegården III	5	1	41	87	15	Møhl (1984)[a, b]
Kongemose	0	4	911	3286	794	Noe-Nygaard (1995)[a, b]
Stationsvej	0	0	138	199	90	Persson (1999)[a]; Enghoff (2011)[b]
Nivå 10	0	0	334	103	68	Enghoff (2011)[a, b]
Nivågård	0	0	1235	905	249	Enghoff (2011)[a, b]
Maglemosegaard	0	0	1042	2049	235	Aaris-Sørensen (1985)[b]; Persson (1999)[a]
Sølager (Ertebølle layer)	0	0	186	42	48	Skaarup (1973)[b]; Persson (1999)[a]
Ølby Lyng	0	1	330	646	137	Møhl (1971)[b]; Persson (1999)[a]
Præstelyngen	0	0	263	778	114	Noe-Nygaard (1995)[a, b]
Lollikhuse	0	0	996	828	214	Friborg (1999)[a, b]; Magnussen (2007)[b]
Smakkerup Huse	0	0	661	720	269	Hede (2005)[b]; Price and Gebauer (2005)[a]
Åkonge	0	0	466	1659	189	Gotfredsen (1998)[b]; Persson (1999)[a]
Vejkonge	0	0	6	167	8	Gotfredsen (1998)[b]; Persson (1999)[a]
Nøddekonge	0	0	115	355	7	Gotfredsen (1998)[b]; Persson (1999)[a]
Muldbjerg I	0	0	116	665	5	Noe-Nygaard (1995)[a, b]

Table 7.2 Faunal remains (NISP) of ungulates from Mesolithic sites in Scania, Sweden, ordered by dating. [a]Reference for dating. [b]Reference for archaeozoological data

	Aurochs	Moose	Roe deer	Red deer	Wild boar	References
Ageröd I:B	6	10	18	72	54	Larsson (1978)[a]; Lepiksaar (1978)[b]
Ageröd I:D	17	11	12	123	34	Larsson (1978)[a]; Lepiksaar (1978)[b]
Ageröd I:HC – lower peat	55	27	24	174	104	Larsson (1978)[a]; Magnell (2006)[b]; Previous unpublished data
Ageröd I:HC – white layer	52	23	84	241	168	Larsson (1978)[a]; Magnell (2006)[b]; Previous unpublished data
Ageröd I:HC – gravel layer	32	11	32	119	57	Larsson (1978)[a]; Magnell (2006)[b]; Previous unpublished data
Ageröd I:HC – culture layer	75	94	117	457	313	Larsson (1978)[a]; Magnell (2006)[b]; Previous unpublished data
Tågerup – phase I	1	41	384	518	490	Eriksson and Magnell (2001a)[a, b]
Ringsjöholm	144	239	400	1015	841	Sjöström (1997)[a]; Previous unpublished data[b]
Segebro	3	22	380	675	424	Lepiksaar (1982)[b]; Persson (1999)[a]
Hög	0	3	40	32	47	Iregren and Lepiksaar (1993)[a, b]
Ageröd III	19	8	8	99	44	Previous unpublished data[a, b]
Ageröd V	0	48	59	294	58	Lepiksaar (1983)[b]; Persson (1999)[a]
Tågerup – phase II	0	0	82	84	58	Eriksson and Magnell (2001a)[a, b]
Bökeberg III	0	120	163	1060	190	Persson (1999)[a]; Eriksson and Magnell (2001b)[b]
Skateholm II	0	5	703	248	281	Jonsson (1988)[b]; Persson (1999)[a]
Skateholm I	0	2	174	182	447	Jonsson (1988)[b]; Persson (1999)[a]
Arlöv I	5	3	54	115	146	Jonsson (1988)[b]; Persson (1999)[a]
Nymölla III	0	0	23	77	31	Wyszomirska (1988)[b]; Persson (1999)[a]
Södra Lindved	0	1	14	64	12	Althin (1954)[a]; Previous unpublished data[b]
Tågerup – phase III	0	0	108	129	81	Eriksson and Magnell (2001a)[b]; Karsten and Knarrström (2003)[a]
Sjöholmen (level 4–6)	7	50	109	563	231	Berlin (1930)[b], quantification by the author, Previous unpublished data[a]
Löddesborg	0	0	32	68	44	Hallström (1984)[b]; Persson (1999)[a]
Bredasten	0	0	167	625	1334	Magnell (2006)[a]; Previous unpublished data[b]

The dating of sites is mainly based on radiocarbon dating, but for eight sites no radiocarbon dates have been available; instead, the dating is based on typology of lithics and pollen (Tables 7.1 and 7.2). The median date has been used as an approximation of the chronology of the faunal remains. The study also includes previously unpublished radiocarbon dates of ungulate bones from: Lundby I (LuS 9085 8585 ± 60 BP, LuS 9086: 8535 ± 60 BP, LuS 9087 7950 ± 60 BP), Sværdborg I – excavation 1918 (LuS 9082: 7920 ± 60 BP, LuS 9083: 8170 ± 60 BP, LuS 9084: 8075 ± 60 BP), Ageröd III (LuS 7904: 6950 ± 60 BP), and Sjöholmen LuS 7337: 5635 ± 55 BP, LuS 7869: 5760 ± 120 BP).

Material from Lundby I was only included in the osteometric analysis. In the osteometric analysis no measurements from Zealand sites dating after 8,500 cal BP have been included to avoid any effects on the size caused by isolation, when the area becomes an island, rather than by climate.

Methods

The analysis of the abundance of ungulate taxa has been based on NISP (number of identified specimens) and the relative frequency of different ungulate taxa. Consequently the quantifications are not based on the total fauna of the sites. There are several problems with quantification based on NISP, such as the fact that many fragments can originate from the same individual, differential fragmentation can occur between taxa and sites, and anatomical differences between taxa can affect the result of the quantification (Grayson 1979; Gilbert and Singer 1982). This quantification method was used, however, because these data are available from most sites, while other measures, such as MNI and MNE are not always published and also because these estimates have their own inherent problems.

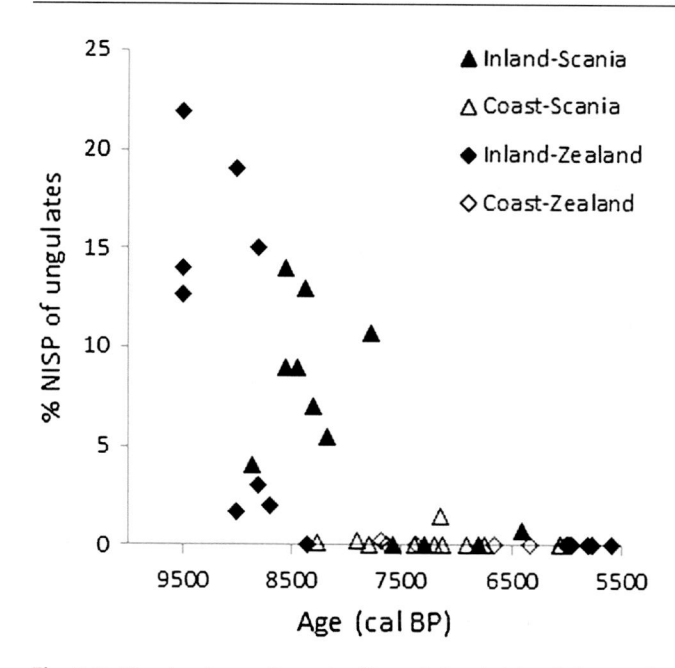

Fig. 7.2 The abundance of aurochs (*Bos primigenius*) in relation to all ungulate taxa from inland and coastal Mesolithic sites on Zealand and Scania in South Scandinavia before and during HTM (7,500–5,500 cal BP)

The osteometric analysis is based on postcranial bones from red deer, roe deer and wild boar, since there are too few bones of aurochs and moose from the HTM. Variations in body size are based on osteometric analysis of measurements according to von den Driesch (1976) and Payne and Bull (1988). The study includes measurements of different bones elements, but the main focus is on osteometric analysis of the astragali: greatest lateral length (GLl) and humeri; breadth of the distal trochlea (BT) for red deer; distal breadth (Bd) for roe deer and diameter of distal trochlea at its narrowest point (HTC) for wild boar. These bone elements provide the largest sample of measurements, and their measurements correlate well with overall body size (Teichert 1969; Magnell 2004).

Results
Abundance of Taxa

The abundance of aurochs is clearly lower in the HTM than during the Early Holocene, both on Zealand and in Scania (Tables 7.3 and 7.4), but the decline of the aurochs population is already apparent in the transition from the Boreal to the Atlantic chronozone, ca 9,000 cal BP (Fig. 7.2). About 8,500 cal BP aurochs were extinct on Zealand and somewhat later they disappear from Southern Sweden. This has been shown previously in other studies (Degerbøl and Fredskild 1970; Aaris-Sørensen 1980, 1999; Ekström 1993). The results presented in this study provide a more detailed description of the decline of aurochs because it is based on a larger data set and on frequencies rather than on presence or absence of taxa. It shows that the extinction on Zealand was a rapid event that probably took place within a few centuries (Fig. 7.2). Further, the results in this study also show that an aurochs population was present in Scania as late as 7,800 cal BP, during a period when the aurochs previously have been assumed to be extinct in southern Sweden (Ekström 1993). Finds and radiocarbon dating of aurochs remains from Arlöv I, a Late Mesolithic site, indicate that a small population was possibly present as late as 6,500 cal BP. During the HTM there are no clear differences in the abundance of aurochs between inland and coastal sites or between Zealand and Scania (Fig. 7.2; Tables 7.5 and 7.6).

The moose population starts to decrease between 9,500–9,000 cal BP and becomes even rarer during the HTM (Fig. 7.3). The frequencies of moose are significantly higher in the period before the HTM in comparison to during the HTM, both in Scania and on Zealand (Tables 7.3 and 7.4). On Zealand the extinction of moose seems to have been a relatively fast event, and by 8,500 cal BP the species has disappeared from the island (Fig. 7.3). In Scania during the HTM, the abundance of moose is, significantly lower in coastal sites in comparison with inland sites (Fig. 7.3; Table 7.5).

There are no differences in the frequency of red deer in the sites from the HTM and the previous period in Scania, but on Zealand there is a significant difference in red deer between the two periods (Fig. 7.4; Tables 7.3 and 7.4). The difference between the two areas could be explained by the fact that there are no sites from Scania dating to the Boreal (i.e., 10,000–9,000 cal BP), a period when the abundance of red deer seems to have been particularly low on Zealand. It is clear that the abundance of red deer increases between 9,500–8,500 cal BP from 9–16 to 42–62%, during the same period as the frequency of aurochs and moose

Table 7.3 Results of Mann-Whitney U test analysis for relative frequencies of NISP from ungulate taxa on sites from the period before (9,500–7,500 cal BP) and during the HTM (7,500–5,500 cal BP) on Zealand, eastern Denmark

	9,500–7,500 cal BP			7,500–5,500 cal BP			Result
	n	Median (%NISP)	Range (%NISP)	n	Median (%NISP)	Range (%NISP)	*P < 0.05
Aurochs (*Bos primigenius*)	11	3	0–22	11	0	0	<0.001*
Moose (*Alces alces*)	11	3	0–28	11	0	0	0.004*
Red deer (*Cervus elaphus*)	11	18	9–66	11	62	15–92	0.019*
Roe deer (*Capreolus capreolus*)	11	24	8–68	11	30	3–67	0.562
Wild boar (*Sus scrofa*)	11	22	10–41	11	10	1–17	<0.001*

Table 7.4 Results of Mann-Whitney U test analysis for frequencies of NISP from different taxa on sites from the period before (9,500–7,500 cal BP) and during the HTM (7,500–5,500 cal BP) from Scania, southern Sweden

	9,500–7,500 cal BP			7,500–5,500 cal BP			Result
	n	Median (%NISP)	Range (%NISP)	n	Median (%NISP)	Range (%NISP)	*P < 0.05
Aurochs (*Bos primigenius*)	12	7	0–14	11	0	0–2	0.003*
Moose (*Alces alces*)	12	6	2–11	11	0	0–8	<0.001*
Red deer (*Cervus elaphus*)	12	45	26–64	11	41	20–70	0.739
Roe deer (*Capreolus capreolus*)	12	14	4–33	11	18	8–57	0.189
Wild boar (*Sus scrofa*)	12	29	13–39	11	25	12–63	0.739

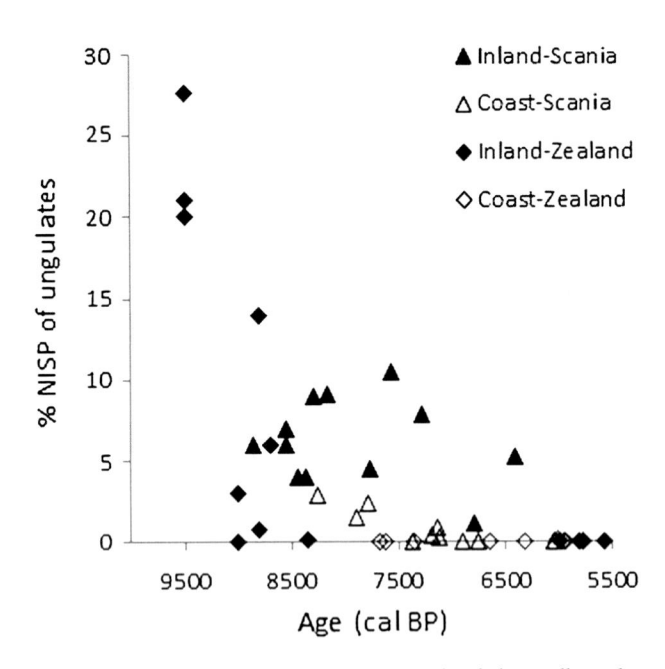

Fig. 7.3 The abundance of moose (*Alces alces*) in relation to all ungulate taxa from inland and coastal Mesolithic sites on Zealand and Scania in South Scandinavia before and during HTM (7,500–5,500 cal BP)

decreases (Figs. 7.2, 7.3 and 7.4) and the frequencies of wild boar and roe deer remain the same (Figs. 7.5 and 7.6). This observation indicates an absolute change in the abundance of red deer and not only a relative due to the decrease of aurochs and moose. During the Late Mesolithic and corresponding to the HTM, a large variation in the frequency of red deer (15–92%) can be noticed during the HTM (Fig. 7.4). The abundance of red deer is also significantly higher in inland sites in comparison with coastal sites during the HTM (Fig. 7.4; Table 7.5).

The abundance of roe deer during the HTM is not significantly different from previous period in both Zealand and Scania (Fig. 7.5; Tables 7.3 and 7.4). The frequency of roe deer varies a lot between different sites (8–67%). Roe deer is also relatively more common in the faunal assemblages from the coastal region as compared to the inland sites (Fig. 7.5; Table 7.5).

The quantification of wild boar indicates different responses of the wild boar populations on Zealand and in Scania between the HTM and the previous period (Fig. 7.6). In Scania no clear differences between the both periods are evident, but on Zealand a significant decrease over time is noticed (Tables 7.3 and 7.4). The decline in wild boar on Zealand from 9,500 to 6,000 cal BP is possibly explained by the isolation of the population when the area becomes an island. Similar to roe deer, the frequencies of wild boar show a large variation between different sites from the HTM (12–63%) in Scania. Wild boar remains are also more abundant on the coastal sites in comparison with inland sites (Fig. 7.6; Table 7.5).

Table 7.5 Results of Mann-Whitney U test analysis for frequencies of NISP from different taxa on inland and coastal sites dating to the HTM (7,500–5,500 cal BP in south Scandinavia (Zealand and Scania)

	Inland sites			Coastal sites			Result
	n	Median (%NISP)	Range (%NISP)	n	Median (%NISP)	Range (%NISP)	*P < 0.05
Aurochs (*Bos primigenius*)	8	0	0–0.7	14	0	0–2	0.867
Moose (*Alces alces*)	4	7	1–11	8	0	0–1	0.004*
Red deer (*Cervus elaphus*)	8	71	59–92	14	40	15–62	<0.001*
Roe deer (*Capreolus capreolus*)	8	15	3–24	14	33	8–67	0.006*
Wild boar (*Sus scrofa*)	8	9	1–24	14	24	7–63	0.006*

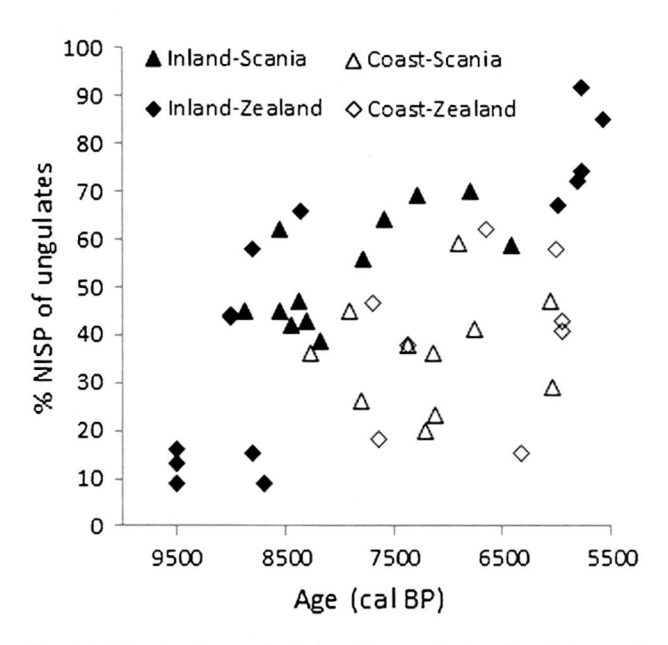

Fig. 7.4 The abundance of red deer (*Cervus elaphus*) in relation to all ungulate taxa from inland and coastal Mesolithic sites on Zealand and Scania in South Scandinavia before and during HTM (7,500–5,500 cal BP)

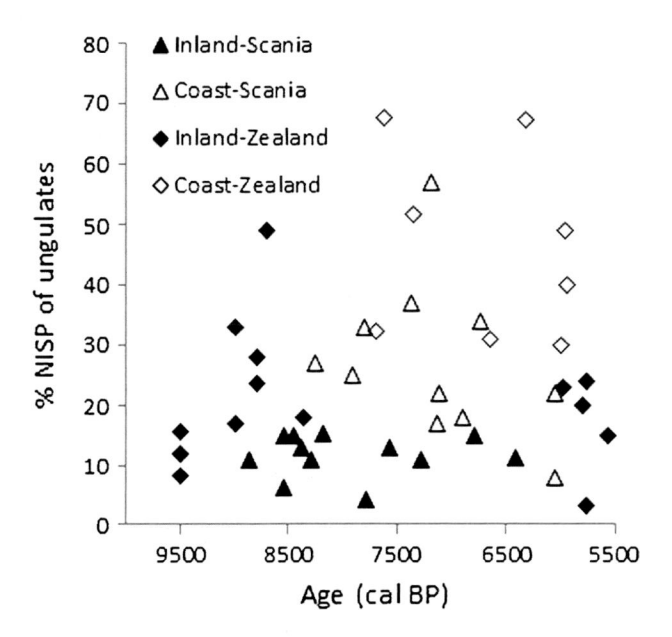

Fig. 7.5 The abundance of roe deer (*Capreolus capreolus*) in relation to all ungulate taxa from inland and coastal Mesolithic sites on Zealand and Scania in South Scandinavia before and during HTM (7,500–5,500 cal BP)

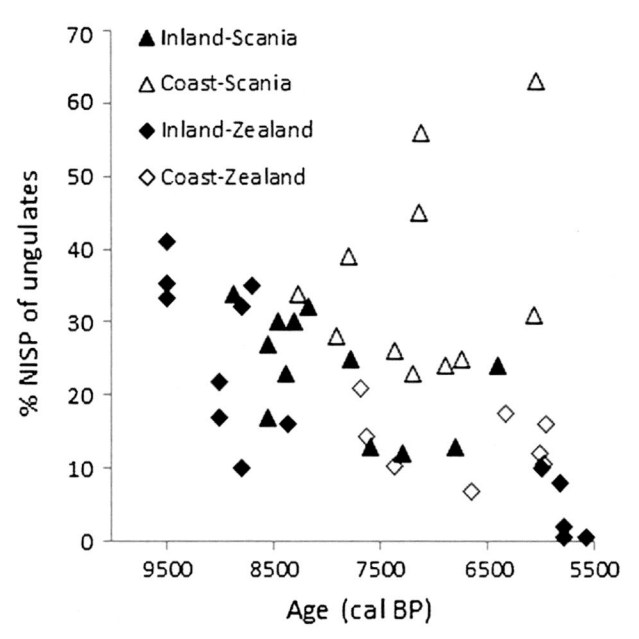

Fig. 7.6 The abundance of wild boar (*Sus scrofa*) in relation to all ungulate taxa from inland and coastal Mesolithic sites on Zealand and Scania in South Scandinavia before and during HTM (7,500–5,500 cal BP)

Osteometry

The body size of red deer seems to be smaller during the HTM than the previous period (Fig. 7.7; Table 7.6). Statistical analysis (t-test) shows that the changes are significant both for astragali ($p < 0.001$) and humeri ($p = 0.008$). The roe deer remains show no change in body size between samples before or during the HTM (Fig. 7.7; Table 7.7). No significant change is evident in the measurements of astragali ($p = 0.842$) and humeri ($p = 0.598$). For wild boar, the osteometric analysis results are ambiguous (Fig. 7.7; Table 7.8). Measurements of the humeri show a significant increase ($p < 0.001$) in size between the period before and during the HTM, while for the astragali ($p = 0.079$) no significant difference is evident.

A possible explanation of the ambiguity in the result of the osteometric analysis, with contradicting results in measurements of the humeri and astragali, could be differences in sex ratios. The upper and anterior limb bones, such as the

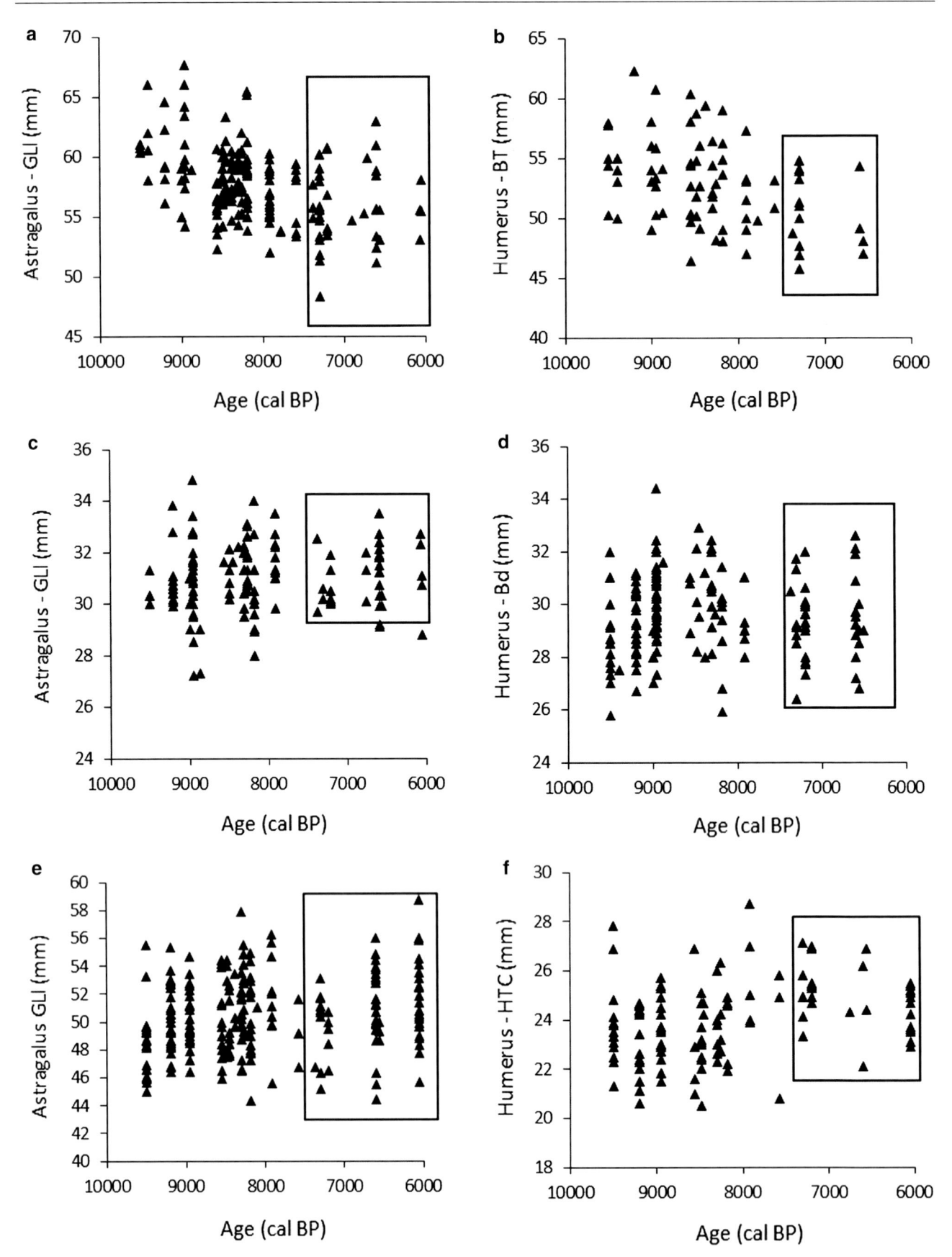

Fig. 7.7 Bone measurements (mm) of red deer (*Cervus elaphus*) (**a**, **b**), roe deer (*Capreolus capreolus*). (**c**, **d**) and wild boar (*Sus scrofa*) (**e–f**) from Mesolithic sites before and during the HTM (rectangle) in South Scandinavia in correlation with dating. Measurements according to von den Driesch (1976) and Payne and Bull (1988). No measurements from Zealand after 8,500 cal BP are included to avoid affects on size caused by isolation due to island formation

Table 7.6 Descriptive statistics of measurements (mm) of red deer (*Cervus elaphus*) before and during HTM in southern Scandinavia. Measurements according to von den Driesch (1976). No measurements from Zealand after 8,500 cal BP are included to avoid affects on size caused by isolation due to island formation

	Astragalus – GLl		Humerus – BT	
	9.500–7.500 cal BP	7.500–6.000 cal BP	9.500–7.500 cal BP	7.500–6.000 cal BP
Mean	58.1	55.8	53.6	50.7
Range	52.0–67.6	51.1–62.9	46.4–60.7	45.7–54.8
n	145	43	64	16
SD	2.846	3.102	4.098	3.184

Table 7.7 Descriptive statistics of measurements (mm) of roe deer (*Capreolus capreolus*) before and during HTM in southern Scandinavia. Measurements according to von den Driesch (1976). No measurements from Zealand after 8,500 cal BP are included to avoid affects on size caused by isolation due to island formation

	Astragalus – GLl		Humerus – Bd	
	9.500–7.500 cal BP	7.500–6.000 cal BP	9.500–7.500 cal BP	7.500–6.000 cal BP
Mean	31.2	31.0	26.9	29.5
Range	27.7–34.8	28.8–33.5	25.8–34.4	26.4–32.6
n	97	35	116	41
SD	1.361	1.148	1.542	1.496

Table 7.8 Descriptive statistics of measurements (mm) of wild boar (*Sus scrofa*) before and during HTM in southern Scandinavia. Measurements according to von den Driesch (1976) and Payne and Bull (1988). No measurements from Zealand after 8,500 cal BP are included to avoid affects on size caused by isolation due to island formation

	Astragalus – GLl		Humerus – HTC	
	9.500–7.500 cal BP	7.500–6.000 cal BP	9.500–7.500 cal BP	7.500–6.000 cal BP
Mean	50.2	50.8	23.6	24.8
Range	44.3–57.9	46.7–58.7	20.5–28.7	20.8–27.1
n	168	68	88	31
SD	2.604	2.806	1.601	1.245

humeri, show more sexual dimorphism in size and bone measurements than the lower posterior limb bones, such as astragali (Payne and Bull 1988). It is possible that the change of size noted in humeri reflects changes in sex ratios in the bone assemblages that not are shown in the less sexually dimorphic measurements of astragali. If this is the case, the size increase in humeri of wild boar would mean a relatively higher proportion of males in the sites from the HTM compared to the previous period.

Studies of changes in the size of red deer have been made by Ahlén (1965), Noe-Nygaard (1995) and Magnell (2006), showing a decrease in size from the Boreal to Late Atlantic, especially on the Danish islands but also in Scania. The body size of roe deer from Denmark has been shown to decrease on the Danish islands but not on Jutland during the Boreal to Atlantic (Jensen 1991). No decrease in size of wild boar during the Mesolithic in South Sweden has previously been observed (Magnell 2004).

Unfortunately, it has not been possible to study the body size in aurochs and moose since there are too few measurements of the two species from the HTM.

Discussion

The decrease in abundance of aurochs correlates rather well with the increasing temperature in the Early Holocene, but it can be questioned whether it is climate or other factors more indirectly linked to climate change, such as sea level and vegetation, that affected the aurochs population. The areal distribution of aurochs shows that South Scandinavia was the northern boundary of their distribution and their presence in the Mediterranean shows that the species was adapted to a warmer climate (Van Vuure 2002). The formation of an outlet of the Baltic Sea in the Stora Belt area about 10.000 cal BP and the Öresund Strait about 8,500 cal BP resulted in an isolation of the Scandinavian aurochs population and made it sensitive to environmental changes. The rapid decrease in abundance of aurochs and extinction on Zealand is most likely an effect of the isolation of the population (Ekström 1993; Yu 2003; Aaris-Sørensen 1999, 2009). The establishment of broad-leaf trees during the Atlantic resulted in a more closed forest with fewer habitats for grazing and possibly a lower population density of the aurochs (Ekström 1993). The isolation and change of biotopes most likely made the aurochs population more sensitive to hunting pressure, resulting in a decrease of abundance and finally extinction of the aurochs. As in other cases of extinction, its dynamics include climatic, geographic and anthropogenic factor (Ochoa and Piper 2017).

The decrease of the moose population by the HTM could possibly be explained by climatic factors. The distribution of moose is closely linked to coniferous forests and pine forests with moors (Skuncke 1949; Peterson 1955). The decrease in moose coincides rather well with a period of warmer climate when pine forests decrease and were replaced by broadleaf forests with oak and elm (Björkman 2007). The fact that increasing temperature also increases the occurrence of pathogens as in recent moose populations could also have affected the moose population during the HTM (Murray et al. 2006). This has previously been suggested to have been a factor in the extinction of moose in Central Europe (Schmölcke and Zachos 2005).

The increase of the red deer population on Zealand during the Early Holocene is probably linked to change of vegetation to temperate broad-leaf forests, but possibly also to less competition for browse from decreasing populations of aurochs and moose. The increase of red deer over time on Zealand during the Late Mesolithic is more evident for inland sites than coastal sites and is possibly explained by specialization in hunting of red deer at inland hunting camps. The reasons for the decrease of wild boar on Zealand are uncertain, but one could be isolation of the population. In Scania no difference in abundance of red deer, roe deer or wild boar could be noticed between the HTM and the previous period. Increasing temperature associated with the onset of the HTM does not seem clearly to have affected the abundance of red deer, roe deer and wild boar in Southern Scandinavia. The large variations in abundance of red deer, roe deer and wild boar on different sites during the HTM is not likely explained by climate; rather, they are better explained by specialisation in hunting and local biotopes at different sites.

The lower abundance of moose and red deer coincident with the higher abundance of roe deer and wild boar in the coastal sites compared to the inland sites is most likely explained by a higher hunting pressure in the coastal region due to denser and more sedentary human populations on the coast. The higher fecundity of wild boar and roe deer in comparison to red deer and moose make the former two species less sensitive to high hunting pressure (Mitchell et al. 1977; Briedermann 1986; Hewison 1996). The differences in hunting strategies between the coast and inland could be explained by optimal foraging theory and the premise of body size as a predictor of prey rank (Smith 1983; Kelly 1995). A decrease of available larger and higher ranked species at the coast may have resulted in a change of prey choice by the Mesolithic hunters from, moose and red deer, towards wild boar and roe deer. On Mesolithic coastal sites from the Northern Adriatic an increasing scarcity of larger prey species such as red deer has also been observed (Pilaar Birch and Miracle 2017). It is possible that the difference in abundance of moose and red deer between the inland and coastal areas is the result of differences in biotopes between the two areas, but reconstruction of vegetation in the coastal and the inland zones during the Atlantic in Scania based on pollen records and soil conditions indicates no major difference in habitat between the two zones (Nilsson 1964; Göransson 1991; Regnell et al. 1995; Regnell and Sjögren 2006).

The decrease in body size of red deer at the HTM could be a response to warmer climate. In Norway geographic differences in size with larger red deer in the northern and colder areas have been shown and have been explained as differences in nutritional quality and digestibility of plants (Langvatn and Albon 1986). A possible explanation for the decrease in body size of red deer at the HTM is that the climate change resulted in lower nutritional quality of plants on which they browsed. Analysis of stable isotopes of red deer from Denmark has also shown a change in the $\delta^{13}C$, indicating a change of the diet. This decrease in body size and change of stable isotope ratios in red deer has been interpreted as the result of lower quality of diet (Noe-Nygaard 1995).

The osteometric analysis shows no indications of an overall decrease in body size of roe deer. Since roe deer are selective browsers of high quality forage, it is possible that roe deer was less affected than red deer by a possible decrease of nutritional quality of plants during the HTM (Cederlund and Liberg 1995). The omnivorous wild boar did not decrease in body size during the HTM, but may have increased in size, though the osteometric analysis is somewhat ambiguous. The tendency towards increased body size of wild boar could be explained by increasing habitat quality in the mixed deciduous forest during the HTM or by hunter-gatherer sex selection of males.

This study shows some long-term changes in the wild game populations in southern Scandinavia at the HTM that could have implications for future environmental conditions brought about by climate change. The consequences for aurochs of climate change are of course of relatively little direct relevance since the aurochs became extinct long ago. However, the study implies that if climate change affects the composition of the forest by increasing broad-leaf forests, a decline in moose population can be expected. The study also shows that increasing temperature does not necessary mean change in abundance of red deer, roe deer and wild boar. One could, for example, have expected that roe deer and wild boar should have been more favoured by warmer climate and milder winters, but this does not seem to have been the case. The result is not a surprise, considering that red deer, roe deer and wild boar are adaptable animals that today have distributions over large parts of Eurasia and are found in different biotopes and climatic zones (Niethammer and Krapp 1986). It can be concluded that these species were able to cope well with climate change and an increasing temperature during the HTM and that they might also cope well with possible climate change in the present.

One could argue that the results and conclusions of this study are irrelevant to nature conservation. Does what happened about 7,000 years ago matter to present or future wild game management? The differences are obvious; the aurochs is extinct, today's cultural landscape and managed forests in South Scandinavia are completely different from vast pristine broadleaf forests of the HTM, the human population density in prehistory was low compared to the present, other species have been introduced by humans, such as fallow deer (*Dama dama*), and predation of carnivores on ungulates is relatively minor in Southern Scandinavia at present. On the other hand, if one is interested in what will happen in the future and in long-term sustainability, what relevance have short-term studies of present populations of wild game? If one considers, for example, the difference between present wild game populations in Scandinavia and conditions 150 years ago, they are completely different with minor populations of roe deer and moose, while wild boar was extinct (Ahlén 1977).

Conclusions

This study shows changes in abundance of different ungulate taxa with a decrease of aurochs as well as moose and wild boar, and an increase of red deer on Zealand. These changes start before the HTM, about 9,000 cal BP. On Zealand, isolation when the area becomes an island ca. 8,500 BP was the most important factor to influence wild game populations. In Scania, aurochs and moose are the only wild game species that show changes between the HTM and the previous period. A decrease in body size of red deer between Early Holocene and the HTM is also shown, while no change in size of roe deer and wild boar could be observed. However, it is not possible to explain these changes only in direct relation to climate. Other important factors are changes in land-bridges, isolation of populations, changes of vegetation and hunting. It also appears that different taxa show different responses to climatic changes. In terms of abundance, moose decreases while red deer increases in Zealand and remains relatively constant in Scania, In terms of body size, red deer decreases while no clear change can be observed for roe deer and wild boar in Scania.

This study brings out both the limitations and benefits of the contribution of archaeozoology to nature conservation and wildlife management. It can be concluded that it thus is difficult to, in detail, predict how the abundance, distribution and body size of wild game in South Scandinavia will react to future global warming and climate change due to altered cultural and natural conditions between past and present populations. However, the study indicates that a decrease of the moose population in relation to deer can be expected. Another possible effect of an increasing temperature could be a decrease in body size of deer, while wild boar may increase in size. This indicates the importance of change in vegetation and habitats caused by climate change.

The study exemplifies how investigations of the fauna of the past can contribute to highlight the complexity of predicting the responses of wild game populations to climate changes in the future. Increasing temperature may affect the abundance and/or body size of certain wild game taxa, while others may be unaffected. Finally, a lesson to be learned from the past is that even though climate change is a factor shaping the wild game populations, other factors such as hunting pressure and changes of habitat quality are important. This means that nature conservation and wild game management, as well as the effects of climate changes, will have an important role in shaping the fauna of the future.

References

Aaris-Sørensen, K. (1976). Zoological investigation of the bone material from Sværdborg I-1943. In B. B. Henriksen (Ed.), *Sværdborg I, Excavations 1943–44. A Settlement of the Maglemose Culture* (pp. 137–148). Copenhagen: Akademisk Forlag.

Aaris-Sørensen, K. (1980). Depauperation of the mammalian fauna of the island of Zealand during the Atlantic period. *Videnskablige Meddelelser fra Dansk Naturhistorisk Forening, 142*, 131–138.

Aaris-Sørensen, K. (1985). An example of taphonomic loss in a Mesolithic faunal assemblage. In B. Hesse & P. Wapnish (Eds.), *Animal bone archaeology: From objections to analysis* (pp. 243–247). Washington D.C.: Taraxacum.

Aaris-Sørensen, K. (1999). The Holocene history of the Scandinavian aurochs (*Bos primigenius* BOJANUS, 1827). *Wissenschaftliche Schriften des Neanderthal Museums, 1*, 49–57.

Aaris-Sørensen, K. (2009). Diversity and dynamics of the mammalian fauna in Denmark throughout the last glacial-interglacial cycle, 115-0 kyr BP. *Fossil and Strata, 57*, 1–59.

Ahlén, I. (1965). *Studies on Red Deer, Cervus elaphus L, in Scandinavia*. Stockholm: Viltrevy.

Ahlén, I. (1977). *Faunavård. Om bevarande av hotade djurarter i Sverige*. Uppsala: Skogshögskolan & naturvårdsverket.

Althin, C.-A. (1954). *The Chronology of the Stone Age Settlement of Scania, Sweden*. Lund: Department of Archaeology, Lund University.

Berlin, H. (1930). Förteckning över benfynd från stenåldersboplatsen vid Sjöholmen. *Meddelanden från Lunds universitets historiska museum, 1929–30*, 40–42.

Bertouille, S. B., & de Crombrugghe, S. A. (1995). Body mass and lower jaw development of the female red deer as indices of habitat quality in the Ardennes. *Acta Theriologica, 40,* 145–162.

Björkman, L. (2007). *Från tundra till skog. Miljöförändringar i norra Skåne under jägarestenåldern.* Stockholm: Riksantikvarieämbetet.

Blankholm, H. P. (1992). *On the track of a prehistoric economy. Maglemosian subsistence pattern in early postglacial south Scandinavia.* Aarhus: Aarhus University Press.

Briedermann, L. (1986). *Schwarzwild.* Berlin: VEB Deutscher Landwirtschaftsverlag.

Cederlund, G., & Liberg, O. (1995). *Rådjuret – Viltet, ekologin och jakten.* Spånga: Svenska Jägarförbundet.

Christensen, J. H., & Hewitson, B. (2007). Regional climate projections. In S. Salmon, D. Qin, M. Manning, M. Marquis, K. Averyt, M. M. B. Tignor, et al. (Eds.), *Climate change 2007. The physical science basis. Contribution of working group I to the fourth assessment report of the intergovernmental panel on climate change* (pp. 847–940). Cambridge: Cambridge University Press.

Clutton-Brock, T. H., & Albon, S. D. (1983). Climatic variation and body weight of red deer. *Journal of Wildlife Management, 47,* 1197–1201.

Cussans, J. E. (2017). Biometry and climate change in Norse Greenland: The effect of climate on the size and shape of domestic mammals. In G. G. Monks (Ed.), *Climate change, human responses: A zooarchaeological perspective* (pp. 197–216). Dordrecht: Springer.

Davis, B. A. S., Brewer, S., Stevensson, A. C., & Guiot, J. (2003). The temperature of Europe during the Holocene reconstructed from pollen data. *Quaternary Research Reviews, 22,* 1701–1716.

von den Driesch, A. (1976). *A guide to the measurement of animal bones from archaeological sites.* Peabody: Peabody Museum of Archaeology, Harvard University.

Degerbøl, M., & Fredskild, B. (1970). *The Urus (Bos primigenius Bojanus) and Neolithic cattle (Bos taurus domesticus Linné) in Denmark.* Copenhagen: Det Konglige Danske Videnskabernes Selskab.

Ekström, J. (1993). *The Late Quaternary History of the Urus (Bos primigenius Bojanus 1827) in Sweden.* Lund: Department of Quaternary Geology, Lund University.

Enghoff, I. B. (2011). *Regionality and biotope exploitation in Danish Ertebølle and adjoining periods.* Copenhagen: The Royal Danish Academy of Sciences and Letters.

Eriksson, M., & Magnell, O. (2001a). Det djuriska Tågerup. Nya rön kring Kongemose- och Erteböllekulturens jakt och fiske. In P. Karsten & B. Knarrström (Eds.), *Tågerup. Specialstudier* (pp. 156–237). Lund: Riksantikvarieämbetet.

Eriksson, M., & Magnell, O. (2001b). Jakt och slakt. In P. Karsten (Ed.), *Dansarna från Bökeberg. Om jakt, ritualer och inlandsbosättning vid jägarstenålderns slut* (pp. 49–77). Lund: Riksantikvarieämbetet.

Frazier, J. (2007). Sustainable use of wildlife: The view from archaeozoology. *Journal for Nature Conservation, 15,* 163–173.

Friborg, M. N. (1999). *En palæoøkologisk undersøgelse af bopladsen Lollikhuse baseret på pattedyrknogler.* M.Sc. Thesis, Institute of Geology, University of Copenhagen.

Gilbert, A. S., & Singer, B. H. (1982). Reassessing zooarchaeological quantification. *World Archaeology, 14,* 21–40.

Gotfredsen, A. B. (1998). En rekonstruktion af palæomiljøet omkring tre senmesolitiske bopladser i Store Åmose, Vestsjælland- baseret på pattedyr- og fugleknogler. *Geologisk Tidsskrift, 2,* 92–104.

Grayson, D. K. (1979). On the quantification of vertebrate archaeofaunas. In M. B. Schiffer (Ed.), *Advances in archaeological method and theory* (Vol. 2, pp. 199–237). New York: Academic Press.

Göransson, H. (1991). *Vegetation and man around Lake Bjärsjöholmssjön during prehistoric time.* Lund: Department of Quaternary Geology, Lund University.

Hallström, A. (1984). Benfynden från Löddesborgsboplatsen. In K. Jennbert (Ed.), *Den produktiva gåvan. Tradition och innovation i Sydskandinavien för omkring 5300 år sedan* (pp. 182–200). Lund: Department of Archaeology, Lund University.

Hede, S. U. (2005). The finds: Mammal, bird, and amphibian bones. In T. D. Price & A. B. Gebauer (Eds.), *Smakkerup Huse. A Late Mesolithic coastal site in northwest Zealand, Denmark* (pp. 91–102). Aarhus: Aarhus university Press.

Henriksen, B. B. (1976). *Sværdborg I, Excavations 1943–44. A Settlement of the Maglemose Culture.* Copenhagen: Akademisk Forlag.

Henriksen, B. B. (1980). *Lundby-holmen. Pladser av Maglemose-typ i Sydsjælland.* Copenhagen: Det Kongelige Nordiske Oldskriftselskab.

Hewison, A. J. M. (1996). Variation in the fecundity of roe deer in Britain: Effects of age and body weight. *Acta Theriologica, 41,* 187–198.

Iregren, E., & Lepiksaar, J. (1993). The Mesolithic site in Hög in a south Scandinavian perspective—a snap shot from early Kongemose culture. *Zeitschrift für Archäologie, 27,* 29–38.

Jensen, P. (1991). Body size trends of roe deer (Capreolus capreolus) from Danish Mesolithic sites. *Journal of Danish Archaeology, 10,* 51–58.

Johnsen, S. J., Dahl-Jensen, D., Gundestrup, N., Steffensen, P., Clausen, H. B., Miller, H., et al. (2001). Oxygen isotope and palaeotemperature records from six Greenland ice-core stations: Camp Century, Dye-3, GRIP, GISP2, Renland and North GRIP. *Journal of Quaternary Science, 16,* 299–307.

Jonsson, L. (1988). The vertebrate faunal remains from the Late Atlantic settlement at Skateholm in Scania, South Sweden. In L. Larsson (Ed.), *The Skateholm Project. I: Man and environment* (pp. 56–88). Lund: Royal Society of Letters.

Karsten, P., & Knarrström, B. (2003). *The Tågerup excavations.* Lund: National Heritage Board.

Kelly, R. L. (1995). *The foraging spectrum: Diversity in hunter-gatherer lifeways.* Washington: Smithsonian Institution Press.

Larsson, L. (1978). *Ageröd I.B-I:D. A study of early Atlantic settlement in Scania.* Lund: Department of Archaeology, Lund University.

Langvatn, R., & Albon, S. D. (1986). Geographic clines in body weight of Norwegian red deer: A novel explanation of Bergmann's rule? *Holarctic Ecology, 9,* 285–293.

Lauwerier, R. C. G. M., & Plug, I. (Eds.). (2004). *The future from the past: Archaeology in wildlife conservation and heritage management.* Oxford: Oxbow Books.

Levinsky, I., Skov, F., Svenning, J.-C., & Rahbek, C. (2007). Potential impacts of climate change on the distributions and diversity patterns of European mammals. *Biodiversity and Conservation, 16,* 3803–3816.

Lepiksaar, J. (1978). Bone remains from the Mesolithic Ageröd I:B and I:D. In L. Larsson (Ed.), *Ageröd I.B-I:D. A study of early Atlantic settlement in Scania* (pp. 234–244). Lund: Department of Archaeology, Lund University.

Lepiksaar, J. (1982). Djurrester från den tidigatlantiska boplatsen vid Segebro nära Malmö. In L. Larsson (Ed.), *Segebro. En tidigatlantisk boplats vid Sege ås mynning* (pp. 105–128). Malmö: Malmö Museum.

Lepiksaar, J. (1983). Animal remains from the Atlantic bog site at Ageröd V in Central Scania. In L. Larsson (Ed.), *Ageröd V, an Atlantic bog site in cCentral Scania* (pp. 159–168). Lund: Department of Archaeology, Lund University.

Liljegren, R., & Lagerås, P. (1993). *Från mammutstäpp till kohage. Djurens historia i Sverige.* Lund: Wallin & Dalholm AB.

Lyman, L. (1996). Applied zooarchaeology: The relevance of faunal analysis to wildlife management. *World Archaeology, 28,* 110–125.

Lyman, L. (2006). Paleozoology in the service of conservation biology. *Evolutionary Anthropology, 15,* 11–19.

Lyman, L., & Cannon, K. P. (Eds.). (2004). *Zooarchaeology and conservation biology.* Salt Lake City, UT: The University of Utah Press.

Magnell, O. (2004). The body size of wild boar during the Mesolithic in South Scandinavia. *Acta Theriologica, 49,* 113–130.

Magnell, O. (2006). Tracking wild boar and hunters. Osteology of wild boar in Mesolithic south Scandinavia. Lund: Department of Archaeology and Ancient History, Lund University.

Magnussen, B. I. (2007). En geologisk og zooarkæologisk analyse af kystbopladsen Lollikhuse på overgangen mellem Mesolitikum og Neolitikum. M.Sc. thesis, University of Copenhagen.

Mitchell, B., Staines, B. W., & Welch, D. (1977). *Ecology of Red Deer. A research review relevant to their management in Scotland.* Banchory: Institute of Terrestrial Ecology.

Murray, D., Cox, E. W., Ballard, W. B., Whitlaw, H. A., Lenarz, M. S., Custer, T. W., et al. (2006). Pathogens, nutritional deficiency, and climatic influences on a declining moose population. *Wildlife Monographs, 166,* 1–30.

Møhl, U. (1971). Oversigt over dyreknoglerne fra Ølby Lyng. En østsjællændsk kystboplads med Ertebøllekultur. *Aarbøger for Nordisk Oldkyndighed og Historia, 1970,* 43–77.

Møhl, U. (1984). Dyreknogler fra nogle af borealtidens senere bopladser i den sjællandske Aamose. *Aarbøger for Nordisk Oldkyndighed og Historia, 1984,* 47–60.

Niethammer, J., & Krapp, F. (1986). *Handbuch der Säugetiere Europas. Paarhufer.* Wiesbaden: AULA-Verlag.

Nilsson, T. (1964). Standardpollendiagramme und C^{14}-datierungen aus dem Ageröds Mosse im mittleren Schonen. *Lunds universitets årsskrift N.F. 59*(7), 1–52.

Noe-Nygaard, N. (1995). Ecological, sedimentary and geochemical evolution of the Late-Glacial to Postglacial Åmose lacustrine basin, Denmark. *Fossils and Strata, 37,* 1–436.

Ochoa, J., & Piper, P. J. (2017). Holocene large mammal extinctions in Palawan Island, Philippines. In G. G. Monks (Ed.), *Climate change and human responses: A zooarchaeological perspective* (pp. 69–86). Dordrecht: Springer.

Parry, M., Canziani, O., Palutikof, J., van der Linden, P., & Hanson, C. (Eds.). (2007). *Climate change 2007. Impacts, adaptation and vulnerability. Contribution of working group I to the fourth assessment report of the intergovernmental panel on climate change.* Cambridge: Cambridge University Press.

Payne, S., & Bull, G. (1988). Components of variations in measurements of pig bones and teeth, and the use of measurements to distinguish wild from domestic pig remains. *Archaeozoologica, 2,* 27–66.

Persson, P. (1999). *Neolitikums början. Undersökningar kring jordbrukets introduktion i Nordeuropa.* Gothenburg: Department of Archaeology, University of Gothenburg.

Peterson, R. L. (1955). *North American moose.* Toronto: University of Toronto Press.

Pilaar Birch, S. E., & Miracle, P. T. (2017). Human response to climate change in the northern Adriatic during the Late Pleistocene and Early Holocene. In G. G. Monks (Ed.), *Climate change, human responses: A zooarchaeological perspective* (pp. 87–100). Dordrecht: Springer.

Price, T. D., & Gebauer, A. B. (2005). *Smakkerup Huse. A Late Mesolithic coastal site in northwest Zealand.* Aarhus: Aarhus University Press.

Regnell, M., Gaillard, M.-J., Bartholin, T. S., & Karsten, P. (1995). Reconstruction of environment and history of plant use during the Late Mesolithic (Ertebølle culture) at the inland settlement of Bökeberg III, southern Sweden. *Vegetation History and Archaeobotany, 4,* 67–91.

Regnell, M., & Sjögren, K.-G. (2006). Vegetational development. In K.-G. Sjögren (Ed.), *Ecology and economy in Stone Age and Bronze Age Scania* (pp. 40–79). Lund: National Heritage Board.

Rosenlund, K. (1971). Zoological material. In E. B. Petersen (Ed.), Sværdborg II. A Maglemose hut from Sværdborg bog, Zealand, Denmark. *Acta Archaeologica 42,* 59–62.

Rosenlund, K. (1980). Knoglematerialet fra bopladsen Lundby II. In B. B. Henriksen (Ed.), *Lundby-holmen. Pladser av Maglemose-typ i Sydsjælland* (pp. 128–142). Copenhagen: Det Kongelige Nordiske Oldskriftselskab.

Schmölcke, U., & Zachos, F. E. (2005). Holocene distribution and extinction of the moose (*Alces alces*, Cervidae) in Central Europe. *Mammalian Biology, 70,* 329–344.

Sjöström, A. (1997). Ringsjöholm. A Boreal-Early Atlantic settlement in central Scania, Sweden. *Lund Archaeological Review, 3,* 5–20.

Skaarup, J. (1973). *Hesselø-Sølager. Jagdstation der südskandinavischen Trichtbergerkultur.* Copenhagen: Akademisk forlag.

Skuncke, F. (1949). *Älgen. Studier jakt och vård.* Stockholm: P.A. Norstedt & Söners Förlag.

Smith, E. A. (1983). Anthropological applications of optimal foraging theory: A critical review. *Current Anthropology, 24,* 625–651.

Snowball, I., Korhola, A., Briffa, K. R., & Koc, N. (2004). Holocene climate dynamics in Fennoscandia and the North Atlantic. In R. W. Battarbee, F. Gasse, & C. E. Stickley (Eds.), *Past climate variability through Europe and Africa* (Vol. 6, pp. 465–494). Dordrecht: Springer.

Tauber, H. (1971). Danske kulstof-14 dateringer af arkæologiske prøver III. *Aarbøger for nordisk oldkyndighed og historia, 1970,* 120–142.

Teichert, M. (1969). Osteometrische Untersuchungen zur Berechnung der Widerristhöhe bei vor- und frühgeschichtlichen Schweinen. *Kühn-Archiv, 83,* 237–292.

Van Vuure, T. (2002). History, morphology and ecology of the aurochs (*Bos primigenius*). *Lutra, 45,* 1–16.

Winge, H. (1904). Oversigt over Knoglematerialet fra Mullerup-bopladsen. In G.F.L. Sarauw, En stenalders boplads i Maglemose ved Mullerup, Sammenholdt med beslægtede fund. *Aarbøger for Nordisk Oldkyndighet og Historie, 1903,* 148–315.

Winge, H. (1919). Oversigt over Knoglematerialet fra Sværdborgbopladsen. In K. Friis-Johansen, En boplads fra den ældste Stenalder i Sværdborg Mose. *Aarbøger for Nordisk Oldkyndighet og Historie, 1919,* 106–235.

Wintemberg, W. J. (1919). Archaeology as an aid to zoology. *The Canadian Field-Naturalist, 33,* 63–72.

Wyszomirska, B. (1988). *Ekonomisk stabilitet vid kusten. Nymölla III. En tidigneolitisk bosättning med fångstekonomi i nordöstra Skåne.* Lund: Department of Archaeology, Lund University.

Yu, S.-Y. (2003). *The Littorina transgression in southwestern Sweden and its relation to mid-Holocene climate variability.* Lund: LUNDQUA Thesis 51, Lund University.

Part III
The Recent Holocene

Chapter 8
Oxygen Isotope Seasonality Determinations of Marsh Clam Shells from Prehistoric Shell Middens in Nicaragua

André C. Colonese, Ignacio Clemente, Ermengol Gassiot, and José Antonio López-Sáez

Abstract Marsh clams (*Polymesoda* sp.) were an important dietary item for pre-Columbian people living along the Caribbean coast of Nicaragua. Their intensive exploitation is synchronous with major cultural changes associated with the emergence of socio-political complexity in Central America. In this paper we present the results of an oxygen isotope seasonality study on archaeological shells retrieved from Karoline, a shell midden site dated to ~2 cal kBP and located along the southern margin of Pearl Lagoon (Caribbean coast of Nicaragua). Modern shells (*Polymesoda arctata*) were also analysed for stable isotopes. The results indicate that archaeological specimens from Karoline may have experienced different hydrological conditions or nutrient supply within the lagoon compared to present day. The seasonal analysis reveals that there were no preferential seasons for the collection of marsh clams during the distinct phases of site formation; instead, exploitation occurred throughout the year.

Keywords Central America • Late Holocene aquatic exploitation • Coastal lagoon • Mollusc shell • Stable isotopes

A.C. Colonese (✉)
BioArCh Department of Archaeology, University of York, York, YO10 5DD, UK
e-mail: andre.colonese@york.ac.uk; andre@palaeo.eu

I. Clemente
Department of Archaeology and Anthropology (IMF-CSIC), AGREST (Generalitat de Catalunya), C/Egipcíaques 15, 08001 Barcelona, Spain
e-mail: ignacio@imf.csic.es

E. Gassiot
Department of Prehistory, Universitat Autònoma de Barcelona Edifici B, Bellaterra, 08193 Barcelona, Spain
e-mail: ermengol.gassiot@uab.cat

J.A. López-Sáez
Archaeobiology Group (CCHS-CSIC), C/Albasanz 26-28, 28037 Madrid, Spain
e-mail: joseantonio.lopez@cchs.csic.es

Introduction

The genus *Polymesoda* (Bivalvia: Cyrenidae) is a typical component of Atlantic coastal lagoons and estuaries of Central America (Severeyn et al. 1996; Rueda and Urban 1998), and an important economic item for artisanal fisheries in the region (de La Hoz-Aristiábal 2009). On the Caribbean coast of Nicaragua *Polymesoda* has been intensively exploited since the Late Holocene (~3.4 cal kBP), as attested by numerous shell middens bordering Pearl Lagoon, the largest lagoon on Nicaragua's Caribbean coast (e.g., Magnus 1978; Clemente et al. 2003, 2009; Gassiot 2005; Clemente and Gassiot 2004). Before ~2.5 cal kBP the archaeological record along the coast of Pearl Lagoon is mainly represented by short-term campsites, probably associated with the specialized exploitation (and field processing) of aquatic resources. During this early period shell middens are composed essentially of shells of *Polymesoda* with very small amounts (if any) of vertebrate faunal remains. Artifacts are absent in these deposits. After ~2.5 cal kBP, shell middens (e.g., Punta Masaya, Sitetaia, Brown Bank, Karoline) are composed of a broader range of aquatic and terrestrial resources (large mammals, mollusc, amphibian, reptiles, and fish) along with pottery, stone tools and archaeobotanical remains (Clemente et al. 2003; Gassiot et al. 2003). This time interval is marked by an increase in the number of settlements and the appearance of monumental architecture in the region (e.g., at Cascal de Flor de Pinos; Clemente and Gassiot 2004). It seems thus that estuarine environments were actively exploited during the emergence of socio-political complexity throughout the Central America land-bridge (e.g., Cooke 2005).

Mollusc shells are the most abundant archaeozoological remains in these sites, and we use oxygen isotope analysis to explore seasonal harvesting patterns that in turn will help us understand patterns of settlement and occupation.

Oxygen isotopic analysis of marine mollusc shell carbonate from archaeological deposits is a technique widely used to investigate season of collection and to infer mobility-settlement patterns (e.g., Kennett and Voorhies 1996; Mannino et al. 2003; Colonese et al. 2009, 2012; Kennett and Culleton 2010; Andrus and Thompson 2012), as well as palaeoenvironmental conditions in coastal areas (e.g., Kennett and Voorhies 1995; Monks 2017). Shell $\delta^{18}O$ is a function of temperature and of the $^{18}O/^{16}O$ ratios of water (e.g., Wefer and Berger 1991). Changes in the oxygen isotopic composition of shell carbonate is influenced by fluctuations of water $\delta^{18}O$ and the temperature as the mollusc grows. In transitional coastal areas (e.g., lagoons, estuaries) the contribution of seawater and freshwater can be assessed because freshwater is markedly ^{18}O-depleted compared to seawater. In low-latitude tropical regions this signal is easier to interpret compared to high-latitude estuaries (e.g., Culleton et al. 2009) because of reduced seasonal temperature variation. In tropical coastal environments, changes in salinity associated with variable seasonal precipitation are expected to be the main factor driving shell $\delta^{18}O$ variability (e.g., Kennett and Voorhies 1995, 1996; Ingram et al. 1996; Stephens et al. 2008).

In this paper we present oxygen isotope data for prehistoric *Polymesoda* sp. shells from Karoline, an early ceramic shell midden site located in the southern margin of Pearl Lagoon on the Caribbean coast of Nicaragua. Modern shells of *Polymesoda arctata* were initially analyzed to assess the effect of local temperature, salinity and ontogenic growth rate based on oxygen isotope ratios within their incremental growth. Archaeological shells were then analysed to establish a preliminary seasonal shellfish harvesting profile for the site of Karoline. Our primary aim is to use this approach to improve our understanding of the economic significance of aquatic resources for pre-Columbian coastal peoples of Central America.

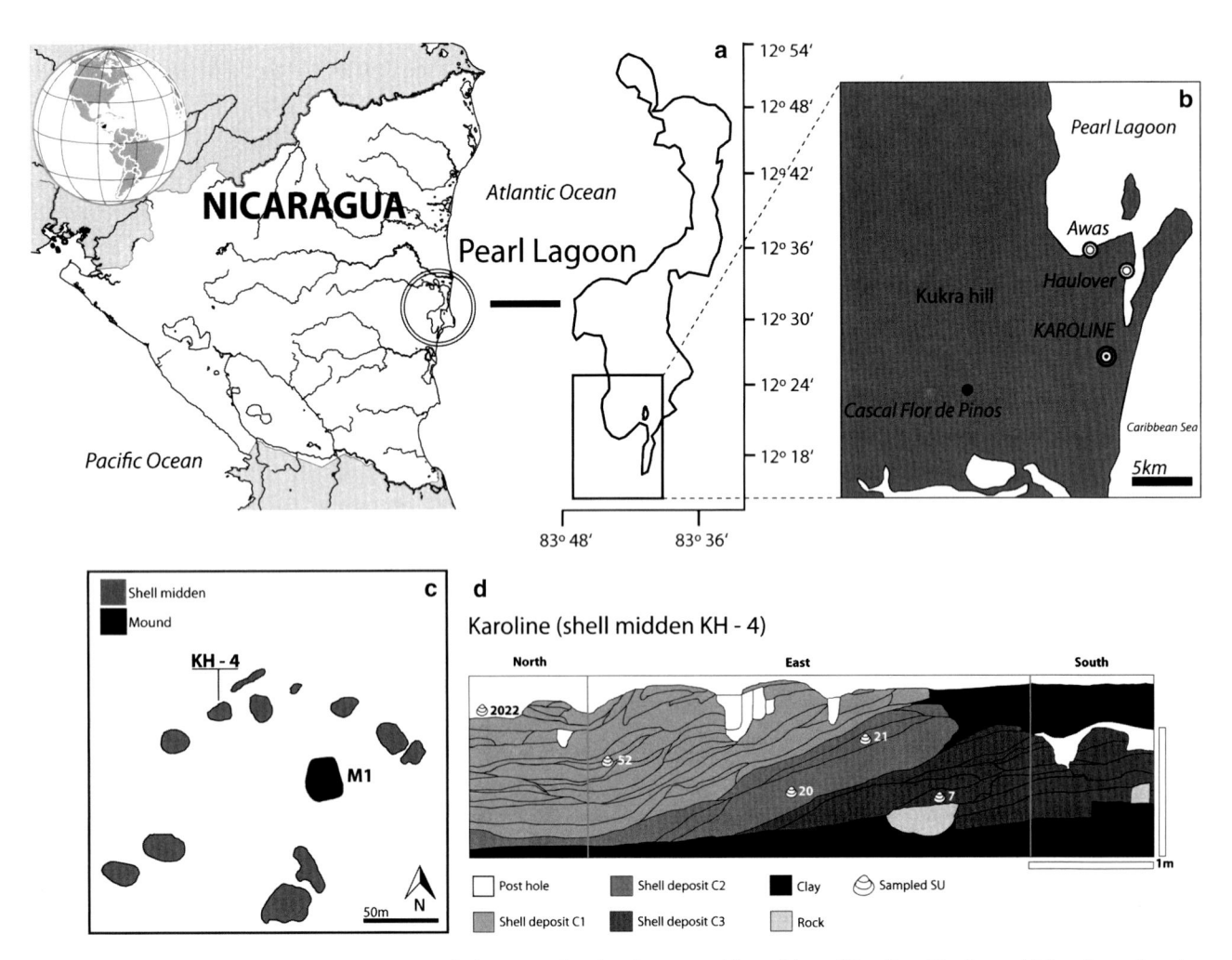

Fig. 8.1 **a** Map of Pearl Lagoon (Caribbean coast of Nicaragua) showing the geographic position of Karoline (**b**), along with locations of modern shells of *P. arctata* (Awas) and measurements of water ST and SS (Haulover, data from 1995−1997). Karoline is formed by different shell middens, including KH-4 (**c**). North, east and south profiles of the archaeological stratigraphy from which sampled shells were taken (**d**)

Environmental and Archaeological Setting
The Study Area

Pearl Lagoon (Fig. 8.1a, b; 571 km²), the largest coastal lagoon on Nicaragua's Caribbean coast, is a shallow estuarine-lagoon system (mean bottom depth 2.5 m), with a microtidal range (mean 0.22 m) (e.g., Brenes and Castillo 1999; Christie et al. 2000; Brenes et al. 2007, and reference therein). Regional climate is influenced by the latitudinal change of the Intertropical Convergence Zone (ITCZ), promoting wet and dry seasons during northward and southward displacement of the ITCZ respectively. Mean daily temperature ranges from ~26 to ~28°C in December and April respectively. The dry and wet season occur between December and April (boreal winter), and between May and November (boreal summer) respectively (e.g., Leduc et al. 2007). Hurricanes are also common in the area, occurring from June to October (Christie et al. 2000). The seasonal distribution of rainfall plays a crucial role in the freshwater input into the study area, controlling the residence time of fresh water and the degree of mixing with seawater which enters through an inlet in the southern sector of the lagoon. River discharges into the lagoon vary considerably, doubling from May to July (from ~500 to ~1040 $m^3 \cdot s^{-1}$) (Brenes et al. 2007). The surface temperature (ST) shows very little variability within the lagoon, whereas the surface salinity (SS) exhibits some variation through the year (Brenes and Castillo 1999). Salinity is mostly controlled by tidal intake of seawater, the amount of fresh discharge from the rivers and wind, and it decreases with distance from the inlet (Brenes and Castillo 1999; Christie et al. 2000). During the wet season the lagoon is dominated by freshwater and very little spatial difference in SS occurs within the lagoon (from ~0 to ~3 PSU in August and September respectively). By contrast, dry season SS variability can be considerable between zones close to the inlet and areas in the innermost part of the lagoon (from ~10 to ~18 PSU between December and April-May respectively). Over a year, little difference occurs within ~10 km from inlet toward the innermost part of the lagoon (Brenes and Castillo 1999). Although hydrological data are not available for historical reconstruction close to the archaeological site, monthly spot measurements of SS and ST carried out under the project DIPAL II between 1995 and 1997 at Haulover (12°18′N 83°37′W; Fig. 8.1b) provide some insights into seasonal SS and ST variability in the study area. In this area the average SS was 24.2 PSU with maximum and minimum values recorded during October-November (6.5 PSU) and May 1997 (38.5 PSU), respectively. Overall, SS decreases from June to November and increases from December to May (Fig. 8.2). In the same area ST shows only minor oscillations, ranging from 25.7°C in December 1996 to 30.3°C in September 1997, with an average of 27.9°C.

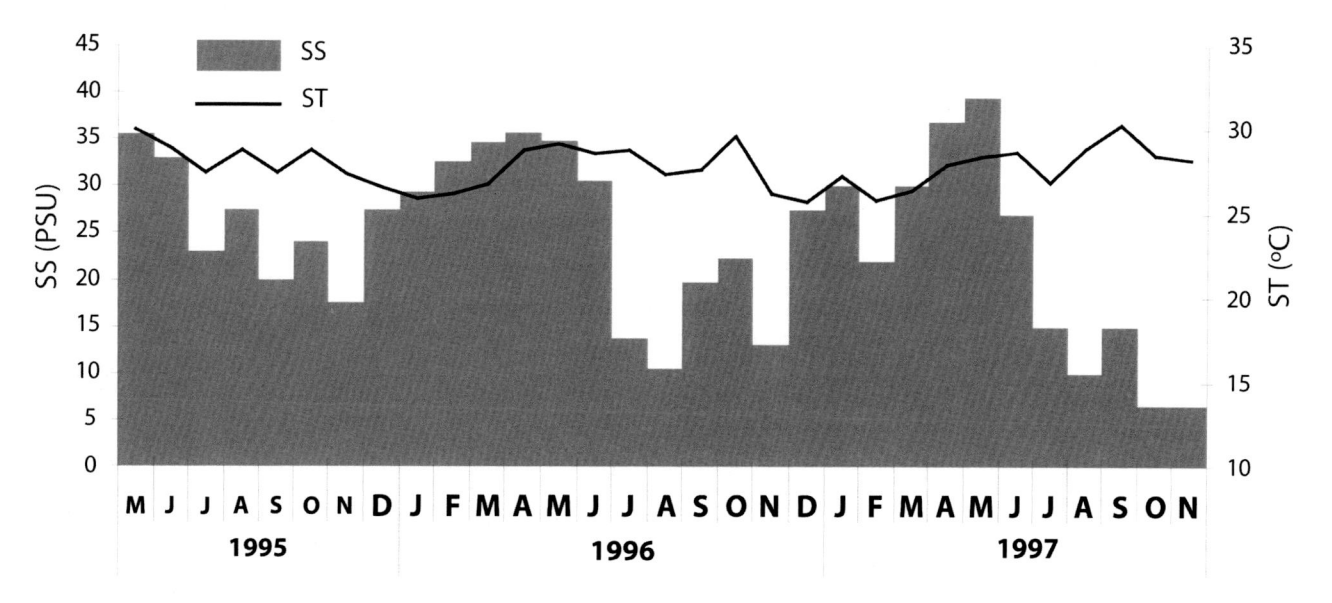

Fig. 8.2 Surface temperature (ST) and salinity (SS) variations from 1995 to 1997 at Pearl Lagoon (data from Haulover)

The Karoline Site (Shell Midden KH-4)

The Karoline site consists of a complex of shell middens located in the south margin of the Pearl Lagoon, at Kukra hill (Región Autónoma del Atlántico Sur: Fig. 8.1b–d). These shell deposits are distributed around an architectonic structure (M1) built on a natural mound, and formed by retaining walls and blocks of different sizes. Pottery and lithic remains retrieved in these shell middens show great affinity between them and with artifacts from this architectonic building (M1), suggesting that these structures were built at the same time.

Since 2002, systematic research has been carried out at shell midden number 4 (KH-4) (Clemente et al. 2009), revealing four distinct phases of shell accumulation (i.e., Surface occupation, C1, C2, C3) preliminarily dated between ~1.6 (Surface) and ~2.2 cal kBP (C3) (Fig. 8.1c, Table 8.1[1]; Clemente and Gassiot 2004, 2005). This time interval, however, does not necessarily imply a gradual site formation over ~600 years. Other evidence suggests that complex shell middens in Central America may represent relatively short accumulation events (e.g., Kennett et al. 2011), thus further radiocarbon data would be required for a more detailed understanding of site formation process. Archaeological evidence includes house floors, stone tools, artifacts made from bone and shell, stratified hearths associated with numerous "post holes" (suggesting the presence of structures for cooking or smoking food), seeds, and terrestrial and aquatic (both marine and freshwater) faunal remains (e.g., Gassiot 2005; Clemente et al. 2009; Zorro 2010). Remains belonging to Caviidae, Dasyproctidae and Muridae dominate the mammal fauna, followed by Cervidae (*Odocoileus virginianus*), Tayassuidae (e.g., *Tayassu pecari*), Felidae (*Herpailurus yagouaroundi*), Cebidae (*Ateles* sp., *Cebus capucinus*), Dasypodidae (*Bradypus* sp., *Dasypus novemcinctus*) and Tapiridae (*Tapirus* sp.). Bird are present but rare. In particular the amount of fish (the subject of an ongoing study), reptile (e.g., Emydidae/Geoemydidae, Chelydridae), amphibian (e.g., *Bufo* sp.) and mollusc (*Donax* sp., *Polymesoda* sp.) debris indicates a persistent exploitation of the transitional environment (e.g., Zorro 2010). Aquatic and terrestrial resources were complemented with agricultural products during the later phases of occupation. This is suggested by *metates*, pottery and pollen data (Gassiot 2005; Zorro 2010). Fossil pollen from the two latest phases of occupation (C1 and C2; Fig. 8.1d) reveals a complex history of forest clearing and domesticated plant use. For the occupation phase C2, pollen records show the local dominance of a pioneer community (*Cecropia*, *Trema*, Piperaceae, Urticales) and the development of both Caribbean mangrove (Rhizophora) and tropical evergreen lowland swamp forest (*Coccoloba*, *Raphia taedigera*, *Acoelorrhaphe wrightii*, *Campnosperma panamensis*, Melastomataceae). This vegetation cover can be associated with both human activity and influence of marine waters. In this phase, pollen data do not record evidence of cultigens. Subsequently, with the occupation phase C1, pollen data attest to the development of a tropical evergreen forest around the site (*Ficus*, *Vochysia*, *Bursera*, *Inga*, *Brosimum*), the early cultivation of maize (*Zea mays*) and probably beans (*Phaseolus vulgaris*) and squash (*Cucurbita* sp.).

Donax denticulatus, *D. estriatus* and *Polymesoda* sp. were the most frequent molluscan taxa in the Karoline assemblage. Stable isotope data reported in this paper were obtained from shells collected in different stratigraphic units (SU) of Karoline.

Materials and Methods
Stable Isotope Analysis of Modern Shells – Polymesoda arctata (Deshayes 1854)

Polymesoda arctata (also known as *P. solida*; Philippi, 1846) inhabit fine-grained sediments in coastal lagoons and estuaries on the western Atlantic coast of Central and South America from eastern Venezuela to Belize (Severeyn et al. 1994). It is usually found in waters with salinity ranging between ~1 and ~20 PSU, although it has been reported to occur in salinities of up to 30 PSU, and at temperature between ~28 and ~31°C (Severeyn et al. 1996; Rueda and Urban 1998; de La Hoz-Aristiábal 2009). The species is particularly tolerant of low salinity waters, and is not very resistant to sharp increases in salinity (de La Hoz-Aristiábal 2009). Previous work suggests that salinity oscillations are a key factor influencing its growth rate, reproduction and density (Rueda and Urban 1998; de La Hoz-Aristiábal 2009). Absolute growth, however, seems to be influenced by both salinity and temperature. Maximum growth rate occurs when salinity is around 9 PSU, and a temperature >30°C is critical for the survival of this species (Severeyn et al. 1996). Growth rate is consistently higher during the first year but decreases with age. Severeyn et al. (1996) report shell secretion up to 10 mm during the first two months of growth, corresponding with ~¼ of maximum shell length (45 mm, 7 years). An important reduction in shell secretion (by 1 to 3 mm/yr) starts after ~35 mm of shell length (>1 yr). The reproductive cycle in *P. arctata* is variable, including both monoecious (hermaphroditic) combined with

---
[1]Calibration was performed with CALPAL_A (advanced) (http://www.calpal.de; Weninger and Jöris 2010). Sample pre-treatments: AAA-method (Alkali-Acid-Alkali) for charcoal; Longin-method for bones.

Table 8.1 AMS radiocarbon data from KH-4 reported as conventional and calibrated BP (CalPal 2007Hulu calibration curve)

Phase	SU	^{14}C yr BP	^{14}C yr cal BP (68%)	^{14}C yr cal BP (95%)	Lab code	Material	δ^{13}C (V-PDB) (‰)	C/N
Surface	2022	1735 ± 25	1660 ± 40	1740–1580	KIA-17978	Bone (*Odocoileus virginianus*)	−23.91	3.3
C3	7	2195 ± 25	2230 ± 60	2350–2110	KIA-17648	Charcoal	−30.76	

protogyny (i.e., it changes from male to hermaphroditic when the shell is about 21 mm length) (Rueda and Urban 1998) and dioecious strategies (i.e., the two sexes can be separated from very small size) (Severeyn et al. 1996), probably due to the combined effect of endogenous and exogenous factors (Rueda and Urban 1998). In some coastal areas of South and Central America this species remains an important part of the fishery and is intensively exploited (Rueda and Urban 1998; de La Hoz-Aristiábal 2009).

Sample Procedure and Isotopic Analysis

Living specimens were collected in April 2004 at the Awas locality (Fig. 8.1b), and of these, two specimens (M1 and M2) were selected for sequential stable isotope analysis of shell carbonate. M1 and M2 differed in size (38 mm and 28.5 mm lengths respectively Fig. 8.3) and were analyzed in order to evaluate the potential influence of ontogenic growth on shell δ^{18}O resolution before selection of archaeological specimens. Shells were cleaned of flesh in the field and then were cleaned more thoroughly in the laboratory using distilled H$_2$O. The left valves were selected, partially embedded in an epoxy resin, sectioned along the axis of maximum growth and polished. Carbonate powders (~150 μg) were then sampled sequentially in the outer crossed-lamellar layer composed of aragonite (OCL) (Taylor et al. 1973; Fig. 8.3), from the edge toward the umbo, working parallel to small-scale periodic microgrowth increments (SCPGI, possibly diurnal) visible in the OCL (e.g., Figure 8.3b). Samples were taken using a manual microdrill with a 0.35 mm bit (Fig. 8.3).

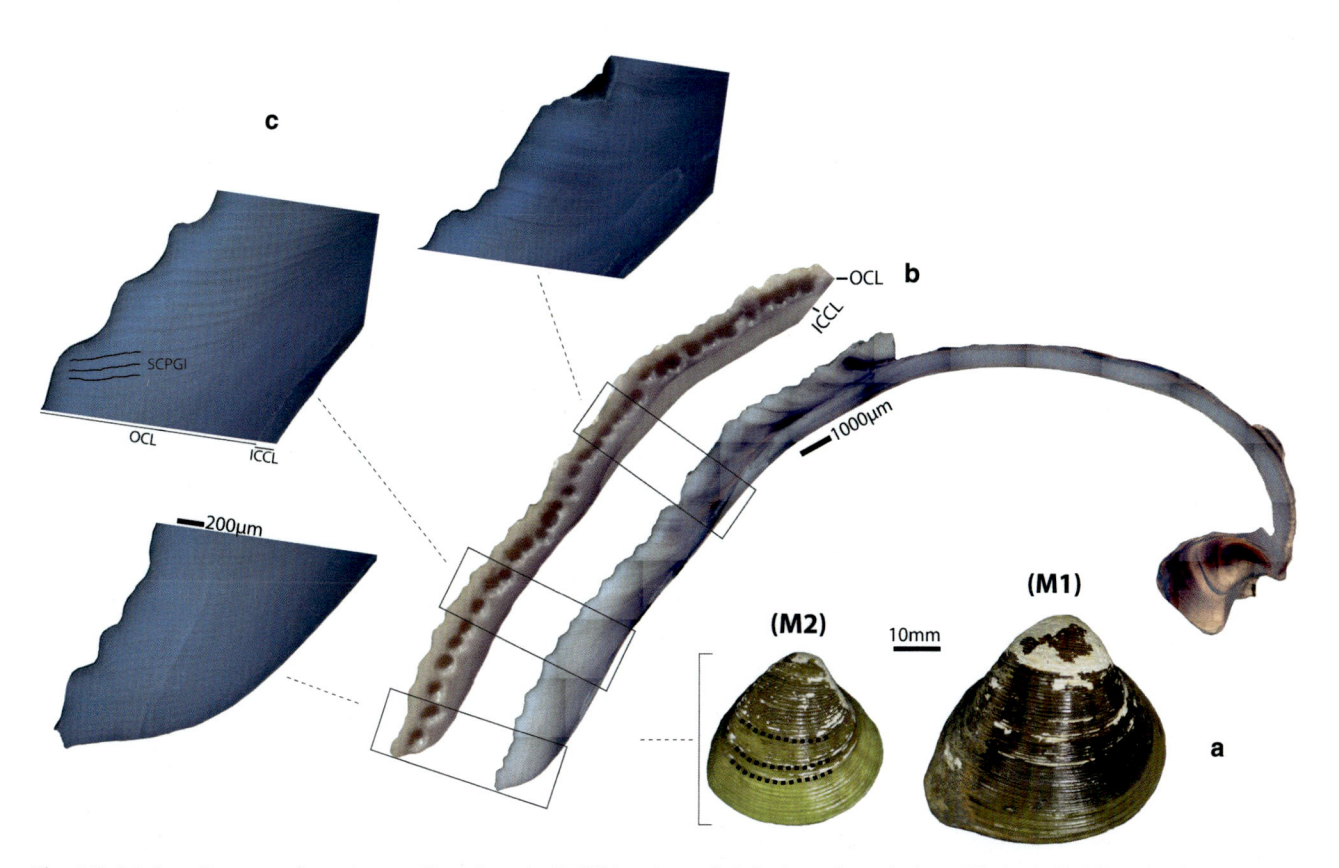

Fig. 8.3 Modern *P. arctata* from Awas collected on April 2004 and sampled for isotopic analysis. **a** Whole shells M1 and M2; **b** Enlarged sections of M2 OCL; and **c** Enlarged areas of M2 OCL showing growth variations. The inner complex crossed-lamellar layer bounded by the trace of the pallial line is also reported (ICCL). Dotted lines on M2 mark the external growth checks

Archaeological specimens from Karoline (N = 14) were recovered during archaeological excavation in 2008. Shells were selected from several stratigraphic units (SU) representing four occupational episodes (Fig. 8.1d). Only the left valves with similar size (~30 mm lengths) were used in order to establish the season of collection. Sample preparation of archaeological shells follows that of modern counterparts, but after being sectioned and rinsed with distilled water, in most of shells only four continuous carbonate samples starting from the shell-edge were obtained in the OCL. The $\delta^{18}O$ value of the first sample (shell-edge) is representative of the period (season) of mollusc collection. In the case of harvesting during intermediate seasons the further three samples behind the shell-edge would provide indications of changes in ST (cooling or warming) and/or ambient water $\delta^{18}O$ ($\delta^{18}Ow$) (dry or wet season). Of these, two shells (2022-1 and 2022-2) were sampled sequentially using a manual microdrill with a 0.35 mm bit. These sequential samples provide the ontogenic (intra-annual) shell $\delta^{18}O$ variability, which is representative of seasonal ST and $\delta^{18}Ow$ variations at the time of shell collection (e.g., Mannino et al. 2003; Colonese et al. 2009, 2012).

Isotopic analyses were performed at the Vrije Universiteit in Amsterdam. Carbonate samples were analysed using a Finnigan MAT 252 equipped with an automated preparation line (Kiel II type). The reproducibility of a routinely analyzed carbonate standard (NBS 19) is better than 0.09‰ for both $\delta^{18}O$ and $\delta^{13}C$. The isotopic composition is expressed using δ (‰) notation. Shell carbonate isotope signatures are related to V-PDB standard (Vienna Pee Dee Belemnite) and water values to V-SMOW (Vienna Standard Mean Ocean Water). X-ray diffraction on powdered carbonate revealed that the OCL of both modern and archaeological shells was entirely composed of aragonite; no evidence of diagenetic alteration was observed.

Results and Discussion
Modern Shells

M1 and M2 had shell $\delta^{18}O$ average values of -3.9 ± 1.3‰ and -3.3 ± 1.5‰ respectively. Sequential shell $\delta^{18}O$ depict very clear sinusoidal trends, with values ranging from -1.7‰ to -6.2‰ for M1 and -1.3‰ to -6.1‰ for M2, revealing intra-shell $\delta^{18}O$ variability ($\Delta^{18}O$) of 4.5‰ and 4.8‰ respectively (Fig. 8.4; Table 8.2). The $\delta^{18}O$ value in biogenic aragonite is the product of both temperature and $\delta^{18}Ow$, as expressed by the follow equation on the temperature dependence of aragonite fractionation in biogenic minerals (Grossman and Ku 1986):

$$T(°C) = 20.86 - 4.69 * (\delta^{18}O_{shell} - \delta^{18}O_{water}) \quad (8.1)$$

This relationship predicts that a shift in ST of 4.69 °C corresponds to a shell $\delta^{18}O$ change of 1‰. Using this relationship on measured ST, SS and shell oxygen isotope values (M1 and M2), we estimate that the local intra-annual ST range of 4.6 °C should account for ~1‰ of $\Delta^{18}O$ in both specimens. After the removal of ST imprinting, the residual shell $\Delta^{18}O$ of ~3.8‰ (M1) and ~3.5‰ (M2) can be associated with oscillations in $\delta^{18}Ow$ linked with the strong seasonal salinity variation (by 33 PSU). In so doing, higher and lower shell $\delta^{18}O$ values are associated with higher and lower SS conditions respectively. No $\delta^{18}Ow$ values are available for the study area, so the $\delta^{18}Ow$ was estimated using shell $\delta^{18}O$ values and the average local ST (27.9 °C) in Eq. 8.1. The ensuing results indicate that M1 and M2 experienced $\delta^{18}Ow$ in the range of ~0 to ~-4.6‰. The following equation proposed by LeGrande and Schmidt (2006) for the Tropical Atlantic area permits, with a general approximation, the conversion of $\delta^{18}Ow$ values into SS (PSU):

$$\delta^{18}O_{water} = 0.15 * SS(PSU) - 4.61 \quad (8.2)$$

Predicted $\delta^{18}Ow$ values suggest that both M1 and M2 experienced salinity oscillations ranging from near seawater (~31 PSU) to freshwater (~0 PSU) values during shell formation, which is quite reasonable for the study area. Although the lowest estimated $\delta^{18}Ow$ (~-4.6‰) falls into the range of continental surface water in Central America (Lachniet and Patterson 2002; Marfia et al. 2004), measured SS data have values substantially above freshwater (~5 PSU; Fig. 8.2). Thus Eq. 8.2 provides only an approximate range of $\delta^{18}Ow$ experienced by M1 and M2; these are not absolute values. However, this exercise reveals that *P. arctata* tolerates a broad range of salinity conditions, thus the salinity record based on its shell $\delta^{18}O$ can be used as indicator of wet and dry seasons, in agreement with previous studies (Kennett and Voorhies 1996). We conclude that the seasonal freshwater/seawater fluctuation accounts for the majority of the observed shell $\delta^{18}O$ variability at Pearl Lagoon (~80% of $\Delta^{18}O$); temperature contribution is less significant due to its lower seasonal variation (~20% of $\Delta^{18}O$).

Oxygen isotopic profiles show that growth rate decreases with age and that ~6 yrs of shell secretion is evident in the record (Fig. 8.4). This result is consistent with field observations about changing growth rate (Severeyn et al. 1996). By combining shell $\delta^{18}O$ fluctuations with the distance of samples from the shell-edge, we estimate faster shell secretion in the first year (~18 mm/yr) followed by a considerable decrease during the second year (~6.2 mm/yr). These two years of growth are then followed by several cyclical $\delta^{18}O$

curves representing ~4 yrs of growth, with much reduced rates (~2 mm/yrs; Fig. 8.4). Pronounced external growth checks (bands) and major internal growth lines coincide with both lower and higher $\delta^{18}O$ values. These growth increments likely reflect interruptions and/or reductions of shell formation due to semi-periodic events, involving physiological (e.g., spawning) and environmental factors (e.g., salinity change, storms). Several spawning events per year have been reported by Rueda and Urban (1998) in populations from Central America, and growth cessation due to strong storms have been also recorded in diurnal microgrowth increments of other Cyrenidae (Fritz and Lutz 1986). In M1, pronounced external growth checks coincide with major microgrowth increments only during the first year of growth. From approximately the second year, the external growths checks are less distinguishable from other external growth increments. Small-scale periodic microgrowth increments are also visible during the first year of growth (see for instance in M2, Fig. 8.3). In M2, the sequential $\delta^{18}O$ values represent ~1 yr of growth and it has a similar growth rate to M1 during the first year (~19 mm/yr). Reduced growth rates are evident during wet season $\delta^{18}O$ values and faster growth coincides with dry season higher $\delta^{18}O$ (Fig. 8.3). Overall these results suggest reduced growth occurring in response to persistent exposure to low salinity water, probably within, or close to, values of freshwater. Here, reduced increment widths suggest distinct short-term episodes of decreased growth rate lasting days or weeks (Fig. 8.3c).

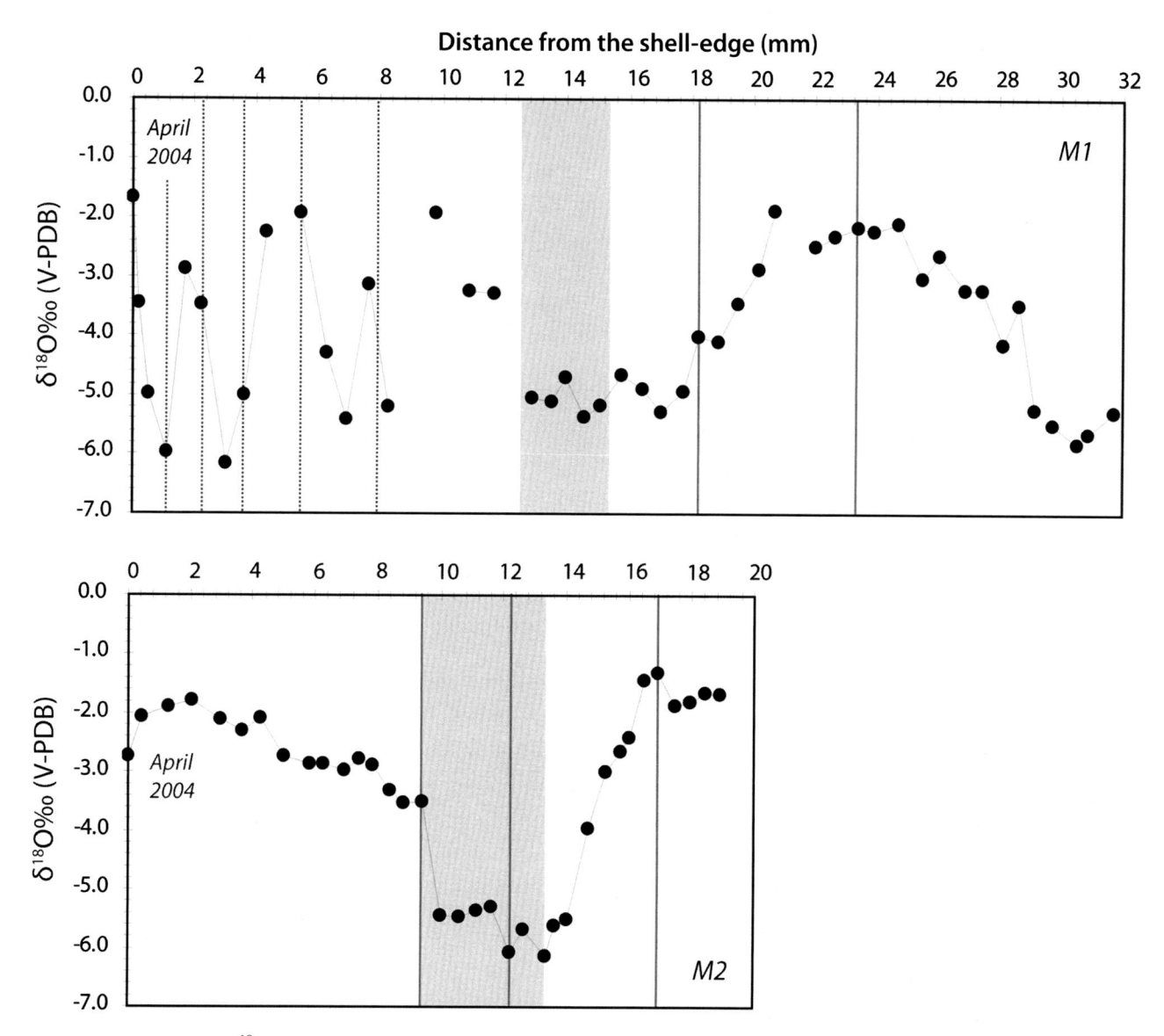

Fig. 8.4 Sequential shell $\delta^{18}O$ of *P. arctata* (M1 and M2) collected on April 2004 plotted versus distance from the shell-edge (mm). The grey bands represent the interval with closely spaced growth increments (reduced growth rate). Vertical dotted lines in M1 represent major internal growth increments; vertical grey lines refer to external growth checks

Table 8.2 Sequential shell $\delta^{18}O$ of modern specimens of *P. arctata* (M1 and M2)

Samples	Modern shells $\delta^{18}O‰$ (V-PDB)	
	M1	M2
1	−1.68	−2.74
2	−3.45	−2.06
3	−4.98	−1.90
4	−5.96	−1.78
5	−2.87	−2.11
6	−3.48	−2.30
7	−6.15	−2.09
8	−4.99	−2.74
9	−2.26	−2.86
10	–	−2.87
11	−1.92	−2.98
12	−4.30	−2.79
13	−5.42	−2.89
14	−3.14	−3.32
15	−5.19	−3.53
16	–	−3.51
17	−1.94	−5.45
18	–	−5.46
19	−3.25	−5.35
20	−3.29	−5.30
21	–	−6.08
22	−5.05	−5.68
23	−5.11	−6.14
24	−4.70	−5.62
25	−5.38	−5.51
26	−5.18	−3.96
27	−4.66	−3.00
28	−4.89	−2.65
29	−5.29	−2.40
30	−4.95	−1.43
31	−4.02	−1.32
32	−4.09	−1.88
33	−3.46	−1.81
34	−2.88	−1.65
35	−1.88	−1.68
36	–	
37	−2.48	
38	−2.32	
39	−2.17	
40	−2.23	
41	−2.09	
42	−3.02	
43	−2.63	
44	−3.22	
45	−3.22	
46	−4.15	
47	−3.48	
48	−5.24	
49	–	
50	−5.50	
51	−5.82	
52	−5.65	
53	−5.28	
Max	−6.15	−6.14
Min	−1.68	−1.32
Mean	−3.92	−3.28
$\Delta\delta^{18}O‰$	4.5	4.8

The effect of decreasing growth rate in the shell sample resolution is also evident in the discrepancy between shell-edge $\delta^{18}O$ values of M1 (−1.7‰) and M2 (−2.7‰), collected at the same time at the end of the dry season (April). This additional feature further emphasizes the need to select animals of approximately one year old (e.g., medium sized specimens, 28.5 mm) for seasonal palaeoenvironmental reconstructions using incremental oxygen isotope measurements.

Archaeological Shells

Our results suggest that *Polymesoda arctata* grow all year round, although with distinctly different growth rates, and that shell oxygen isotope composition is informative of seasonal SS and $\delta^{18}Ow$ variability in the study area. This in turn is controlled by the seasonal distribution of precipitation induced by the northward/southward shift of ITCZ.

Table 8.3 Sequential shell $\delta^{18}O$ of archaeological specimens of *Polymesoda* sp

Samples	Archaeological shells $\delta^{18}O‰$ (V-PDB)	
	C2022-1	C2022-2
1	−5.39	−5.26
2	−3.65	−5.25
3	−3.94	−4.89
4	−6.04	−5.14
5	−6.13	−5.22
6	−2.85	−4.74
7	−2.57	−2.19
8	−3.80	−2.60
9	−5.10	−3.18
10	−5.08	−3.42
11	−4.66	−3.38
12	−5.19	−3.60
13	−5.28	−4.27
14	−5.34	−3.12
15	−4.18	−4.75
16	−2.33	−4.58
17	−2.32	−4.67
18	−1.53	−4.06
19	−1.88	−5.99
20	−2.04	−6.48
21	−1.97	−5.80
22	−2.21	−6.59
23		−5.27
24		−2.67
25		−3.39
26		−4.15
Max	−6.13	−6.59
Min	−1.53	−2.19
Mean	−3.79	−4.41
$\Delta\delta^{18}O‰$	4.61	4.40

However, in contrast to previous work (e.g., de La Hoz-Aristiábal 2009), we demonstrate that slow growth rates in *P. arctata* occur during low salinity conditions that were close to fresh water. Therefore the annual growth rates of medium sized specimens (28.5 mm) provide a more extensive and detailed record of recent environmental conditions compared with older slow growing specimens (38 mm) that have reduced annual growth rates later in their life cycle. As a consequence, archaeological shells of similar size (30 mm length were) were used for seasonality studies.

Our results provide compelling evidence that shell $\delta^{18}O$ of *P. arctata* record seasonal hydrological conditions that are in agreement with previous studies on bivalve shells from transitional environments (e.g., Kennett and Voorhies 1995, 1996; Stephens et al. 2008). The same isotopic pattern is assumed for archaeological shells from Karoline: shell ^{18}O-depletion reflect enhanced freshwater inputs and reduced SS during the wet season. By contrast, ^{18}O-enrichment in shells indicates more saline conditions during the dry season when the flow of continental waters is reduced.

Past climate conditions may have affected local SS (and $\delta^{18}O_w$) consequently, the intra-shell $\delta^{18}O$ variability of modern specimens might not provide an appropriate isotopic baseline for archaeological counterparts. We examined this with two archaeological shells from the most recent occupation episode (2022-1 and 2022-2). These shells were sampled sequentially in order to obtain intra-annual $\delta^{18}O$ values as indirect evidence of past seasonal SS and $\delta^{18}O_w$ variability at the time of shell collection (Fig. 8.5; Table 8.3). Palaeoclimatic records from the southern Caribbean suggest that at the time of KH-4 formation (~ 2 cal kBP) climatic conditions were more stable and wetter compared with today (Malaizé et al. 2011, and references therein). Relatively stable conditions enable us to use the shell $\delta^{18}O$ amplitudes of 2022-1 and 2022-2 as references for shell-edge $\delta^{18}O$ values from other stratigraphic units.

The $\delta^{18}O$ profiles of 2022-1 and 2022-2 both show cyclical oscillations similar to the modern specimens. Incremental $\delta^{18}O$ values in 2022-1 reveal approximately two years of shell growth, whereas in 2022-2 they encompass one and a half years. Specimen 2022-1 displays $\delta^{18}O$ values ranging from −1.5 to −6.1‰ ($\Delta^{18}O = 4.6$‰) and an average of −3.8 ± 1.5‰. Specimen 2022-2 shows sequential $\delta^{18}O$ values ranging from −2.2 to −6.6‰ ($\Delta^{18}O = 4.4$‰) and an average of −4.4 ± 1.2‰. According to modern observations, shell-edge $\delta^{18}O$ values of 2022-1 and 2022-2 (−5.4‰ and −5.3‰ respectively) indicate that both specimens were collected during the wet season. Taken as a whole, amplitudes of archaeological $\delta^{18}O$ do not differ much from modern counterparts. However it is worth noting that archaeological shells display more $\delta^{18}O$ cycles than M2. In spite their similar size, $\delta^{18}O$ profiles from 2022-1 and 2022-2 show three and two $\delta^{18}O$ curves respectively, whereas M2 only provides a unique cycle of higher and lower values. Moreover, 2022-2 exhibits an average shell $\delta^{18}O$ (−4.4‰) that is ~ 1‰ lower than the modern M2 (−3.3‰). Although this $\delta^{18}O$ discrepancy falls within the intra-specific variability observed in previous studies on the genus *Polymesoda* from the Central America (Kennett and Voorhies 1995, 1996), it could also imply that archaeological specimens experienced different environmental conditions or distinct growth rates compared with M2. In particular, in both 2022-1 and 2022-2 the transitions from higher to lower $\delta^{18}O$ values occur more abruptly than in M2 and in the first two years of M1. Such shifts in $\delta^{18}O$ values could reflect abrupt changes in salinity that may have been promoted by short-term atmospheric events, e.g., hurricanes. Modern and past strong tropical cyclones have been widely documented over the study area (Christie et al. 2000; Urquhart 2009) and palaeoclimatic records suggest that they occurred with different frequency and intensity in the recent past (Donnelly and Woodruff 2007; Urquhart 2009; Malaizé et al., 2011). An alternative explanation is that isotopic discrepancies are associated with exploitation in distinct zones within the lagoon, with different salinity distribution (e.g., Lesure et al. 2010; Andrus and Thompson 2012). However, both archaeological and modern shells show

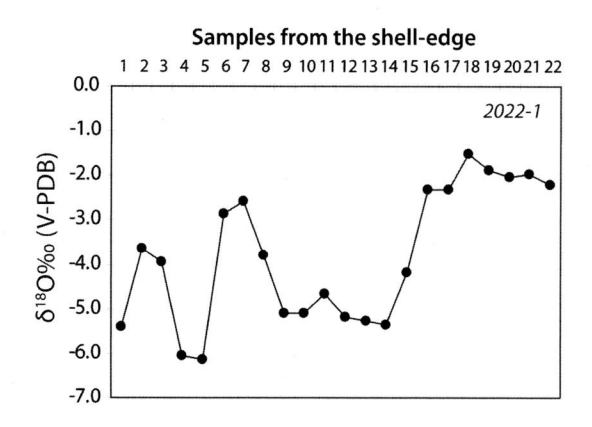

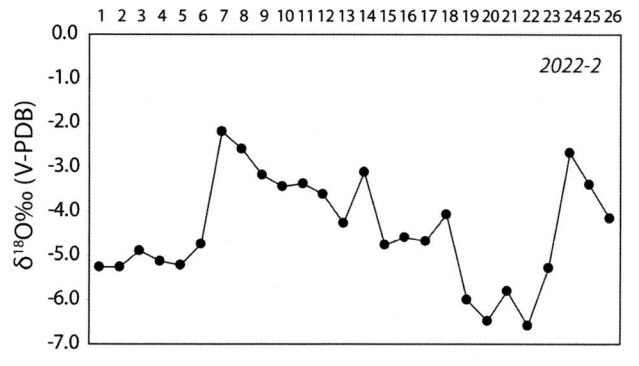

Fig. 8.5 Sequential shell $\delta^{18}O$ values of archaeological specimens of *Polymesoda* sp. (2022-1 and 2022-2) plotted by samples from the shell-edge

compatible $\delta^{18}O$ amplitudes, suggesting similar seasonal salinity oscillation. Furthermore, salinity variations within the lagoon are not particularly pronounced at a distance of ~10 km from the inlet (Brenes and Castillo 1999). Karoline is situated close to this entrance, thus if molluscs were exploited within this area we could expect that the effect of different salinity distribution within the lagoon is negligible on our shell $\delta^{18}O$ results.

Modern increased population and consequent land use changes in and around Pearl Lagoon might additionally explain shell growth variability between modern and past populations. The effect of anthropogenic eutrophication within Pearl Lagoon (from organic sewage or fertilizer runoff) has been previously postulated for the seagrass decline in this environment (Schuegraf 2004). High nutrient levels triggering phytoplankton blooms may enhance growth rates of suspension feeders such as bivalves (e.g., Kirby and Miller 2005), which could explain the difference between modern and archaeological shells. Nevertheless the size of sample is not large enough to address this issue in any detail, but highlights avenues of future research.

Seasonal Harvesting Profiles at Karoline

Sequential $\delta^{18}O$ values of 2022-1 and 2022-2 were pooled and then grouped into quartiles in order to obtain a more robust picture of shell $\delta^{18}O$ distribution as a function of local seasonal hydrological balance (Fig. 8.6a). Quartile distribution organizes intra-shell $\delta^{18}O$ values into four different groups of equal parts. The lowest $\delta^{18}O$ values correspond to the wet season and are grouped in the lower quartile (lowest 25% of data). The highest $\delta^{18}O$ values correspond with the dry season and are grouped into the upper quartile (upper 25%). Intermediate $\delta^{18}O$ values representing conditions between the two are grouped into the interquartile range (>25% and <75%), with 50% of the $\delta^{18}O$ data falling into this range (Mannino et al. 2007; Colonese et al. 2009, 2012).

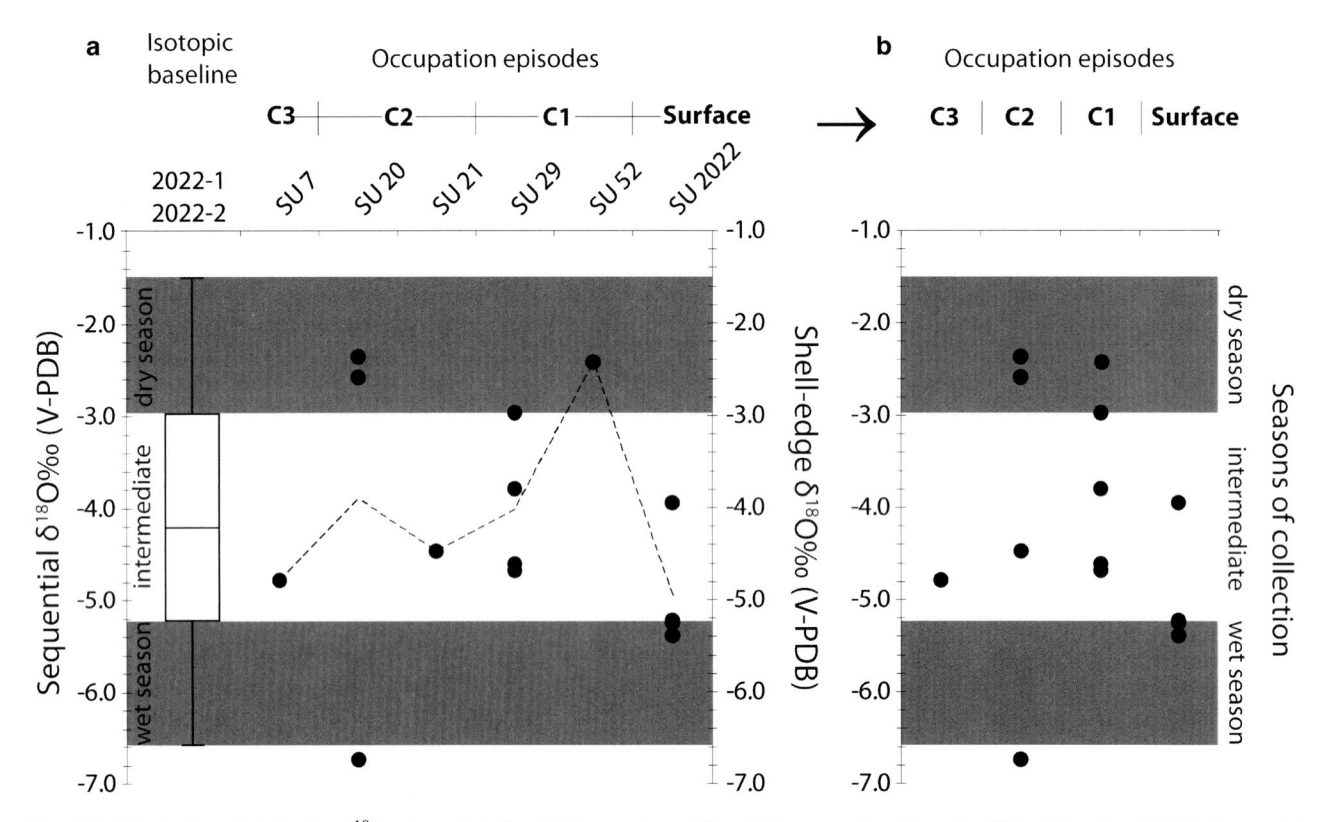

Fig. 8.6 Distribution of shell-edge $\delta^{18}O$ values of shells of *Polymesoda* sp. (N = 14) from stratigraphic units (SU) of Karoline (KH-4). Sequential shell $\delta^{18}O$ of archaeological specimens (2022-1 and 2022-2) were grouped in a boxplot and reported for comparison (**a**). Note that shell-edge $\delta^{18}O$ values are distributed from the upper to the lower quartile of sequential $\delta^{18}O$ values. Such distribution denotes exploitations in different periods of the year during different period of site occupation (**b**)

Shell-edge $\delta^{18}O$ values of archaeological specimens from different stratigraphic units at Karoline range from −2.4‰ (SU 20) to −6.7‰ (SU 20), with an average $\delta^{18}O$ value of −4.2‰ ($\Delta^{18}O$ = 4.3‰). Considerable shell-edge $\delta^{18}O$ amplitudes are detected within occupation episodes. Inherent shell $\delta^{18}O$ variability due to intra-specific variations in growth rate alone does not explain these large isotopic ranges (Kennett and Voorhies 1996). In addition, the similar amplitude of sequential $\delta^{18}O$ in both archaeological and modern shells suggests that they experienced similar seasonal salinity oscillation. Assuming that molluscs were collected within a distance ≤10 km from the site, these data would indicate exploitation during different times of the year. More than one third of the analysed specimens provide isotopic signatures compatible with intermediate seasons (N = 6; Fig. 8.6a; Table 8.4). The residual specimens had either a dry (N = 4) or wet (N = 4) season isotopic signal. A more detailed picture can be suggested for occupations C2, C1 and Surface. Shells from occupation C2 were collected during wet, intermediate and dry seasons. During occupation C1 they show isotopic signature consistent with collection in dry and intermediate seasons. Finally, in the most recent phase (Surface) shells were mainly collected toward the onset or cessation of the wet season, but collections also occur in the intermediate period (Fig. 8.6b). Overall, the data suggest that there were no preferential seasons of collections of *Polymesoda* sp. during the distinct phases of site formation; collection of these animals is recorded in different periods of the year. Nevertheless, a more confined exploitation during the wet season seems to be detected in the uppermost deposits (Surface), contemporaneous with agricultural practices in the site. It is worth noting that on the Pacific coast of south Mexico (Acapetahua Estuary), Kennett and Voorhies (1996) observed a clear shift from year-round to wet season use of coastal environments through the late Archaic Period (∼3000–1800 BC), at the time of development of maize agriculture in the region. The authors suggest that farming activities became increasingly greater during the dry season, resulting in a more focused exploitation of coastal resources in the wet season. Although a detailed reconstruction of seasonal exploitation patterns at Karoline is precluded by the small sample size, our results suggest that similar seasonal use of lagoon resources also may have occurred in the Caribbean side of Central America. However, possible relationships with agricultural practices need to be tested with future extensive studies.

Stable oxygen isotopes at ∼2.0 cal kBP in *Polymesoda* sp. shells from Karoline reveal that humans exploited aquatic resources at Pearl Lagoon under distinct seasonal hydrological conditions. As observed in other regions of Central America, from the Archaic to the Late Formative period (e.g., Blake et al. 1992; Kennett et al. 2006; Rosenswig 2006), our results suggest that transitional environments may have played an important role in the intra-annual subsistence systems of peoples inhabiting the Caribbean coast of Nicaragua during the Late Formative Period. This observation is consistent with other archaeozoological records from the same occupation phases (Zorro 2010). Although further

Table 8.4 $\delta^{18}O$ results of shells selected for seasonal analysis. Shell-edge $\delta^{18}O$ is indicative of the season of collection. The further three samples behind the shell-edge provide indications of SS trend (dry or wet). *Archaeological shells sampled sequentially (2022-1 and 2022-2)

Occupation episodes	SU-Shells	$\delta^{18}O$‰ (V-PDB)				Season of collection
		1st sample (shell-edge)	2nd sample	3rd sample	4th sample	
Surface	US2022 – 1*	−5.39	−3.65	−3.94	−6.04	Wet
Surface	US2022 – 2*	−5.26	−5.25	−4.89	−5.14	Wet
Surface	US2022 – 3	−3.96	−4.18	−4.56	−2.42	Intermediate
Surface	US2022 – 4	−5.23	−2.77	−6.36	−4.99	Wet
C1	US52 – 1	−2.43	−4.99	−3.18	−3.37	Dry
C1	US29 – 1	−3.80	−3.97	−3.79	−4.18	Intermediate
C1	US29 – 2	−4.62	−3.91	−4.45	−4.87	Intermediate
C1	US29 – 3	−4.69	−4.36	−2.03	−5.95	Intermediate
C1	US29 – 4	−2.97	−2.81	−4.21	−4.09	Dry
C2	US21 – 1	−4.48	−5.24	−4.06	−5.27	Intermediate
C2	US20 – 2	−6.74	−5.09	−3.74	−5.65	Wet
C2	US20 – 3	−2.37	−3.98	−4.43	−4.17	Dry
C2	US20 – 1	−2.59	−5.53	−3.96	−2.16	Dry
C3	US7 – 1	−4.79	−5.98	−4.40	−3.46	Intermediate

extensive studies are required, previous research recognizes possible changes in aquatic (lagoon and marine) exploitation starting after ~2.5 cal kBP. Before this interval, aquatic food remains seems to be limited to short-term campsites, with no relation to residential features. From ~2.5 cal kBP, the archaeological records from Karoline reveal a broad range of aquatic resources in association with residential activities and, later, with agricultural products (Clemente et al. 2003; Gassiot et al. 2003). This economic picture is chronologically consistent with a period of greater stability of settlements and social investment in monumental architecture in the area (Clemente et al. 2003). This occurs at a time of accelerated social divergence and complexity marked by the emergence of social inequality and hierarchical political institutions in Central America (Cooke 2005; Hoopes 2005). It is within this context that $\delta^{18}O$ data of *Polymesoda* sp. shells suggest that peoples were exploiting aquatic resources at Pearl Lagoon in all seasons, a pattern consistent with other indications for greater sedentism. More restricted collections during the wet season in the most recent phase of site occupation seems to parallel the rescheduling of estuarine resource exploitation in the Pacific coast during development of maize agriculture, but further studies are required to address this hypothesis.

Conclusion

The $\delta^{18}O$ values of *Polymesoda arctata* shells in the Pearl Lagoon of Nicaragua are influenced by the balance between fresh and seawater that changes seasonally in response to monsoonal rains. Pre-colonial shell middens sites are abundant along the coast of Central America and preserve invaluable snapshots of subsistence strategies involving both aquatic and terrestrial resources and agricultural products. *Polymesoda* sp. is usually dominant in these archaeological contexts and is highly suitable for studying past seasonal hydrological changes in this region, and it also provides a proxy for determining the season of shellfish harvesting. The $\delta^{18}O$ profiles of archaeological specimens from Karoline (shell middens KH-4) suggest that they experienced environmental conditions at ~2 kBP (reduced level of nutrient input, salinity) not comparable with modern counterparts. The results also indicate that people were exploiting aquatic resources during dry, wet and intermediate seasons, a pattern that is consistent with other evidence (houses, greater faunal and artifact diversity, and monumental architecture) suggesting greater sedentism starting after ~2.5 cal kBP along the Caribbean coast of Nicaragua.

The Caribbean coast of Nicaragua is highly susceptible to extreme weather events like hurricanes and storms, and palaeoclimatic records reveal the frequency of these events also in the recent past (e.g., Donnelly and Woodruff 2007). Hurricanes and strong storms have significant impact on human life, property and production systems in the region (e.g., Christie et al. 2000), but how past peoples withstood the effects of these natural hazards is still a matter of debate (Cooper and Peros 2010). Dynamic and small-scale subsistence systems (combining gathering, fishing, hunting, and agriculture) that were built on different environments (e.g., coast, savannah, rainforest and mangroves) offer some advantages in withstanding dramatic atmospheric events in this region (e.g., Kronike and Verner 2010). Similar economic strategies may have occurred in the past in different areas of Central America (Cooper and Peros 2010). The elaborated economic system observed at Karoline relied on distinct habitats (terrestrial and aquatic) and resource items (wild and domestic products) in different seasons, it may, therefore, represent a local response to past climatic condition in this tropical coastal area.

Acknowledgments Authors are particularly grateful to Gregory Monks (University of Manitoba, Canada), John Currey, Matthew Collins, Igor Gutiérrez Zugasti and Annika Burns (Department of Archaeology, University of York, UK) for their support in the preparation of the manuscript. We also thank WoRMS Team for their assistance on the taxonomic status of *Polymesoda solida* (Philippi 1847). We are very grateful to Rodolfo Sánchez Barquero for kindly providing hydrological data for the study area (DIPAL II, Proyecto para el Desarrollo Integral de la Pesca Artesanal en la Región Autónoma Atlántico Sur, Nicaragua). We are grateful to Douglas J. Kennett and to the other two anonymous referees for their constructive comments and suggestions on the manuscript.

References

Andrus, C. F. T., & Thompson, V. D. (2012). Determining the habitats of mollusk collection at the Sapelo Island shell ring complex, Georgia, USA using oxygen isotope sclerochronology. *Journal of Archaeological Science, 39*, 215–228.

Blake, M., Chisholm, B. S., Clark, J. E., Voorhies, B., & Love, M. W. (1992). Prehistoric subsistence in the Soconusco region. *Current Anthropology, 33*(1), 83–94.

Brenes, R. C., & Castillo, V. E. (1999). Caracterizacion hidrografica de la Laguna de Perlas, Nicaragua. *Proyecto para el Desarrollo Integral de la Pesca Artesanal en la Region Autonomia Atlantico Sur, Nicaragua (DIPAL II)*.

Brenes, C. L., Hernández, A., & Ballestero, D. (2007). Flushing time in Perlas Lagoon and Bluefields Bay, Nicaragua. *Invest. Mar., Valparaíso, 35*(1), 89–96.

Christie, P., Bradford, D., Garth, R., Gonzalez, B., Hostetler, M., Morales, O. et al. (2000). *Taking care of what we have: Participatory natural resource management of the Caribbean Coast of Nicaragua*. Center for Research and documentation of the Atlantic Coast AP A-186 Managua, Nicaragua and the International Development Research Centre, Ottawa, ON, Canada. K1G 3H.

Clemente, I. C., Gassiot, B. E., & Lechados, L. (2009). Shellmiddens of the Atlantic coast of Nicaragua: Something more than mounds. In M. C. Afonso & G. Bailey (Eds.), *Evolutions and environment—C62—coastal geoarchaeology: The research of shellmounds* (pp. 119–125). Oxford: BAR International Series.

Clemente, I. C., Gassiot, E. B., Lechado, L. R., & Oltra, J. P. (2003). *Las plataformas del sitio "El Cascal de Flor de Pino": Nuevas evidencias de arquitectura prehistórica monumental en las tierras bajas de la Costa Atlántica de Nicaragua* (p. 2003). Santo Domingo: Acts of the XX International Association of Caribbean Archaeology Congress.

Clemente, I. C., & Gassiot, E. B. (2004/2005). ¿En el camino de la desigualdad? El litoral de la Costa Caribe de Nicaragua entre el 500 cal ANE y el 450 cal NE. *Revista Atlántica Mediterránea de Prehistoria y Arqueología Social 7*, 109–130.

Colonese, A. C., Troelstra, S., Ziveri, P., Martini, F., Lo Vetro, D., & Tommasini, S. (2009). Mesolithic shellfish exploitation in SW Italy: Seasonal evidence from the oxygen isotopic composition of Osilinus turbinatus shells. *Journal of Archaeological Science, 36*, 1935–1944.

Colonese, A. C., Verdún-Castelló, E., Álvarez, M., Godino, I. B., Zurro, D., & Salvatelli, L. (2012). Oxygen isotopic composition of limpet shells from the Beagle Channel: Implications for seasonal studies in shell middens of Tierra del Fuego. *Journal of Archaeological Science,*. doi:10.1016/j.jas.2012.01.012.

Cooke, R. (2005). Prehistory of Native Americans on the Central American land bridge: Colonization, dispersal, and divergence. *Journal of Archaeological Research, 13*(2), 129–187.

Cooper, J., & Peros, M. (2010). The archaeology of climate change in the Caribbean. *Journal of Archaeological Science, 37*, 1226–1232.

De La Hoz-Aristiábal, M. V. (2009). Densidad, estructura de tallas y explotación pesquera del bivalvo Polymesoda solida en un sistema lagunar del Caribe colombiano. *Boletín del Centro de Investigaciones Biológicas, 43*(1), 1–27.

Donnelly, J. P., & Woodruff, J. D. (2007). Intense hurricane activity over the past 5,000 years controlled by El Niño and the West African monsoon. *Nature, 447*, 465–468.

Fritz, L. W., & Lutz, R. A. (1986). Environmental perturbations reflected in internal shell growth patterns of Corbicula fluminea (Mollusca: Bivalvia). *The Veliger, 28*(4), 401–417.

Gassiot, E. B., Clemente, I. C., & Palomar, B. P. (2003). *Entre lagunas y manglares: Poblamiento y explotación del litoral en la Costa Caribe de Nicaragua (1,400 cal AC a 1,000 cal DC)* (p. 2003). Santo Domingo: Acts of the XX International Association of Caribbean Archaeology Congress.

Gassiot, E. B. (2005). Shell middens in the Caribbean coast of Nicaragua: Prehistoric patterns of mollusk collecting and consumption. In D. Bar-Yosef (Ed.), *Archaeomalacology: Molluscs in former environments of human behaviour* (pp. 40–53). Oxford: Oxbow Books.

Grossman, E. L., & Ku, T.-L. (1986). Oxygen and carbon isotope fractionation in biogenic aragonite; temperature effects. *Chemical Geology, 59*, 59–74.

Hoopes, J. W. (2005). The emergence of social complexity in the Chibchan world of southern Central America and northern Colombia, AD 300–600. *Journal of Archaeological Research, 13*(1), 1–46.

Ingram, B. L., Conrad, M. E., & Ingle, J. C. (1996). Stable isotope and salinity systematics in estuarine waters and carbonates: San Francisco Bay. *Geochimica et Cosmochimica Acta, 60*(3), 455–467.

Kennett, D. J., & Voorhies, B. (1995). Middle Holocene periodicities in rainfall inferred from oxygen and carbon isotopic fluctuations in prehistoric tropical estuarine mollusc shells. *Archaeometry, 23*(1), 157–170.

Kennett, D. J., & Voorhies, B. (1996). Oxygen isotopic analysis of archaeological shells to detect seasonal use of wetlands on the southern Pacific coast of Mexico. *Journal of Archaeological Science, 23*, 689–704.

Kennett, D. J., Voorhies, B., & Martorana, D. (2006). An evolutionary model for the origins of agriculture on the Pacific coast of southern Mexico. In D. J. Kennett & B. Winterhalder (Eds.), *Behavioral ecology and the transition to agriculture* (pp. 103–136). Berkeley: University of California Press.

Kennett, D. J., & Culleton, B. J. (2010). Shellfish harvesting strategies at El Varal. In R. G. Lesure (Ed.), *Settlement and subsistence in Early Formative Soconusco: El Varal and the problem of inter-site assemblage variation* (pp. 173–178). Los Angeles: Cotsen Institute of Archaeology.

Kennett, D. J., Culleton, B. J., Voorhies, B., & Southon, J. R. (2011). Bayesian analysis of high precision AMS 14Cd Dates from a prehistoric Mexican shellmound. *Radiocarbon, 53*(2), 101–116.

Kirby, M. X., & Miller, M. H. (2005). Response of a benthic suspension feeder (Crassostrea virginica Gmelin) to three centuries of anthropogenic eutrophication in Chesapeake Bay. *Estuarine, Coastal and Shelf Science, 62*, 679–689.

Kronik, J., & Verner, D. (2010). *Peoples and climate change in Latin America and the Caribbean*. The World Bank: Washington D.C.

Lachniet, M. S., & Patterson, W. P. (2002). Stable isotope values of Costa Rica surface waters. *Journal of Hydrology, 260*, 135–150.

Leduc, G., Vidal, L., Tachikawa, K., Rostek, F., Sonzogni, C., Beaufort, L., et al. (2007). Moisture transport across Central America as a positive feedback on abrupt climatic changes. *Nature, 445*, 908–911.

LeGrande, A. N., & Schmidt, G. A. (2006). Global gridded data set of the oxygen isotopic composition in seawater. *Geophysical Research Letters, 33*, L12604. doi:10.1029/2006GL026011.

Magnus, R. (1978). The prehistoric and modern subsistence patterns of the Atlantic coast of Nicaragua: A comparison. In B. Stark & B. Voorhies (Eds.), *Prehistoric coastal adaptations. The economy and ecology of maritime Middle America* (pp. 61–80). New York: Academic Press.

Malaizé, B., Bertran, P., Carbonel, P., Bonnissent, D., Charlier, K., Galop, D., et al. (2011). Hurricanes and climate in the Caribbean during the past 3700 years BP. *The Holocene, 21*(6), 911–924.

Mannino, M. A., Spiro, B. F., & Thomas, K. D. (2003). Sampling shells for seasonality: Oxygen isotope analysis on shell carbonates of the inter-tidal gastropod Monodonta lineata (da Costa) from populations across its modern range and from a Mesolithic site in southern Britain. *Journal of Archaeological Science, 30*, 667–679.

Mannino, M., Thomas, K. T., Leng, M. J., Piperno, M., Tusa, S., & Tagliacozzo, A. (2007). Marine resources in the Mesolithic and Neolithic at the Grotta dell'Uzzo (Sicily): Evidence from isotope analyses of marine shells. *Archaeometry, 49*(1), 117–133.

Marfia, A. M., Krishnamurthy, R. V., Atekwana, E. A., & Panton, W. F. (2004). Isotopic and geochemical evolution of ground and surface waters in a karst dominated geological setting: A case study from Belize, Central America. *Applied Geochemistry, 19*, 937–946.

Monks, G. G. (2017). Evidence of changing climate and subsistence strategies among the Nuu-chah-nulth of Canada's west coast. In G. G. Monks (Ed.), *Climate change and human responses: A zooarchaeological perspective* (pp. 173–196). Dordrecht: Springer.

Rosenswig, R. M. (2006). Sedentism and food production in early complex societies of the Soconusco. *Mexico. World Archaeology, 38*(2), 330–355.

Rueda, M., & Urban, H.-J. (1998). Population dynamics and fishery of the fresh-water clam Polymesoda solida (Corbiculidae) in Cienaga Poza Verde, Salamanca Island, Colombian Caribbean. *Fisheries Research, 39*, 75–86.

Schuegraf, M. J. (2004). *Establishment of seagrass decline and causative mechanisms in Pearl Lagoon, Nicaragua through use of traditional ecological knowledge, sediment coring and direct visual census*. FES Outstanding Graduate Student Paper Series: University of New York.

Severeyn, H. J., Garcia de Severeyn, Y., & Ewald, J. J. (1994). Taxonomic revision of Polymesoda solida (Philippi 1846) (Bivalvia: Corbiculidae), a new name for Polymesoda arctata, the estuarine clam of Lake Maracaibo and other estuaries of the tropical Atlantic coasts of America. *Ciencia, 2*(2), 53–65.

Severeyn, H. J., García, Y., Ewald, J. J., & Morales, F. (1996). Efectos de parámetros ambientales y la talla inicial sobre el crecimiento de la almeja comercial *Polymesoda solida* (Philippi 1846) (Bivalvia: Corbiculidae) en condiciones naturales. *Rev. Fac. Agron., 13*, 341–356.

Stephens, M., Mattey, D., Gilbertson, D. D., & Murray-Wallace, C. V. (2008). Shell-gathering from mangroves and the seasonality of the Southeast Asian monsoon using high-resolution stable isotopic analysis of the tropical estuarine bivalve (*Geloina erosa*) from the Great Cave of Niah, Sarawak: Methods and reconnaissance of molluscs of Early Holocene and modern times. *Journal of Archaeological Science, 35*, 2686–2697.

Taylor, J. D., Kennedy, W. J., & Hall, A. (1973). The shell structure and mineralogy of the Bivalvia. II Lucinacea-Clavagellacea, conclusions. *Bulletin of the British Museum (Natural History) Zoology, 22*, 255–294.

Urquhart, G. R. (2009). Paleoecological record of hurricane disturbance and forest regeneration in Nicaragua. *Quaternary International, 195*, 88–97.

Wefer, G., & Berger, W. H. (1991). Isotope paleontology: Growth and composition of extant calcareous species. *Marine Geology, 100*, 207–248.

Weninger, B., & Jöris, O. (2010). Glacial radiocarbon calibration: The CalPal program. In T. Higham, C. Bronk, C. Ramsey, & C. Owen (Eds.), *Radiocarbon and archaeology* (pp. 9–15). Oxford: Oxford University School of Archaeology.

Zorro, C. L. (2010). Étude del Tétrapodes d'un échantillon de faune de l'amas coquillier 4 (KH-4) du site de Karoline (Kukra Hill, Nicaragua). Ph.D. Dissertation. Muséum National d'Histoire Naturelle.

Chapter 9
Climatic Changes and Hunter-Gatherer Populations: Archaeozoological Trends in Southern Patagonia

Diego Rindel, Rafael Goñi, Juan Bautista Belardi, and Tirso Bourlot

Abstract Archaeozoological studies in Patagonia have tended in the past to focus on evidence provided by rock-shelters. However, a regional perspective, such as the one employed in this paper, allows us to identify trends and patterns during the Late Holocene (last 2,500 years) that could remain in the shadows if a microregional scale alone were used. Climatic changes occurred during the Late Holocene and specifically during the Medieval Climatic Anomaly (MCA), *ca.* 900 BP that were very different from the preceding times. It was proposed that lower lacustrine basins (lowlands) were residentially used by hunter-gatherer populations while the high basaltic plateaus (highlands) show an archaeological signal related to a seasonal logistic strategy. Consequently, it is expected that regional archaeozoological records obtained in different type of basins should follow these archaeological patterns. Spatial distribution of the zooarchaeological record present important differences between the Middle and Late Holocene in terms of skeletal part frequencies and processing evidence that is in agreement with the proposal.

Keywords Late Holocene • Guanaco • Southern South America • Lake basins • Basaltic Plateaus • Regional scale • Rockshelters and open air sites

D. Rindel (✉) · T. Bourlot
CONICET-INAPL, UBA, 3 de Febrero 1370, C1426BJN Capital Federal, Argentina
e-mail: drindelarqueo@yahoo.com

T. Bourlot
e-mail: tjbourlot@hotmail.com

J.B. Belardi
UNPA-CONICET, Av. Lisandro de La Torre 1070 (9400), Río Gallegos, Santa Cruz, Argentina
e-mail: juanbautistabelardi@gmail.com

R. Goñi
INAPL, UBA, 3 de Febrero 1370, C1426BJN Capital Federal, Argentina
e-mail: rafaelagustingoni@gmail.com

A Model of Late Holocene Human Peopling

A model of hunter-gatherer colonization during the Late Holocene has been proposed for southern Santa Cruz Province, Argentina (Goñi 2010). This model proposes that in the last 2,500 years residential mobility underwent a drastic reduction as a result of the environmental desiccation suffered by the region during that period (Goñi 2000; Goñi et al. 2000–2002, 2005, among others).

The main assumption that has guided research has been that human occupation in southern Patagonia is highly dependent on climatic and environmental variables which have changed throughout the Holocene. Climatic changes are initially defined by changes in the direction (begun in the Middle Holocene *ca.* 6,000 BP) and intensity (*ca.* 1,800 BP) of the westerly winds or *southern westerlies* (Gilli 2003; Gilli et al. 2005). Their consequences were manifested in new environmental (ecological and geographical) conditions, with tendencies to progressive desiccations during the Late Holocene (Stine and Stine 1990). The height of this progressive desiccation was reached during the so-called Medieval Climatic Anomaly – MCA – (AD 1,021 to 1,228) (Stine 1994, 2000). The new ecological characteristics involve the widening of shrub-like steppes featuring the presence of molle/pepper trees (*Schinus marchandii*) (Paruelo et al. 1992) and a decline in the water level of various basins in the region, especially lake basins, which fluctuated significantly, e.g., lakes Cardiel (Stine and Stine 1990) and Belgrano/Burmeister (González 1992). Equivalent changes, still under study, appear also to have occurred in the area of lakes Salitroso and Posadas, indicating that the possibility of any record of human occupation prior to 3,000–4,500 BP in the low-lying sectors of these basins would be highly improbable because these areas would have previously been covered by different types of lake basins or river-beds (Pereyra 1997; Horta and Gonella 2009). To summarize, the distribution of water as a critical resource in Patagonia varied drastically in regional and temporal terms.

Thus, a general premise was postulated: water and its spatial distribution must have been the critical resource for the region during the Holocene. Climatic changes occurred especially during the Late Holocene (last 2,500/2,000 years) and affected the distribution and availability of water and hence of different resources, especially the main prey: guanaco (Goñi 2000).

The new ecological conditions that were established can be synthesized into new primary productivity relationships (Binford 2001), given the regional variability of effective temperatures which persists to the present day. Lowland broader and open sand-dune landscapes, with predominantly shrub-like vegetation and seasonal pastures in highlands, was the geographic landscape in which the colonizing dynamics of the hunters of the Late Holocene unfolded. In this sense, effective temperatures (ET) obtained for each subregion present an internal variability in the entire region under study (Posadas/Salitroso: 12, 60; Cardiel: 11, 59; Parque Nacional Perito Moreno (PNPM): 10,41). The PNPM and the lake Cardiel basin, displaying lower ETs, present an alternative to Salitroso/Posadas for environmental productivity (Binford 2001). Thus, the ET of SalitrosoPosadas resembles the ET threshold and environments that accept an intensification of resources for other regions, such as northern Patagonia (Johnson et al. 2009).

Under the new climatic and environmental conditions present in Late Holocene, a reduction in mobility would only be expected in the face of intensification in the use of two types of resources: aquatic (see Binford 2001, Proposition 7.01) or plant resources. According to Johnson et al. (2009: 31), this situation is not foreseeable in the region here analyzed, given the absence of this kind of resources. Additionally, quasi sedentary groups that rely on land animals for their subsistence are very rare or non-existent according to Generalization 7.16 and Proposition 7.03 made by Binford (2001).

Therefore, if there was no reduction of mobility as a result of intensification processes, as could be expected based on the generalizations and propositions made by Binford (2001), a reduction in mobility should be explained by a different behavior. This process can be understood as *extensification* (Binford 2001).

According to the actualistic data provided as frames of reference by Binford (2001), the initial conditions in the environments studied show that the prevalent ETs, each taken separately, would not allow the possibility of reduced mobility. Nevertheless, if all the subregions considered are viewed complementarily, this possibility begins to make sense. This argument is considered possible for periods prior to contact with Europeans, although the exact aspects of mobility that changed with the adoption of horses as a means of transport and mobility (Goñi 2000) are still poorly understood.

Then, according to archaeological evidence, severe changes are presumed to have occurred in the organizational systems of the local hunter-gatherer societies (see previous conditions and subsequent consequences in Goñi 2000). These changes have implied: (a) a marked decline in residential mobility, focused in the low-lying basins of lakes Posadas-Salitroso and Cardiel (b) a potential spatial constraint, leading to a greater concentration of population in clusters (i.e., differential demographic increase in these basins and in certain sectors of the region), and (c) a process of *extensification* with the ensuing expansion of the ranges of action in logistic and seasonal terms (*sensu* Binford 1980). This extensification came to include the surrounding high-lying areas (e.g., the plateaus of Strobel, Cardiel Chico Pampa del Asador/Guitarra and high-lying basins of the Perito Moreno National Park), taking advantage of the different primary environmental productivities (sectors with different effective temperatures), in line with their intrinsic resource structures. According to this proposition, the colonization and effective occupation (*sensu* Borrero 1989–1990) of the area studied, in particular Posadas/Salitroso and Cardiel lake basins, would have occurred during the Late Holocene (Belardi et al. 2003; Goñi and Barrientos 2004). What follows is a brief summary of the spatial distribution of the archeological record and the functional characterizations assigned to the different sectors of the study area (Fig. 9.1).

The Strobel Plateau has been considered a special ecological environment allowing the convergence of population groups at a regional level (Belardi and Goñi 2006). The high frequencies of hunting blind structures and projectile points (Belardi et al. 2005), in addition to the elevated variety/quantity of engraved designs and motifs, support the idea of higher populations sharing the same area (Re et al. 2009). Furthermore, human burials, mostly concentrated in the basin of Salitroso Lake (García Guraieb 2006), an environment that offered optimal conditions for year-round habitation, support the idea that this could have been a central residential area for populations during Late Holocene times (Goñi and Barrientos 2004). Also, high pottery frequencies and grinding artifacts support this interpretation (Cassiodoro 2008). A similar case of suitable annual habitation conditions is offered by the Lake Cardiel basin, which shows a strong late signature for human occupations (Belardi et al. 2003; Goñi et al. 2005). This area is characterized by occupations in sand-dunes, abundant archeofaunal evidence (Bourlot 2009), a variety of designs in lithic technology and expressions of rock art in rockshelters (Ferraro and Molinari 2001) that exhibit a painting technique that is different from the engravings found on the nearby Strobel and Cardiel Chico plateaus (Belardi and Goñi 2002, 2003). Similarly, the

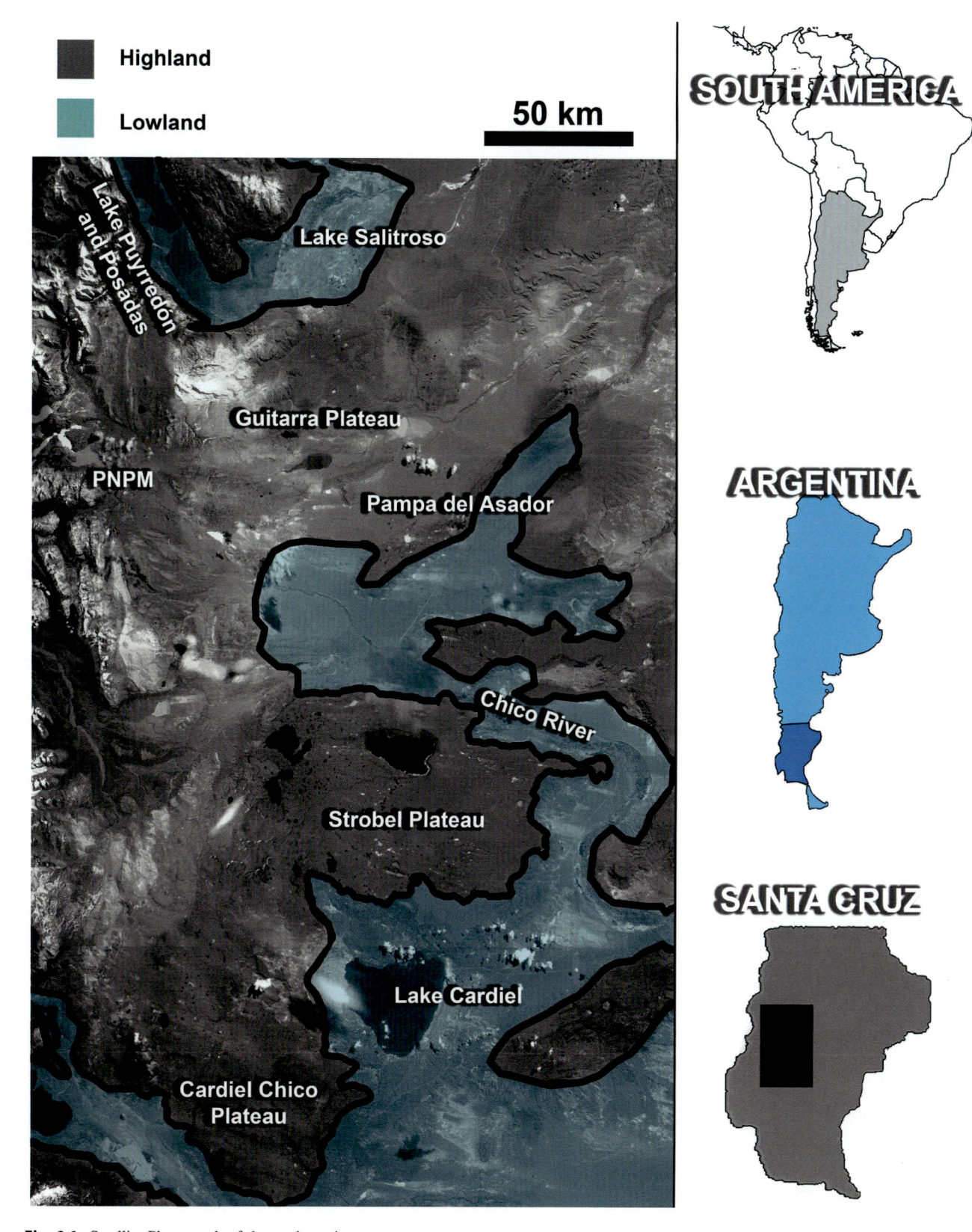

Fig. 9.1 Satellite Photograph of the study region

Pampa del Asador plain stands out as a very specific case of an environment that concentrates activities associated with the procurement of lithic raw materials (obsidian and silex) (Espinosa and Goñi 1999; Cassiodoro 2008), seasonal hunting (Rindel 2009) with the use of blinds in certain strategic locations (Aragone and Cassiodoro 2005–2006) and expressions of rock art that connect motifs from the north and south of the Lake Guitarra Plateau within the Pampa del Asador (Goñi et al. 2010). The Perito Moreno National Park, a high-lying basin that combines forest and steppe environments, presents an abundant late lithic and faunal record whose extensive distribution in open air sites, rockshelters and caves indicates a generalized seasonal utilization, with few hunting blind structures and almost no pottery or funerary structures or "*chenques*" (Cassiodoro 2001; Aschero et al. 2005; Rindel 2009).

The Study Area: Landscape and Fauna

The environment can be characterized as a shrub and herbaceous steppe (Oliva et al. 2001) located in the Southern tip of the American Continent (Province of Santa Cruz, Argentina). The landscape is characterized by inland basaltic plateaus between 900 and 1,400 meters above sea level (m asl) with few areas under 300 m asl (Mazzoni and Vázquez 2004).

The study region runs along 250 km from the Salitroso and Posadas lake basins to the North, to the Tar and San Martin lake basins to the South, and 120 km from the Andes mountain range to the Pampa del Asador and the Cardiel Lake basin to the east (see Fig. 9.1). The spatial scale utilized in this study is a mesoregional one (Dincauze 2000).

It is a landscape of lacustrine, glacial and tectonic basins (Gilli 2003) separated by wide basaltic Miocene plateaus (Ramos 2002), with the Chico River valley as well as the Lista and Belgrano Rivers dividing them from the Northwest to the Southeast.

For the purposes of this paper, the following subregions are considered: (a) Perito Moreno National Park (PMPM) (highland area), (b) Salitroso and Posadas lake basins (lowland area), (c) Pampa del Asador and Cerro Pampa (highland area), (d) Cardiel and Strobel basins (lowland area) and their surrounding plateaus – Strobel plateau and Cardiel Chico plateau- (highland areas) and (e) Tar basin (lowland area). Figure 9.2 shows the general landscape of the different areas.

Fig. 9.2 General view of the different areas. Above, left to right: Parque Nacional Perito Moreno, Salitroso Lake and Pampa del Asador below, left to right: Cardiel Lake, Cardiel Chico plateau (with hunting blinds) and Tar Lake

The Main Prey: The Guanaco

The broad regional scale allows for monitoring of the relationship between the lowlands and the highlands in which environmental, ecological and geographical differences have been taken into account. A key element of this model is the differential availability of the main prey throughout the annual cycle. For this reason a brief description of this species is given.

The guanaco (*Lama guanicoe*) is a camelid, and was the most important mammal for inland hunter-gatherer-subsistence. This animal has the ability to exploit the structure and the ecological diversity of the region. Indeed, the archaeological data show a pattern of high representation of the taxon in most of the macroregional records, and in the assemblages introduced here it averages 92% of the NISP.

As a generalist herbivore, the guanaco feeds on pastures, but it can also browse according to the availability of food at different times of the year (Raedeke 1978). It normally lives in open spaces occupying areas of steppe or prairie, though it may also live in or use the forest that are present at the Andean foothills (Cajal and Ojeda 1994; Montes et al. 2000). Its behavior is seasonally territorial, being related to sexual selection repertoire or resource distribution; however, this behavior could be modified when different groups share water (Casamiquela 1983; Oporto 1983). As a social species, according to the harsh and changing environment, the guanaco responds through modifications in the composition and size of the different groups. In guanaco wild populations, family groups, male groups and solitary individuals may be identified. Family groups are composed of an adult male with females and young. These groups are generally composed of a variable number of animals (5 to 13 members), with an average of 8.2 individuals (Puig and Videla 1995). In the male groups the age range is wider and there isn't a clear leadership behavior. The size of these groups is slightly bigger that the one recorded for family groups (7 to 20 individuals) with an average of ten animals (Puig and Videla 1995). Solitary individuals have been described as adult males who may or may not present territory. These are usually young or senile animals (Puig and Videla 1995). There are also other structures described in some populations: female groups with calves and mixed groups. In the case of female groups with calves, they might be composed of juvenile or old females with or without youngsters that leave their territory, probably due to the expulsion of the juveniles by the family male (called *relincho*). The average size of these groups is 2.6 animals, and they are composed of females and youngsters in similar proportions (Puig and Videla 1995). In several cases, mixed groups were described. These social structures consist of several kinds of groups that are formed towards the end of autumn and migrate towards the winter areas, separating again at the beginning of spring (October/November) when reproduction time approaches (Puig 1986; Puig and Videla 1995). The mixed groups are very changeable in size, sometimes reaching 150 individuals (Franklin 1982), even though the available estimates show lower average numbers: 38.7 individuals in La Payunia and 25.3 members in Torres del Paine (Puig and Videla 1995). Thus, several authors (see revision in Puig and Videla 1995) have highlighted these seasonal variations in the composition and size of the population parameters and the differences between populations. Also, it is important to note that in the past, and recorded by many visitors of Patagonia in 19th century, the herd size of guanacos could be significantly higher than in the present (Darwin 1951; Musters 1997; Lista 1998; among others).

Given these environmental, social and seasonal characteristics, Patagonian guanaco hunting required isomorphic residential mobility with seasonal guanaco movements in places where this animal was migratory. Winter is the season in which guanacos concentrate in the lowlands, and in summer they migrate to new pastures in the highlands, which are covered by snow during cold seasons. Goñi (2010) has proposed that human mobility during the Late Holocene was facilitated by a new technology: the use of leather shelters called *toldos*.

The Zooarchaeological Evidence

The main objective of this paper is to discuss, in terms of the model proposed on the basis of artefacts, features and site locations, the distribution of zooarchaeological remains in the region with reference to changes that occurred during the Late Holocene. We propose that differences in the use of lowlands (less than 275 m asl) versus the use of highlands (over 900 m asl)[1] will be observed and that the zooarchaeological data will reveal general distributional patterns at the regional scale that are not observable at the microregional scale.

The zooarchaeological discussions in southern Patagonia have been focused mainly on the evidence recovered from caves and rockshelters (Mengoni Goñalons and Silveira 1976; Silveira 1979; Mena 1983; Herrera 1988; Mena and Jackson 1991; De Nigris 1994; Gradín and Aguerre 1994; Miotti 1996, 1998; Muñoz 1997, 2000; Mengoni Goñalons 1999; Mengoni Goñalons and De Nigris 1999; De Nigris 1999a, b, 2000, 2003, 2004; De Nigris and Mengoni Goñalons 2000, 2002, 2004; Mena et al. 2000, 2003; Cassiodoro et al. 2000; Catá 2003; Rindel 2003, 2004; De Nigris and Catá 2005, among

---

[1]Most of the space between 275 and 900 m asl corresponds to the narrow and steep slope to the plateaus.

others; also, see discussion in Borrero 1989). Beyond the indubitable contribution of these works, several avenues of inquiry were opened by considering the "Monitoring perspective" (Thomas and Mayer 1983). Consistent with this perspective, we provide a novel approach in extending the contexts of sites analyzed, particularly those of the Late Holocene. Thus, we consider not only sites in rockshelters but also open air sites. The analysis of open air sites is important because rockshelters and caves in the Patagonian region have been used for specific purposes (Goñi 1995; Cassiodoro et al. 2000; Rindel 2003; among others) or as residential bases (Mengoni Goñalons and Silveira 1976; Silveira 1979; De Nigris 2004, among others). Considering that the processing of a carcass consists of several stages (Binford 1981; Lyman 1994) in which we can distinguish activities involving the procurement, initial dismemberment, secondary processing, preparation, final consumption and disposal of remains, caves and rockshelters usually show evidence of the last stages. One implication of this situation is that our knowledge of the Patagonian faunal record is biased because, as Borrero and colleagues (1985) noted, there are few kill and initial processing sites recorded. This should guide us to searching other settings besides caves and rockshelters.

Middle Holocene Zooarchaeological Record

For the purpose of allowing a comparison and discussion of the faunal record of the Late Holocene, it is important to present the main features of this record during the Middle Holocene in the study area. As we will see, it is important to note that in the stratified sites that have been taken into account in our analysis, there are substantial changes in the pattern observed for the Late Holocene (Rindel 2009, 2012). Also, it must be emphasized that there are few open air sites in Southern Patagonia with Middle Holocene faunal assemblages, and the basic chronology for basalt plateaus correspond to the last 2,000 years (Goñi 2010). Thus, the record has an uneven distribution in the Middle Holocene in the study area. In the PNPM there is a high occupational intensity while in Cerro Pampa the first evidence of planned use of the area is dated only in the Late Holocene; however, the presence of obsidian from Pampa del Asador in a neighboring area, has been known since early times indicating that these spaces were traveled but not colonized earlier (Goñi et al. 2010). In Cardiel Lake basin, the first signs of human occupation are dated in the Middle Holocene, but they are very weak. In contrast, the archaeological evidence for the Late Holocene is very strong (Goñi et al. 2004). At Strobel Plateau and the Salitroso and Posadas basins, there are several radiocarbon dates for the late Middle Holocene (4,000–3,000 BP) (De Nigris et al. 2004), but again, the bulk of occupations occur during the Late Holocene. The same situation holds for the Tar and San Martin lake basins.

Some general trends in the use of animal prey characterize the study area during the Middle Holocene. First, there has been a consistent and predominant use of guanaco. A small number of additional taxa were minimally exploited, e.g., the huemul (*Hippocamelus bisulcus*). Second, there is a low frequency of damage caused by natural agents and processes (rodents, carnivores, weathering), possibly linked with increased occupational redundancy at the Late Holocene.

Guanaco remains show a high degree of completeness in the Middle Holocene assemblages as evidenced by the ratio of axial to appendicular bones, anatomical completeness indices, the ratio of epiphysis to diaphysis fragments and MNE frequencies (Rindel 2009, 2012). This completeness can be interpreted as an indicator of relatively complete carcasses entering the sites. With regard to the frequency of anthropogenic changes, Middle Holocene assemblages exhibit a high ratio of bones with processing evidence in both axial and appendicular skeleton (Rindel 2009, 2012). This trend appears, with few variations, in each of the data sets that were analyzed. Consideration of cut and percussion marks is consistent with disarticulation, defleshing and fracture of long bones in order to obtain the bone marrow. There is also a large quantity of burned bones, regardless of the site function (base camp or limited activities sites). This thermal alteration can be linked with final consumption activities. These tasks, with minor variations, were carried out at all sites in a similar manner, resulting in a high degree of interassemblage similarity (Rindel 2009, 2012). It indicates that all aspects of carcass processing appear to have been done at the sites of this period. This also underscores the domestic character of the activities carried out at different sites during this period.

The age structure of archaeological guanacos and the presence of certain species in the assemblages can be considered as indicators of the seasonality of hunting, and possibly residence, at the sites. The case of the exploitation of huemul raises the possibility of using the PNPM area during the Early and Middle Holocene under winter conditions. Although the mere presence of this animal is not an indicator of seasonality *per se*, it is interesting to note that Herrera (1988) postulated a predation strategy in terms of the altitudinal movements of this species throughout the annual cycle. Analyses on the fibers recovered from this animal are also consistent with winter exploitation (Reigadas 2005, 2007). According to this authors (Herrera 1988; Reigadas 2005, 2007) it is suggested a preferential use of this animal on a seasonal basis, possibly into the fall and winter. Further evidence of winter predation on guanacos is the presence of fetal remains in PNPM sites (De Nigris 2003). One possible explanation for this pattern is that the exploration and colonization of the northwest Santa Cruz province took place in a landscape profoundly different from that observed today, with a more widespread distribution of water. Lakes, lagoons, *vegas* and streams should be more ubiquitously distributed than today (Aschero et al. 1992, 2005). The impact of water distribution in space must have been considerable, mainly in relation to human and animal circulation routes.

The Studied Sample

The studied assemblages belong to the Late Holocene (last 2,500 years), given that it is in this period when the rearrangement of human occupation related to modifications of surface water distribution took place.

The units of analysis we use are: (a) each I individual bone or identifiable bone fragment so as to observe the

Table 9.1 Lowlands NISP per site. The chronology of sites is distributed temporally over the past 2,500 years. The assemblage or site unit is taken into account to get spatial and temporal information. Different sites show more than one layer (CI) or more than one concentration (MS1S3, MS1S4)

Site	Acronym	NISP[a]	Site type	References
Cerro de los Indios 1 capa 3a	CI1 3a	553	Rockshelter/stratigraphy	Mengoni Goñalons (1999)
Cerro de los Indios 1 capa 3b	CI1 3b	447	Rockshelter/stratigraphy	Mengoni Goñalons (1999)
Cerro de los Indios 1AE2-C1	CI1AE2-C1	2430	Rockshelter/stratigraphy	Catá (2003)
Cerro de los Indios 1AE2-C6	CI1AE2-C6	611	Rockshelter/stratigraphy	De Nigris y Catá (2005)
Grippa Si Lito No	GSLN	833	Open air/surface-stratigraphy	Bourlot (2009)
Médanos Sur 1 Sector 4	MS1S4	244	Open air/stratigraphy	Bourlot (2009)
Médanos Sur 1 Sector 3	MS1S3	409	Open air/stratigraphy	Bourlot (2009)
La Primera Argentina capa 1	LPA1-C1	250	Open air/stratigraphy	Bourlot (2009)
La Siberia 2	LS2	154	Open air/surface	Bourlot (2009)
La Siberia 3 capa 1	LS3c1	292	Open air/stratigraphy	Bourlot (2009)
La Siberia 5	LS5	199	Open air/surface	Bourlot (2009)
Lago Tar 1	TAR-1	182	Open air/stratigraphy	this paper
Lago Tar 2	TAR-4	306	Open air/stratigraphy	this paper
Lago Tar 5	TAR-5	229	Open air/stratigraphy	this paper
Total		7139		

[a]Excluding teeth

Table 9.2 Highlands NISP per site. The chronology of sites is distributed temporally over the past 2,500 years. The assemblage or site unit is taken into account to get spatial and temporal information. Different sites show more than one layer (ADG, ADO and AGV) or more than one concentration (CP2)

Site	Acronym	NISP[a]	Site type	References
Alero Destacamento Guardaparque capa 4	ADG C4	291	Rockshelter/stratigraphy	Rindel (2009)
Alero Destacamento Guardaparque capa 3	ADG C3	136	Rockshelter/stratigraphy	Rindel (2009)
Alero Destacamento Guardaparque capa 2	ADG C2	16	Rockshelter/stratigraphy	Rindel (2009)
Istmo Lago Belgrano	ILB	651	Open air/surface	Rindel (2009)
Alero Gorra de Vasco-capa 5	AGV C5	59	Rockshelter/stratigraphy	Rindel (2009)
Alero Gorra de Vasco capa 4	AGV C4	109	Rockshelter/stratigraphy	Rindel (2009)
Alero Gorra de Vasco capa 3	AGV C3	255	Rockshelter/stratigraphy	Rindel (2009)
Alero Gorra de Vasco capa 2	AGV C2	162	Rockshelter/stratigraphy	Rindel (2009)
Alero Gorra de Vasco capa 1	AGV C1	18	Rockshelter/stratigraphy	Rindel (2009)
Alero Dirección Obligatoria capa 3	ADO C3	524	Rockshelter/stratigraphy	Rindel (2009)
Cerro Pampa 2 Ojo de Agua	CP2 OA	581	Open air/surface	Rindel (2009)
Cerro Pampa 2 Parapeto 4 estratigrafía	CP2 par 4 est	133	Open air/stratigraphy	Aragone (2007), Rindel et al. (2007)
Cerro Pampa 2 Parapeto 4 Superficie	CP2 par 4 sup	18	Open air/surface	Aragone (2007), Rindel et al. (2007)
Cerro Pampa 2 Parapeto 4 estratigrafía	CP2 par 2 est	9	Open air/stratigraphy	Aragone (2007), Rindel et al. (2007)
Cerro Pampa 2 Parapeto 2 superficie	CP2 par 2 sup	1	Open air/surface	Aragone (2007), Rindel et al. (2007)
Cerro Pampa 2c Parapeto 3 subactual	CP 2c par 3 sub	45	Open air/stratigraphy	Aragone (2007), Rindel et al. (2007)
Cerro Pampa 2a Parapeto 3 superficie	CP2a par 3 sup	4	Open air/surface	Aragone (2007), Rindel et al. (2007)
Cerro Pampa 2a Parapeto 5 superficie	CP2a par 5 sup	29	Open air/surface	Aragone (2007), Rindel et al. (2007)
Cerro Pampa 2a Parapeto 6 superficie	CP2a par 6 sup	2	Open air/surface	Aragone (2007), Rindel et al. (2007)
Cerro Pampa 2a Parapeto 7 superficie	CP2a par 7 sup	10	Open air/surface	Aragone (2007), Rindel et al. (2007)
Total		3053		

[a]Excluding teeth

Table 9.3 Lowland MNE per Skeletal element by assemblage and site

Part	CI1 3a	CI1 3b	CI AE2 C1	CI AE2 C6	GSLN	LPA1	LS3 c1	LS 5	LS 2	MS1S3	MS1M4	TAR-1	TAR-4	TAR-5	MNE total
Skull	3	2	9	4	3	3	3	2	2	2	1	0	0	1	35
Mandible	9	4	6	2	4	2	4	1	1	2	1	1	0	1	38
Atlas	2	0	4	1	3	1	0	2	1	0	0	0	0	0	14
Axis	3	1	9	2	3	1	1	1	0	0	0	1	0	0	22
Indeterminate Vertebrae	0	0	0	0	16	5	4	7	2	5	3	4	20	1	67
Cervical Vertebrae	7	7	54	14	20	3	4	6	4	1	1	4	6	1	132
Thoracic Vertebrae	12	7	30	14	23	7	2	21	6	2	1	6	4	8	143
Lumbar Vertebrae	20	14	25	13	7	5	2	1	3	8	4	0	1	1	104
Sacrum	3	1	3	1	3	1	0	0	1	1	1	1	1	0	17
Caudal Vertebrae	0	0	23	0	1	0	0	0	0	0	0	0	0	0	24
Pelvis	3	1	3	3	5	1	1	2	2	3	5	3	1	2	35
Rib	20	14	27	18	19	12	5	20	6	8	6	5	5	8	173
Sternebrae	1	0	3	0	5	0	1	5	1	0	0	2	0	1	19
Scapula	3	2	7	5	10	1	2	6	1	1	0	3	0	3	44
Humerus Proximal	0	2	0	0	11	2	0	4	0	0	0	5	3	0	27
Humerus Shaft	6	5	10	6	3	3	4	1	4	1	1	1	1	4	50
Humerus Distal	0	0	0	0	17	3	1	4	3	2	0	11	5	1	47
Radioulna Proximal	4	2	0	0	15	3	2	12	5	2	0	15	3	4	67
Radioulna Shaft	4	6	0	4	5	1	4	0	2	1	1	0	1	3	32
Radioulna Distal	3	2	10	0	13	3	1	3	4	1	1	7	6	5	59
Carpals	18	10	0	0	26	16	4	1	6	8	2	3	24	22	140
Metacarpal	2	4	0	0	3	3	2	0	0	4	0	2	7	3	30
Femur Proximal	4	0	0	0	9	2	2	10	2	4	1	5	8	2	49
Femur Shaft	6	8	14	3	8	2	3	1	1	3	1	1	3	3	57
Femur Distal	1	2	0	0	11	4	1	2	1	2	2	9	3	1	39
Patella	4	3	0	0	9	1	0	1	2	2	0	2	1	5	30
Tibia Proximal	2	2	0	0	9	1	3	7	0	5	2	8	3	3	45
Tibia Shaft	3	7	0	5	5	1	7	0	1	3	0	1	1	5	39
Tibia Distal	2	5	13	0	9	4	2	5	4	5	0	7	8	5	69
Tarsals	7	6	0	0	16	11	3	1	1	1	2	5	13	4	70
Astragalus	7	2	9	1	6	3	4	1	1	2	2	8	7	1	54
Calcaneus	4	6	26	3	9	7	2	0	1	7	2	4	6	2	79
Metatarsal	2	3	0	0	2	7	0	1	1	0	1	0	5	2	24
Metapodial Shaft	11	11	0	0	8	3	4	2	1	6	4	2	3	4	59
Metapodial Distal	11	11	64	16	17	3	4	5	2	8	11	6	9	14	181
Phalange 1	19	20	208	31	26	6	10	2	4	33	44	9	7	11	430
Phalange 2	17	13	119	25	19	2	9	1	4	16	28	1	4	7	265
Phalange 3	11	4	53	14	10	0	8	0	1	5	1	0	1	2	110
Sesamoid	0	0	0	0	16	0	5	1	2	16	10	0	1	0	51
Total	234	187	729	185	404	133	114	139	83	170	139	142	171	140	2970

Table 9.4 Highland MNE per Skeletal element by assemblage and site

Part	ADG 4	ADG 3	ADG 2	ILB	CP 2 OA	AGV 5	AGV 4	AGV 3	AGV 2	AGV 1	ADO 3	TOT HB[a]	MNE total
Skull	2	2	0	2	3	1	2	2	3	1	3	4	25
Mandible	1	1	0	10	2	0	0	6	2	0	6	3	31
Atlas	0	0	0	0	3	0	0	0	0	0	2	2	7
Axis	1	0	0	1	4	0	1	0	0	0	0	1	8
Indeterminate Vertebrae	1	2	0	2	4	1	0	0	0	0	0	0	10
Cervical Vertebrae	4	2	1	13	23	2	3	7	3	1	7	2	68
Thoracic Vertebrae	5	3	0	30	28	1	2	6	2	0	7	3	87
Lumbar Vertebrae	6	1	0	16	23	1	2	5	4	1	5	6	70
Sacrum	0	0	0	5	1	0	2	0	0	0	1	0	9
Caudal Vertebrae	3	0	0	0	0	0	0	0	2	0	0	0	5
Pelvis	1	1	0	22	24	0	1	3	1	1	2	2	58
Rib	10	5	0	12	11	2	5	5	0	0	13	4	67
Sternebrae	2	1	0	2	0	0	0	0	0	0	1	0	6
Scapula	4	3	0	28	39	1	2	2	2	0	1	1	83
Humerus Proximal	0	1	0	8	0	0	0	0	0	0	0	0	9
Humerus Shaft	3	4	0	17	0	2	2	3	3	0	7	0	41
Humerus Distal	1	1	0	44	31	0	0	0	0	0	0	3	80
Radioulna Proximal	3	0	0	31	0	0	0	0	0	0	0	3	37
Radioulna Shaft	4	6	1	12	0	0	4	6	2	2	9	5	51
Radioulna Distal	1	0	0	31	23	0	0	0	0	0	0	0	55
Carpals	6	3	0	5	5	0	0	0	2	0	0	3	24
Metacarpal	1	3	0	12	6	1	0	4	2	0	5	1	35
Femur Proximal	0	0	0	18	24	0	3	0	0	0	0	2	47
Femur Shaft	4	4	0	6	0	4	0	3	3	1	10	5	40
Femur Distal	0	0	0	27	0	0	0	0	0	0	0	2	29
Patella	1	0	0	6	4	0	2	3	1	0	1	0	18
Tibia Proximal	0	0	0	16	0	1	0	0	0	0	0	2	19
Tibia Shaft	2	4	0	3	0	0	3	6	4	0	9	6	37
Tibia Distal	3	0	1	25	24	0	0	0	0	1	0	1	55
Tarsals	4	5	1	12	6	0	0	0	3	1	0	3	35
Astragalus	4	2	0	13	25	0	1	2	0	0	7	0	54
Calcaneus	8	1	1	12	12	2	2	3	6	0	11	1	59
Metatarsal	4	2	0	10	6	1	0	4	4	0	8	1	40
Metapodial Shaft	5	5	1	8	4	0	0	0	5	0	0	4	32
Metapodial Distal	5	5	1	29	39	0	3	0	2	0	0	4	88
Phalange 1	12	7	5	16	25	5	7	13	13	1	22	9	135
Phalange 2	6	1	0	4	1	3	2	3	4	1	23	5	53
Phalange 3	2	3	0	1	0	0	0	1	0	0	19	7	33
Sesamoid	2	3	0	0	0	0	0	0	0	0	0	4	9
Total	121	81	12	509	400	28	49	87	73	11	179	99	1649

[a]TOT HB = Sum of nine assemblages from hunting blinds

distribution of the different skeletal parts within their spatial context. This unit was chosen in order to see if a differential distribution of certain bones across the region could have existed. The logic of the argument stands on the availability of different food resources and raw material associated with each anatomic portion of the main prey (the guanaco – *Lama guanicoe*) then each bone would be the expression of the use of such associated resources[2]; then (b) the archaeological sites that have been taken into account, with the aim of linking the bone sets in space and time to achieve a regional perspective.

The analyzed lowland assemblages come from Cardiel Lake (seven assemblages from seven sites: GSLN, LPA1, LS2, LS3-C1, LS5, MS1S3 and MS1M4), Posadas-Salitroso Lakes (four assemblages from one stratified site: CI 3a, 3b, CI AE2 C1, AE2 C6), Tar Lake (three assemblages, TAR-1, TAR-4 and TAR-5), while the highlands assemblages come from PNPM (ten assemblages from four sites: ADG, AGV, ADO and ILB) and from Cerro Pampa (ten assemblages from eight sites: CP 2 OA, CP 2 hunting blind 4 stratigraphy and surface, Hunting blind (historic) 3, CP 2 hunting blind 2 stratigraphy and surface, CP 2 hunting blind 3, 5, 6 and 7 surface). The total analyzed sample NISP is 7,139 bones for the lowlands and 3,053 bones for the highlands (Tables 9.1 and 9.2), making a total sample of 10,192 guanaco elements.

The variables considered were the frequency of skeletal parts, the degree of fragmentation, and the evidence of anthropogenic modifications. These observations will allow us to evaluate differential use of the landscape in relation to the differential availability of resources.

The measures of anatomical abundance we have used are MNE, MAU and %MAU (Binford 1984) because they are the most commonly used in archaeozoology and they allow assemblages with different sample sizes to be standardized, a necessary condition for the comparisons required. In the case of the samples from hunting blinds, the sum of the different assemblages was obtained, then the MNE and a total MAU were derived (Tables 9.3 and 9.4). This does not affect the total highland value.

In the case of the processing evidence, due to the lack of detailed data for processing marks for all the cases, the total NISP considered was 6,224 specimens, 3,438 bones for lowlands and 2,786 bones for highlands.

Spatial Distribution of the Archaeozoological Record: %MAU and Correspondence Analysis for Lowlands and Highlands

MAU and %MAU values were obtained, and they are compared in Fig. 9.3. All bones were taken into account. These criteria produced a sample of 14 lowland assemblages and 9 for the highlands.

In this section the distribution of the different bones for the lowlands and the highlands is introduced. For this analysis, the %MAU was used because it facilitates bone assemblage standardization and comparability.

As shown in Fig. 9.3, the recorded trends reveal similarities and differences in skeletal representation patterns. The similarities appeared in the first place in the skull and the jaw, well represented in the lowlands and the highlands. Also, the representation of upper and mid-limb bones (humerus, radio-cubitus, femur and tibia) are similar, although, in the sample of the lowlands there is a higher shaft frequency, while the highland assemblages have a high amount of proximal and distal epiphyses (due to the amount of flesh and marrow they have and the ease of transport and processing of these bones). On the other hand, the lowland assemblages show a high frequency of neck elements (atlas, axis and cervical vertebrae), double the frequency of ribs and sternum recorded in the highlands and a high frequency of lower limb bones (metapodial, carpals/tarsals and phalanges). These latter elements would be related to ease of transportation, high quality of marrow, and as by-products in leather production for making windbreaks (Goñi 2010).

In the highland, there is a high frequency of those elements poorly represented in the lowlands, especially the scapulae and pelves which are difficult to process and transport. It is important to highlight the representation of epiphyses, which is linked to a high frequency of percussion and cut marks closely related to transversal and perimetral marking fracture techniques.[3]

Synthesizing these observations, the first point to note is that, given the large sample size, the large number of sites and the

---

[2]The archaeozoological sample used to carry out this research belongs to a series of assemblages that were previously studied separately (see Bourlot 2009; Catá 2003; Mengoni Goñalons 1999; De Nigris and Catá 2005; Espinosa et al. 2010 and Rindel 2009). In their original studies, they were evaluated also from a taphonomic and site formational perspective; consequently, these discussions do not appear in this paper. It is worth mentioning that in this paper we decided – due to low sample size or/and taphonomic conditions – that several of the assemblages should be discarded as unsuitable for this general analysis.

[3]The main feature of this type of fracture is the presence of straight edges transversal to the long axis of the bone, along with the separation of the articular ends from their diaphyseal segments. This definition is applied to the case of transverse fractures, while we consider the perimetral marking fracture as a subset of these, involving the preparation of the bone by a complete or incomplete groove around the surface of the bone. While it is a processing technique that is very common in Southern Patagonia (Mengoni Goñalons and Silveira 1976; Mengoni Goñalons 1982; Muñoz and Belardi 1998; Mena and Jackson 1991; Hajduk and Lezcano 2005; Bourlot et al. 2008), its use has also been recorded in human groups from other regions of the world, such as the Dassanecht of Kenya (Gifford-González 1989) and the Hadza of Tanzania (Lupo 1994).

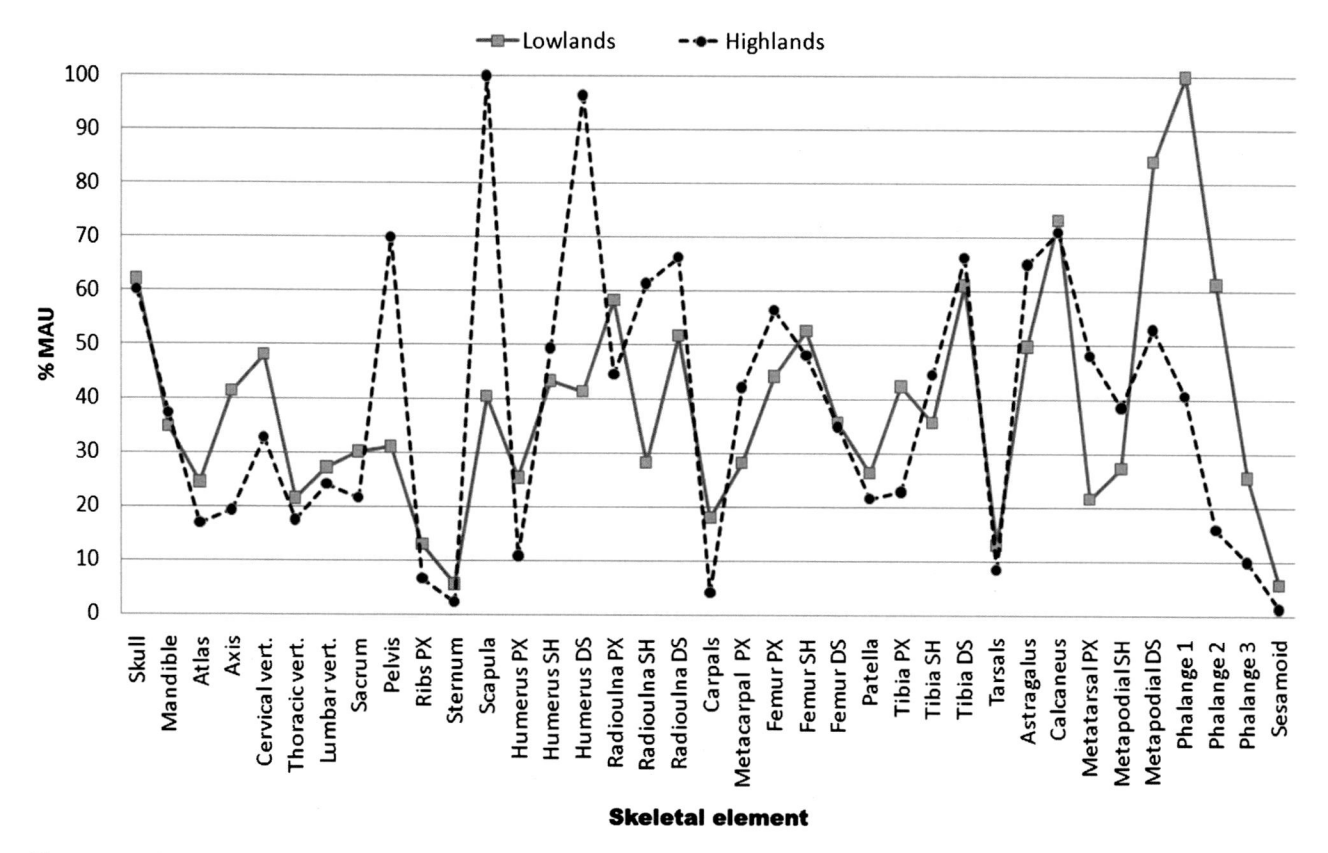

Fig. 9.3 Lowlands and highlands guanaco %MAU

different landscapes, there is a high degree of variability in the archaeozoological record at the regional scale that is not evident in micro-regional analyses, as is noted by the distribution of three main groups in Fig. 9.4. The second point is that a trend to spatial differences between epiphyses and diaphysis, lower limbs and upper limbs, and axial and appendicular skeleton appeared. Lastly, this diversity is related to the differential processing of the prey according to the functional categories of the sites involved, which is explained by the model in terms of seasonality and logistic versus residential mobility strategies.

Correspondence analysis (Fig. 9.4) of bones grouped by assemblage and site shows interesting trends. We selected those samples with an NISP higher than 100, with the objective of assuring their comparability.

Axis 1 orders assemblages by the type of site they come from (open air – ILB, GSLN, MS4, MS1S3, LPAc1, LS2, LSIIIc1, LPAc2, SAC 25, TAR1, 4 and 5, CP2OA, and hunting blinds –, and rock-shelter – CI1 3a, CI1 3b and CI1 AE2, ADGc4, ADGc3, ADOc3, AGVc4, AGVc3 and AGVc2). Open air sites, disregarding their altitude, show a high frequency of articular ends (epiphysis) that are generally associated with a particular kind of percussion technique, the transversal or perimetral fracture (Bourlot et al. 2008) plus scapula and pelvic girdle, sternebrae and vertebrae. This is the case of sites like ILB (17), CP2 OA (18), LS2 (6), LPA-c1 (9) and GSLN (5). In contrast, most assemblages recovered in rock-shelters have a high frequency of segments belonging to long bone diaphysis plus crania/mandible and ribs, as it is evident in sites AGV layers 2 to 4 (20, 21 and 22), ADG layers 4 to 3 (15, 16), ADO (19) and Cerro de los Indios 3a and 3b (1, 2 and 3). Those differences might correspond to differential preservation of the long bones segments (see Belardi et al. 2010a). It is very interesting to observe, though, that the assemblages made of samples from hunting blinds also correspond with those of the rock-shelters, even though they are strictly open air sites (see point 23, left above in Fig. 9.4). Thus, the abundance of long bone segments might be related to the stages of the animal processing carried out at both these sites, rather than bone preservation. Axis 2 corresponds to segments that belong to lower limbs, especially metapodials and phalanges, in lowland rockshelters and open air sites: MS4 (7), SAC 25 (4), MS1S3 (8), CI1 AE2 (3).

As we noted above, bone assemblages grouped according to site show that those from rock-shelters have a close

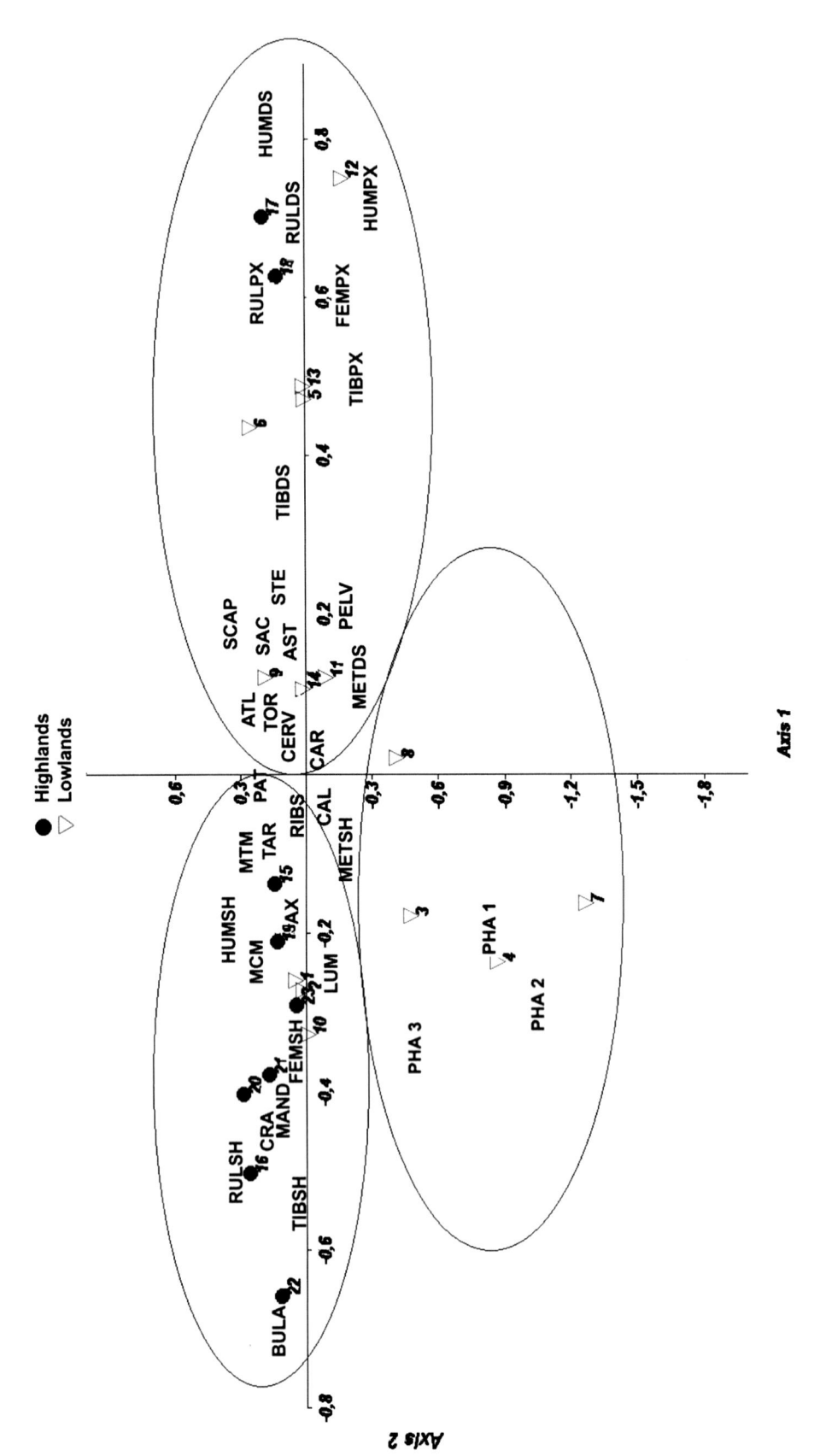

Fig. 9.4 Bone correspondence analysis in lowland (open triangles)/highland (black circles) sites; Ref.: 1 = CI13a; 2 = CI13b; 3 = CIIAE2; 4 = SAC25; 5 = GSLN; 6 = LS2; 7 = MS4; 8 = MS1S3; 9 = LPAc1; 10 = LSIIIc1; 11 = LPAc2; 12 = TAR1; 13 = TAR4; 14 = TAR5; 15 = ADGc4; 16 = ADGc3; 17 = ILB; 18 = CP2OA; 19 = ADOc3; 20 = AGVc4; 21 = AGVc3; 22 = AGVc2; 23 = total hunting blinds

correspondence among them. At the same time, two big groups appear in most open air sites: those related specifically to lower limbs (left below in Fig. 9.4) and those that have pieces of long bone articular ends (right above in Fig. 9.4). It is also noteworthy that the two highland open air sites (ILB and CP2 OA), whose assemblages are extremely similar, are 60 km apart, and both assemblages have similar chronologies (*circa* 1,400 BP).

Finally, a higher degree of variability is found in lowland sites, which is consistent with the expectation of a mainly residential site function in these basins. On the other hand, the highland sites are more homogeneous, due to their specific function of prey processing.

Cultural Processing Marks

With the objective of evaluating aspects related to prey processing in different areas, the assemblages were also analyzed in relation to the frequency of processing marks, both cut and percussion (Fig. 9.5). The observed trend indicates higher processing evidence in the highland assemblages.

The sample was subdivided in cut mark (Fig. 9.6) and percussion mark frequencies (Fig. 9.7). The trend mentioned above remains in both areas; however, the cut mark frequencies are more homogeneously distributed between areas, while the percussion mark frequencies are considerably higher in certain sites of the highlands than in the lowlands.

The visibility of marks in the highlands is high and results from initial transport and immediate consumption, i.e., minimal processing, of prey. Conversely, in the lowlands, the percentage of cultural marks is lower because bones are intensively processed, losing marks in the final stage of consumption, as we note below.

Bone Fragmentation

Bone fragmentation was measured by the NISP/MNE ratio. Results indicate that in highland assemblages, values of bone fragmentation are lower than those in the lowlands (Fig. 9.8).

The observed pattern might be related to the residential use of the lowland landscape (showing a higher degree of variability in Fig. 9.8), where the availability of a bigger array of resources (shelter and fire wood), would allow a more intensive final processing of carcasses. The greater fragmentation of guanaco bones in lowland sites is possibly related to winter occupation, probably by relatively large groups of people. Under such circumstances, intensive processing might signal a strain of the human population against the environment's carrying capacity. Greater fragmentation might also indicate more

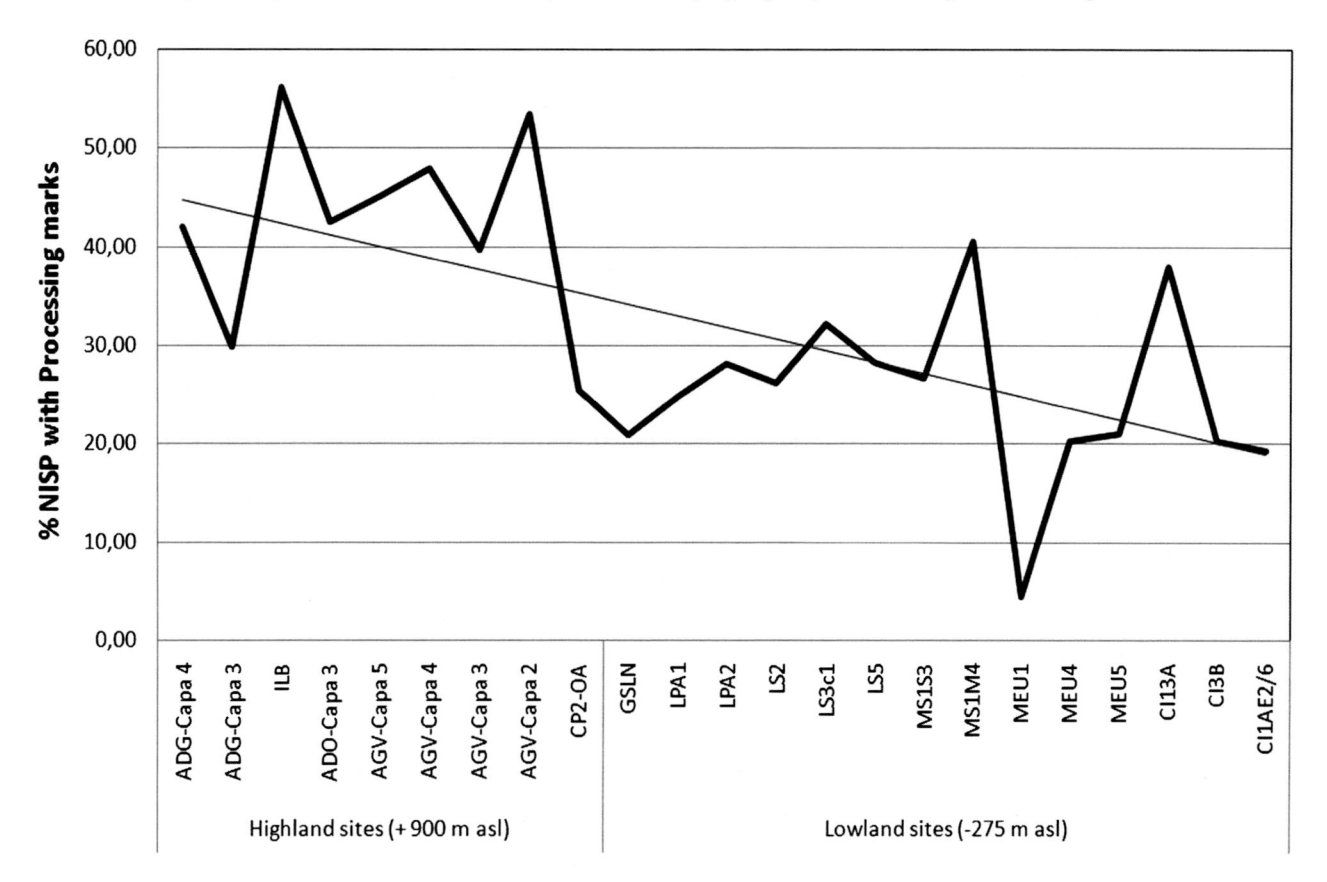

Fig. 9.5 General sample of processing marks in lowland/highland sites (%NISP with processing marks)

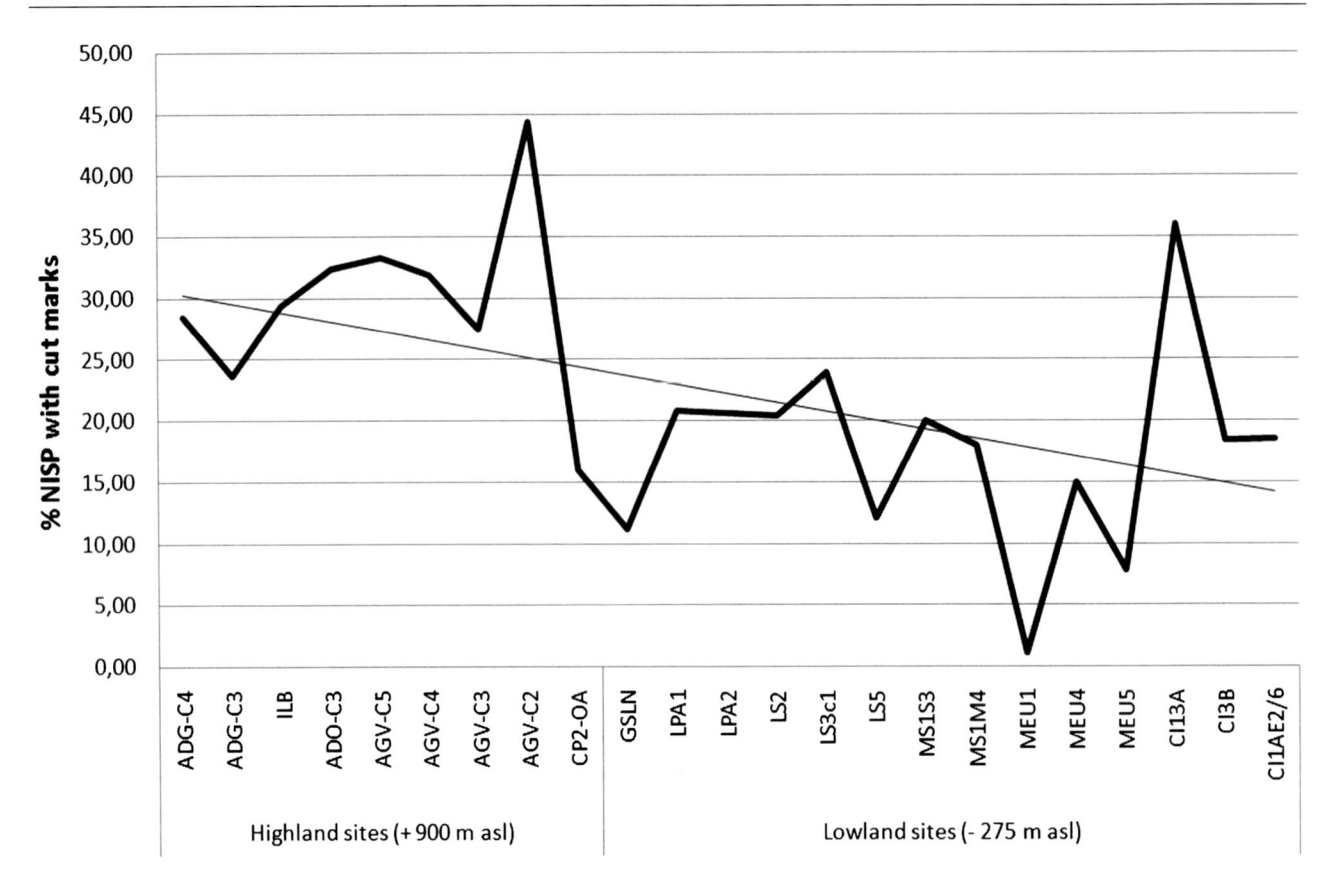

Fig. 9.6 Cut marks in lowland and highland samples (%NISP with cut marks)

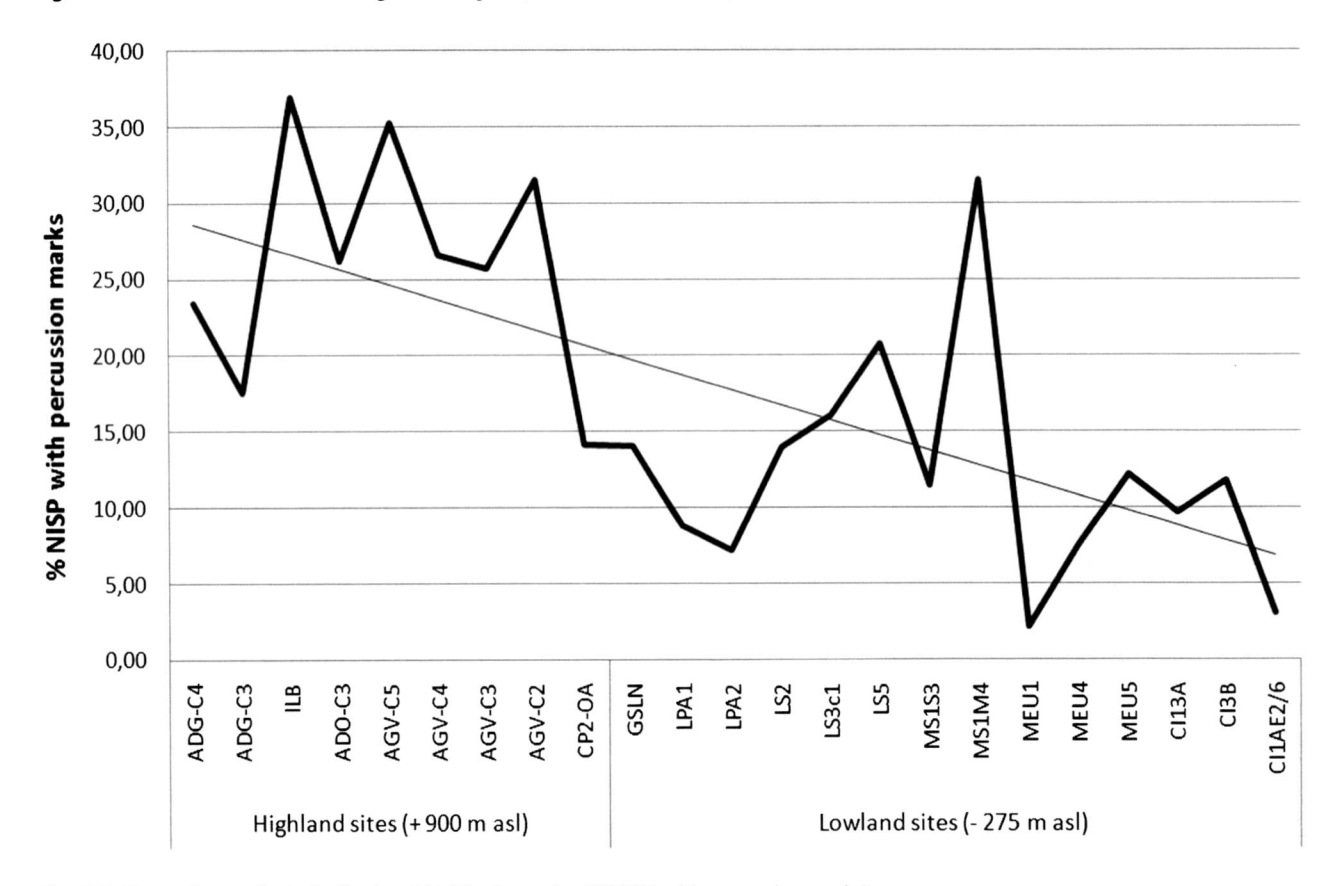

Fig. 9.7 Percussion marks in lowland and highland samples (%NISP with percussion marks)

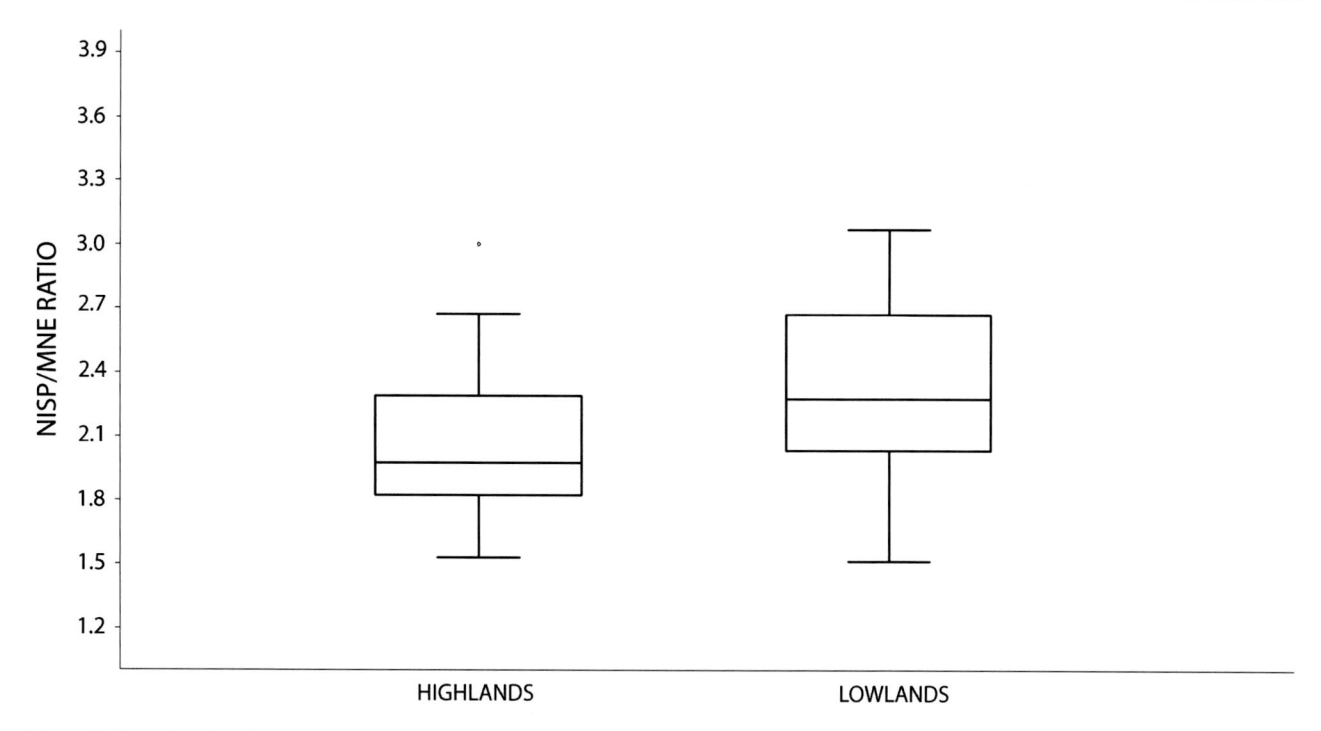

Fig. 9.8 Box plot showing the fragmentation degree of the samples calculated by NISP/MNE

marrow and bone grease extraction to obtain fats. It may also indicate more time to process carcasses under more seasonally sedentary conditions (as opposed to mobile highland hunting in summer). Conversely, the seasonal (summer) and logistic use of the plateaus, mainly for hunting activities, would show initial prey processing, with later transport of skeletal parts towards other areas, as can be seen in the complementary frequencies of skeletal parts between lowlands and highlands (see Fig. 9.3). We observed that as the number of bone processing stages increases so does fragmentation and loss of cultural marks. This is also confirmed by the characteristics of the artifact assemblages found in lowland site, e.g., ceramic sherds, grinding stones, heated stones, hammer and anvil stones, which indicate their possible relation to bones fracture and marrow obtaining activities (Belardi et al. 2010b; Goñi 2010; Cassiodoro 2011).

Discussion and Conclusions

The following discussion refers to: (a) the distribution of the archaeological record in space and time, in a macro scale; (b) the concept of mobility monitored archaeologically; and (c) how this concept of mobility operates at the regional scale in the Late Holocene and how it compares with previous work in the Middle Holocene.

(a) Figure 9.4 of this study shows the main trends in the distribution of the zooarchaeological record throughout the study region. A sample of over 10,000 bone specimens, from an area the size of Sicily (mesoregional scale sensu Dincauze 2000), can be considered indicative of human behaviour in terms of resource use, related mobility and environmental variability. Thus, the implementation of a large spatial scale, rather than the individual archaeological site, has been extremely useful to understand the distribution of different skeletal parts. Considering each bone as an analytical unit allows us to find major trends related to human mobility patterns, beyond the particularities of each site. This methodological approach has recorded new patterns emerging from the bone assemblages. Thus, the use of these scales and derived trends are the appropriate way to study the archaeological record in terms of questions about social dynamics of the past, separating the discussion from the narrow concept of *archaeological site*. In short, the approach taken here allows us to study processes occurring in large scales, which is particularly important to interpret patterns and trends in the long term. We consider that this is the appropriate scale to monitor climate change and cultural change. These patterns underscore not only the importance of the *presence* of certain variables in the archaeological record, but they are also crucial in interpreting absences that are not manifest at the scale of individual sites. A good example is phalanges which relate to the use of skins for toldos in the lowland steppes, and which occur in low frequencies in the highlands (see Fig. 9.3).

(b) The current concept of mobility (Binford 1980; Wendrich and Barnard 2008) has its methodological roots in ethnoarchaeological and ethnographic work. However, this is a category that also can be tested in the archaeological record. The basis of this process is to assume that mobility patterns are manifested in derived archaeological patterns, or, to put it another way, the archaeological patterns can be assumed to be derived from patterns of social dynamics (Binford 1992). Thus, the pattern described here in terms of the spatial distribution of the zooarchaeological record fulfils minimum requirements to get compatible scales between ethnography and archaeology. The spatial scale included different microenvironments, i.e., a regional scale, in order to ensure environmental variability. The time scale, on the other hand, sets the boundaries to certain climatic conditions governing environmental variability. Our analyses of Late Holocene zooarchaeological patterns allow us to compare them with other analyses of the Middle Holocene to establish *what* changed with time in terms of both climate/environment and human behaviour patterns. This variable, the space use over time, generates an accumulative archaeological landscape (Binford 1992) that exceeds a particularistic ethnographic scale. As a corollary, we can add that from an archaeological perspective, a series of questions can be derived from ethnoarchaeological and ethnographic models (Binford 1980; Wendrich and Barnard 2008) that refer specifically to distributional patterns at regional scales. The combination of both approaches, ethnographic and archaeological, provides cross-questions that enrich explanations about the past.

(c) We can now establish the general trends in the use of fauna and its implications in terms of mobility and change for the Late Holocene in the study area.

During the Late Holocene (the last 2,500 years) changes in the landscape were generated in Southern Patagonia in terms of a new distribution of the critical resource: water. These environmental modifications led to changes in both prey distribution and hunter-gatherer mobility. In these new conditions low steppe basins played a different role from the highlands in the subsistence and settlement systems. The archaeozoological implication of these changes in terms of mobility is that the residential character of the lowlands was manifested as a greater skeletal part representation related to increased sedentism (year round), greater processing intensity and discard of bone and to logistic seasonal exploitation of the highlands. In the latter, less variety of skeletal part representation would be associated with transport strategies. The implementation of these transport strategies is observed in the differential representation of skeletal parts between both areas. At lowland sites high frequencies of lower limbs (metapodials and phalanges), are possibly related to consumption and to leather processing in residential areas. Finally, the higher degree of fragmentation recorded in lowland lacustrine assemblages is also consistent with the residential use of these areas where more intensive carcass processing and final consumption would have occurred. This residential pattern is also indicated by the presence of site furniture, human burials, mild winter climate, and critical resources such as water, firewood, and available prey.

This Late Holocene pattern contrasts with the information obtained for the Middle Holocene. As previously stated, the Middle Holocene is linked to high residential redundancy in the highlands (PNPM). This is evidenced by a set of characteristics: (1) a low frequency of natural agents and processes, (2) a high degree of anatomical completeness, (indicator of near complete carcass transport to the residential sites), (3) a high frequency of processing evidence such as cut marks and marrow extraction related fractures, and (4) high frequencies of burned bones that emphasize the domestic nature of the highland settlements in the Middle Holocene. In PNPM there is also evidence interpreted as winter occupation, such as remains of foetal guanacos, as well as evidence of occupation during spring/summer, such as the find of neonate guanacos (De Nigris 2004). As a consequence, in the Middle Holocene we do not observe a marked seasonal use such as that of the Late Holocene. In addition, basaltic plateaus have shown a chronology concentrated in the Late Holocene with almost no occupation dates for Middle Holocene (Goñi 2010).

The Late Holocene archaeological pattern, on the contrary, clearly presents a contrast between lowlands and highlands. The differential distribution of sites and zooarchaeological remains has been explained in terms of residential and logistical/seasonal mobility (Goñi 2010), which differs from the Middle Holocene. This explanation not only took into account the archaeological record, but it also climatic and environmental determinants that generated an entirely new geographical landscape due to desiccation that occurred at the end of Holocene, especially, but not only, during the MCA. As desiccation increased, new large shrub steppes became established in spaces previously occupied by lakes and rivers or by other steppes types. The expansion of hunting grounds followed from the expansion of lowland steppe area, but it also added cost in terms of residential mobility and prey search and handling time. This cost could be minimized by logistical seasonal exploitation of the highlands, combined with stable residence in optimal locations for habitation in the lowlands. The result was complementary strategies of initial processing, transport and final consumption in the highlands and the lowlands. The analyses presented here and previous work (Goñi 2010) shows that after 2,500 BP, climatic and environmental changes opened the possibility of exploiting the highlands seasonally on a logistic basis while maintaining a more sedentary occupation pattern in the lowlands.

In sum, this paper has presented an analysis of the zooarchaeological evidence of a large section of Southern Patagonia. The regional approach allowed us to make arguments about human mobility in relation to climatic conditions established during the Late Holocene, and other archaeological evidence supports the results obtained here (see Goñi 2010). A variety of causes can lead to differential patterns in the archaeological record. However, as we propose in this study, the climatic and environmental factors of the Late Holocene exerted a major influence on behaviour among hunter-gatherer populations in Southern Patagonia, generating profound changes in the archaeological record.

Acknowledgments We thank the ICAZ 2010 Meeting Organizers, and we thank Greg Monks for inviting us to publish this paper and for his suggestions that greatly improved the manuscript. Also, we thank the reviewers of this paper for their very useful suggestions. Thanks to Florencia Gordón for reading the paper and for helping us with the graphics. This research was conducted with the support of CONICET (PIP 6405, PIP 0122, D325/10), ANPCyT (PICT 26295, PICT 1247), Universidad de Buenos Aires (UBACyT n° 20020100100441), Universidad Nacional de la Patagonia Austral (UNPA-UARG A/183/2, A/213-1 and 29/A245-1), and the National Ministry of Culture.

References

Aragone, A. (2007). Análisis comparativo entre los conjuntos óseos de médanos del lago Posadas y parapetos de Pampa del Asador. Licenciatura Thesis in Anthropological Sciences. Universidad de Buenos Aires.

Aragone, A., & Cassiodoro, G. (2005–2006). Los parapetos de Cerro Pampa: registro arqueofaunístico y tecnológico (noroeste de la provincia de Santa Cruz). *Arqueología 13*, 131–154.

Aschero, C. A., Bellelli, C., & Goñi, R. A. (1992). Avances en las investigaciones arqueológicas del Parque Nacional Perito Moreno (provincia de Santa Cruz, Patagonia argentina). *Cuadernos del Instituto Nacional de Antropología y Pensamiento Latinoamericano, 14*, 143–170.

Aschero, C., Goñi, R., Civalero, M., Molinari, R., Espinosa, S., Guráieb, A., et al. (2005). Holocenic Park: arqueología del PNPM. *Anales de la Administración de Parques Nacionales, 17*, 71–119.

Belardi, J., & Goñi, R. (2002). Distribución espacial de motivos rupestres en la cuenca del lago Cardiel (Patagonia Argentina). *Boletín SIARB, 16*, 29–38.

Belardi, J., & Goñi, R. (2003). Motivos rupestres y circulación de poblaciones cazadoras-recolectoras en la meseta del Strobel (Santa Cruz, Patagonia argentina). *CD publication from VI Simposio Internacional de Arte Rupestre* (pp. 186–195). Jujuy, Argentina.

Belardi, J., & Goñi R. (2006). Representaciones rupestres y convergencia poblacional durante momentos tardíos en Santa Cruz (Patagonia argentina). El caso de la meseta del Strobel. In D. Fiore & M. Podestá (Eds.), *Tramas en la Piedra* (pp. 85–94). Buenos Aires: WAC, SAA and AINA.

Belardi, J., Goñi, R., Bourlot, T., & Aragone, A. (2003). Paisajes arqueológicos en la cuenca del Lago Cardiel (Provincia de Santa Cruz, Argentina). *Magallania, 31*, 95–106.

Belardi, J. B., Espinosa, S., & Cassiodoro, G. (2005). Un paisaje de puntas: las cuencas de los lagos Cardiel y Strobel (Provincia de Santa Cruz, Patagonia argentina). *Werken, 7*, 57–76.

Belardi, J., Bourlot, T. J., & Rindel, D. D. (2010a). Representación diferencial de diáfisis y epífisis de huesos largos de guanaco (Lama guanicoe) en contextos arqueológicos de médanos en Patagonia austral: el sitio Río Meseta 1 (lago Tar, provincia de Santa Cruz). In M. Gutiérrez, M. De Nigris, P. Fernández, M. Giardina, A. Gil, A. Izeta., et al. (Eds.), *Zooarqueología a principios del siglo XXI: aportes teóricos, metodológicos y casos de estudio* (pp. 119–131). Buenos Aires: Ediciones del Espinillo.

Belardi, J., Espinosa, S., Carballo Marina, F., Barrientos, G., Goñi, R., Súnico, A., et al. (2010b). Las cuencas de los lagos Tar y San Martín (Santa Cruz, Argentina) y la dinámica del poblamiento humano del sur de Patagonia: integración de los primeros resultados. *Magallania, 38*(2), 137–159.

Binford, L. (1980). Willow smoke and dog's tails: hunter-gatherer settlement system and archaeological site formation. *American Antiquity, 45*, 4–20.

Binford, L. (1984). *Faunal remains from Klasies River mouth*. New York: Academic Press.

Binford, L. (1992). Seeing the present and interpreting the past—and keeping things straight. In J. Rossignol & L. Wandsnider (Eds.), *Space, time, and archaeological landscapes* (pp. 43–59). New York: Plenum Press.

Binford, L. (2001). *Constructing frames of reference. An analytical method for archaeological theory building using ethnographic and environmental data sets*. Berkeley, Los Angeles and London: University of California Press.

Borrero, L. (1989a). Replanteo de la arqueología patagónica. *Interciencia, 14*(3), 127–135.

Borrero, L. (1989–1990). Evolución cultural divergente en la Patagonia austral. *Anales del Instituto de la Patagonia 19*, 133–139.

Borrero, L. A., Casiraghi, M., & Yacobaccio, H. D. (1985). First guanaco-processing site in southern South America. *Current Anthropology, 26*(2), 273–276.

Bourlot, T. (2009). Zooarqueología de sitios a cielo abierto en el lago Cardiel, Provincia de Santa Cruz: Fragmentación ósea y consumo de grasa animal en grupos cazadores-recolectores del Holoceno Tardío. Ph.D. Dissertation, Universidad de Buenos Aires.

Bourlot, T., Rindel, D., & Aragone, A. (2008). La fractura transversa/marcado perimetral en sitios a cielo abierto durante el Holoceno tardío en el noroeste de Santa Cruz. In M. Salemme, F. Santiago, M. Álvarez, E. Piana, M. Vázquez & M. Mansur (Eds.), *Arqueología de la Patagonia. Una mirada desde el último confín* (pp. 693–705). Ushuaia, Tierra del Fuego: Editorial Utopías.

Cajal, J., & Ojeda, R. (1994). Camélidos Silvestres y Mortalidad por tormentas de nieve en la cordillera frontal de la Provincia de San Juan, Argentina. *Mastozoología Neotropical, 1*(1), 81–88.

Casamiquela, R. (1983). La significación del guanaco en el ámbito pampeano-patagónico; aspectos cronológicos, ecológicos, etológicos y etnográficos. *Mundo Ameghiniano, 4*, 20–46.

Cassiodoro, G. (2001). Variabilidad de la tecnología lítica en el sitio Alero Destacamento Guardaparque (Santa Cruz): análisis de los instrumentos formatizados. Licenciatura Thesis in Anthropological Sciences, Universidad de Buenos Aires.

Cassiodoro, G. (2008). Movilidad y uso del espacio de cazadores-recolectores del Holoceno tardío: estudio de la variabilidad del registro tecnológico en distintos ambientes del noroeste de la provincia de Santa Cruz. Ph.D. Dissertation, Universidad de Buenos Aires.

Cassiodoro, G. (2011). *Movilidad y uso del espacio de cazadores-recolectores del Holoceno tardío. Estudio de la variabilidad del registro tecnológico en distintos ambientes del noroeste de la provincia de Santa Cruz*. Oxford: BAR International Series 2259, South American Archaeology Series No. 13.

Cassiodoro, G., Lublin, G., Piriz, M., & Rindel, D. (2000). Los primeros pasos del Alero Destacamento Guardaparque: análisis lítico y faunístico (N.O. Provincia de Santa Cruz, Argentina). In *Desde el País de los Gigantes. Cuartas Jornadas de Arqueología de la Patagonia*. Río Gallegos, Argentina.

Catá, M. P. (2003). Aspectos metodológicos relacionados con el concepto de grano. Un caso de aplicación al registro arqueofaunístico: Cerro de los Indios 1. Licenciatura Thesis in Anthropological Sciences, Universidad de Buenos Aires.

Darwin, C. [1839] (1951). *Viaje de un naturalista alrededor del mundo*. Buenos Aires, Argentina: Editorial El Ateneo.

De Nigris, M. (1994). Patrones de fragmentación de huesos lagos en el sitio Cerro de los Indios 1, Lago Posadas. Licenciatura Thesis in Anthropological Sciences, Universidad de Buenos Aires.

De Nigris, M. (1999a). Lo crudo y lo cocido: sobre los efectos de la cocción en la modificación ósea. *Arqueología, 9*, 239–264.

De Nigris, M. (1999b). De fracturas y otros huesos: consumo de médula en Patagonia meridional. *Resúmenes del XIII Congreso Nacional de Arqueología Argentina* (pp. 392–393, Córdoba).

De Nigris, M. (2000). Procesando para el consumo: dos casos de Patagonia Meridional. In *Desde el país de los gigantes: perspectivas arqueológicas en Patagonia. Cuartas Jornadas de Arqueología de la Patagonia*. Río Gallegos.

De Nigris, M. (2003). Procesamiento y consumo de ungulados en contextos arqueológicos de Patagonia Meridional: el caso de Cerro Casa de Piedra Cueva 7. Ph.D. Dissertation, Universidad de Buenos Aires.

De Nigris, M. (2004). *El consumo en grupos cazadores recolectores. Un ejemplo zooarqueológico de Patagonia meridional*. Buenos Aires: Sociedad Argentina de Antropología.

De Nigris, M., & Catá, M. (2005). Cambios en los patrones de representación ósea del guanaco en Cerro de los Indios 1 (Lago Posadas, Santa Cruz). *Intersecciones en Antropología, 6*, 109–119.

De Nigris, M., & Mengoni Goñalons, G. (2000). Patrones y tendencias generales de los conjuntos faunísticos en Cerro de los Indios 1. *Arqueología, 10*, 227–243.

De Nigris, M., & Mengoni Goñalons, G. (2002). The guanaco as a source of meat and fat in the southern Andes. In J. Mulville & A. Outram (Eds.), *The zooarchaeology of milk and fats* (pp. 160–166). London: Oxbow Books.

De Nigris, M., & Mengoni Goñalons, G. (2004). El guanaco como fuente de carne y grasas en Patagonia. In M. Civalero, P. Fernández & A. Guraieb (Eds.), *Contra Viento y Marea: Arqueología de Patagonia* (pp. 469–476). Buenos Aires: Edited by INAPL.

De Nigris, M., Figuerero Torres, M., Guráieb, A., & Mengoni Goñalons, G. (2004). Nuevos fechados radiocarbónicos en la localidad de Cerro de los Indios 1 (Santa Cruz) y su proyección areal. In M. Civalero, P. Fernández & A. Guraieb (Eds.), *Contra Viento y Marea: Arqueología de Patagonia* (pp. 537–544). Buenos Aires: Edited by INAPL.

Dincauze, D. (2000). *Environmental archaeology, principles and practices*. Cambridge: Cambridge University Press.

Espinosa, S., & Goñi R. (1999). ¡Viven! Una fuente de obsidiana en la Provincia de Santa Cruz. In *Soplando en el viento… Actas de las III Jornadas de Arqueología de la Patagonia* (pp. 177–189). Bariloche: Universidad Nacional del Comahue e INAPL.

Ferraro, L., & Molinari, R. (2001). ¡Último momento! El arte de los cazadores recorre el lago Cardiel y se dirige al Strobel. In *Actas del XIV Congreso Nacional de Arqueología Argentina* (pp. 70–81, Rosario).

Franklin, W. L. (1982). Biology, ecology, and relationship to man of the South American camelids. In H. Mares & M. G. Genoways (Eds.), *Mammalian biology in South America* (Vol. 6, pp. 457–488). Pittsburg: University of Pittsburgh, Special Publication Series. Pymatuning. Laboratory of Ecology.

García Guráieb, S. (2006). Salud y enfermedad en cazadores recolectores del Holoceno tardío en la cuenca del lago Salitroso (Santa Cruz). *Intersecciones en Antropología, 7*, 37–48.

Gifford-González, D. (1989). Ethnographic analogues for interpreting modified bones: Some cases from East Africa. In *Bone modification* (pp. 179–246). Center for the Study of the First Americans, Orono: University of Maine.

Gilli, A. (2003). Tracking late quaternary environmental change in southernmost South America using lake sediments of Lake Cardiel (49°S), Patagonia, Argentina. Ph.D. Dissertation, Swiss Federal Institute of Technology Zurich.

Gilli, A., Ariztegui, D., Anselmetti, F., Mckenzie, J., Markgraf, V., Hajdas, I., et al. (2005). Mid-Holocene strengthening of the southern westerlies in South America-sedimentological evidences from Lago Cardiel, Argentina (49°S). *Global and Planetary Change, 49*, 75–93.

González, M. (1992). *Paleoambientes del Pleistoceno Tardío/Holoceno Temprano en la cuenca de los lagos Belgrano y Burmeister (47° 40' Sur, 72° 30' Oeste, Santa Cruz)*. Fundación Carl C: Zon Caldenius. Informe Técnico no. 9: 1–7.

Goñi, R. (2000). Arqueología de Momentos Históricos fuera de los Centros de Conquista y Colonización: un análisis de caso en el sur de la Patagonia. In *Desde el País de los Gigantes. Perspectivas Arqueológicas en Patagonia* (pp. 283–296). Río Gallegos: UNPA.

Goñi, R. (2010). Cambio climático y poblamiento humano durante el Holoceno tardío en Patagonia Meridional. Una perspectiva arqueológica. Ph.D. Dissertation, Universidad de Buenos Aires.

Goñi, R., Barrientos, G., & Cassiodoro, G. (2000–2002). Las condiciones previas a la extinción de las poblaciones humanas del sur de Patagonia: una discusión a partir del análisis de la estructura del registro arqueológico de la cuenca del Lago Salitroso. *Cuadernos del Instituto Nacional de Antropología y Pensamiento Latinoamericano, 19*, 249–266.

Goñi, R., & Barrientos, G. (2004). Poblamiento tardío y movilidad en la cuenca del lago Salitroso. In M. Civalero, P. Fernández & A. Guraieb (Eds.), *Contra Viento y Marea. Arqueología de la Patagonia*, (pp 313–324). Buenos Aires: INAPL.

Goñi, R., Belardi, J., Espinosa, S., & Savanti, F. (2004). Más vale tarde que nunca: cronología de las ocupaciones cazadoras- recolectoras en la cuenca del lago Cardiel (Santa Cruz, Argentina). In M. Civalero, P. Fernández & A. Guraieb (Eds.), *Contra Viento y Marea. Arqueología de la Patagonia* (pp. 237–248). Buenos Aires: INAPL.

Goñi, R., Cassiodoro, G., Re, A., Guichón, F., Flores Coni, J., & Dellepiane, J. (2010). Arqueología de la Meseta del lago Guitarra (Santa Cruz). *Libro de resumenes expandidos del XVI Congreso Nacional de Arqueología Argentina* (pp. 174–181). Jujuy: Editorial de la Universidad Nacional de Jujuy.

Goñi, R., Espinosa, S., Belardi, J., Molinari, R., Savanti, F., Aragone, A.., et al. (2005). Poblamiento de la estepa patagónica: cuenca de los Lagos Cardiel y Strobel. In *Actas del XIII Congreso Nacional de Arqueología Argentina* (Tomo 4, pp. 1–18). Córdoba: Universidad de Río Cuarto.

Gradín, C., & Aguerre, A. (Eds.). (1994). *Contribución a la Arqueología del Río Pinturas*. Concepción del Uruguay: Ediciones Búsqueda de Ayllu.

Hajduk A. J., & Lezcano, M. J. (2005). Un "nuevo viejo" integrante del elenco de instrumentos óseos de Patagonia: los machacadores óseos. *Magallania, 33*(1), 63–80.

Herrera, O. (1988). Arqueofaunas del sitio Cerro de Piedra 5. *Precirculados del IX Congreso Nacional de arqueología* (pp. 60–72). Bs. As, Argentina.

Horta, L., & Gonella, C. (2009). *Análisis paleoambiental y reconstrucción paleoclimática mediante isótopos estables en el área del lago Pueyrredón, Provincia de Santa Cruz–Argentina*. Área 3: Geología: Universidad Nacional de Tucumán.

Johnson, A., Gil, A., Neme, G., & Freeman, J. (2009). Maíces e intensificación: explorando el uso de los marcos de referencia. In G. López & M. Cardillo (Eds.), *Arqueología y Evolución. Teoría, metodología y casos de estudio* (pp. 23–47). Buenos Aires: Editorial SB, Colección Complejidad Humana.

Lista, R. (1998). *Obras*. Volúmenes 1 y 2., Buenos Aires: Editorial Confluencia.

Lupo, K. D. (1994). Butchering marks and carcass acquisition strategies: Distinguishing hunting from scavenging in archaeological contexts. *Journal of Archaeological Science, 21*, 827–837.

Lyman, R. L. (1994). *Vertebrate taphonomy.* Cambridge University Press.

Mazzoni, E., & Vázquez, M. (2004). *Ecosistemas de mallines y paisajes de la Patagonia Austral (Provincia de Santa Cruz).* Buenos Aires: Ediciones INTA.

Mena, F. (1983). Excavaciones arqueológicas en la cueva Las Guanacas (RI 16), XI Región de Aisén. *Anales del Instituto de la Patagonia, 14*, 67–75.

Mena, F., & Jackson, D. (1991). Tecnología y subsistencia en Alero Entrada Baker, Región de Aisén, Chile. *Anales del Instituto de la Patagonia (Serie Ciencias Sociales), 20*, 169–203.

Mena, F., Lucero, V., Reyes, O., Trejo, V., & Velásquez, H. (2000). Cazadores tempranos y tardíos en la Cueva Baño Nuevo-1, margen occidental de la estepa centropatagónica (XI Región de Aisén, Chile). *Anales del Instituto de la Patagonia (Serie Ciencias Históricas), 28*, 173–195.

Mena, F., Velásquez, H., Trejo, V., & Torres Mura, J. (2004). Aproximaciones zooarqueológicas al pasado de Aisén continental (Patagonia central chilena). In Mengoni Goñalons, G. (Ed.), *Zooarchaeology of South America* (pp. 99–122). Oxford: BAR International Series 1298.

Mengoni Goñalons, G. L. (1999). *Cazadores de guanacos de la estepa patagónica.* Buenos Aires: Sociedad Argentina de Antropología.

Mengoni Goñalons, G., & De Nigris, M. (1999). Procesamiento de huesos largos de guanaco en Cerro de los Indios 1 (Santa Cruz). In *Soplando en el viento… Actas de las Terceras Jornadas de Arqueología de la Patagonia* (pp. 461–475). Neuquén: Universidad del Comahue e INAPL.

Mengoni Goñalons, G., & Silveira, M. (1976). Análisis e interpretación de los restos faunísticos de la Cueva de las Manos, Estancia Alto Río Pinturas (Prov. de Santa Cruz). *Relaciones Sociedad Argentina de Antropología X*: 261–270.

Miotti, L. (1996). Piedra Museo. Nuevos datos para la ocupación pleistocénica en Patagonia. In J. Gómez Otero (Ed.), *Arqueología. Solo Patagonia* (pp. 27–38). Puerto Madryn: CENPAT CONICET.

Miotti, L. (1998). *Zooarqueología de la meseta central y costa de Santa Cruz. Un enfoque de las estrategias adaptativas aborígenes y los paleoambientes.* Museo de Historia Natural. Secretaria de Gobierno, Departamento de San Rafael, Mendoza.

Montes, C., De Lamo, D., & Zavatti, J. (2000). Distribución de abundancias de guanaco (Lama guanicoe) en los distintos ambientes de Tierra del Fuego, Argentina. *Mastozoología Neotropical, 7*(1), 5–14.

Muñoz, A. (1997). Explotación y procesamiento de ungulados en Patagonia meridional y Tierra del Fuego. *Anales del Instituto de la Patagonia (Serie Ciencias Históricas), 25*, 201–222.

Muñoz, A. (2000). El procesamiento de guanacos en Tres Arroyos 1, Isla Grande de Tierra del Fuego. In *Desde el País de los Gigantes. Perspectivas Arqueológicas en Patagonia* (pp. 499–515). Río Gallegos: UNPA.

Muñoz, A., & Belardi, J. B. (1998). El marcado perimetral en los huesos largos de guanaco de Cañadón Leona (Colección Junius Bird): implicaciones arqueofaunísticas para Patagonia Meridional. *Anales del Instituto de la Patagonia, 26*, 107–118.

Musters, G. (1997). *Vida entre los Patagones.* Buenos Aires: El Elefante Blanco.

Oliva, G., González, L., Rial, P., & Livraghi, E. (2001). El ambiente en la Patagonia Austral. In P. Borrelli & G. Oliva (Eds.), *Ganadería Ovina Sustentable en la Patagonia Austral. Tecnologías de Manejo Extensivo* (pp. 19–82). Buenos Aires: Ediciones INTA.

Oporto, N. (1983). Contribución al conocimiento sobre el comportamiento de guanaco y posibles aplicaciones. *Mundo Ameghiniano, 4*, 1–19.

Ortega, I. (1985). Social organization and ecology of migratory guanaco population in Southern Patagonia. M.A. Thesis, Iowa State University.

Paruelo, J., Aguilar, M., Golluscio, R., & León, R. (1992). La Patagonia extrandina. Análisis de la estructura y el funcionamiento de la vegetación a distintas escalas. *Ecología Austral, 2*, 123–136.

Pereyra, F. (1997). *Informe. Geología Alero Cerro de los Indios y Zona de Lago Posadas, Prov. de Santa Cruz.* Ms.

Puig, S. (1986). Ecología poblacional del guanaco (Lama guanicoe, Camelidae, Artiodactyla) en la Reserva Provincia de La Payunia (Mendoza). Ph.D. Dissertation, Universidad de Buenos Aires.

Puig, S., & Videla, F. (1995). Comportamiento y organización social del guanaco. In S. Puig (Ed.), *Técnicas para el manejo del Guanaco* (pp. 97–118). Gland, Switzerland: UICN.

Raedeke, K. (1978). *El guanaco de Magallanes, Chile. Su distribución y biología.* Chile: CONAF, Publicación Técnica 4 Ministerio de Agricultura.

Ramos, V. (2002). El magmatismo neógeno de la cordillera patagónica. In M.J. Haller (Ed.), *Geología y Recursos Naturales de Santa Cruz. Relatorio del XV Congreso Geológico Argentino (Tomos I–XIII*: 187–199). El Calafate, Santa Cruz.

Re, A., Belardi, J., & Goñi, R. (2009). Dinámica poblacional tardía en Patagonia meridional: su discusión y evaluación a través de la distribución de motivos rupestres. En: M. Sepúlveda, L. Briones & J. Chacama (Eds.), *Crónicas sobre la piedra. Arte rupestre de las Américas* (pp. 293–309). Ediciones Universidad de Tarapacá, Arica.

Reigadas, M. (2005). Fibras arqueológicas de origen animal. Análisis microscópico de muestras de fibras de Cerro Casa de Piedra-CCP7 y CCP5 (Santa Cruz, Argentina). *Relaciones de la Sociedad Argentina de Antropología XXX*, 235–243.

Reigadas, M. (2007). Cazadores del Holoceno y los recursos faunísticos. Estudio de fibras animales de Cerro Casa de Piedra-CCP7 y CCP5 (Santa Cruz). In F. Morello, M. Martinic, A. Prieto y G. Bahamonde (Eds.), *Arqueología de Fuego-Patagonia. Levantando piedras, desenterrando huesos…y develando arcanos* (pp. 663–674). Punta Arenas: CEQUA.

Rindel, D. (2003). Patrones de procesamiento faunístico durante el Holoceno medio y tardío en el sitio Alero Destacamento Guardaparque (Parque Nacional Perito Moreno, Provincia de santa Cruz, argentina). Licenciatura Thesis in Anthropological Sciences, Universidad de Buenos Aires.

Rindel, D. (2004). Patrones de procesamiento faunístico en el sitio Alero Destacamento Guardaparque durante el Holoceno Medio. In M. Civalero, P. Fernández & G. Guraieb (Eds.), *Contra Viento y Marea. Arqueología de la Patagonia* (pp. 263–276). Buenos Aires: SAA y AINA.

Rindel, D. (2009). Arqueología de momentos tardíos en el noroeste de la Provincia de Santa Cruz (Argentina): una perspectiva faunística. Ph.D. Dissertation, Universidad de Buenos Aires.

Rindel, D. (2012). Arqueofaunas del Holoceno medio y tardío de sectores altos del noroeste de la Provincia de Santa Cruz. Paper delivered at the *VIII Jornadas de Arqueología de la Patagonia*. In evaluation.

Rindel, D., Cassiodoro, G., & Aragone, A. (2007). La utilización de las mesetas altas durante el Holoceno tardío: el sitio Cerro Pampa 2 Ojo de Agua (Santa Cruz)). In F. Morello, M. Martinic, A. Prieto y, G. Bahamonde (Eds.), *Arqueología de Fuego-Patagonia. Levantando piedras, desenterrando huesos…y develando arcanos* (pp. 649–662). Punta Arenas: CEQUA.

Silveira, M. (1979). Análisis e interpretación de los restos faunísticos de la Cueva Grande de Arroyo Feo (Provincia de Santa Cruz). *Relaciones de la Sociedad Argentina de Antropología XIII*: 229–247.

Stine, S. (1994). Extreme and persistent drought in California and Patagonia during mediaeval time. *Nature, 369*, 546–549.

Stine, S. (2000). On the medieval climatic anomaly. *Current Anthropology, 41*(4), 627–628.

Stine, S., & Stine, M. (1990). A record from lake cardiel of climate change in southern South America. *Nature, 345*, 705–708.

Thomas, D., & Mayer, D. (1983). Behavioral faunal analysis of selected horizons. In D. Thomas (Ed.), *The archaeology of the Monitor Valley 2: Gatecliff Shelter* (pp. 353–380). New York: Anthropological Papers of the American Museum of Natural History 59(1).

Wendrich, W., & Barnard, H. (2008). The archaeology of mobility: Definitions and research approaches. In H. Barnard & W. Wendrich (Eds.), *The archaeology of mobility. Old World and New World nomadism* (Chap. 1: 1–21). Los Angeles: Cotsen Institute of Archaeology, UCLA.

Chapter 10
Evidence of Changing Climate and Subsistence Strategies Among the Nuu-chah-nulth of Canada's West Coast

Gregory G. Monks

Abstract Zooarchaeological data from Canada's west coast are presented which show a shift from a greater abundance of rockfish (genus *Sebastes*) during the Medieval Climatic Anomaly to a greater abundance of salmon (genus *Oncorhynchus*) during the Little Ice Age. Measurements of rockfish hyomandibulars and $\delta^{18}O$ analysis of *Saxidomus gigantea* shells are used within an optimal foraging framework as proxy measures to evaluate the hypothesis that human subsistence strategies changed in association with climate-driven environmental change. As well, suites of other marine resources that were preferred during each climatic period are presented and the cultural implications of the data are discussed. These data suggest that there was a change in subsistence strategy associated with this climatic transition, but a causal link is not demonstrated.

Keywords Butter clam (*Saxidomus gigantea*) · Little Ice Age · Medieval Climatic Anomaly · Northwest Coast of North America · Optimal foraging theory · Oxygen isotope analysis · Rockfish (*Sebastes* spp.) · Salmon (*Oncorhynchus* spp.)

Introduction

Indigenous cultures of Canada's west coast (the Northwest Coast, or NWC for short) are renowned as examples of complex hunter-fisher-gatherers (Kelly 1995; Ames 2003; see also Colonese et al. 2017). Much of that complexity has been attributed to the vast numbers of salmon that were harvested in large numbers and stored for later consumption (Boas 1966; Matson 2006). In an attempt to bring nuance to generalizations about food resource abundance and the opulent cultures that resulted from it, Suttles (1962) proposed that four sources of environmental variation should be considered: local, seasonal, annual and unpredictable. Subsequent research has shown the cultural effects of local as well as annual salmon availability (Donald and Mitchell 1975, 1994), and cultural response to seasonal variability has been exemplified by several researchers (Monks 1987; Cannon 2002). Although Suttles framed unpredictable variation in terms of the environment, the unpredictability of human decision-making has broadened the scope of this term (Cannon 1998). There has been a welcome recognition by archaeologists in the last several decades that environmental and cultural variability, along with agency and contingency, are the rule among Northwest Coast indigenous populations. Specifically, long-term variability is a dimension not addressed by Suttles, although, not surprisingly, archaeologists have recognized its importance (e.g., Ames 1991, 2005; Cannon 1998, 2002; Matson 2003: 5). Causative agents in long-term variability have been identified as environmental change, resource depletion, population pressure and technological development, to name the more common ones. In this chapter, I wish to focus on the relationship between environmental change and cultural responses in the recent prehistory of the Nuu-chah-nulth of western Vancouver Island, British Columbia (see Fig. 10.1). Specifically, I will present zooarchaeological data from two sites in the territory of one local group of the Nuu-chah-nulth, namely the Toquaht of Barkley Sound, for the past 1,200 years, and I will present relevant climatic and environmental data from the Medieval Climatic Anomaly (MCA) and the Little Ice Age (LIA) as it was experienced on the northeast Pacific Ocean coast using an optimal foraging framework to analyze and interpret these data.

G.G. Monks (✉)
Department of Anthropology, University of Manitoba,
15 Chancellors Circle, Winnipeg, MB R3T 5V5, Canada
e-mail: Gregory.Monks@umanitoba.ca

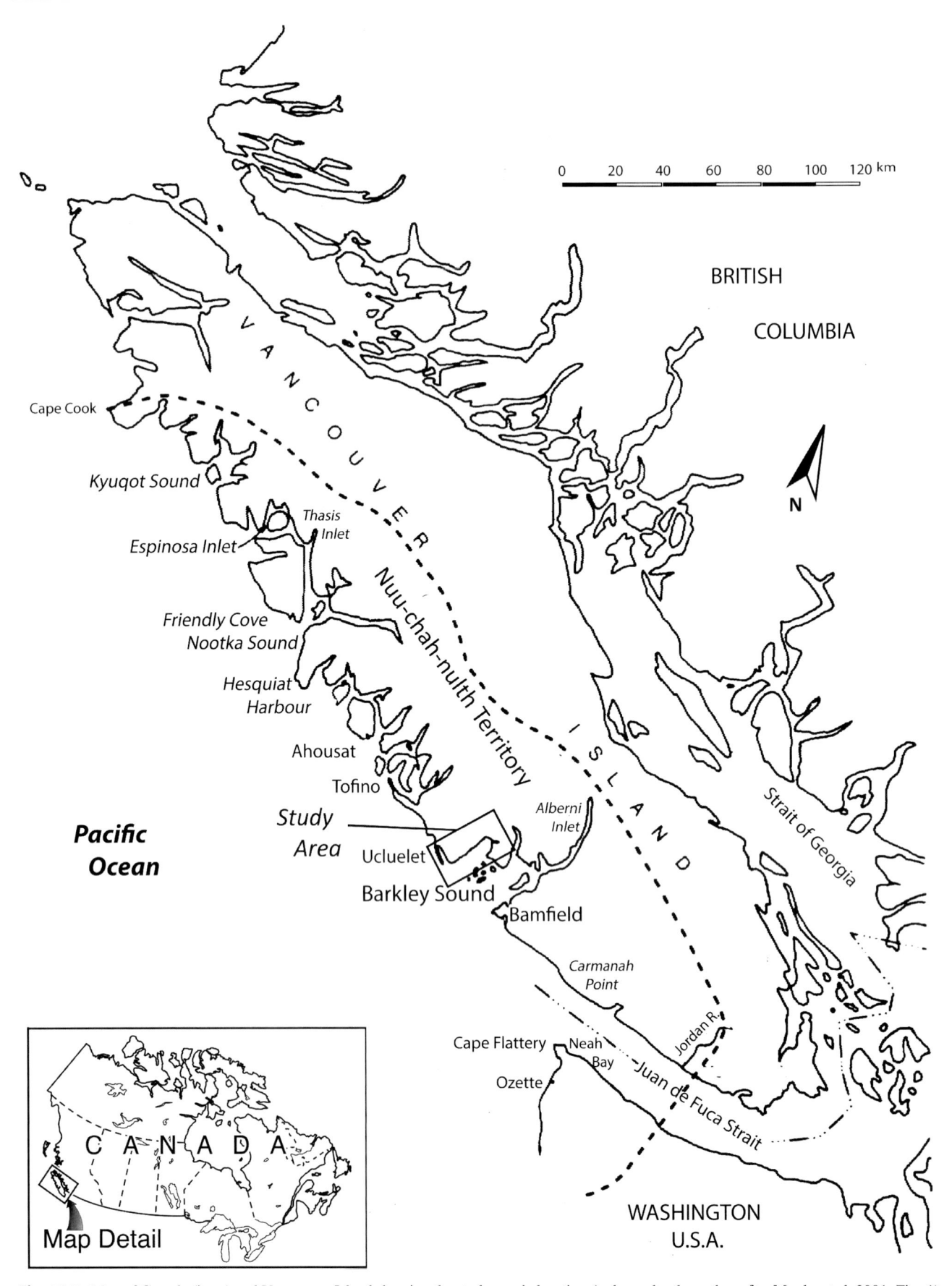

Fig. 10.1 Map of Canada (inset) and Vancouver Island showing the study area's location (redrawn by the author after Monks et al. 2001: Fig. 1)

Climate Change

Was there, in fact, a change in climate between the MCA and the LIA?

Millennial Scale Variation

There is not universal agreement on this question (Bradley et al. 2003), but the weight of evidence suggests that such a phenomenon occurred (North 2006). Sometimes called the A.D. 1300 Event, this transition has been recorded for Europe (Lamb 1965), the north Atlantic (Trouet et al. 2009), and the Pacific Ocean (Nunn 2007). Multi-proxy climatic records are deemed to be reliable after ca. A.D. 1,600, but they are only moderately reliable from A.D. 900–1,600 because of the accuracy of individual proxy measures, to issues related to proxy data syntheses and to variability in climatic effects from region to region (North 2006: 118).

The MCA onset is estimated variously according to the area where evidence has been reported and to the kind of proxy indicator(s) used, and warming of sea surface temperatures in the Indian Ocean and the western Pacific Ocean is thought to be an important causative agent for the global effects of the MCA (Graham et al. 2011). Nunn (2007: 3–5) cites onset dates of A.D. 950–1,220 and A.D. 800–1,400 in California and a terminal date of A.D. 1,200 in the Canadian Rockies. The MCA transition into the LIA in North America is generally thought to have begun approximately A.D. 1,300 while the latter period continued until the late 19th century (Ahlenius 2007).

During the MCA, climatic effects were complex and varied by region. In California, for example, cool arid conditions prevailed in inland areas (Stine 1994), and ocean temperatures were cooler than the Late Holocene median, resulting in greater upwelling and high but variable marine productivity, in contrast to earlier research that indicated ocean temperatures were higher during this period (Kennett and Kennett 2000: 381, 384). Discussing possible future effects of current global warming on Canada's west coast, Clague and Turner suggest that warmer, wetter conditions may prevail and that there may be increased risk of flooding as a result of increased El Niño positive phases (Clague and Turner 2000: 116–118). As a general model, the same effects were likely felt during the MCA in coastal British Columbia. How warm the coast became during the MCA is unclear, but, as it ended, temperatures at the Columbia Ice Sheet in the Rocky Mountains fell ca. 1.2°C around A.D. 1,290 (Luckman et al. 1997; see also Luckman and Wilson 2005: 140–141).

The LIA in Prince William Sound, southern Alaska, has been characterized on the basis of dendrochronology (Barclay et al. 1999). These proxy data showed that coastal forests of the MCA were killed by glacial advances of the LIA (Barclay et al. 1999: 79) and that there were three cold and three warm intervals during the LIA (Barclay et al. 1999: 83). These data conform to Nunn's observation that the LIA was a time of greater climatic variability and possibly greater storminess in the Pacific (Nunn 2007: 6; Ahlenius 2007).

Koch and Clague (2011), however, present evidence of several glacial advances in northwestern North America during the MCA as well. They argue that there is much regional variation in the overall effects of the MCA and that precipitation, rather than temperature, may have a greater effect on these variations (Koch and Clague 2011: 608). A case in point are the lake sediment cores from Pine Lake, Alberta, which showed that grain size, as a proxy of environmental humidity, indicated notably less moisture during the MCA and notably more humidity during the LIA (Campbell 2002: 100). Similarly, Liard et al. (2003) analyzed diatom assemblages from lakes in northwestern North America and demonstrated that major changes in Canadian sites occurred near the beginning of the MCA while similar changes occurred in U.S. lakes near its termination. Pollen data with associated radiocarbon dates also indicate that a major vegetation transition began ca. 1,600 BP and reached its apex during the MCA, and another transition began ca. 600 BP, culminating in the LIA (Viau et al. 2002). Research on Fire Return Intervals (FRI) in the Pacific Northwest, and particularly on western Vancouver Island, indicates such events occur too infrequently to differentiate the MCA from the LIA (McKenzie 1998; Lertzman, et al. 2002; Whitlock et al. 2010; Wang et al. 2009), although Gavin et al. (2003: 582) found evidence of much reduced local fire incidence on the west coast of Vancouver Island at the onset of the LIA.

Multi-Decadal Scale Variation

The northeast Pacific appears to be more directly affected by the Pacific Decadal Oscillation (PDO) than by the El Niño/southern oscillation (ENSO) (Mantua 2002: 592). The ENSO is a coupled atmospheric-oceanic observation for the tropical Pacific Ocean that is measured by oceanic variation (SST) in the eastern Pacific and atmospheric variation (sea surface air pressure – SLP) in the tropical western Pacific, or a by a combination of a number of variables, e.g., the Multivariate ENSO Index (MEI). The PDO is a measure of SST and atmospheric sea surface pressure, and it is applicable to the north Pacific Ocean. The two are therefore related but are not identical. Climate in coastal California and Mexico is relatively more affected by the ENSO, whereas that of Alaska and western Canada is more directly affected by the PDO. While ENSO events can change at an annual or sub-annual rate, PDO events in the 20th century tend to have a duration of two or three decades (Mantua 2002: 592). The PDO normally centers over the central north Pacific Ocean between the Aleutian Islands and the Kamchatka Peninsula (Mantua 2002: Fig. 1). Oceanic SST and

Sea Level Pressure (SLP) in this central area are inversely related to corresponding index values along the North American coast (Mantua 2002: 592), thus cool temperature and low pressure regimes in the core area bring warmer, dryer average mid-winter climatic conditions (i.e., warm-phase PDOs) to western coastal Canada, southeastern Alaska and the northwestern coast of the United States (Mantua 2002: 593, Table 1; Gedalof et al. 2002). Particularly relevant is the tendency for spring snow pack and flood risk to be low for the Northwest Coast during PDO warm phases and the opposite during cold phases (Mantua 2002: 593, Table 1). The most extreme expressions of the PDO effects are felt when the PDO and ENSO are in phase, i.e., warm PDO and El Niño or cool PDO and La Niña (Minobe and Mantua 1999; Mantua 2002: 593), and it is the combination of these two environmental phenomena that provides the greatest empirical accuracy in predicting North American climate (Mantua 2002: 593).

Tree-ring data from California and Alberta were compiled for the period A.D. 993–1,996. They showed a 50–70 year periodicity of the PDO and that very negative PDO index values between A.D. 993 and 1,300 were often associated with "mega-drought" conditions in western North America. Little PDO fluctuation was observed between A.D. 1,600–1,800, but noticeable PDO fluctuations were observed between A.D. 1,000–1,200 and A.D. 1,300–1,500 (MacDonald and Case 2005), i.e., approaching the onset and the coldest parts of the LIA.

The southern Northwest Coast experiences relatively warm winters and reduced precipitation, including less snowfall, during El Niño winters. Coastal waters are warmer than normal, and southern marine species are relatively more abundant during these events. La Niña events bring cool marine waters to the coast and precipitation, including snow, tends to be above normal (El Niño and La Niña Effects on Canada 2010). If present El Niño effects provide a model of climatic effects during the MCA and La Niña effects provide a model of climatic effects during the LIA, then warmer air temperatures, warmer sea temperatures reduced precipitation and more southerly marine species relative to present characterized the MCA whereas the LIA was characterized by greater fluctuations between these conditions and those in which a cold SST, cooler air temperatures and greater precipitation were the norm.

In summary, there is convincing evidence of a global climatic anomaly, the MCA, between approximately A.D. 800/900–1,200/1,300 followed by another anomaly, the LIA, that ended in the late 1800's and is frequently described as one of the coldest periods since the onset of the Holocene (Bradley et al. 2003). Air temperatures on Canada's west coast during the MCA tended to be warmer, possibly by 1–2°C, than those of the LIA, and relatively increased climatic variability characterized this area during the LIA as a result of climatic linkage with the PDO. Similarly, SST and SLP variability on the west coast were also linked to PDO fluctuations, as they are now. This variability translates, by inference, into periods of greater and lesser precipitation and corresponding variations in surface run-off, snow pack and spring meltwater run-off during the LIA.

If there was a change, what were the environmental conditions during each period?

Modern "maritime" forests dominated by *Tsuga heterophylla* and Cupressaceae were established on the north coast of Vancouver Island and the west coast of Washington State by ca. 3,000 BP (Hebda 1982: 3191). By extension, the intervening west coast of Vancouver Island would have exhibited this biome as well. While the forest community apparently underwent little change in response to multi-decadal or century-long variations in temperature and humidity, it is likely that precipitation, run-off volume, run-off temperature and SST were more sensitive to longer term changes and variability therein. In Saanich Inlet on the east coast of Vancouver Island, for example, varve analysis of sediment cores revealed a wet climate over the last 500 years, including the LIA (Nederbragt and Thurow 2001). These latter sources of variability may therefore have had fewer repercussions for the terrestrial animals, on which indigenous populations relied relatively less heavily, than for the aquatic resources that were more heavily exploited.

Freshwater volume and temperature are critical to coastal biological productivity in general, including salmon spawning success (Royer et al. 2001). The PDO, characterized by 20–30 year regimes and abrupt regime shifts, has a direct effect on salmon productivity (Mantua et al. 1997; Hare et al. 1999). Finney et al. (2002: 732), for example, used $\delta 15 N$ and diatom data from sediment cores to show that sockeye spawning in lakes on Kodiak Island, Alaska, were more abundant after ca. A.D. 1,200 during a period of glacial advances in Alaska and the Canadian Rockies. However, the Subarctic Current, which directly affects waters in the Gulf of Alaska, i.e., Kodiak Island, and the west coast of North America encounters the west coast of North America roughly around Haida Gwaii where it divides into the north-flowing Alaska Current and the south-flowing California Current (Francis and Hare 1994: Fig. 9; Finney et al. 2002: 732). Thus, relatively small deviations in the latitudinal meeting of these two major current regimes, such as might occur with an El Niño incursion or a La Niña event, may or may not include western Vancouver Island. Given greater variability in the PDO during the LIA than the MCA, and assuming that the PDO is a good proxy for climate variation in the study area, marine SST and freshwater temperatures and run-off volume in streams and rivers of western Vancouver Island are likely to have fluctuated more during the more recent period than in the earlier one.

The main salmon species in the relatively small rivers of western Vancouver Island, especially those that do not emanate from inland lakes, is chum salmon (*Oncorhynchus keta*), and this is the main spawning salmon, followed distantly by Chinook salmon (*Oncorhynchus tshawytscha*) and Coho salmon (*Oncorhynchus kisutch*), in the Toquart River (McDougall 1987: Appendix I) at the heart of traditional Toquaht territory in northwest Barkley Sound. The spawning run of chum salmon begins in September, peaks in October, and ends in November (U.S. Department of Commerce 2011). Spawning females use their tails to create a redd (nest) in river gravels then deposit their eggs. Males fertilize these eggs, then the female covers the redd with gravel. The eggs require cold, clear, oxygen-rich water for incubation and hatching; warm water can provide insufficient oxygen for hatching, and sediment-laden water can suffocate the nest. Avelins appear in the spring and emerge from the gravel after several weeks, at which time they migrate downstream to estuaries where they school and gradually adapt to salt water before moving offshore by fall. Relatively low water levels and relatively warm water temperatures during El Niño years or under El Niño-like conditions would therefore not be optimal for chum salmon reproduction. By contrast, La Niña years or La Niña-like conditions would bring relatively greater water volumes (less chance of sedimentation) and relatively low water temperatures, i.e., more optimal reproduction conditions. During the MCA, the Toquart River was likely to have been under more consistent El Niño-like conditions whereas during the LIA it was likely to have varied frequently between El Niño-like and La Niña-like conditions, although the magnitude of variations within these conditions may have varied over time (Carré et al. 2005).

Local and regional SST conditions, rather than Pacific basin-wide conditions, are directly correlated with survival rates of pelagic juvenile chum salmon north of Yakutat Bay (Mueter et al. 2002: Fig. 1b), but south of that point the correlation is not significant (Pyper et al. 2002; Mueter et al. 2002: 459, Fig. 2b), suggesting that factors other than SST may influence juvenile chum survival. Spawning and incubation conditions are also important to the survival of all salmon species. Warmer, dryer MCA conditions that were generally consistent over time would have the effect of reducing water volume in the Toquart River and increasing the water temperature. Cool periods during the LIA, by contrast, would have increased stream flow and lowered water temperatures. Variation between warmer and cooler periods during the LIA would make salmon spawning conditions in the Toquart River relatively unpredictable, while generally warmer temperatures during the MCA would have reduced water flow and increased water temperature relatively consistently, making salmon spawning conditions in the Toquart River more consistently sub-optimal. Precipitation, as well as just temperature, may be a more appropriate way to characterize the MCA (Koch and Clague 2011: 594, 608), and, by extension, the LIA also.

SST, affected by both run-off temperature and volume and by shifts in current upwelling, plays a major role in the character of the food chains that are locally and regionally supported (Royer et al. 2001). Mueter et al. (2002: 461) make the same observation in suggesting that SST is a proxy for other factors that relate to salmon survival; nevertheless, the cold downwelling regime of the north and the warm upwelling regime of the south influence juvenile salmon survival in opposite ways when SST rises or falls. The boundaries they identify are based on observations over a maximum 47 year period between 1948–1996, so there is reason to think that, despite the PDO being a less accurate predictor of salmon survival than SST, the latter still would be influenced by the former over longer periods of time (Finney et al. 2002: 732). Since the PDO does shift over multi-decadal and multi-centennial time spans, the SST regimes might also be expected to vary within regions over time, as might the boundary locations between northern and southern current domains that Mueter et al. discuss.

Salmon are not the only taxon affected by SST. Warm El Niño waters tend to bring more southerly adapted marine regimes northwards (Strom et al. 2004: 4). Sardines (*Sardinops sagax*) and anchovies (*Engraulis mordax*) were found to be out of phase with sockeye salmon (*Oncorhynchus nerka*) in Alaska (Finney et al. 2002: 732), and sardines were found in British Columbia waters for the first time in 50 years during the warm waters of the 1990s (Strom et al. 2004: 4). This latter observation is consistent with SST fluctuations in coastal Peru (Sandweiss et al. 2004: 332). Personal observation of the summer 1991 El Niño event in Barkley Sound identified massive schools of mackerel (*Scomber japonicus*) on the west coast of Vancouver Island, a phenomenon consistent with more widespread El Niño consequences (Moss et al. 2007: 517). This species preys on juvenile salmon, and it damaged the commercial and sport fishery in the area over a number of subsequent years. The change in food webs that are brought about by changes in SST can therefore be significant. A larger relative of the mackerel, the Pacific bluefin tuna (*Thynnus orientalis*), was actively hunted by the Nuu-chah-nulth on the rare occasions on which it appeared (McMillan 1979; Crockford 1994, 1997), and the remains of these fish have been identified in low numbers in the faunal assemblages of all major Nuu-chah-nulth sites, including those in Toquaht territory.

Rockfish (genus *Sebastes*) vary in diversity along the west coast of North America. In California waters, approximately 46–60 species of this genus occur, whereas in coastal British Columbia there are only approximately 31–35 species (Love et al. 2002: Fig. 4.1). The inshore species of this genus prefer rocky substrates and kelp forests, and these forests prefer warm water. One might expect that shallow inshore waters would tend to be relatively warm relatively consistently during the MCA, thereby fostering the development of extensive, taxonomically diverse kelp forests and an increase in the number of *Sebastes* species inhabiting them as part of the northward expansion of a southern assemblage of zooplankton during warm intervals (Strom et al. 2004: 4). Conversely, cooler and more variable SST conditions during the LIA would be expected to reduce the extent of kelp forests, their diversity, and the numbers of rockfish species inhabiting them. Analysis of rockfish otoliths using a dendrochonological method has shown that yelloweye rockfish (*Sebastes ruberrimus*) growth is positively correlated with warming conditions at the north end of Vancouver Island and in Haida Gwaii and is positively correlated with cooling conditions in the California current (Black et al. 2008: 375). A similar conclusion was reached in a study of Bocaccio rockfish (*Sebastes paucispinis*) in California, and the same study also indicated that good recruitment years must occur at least 15% of the time for the population to remain at a constant level (Tolimieri and Levin 2005: Fig. 3, 466). Different species, however, have somewhat different responses to changing SST conditions (Moser et al. 2000). Inshore rockfish tend to be territorial and they also have relatively long life spans (multiple decades at least). Interannual changes in SST are thus likely to affect hatching success but are unlikely to affect population size of a given species. In contrast, major regime shifts such as those of the PDO within both the MCA and LIA would likely have had an affect on the population size of individual species and on the diversity of species within the genus. Greater variability in SST regimes during the LIA would therefore inhibit rockfish reproduction and diversity to a greater extent than more consistently warm conditions during the MCA. Changes in the distribution and extent of kelp habitat would also affect species diversity and abundance of rockfish species.

In summary, the west coast of Vancouver Island during the MCA probably supported the same forest regime as it presently does. Relatively warm air temperatures and relatively reduced precipitation would mean lower water volumes and higher water temperatures in rivers, although conditions may have resembled cooler LIA intervals during periods of glacial advances on the mainland. SST was also generally warm throughout the period. Chum salmon may have faced less-than-ideal conditions for spawning, hatching and seaward migration because of freshwater conditions in the Toquart River, and SST conditions may not have been ideal for their survival after entering the sea because of the northward push of warm water. Kelp forests and rockfish species, though, likely benefited from consistently warm conditions. The LIA, with its greater climatic variability, would have exhibited many of the features of the MCA during its warm phases, but during its cold phases the increase in precipitation likely raised river volumes and lowered water temperatures thereby improving chum salmon spawning, hatching and escapement. Cooler SST would also have proven beneficial to juvenile salmon survival. On the other hand, kelp forests and associated rockfish species would have diminished in extent and diversity.

Subsistence Strategies

Recent analysis of global fisheries catch data in relation to climatic indices suggests that there is a causal link in the extra-tropical Pacific between climatic variability and fisheries productivity that has been formulated as the remote synchrony hypothesis (Castro-Ortiz and Lluch-Beldaa 2007). This hypothesis applies at the multi-annual level, i.e., fisheries catch statistics reflect climatic changes in the preceding one or more years. The PDO, however, operates at the multi-decadal level, and variations during the LIA are likely to have been responses to this phenomenon. The MCA, with relatively less internal climatic variability, and the LIA, with its greater internal variability, themselves reflect multi-centennial events within the overall cooling trend of the Holocene. Fisheries production, then, can be seen as a complex phenomenon that varies according to the time-frame within which it is viewed, and the causative climatic factors also vary according to the relevant time frame. As well, different taxa in a given location respond differently to climate, precipitation, SST and changes in their food chains.

The Nuu-chah-nulth and their ancestors faced the challenges of these shifting fisheries resource regimes throughout at least the last 5,000 years, including the MCA and the LIA. This chapter focuses on the transition from the former to the latter in terms of marine subsistence resources. Ethnographic and ethnohistoric accounts of the Nuu-chah-nulth begin at the end of the 18th century and continue into the 20th. The earliest accounts, e.g., Cook (1821), Jewitt (1896), describe Nuu-chah-nulth culture in its original state around the end of the LIA. Later documentation, e.g., Sproat (1868), describes a culture in a state of transition, although daily practice and cultural memory were still closely linked to pre-European times. Nevertheless, climatic amelioration subsequent to the end of the LIA would have begun by then. Finally, ethnographic reports of the 20th century, e.g., Drucker (1951), Sapir and Swadesh (1955) (based on Sapir's fieldwork notes 1910–1914) and St. Claire (1991), would all be set in a modern climate and be based on oral tradition, greatly altered daily practice, and cultural memory. Ethnographic and ethnohistoric descriptions of Nuu-chah-nulth culture are presented first, followed by archaeological data in the zooarchaeology section.

These ethnographic and ethnohistoric sources consistently indicate that salmon, especially chum salmon

(*Oncorhynchus keta*), was the most important resource, and herring (*Clupea harengus pallasii*) is described as very important. Halibut are described by Sproat (1868: 225) as being very common, although almost 100 years later Drucker (1951: 36) indicates that they were more important to some Nuu-chah-nulth groups than to others. Various species of fish described as "cod" were available year-round. Sproat (1868: 229) describes two species of "cod"; one he calls "Toosh-ko" and one similar to a Scottish fish called "Tusk". The former appears to correspond to ling cod (*Ophiodon elongatus*, or *tuškuuḥ*) (Nuu-chah-nulth Language; McMillan and St. Claire 2005: 30) (order Scorpaeniformes) while the latter is *Brosme brosme* (Richard Shelton 2012: personal communication) (order Gadiformes) and is thus most closely related to Pacific cod (*Gadus macrocephalus*). Another type of rockfish called Red Snapper (most likely *Sebastes miniatus*) bears the Nuu-chah-nulth name *ƛ'ihapiiḥ* (Nuu-chah-nulth Language 2012). Sapir and Swadesh (1955: 40) provide informant testimony that "cod" (understood in a general sense) were caught both well offshore as well as near shore, and that an abundance was obtained in all cases regardless of the techniques used. Drucker (1951: 9, 36), however, states that "cod" were ubiquitous but were not used except in times of shortage or lack or other resources. No specific mention is made in ethnographic and ethnohistoric documents of rockfish (genus *Sebastes*), although it may have been subsumed under the general label "cod". Nevertheless, Sapir's informant referred to near shore acquisition of "black bass", which is one of the common names of one species of rockfish (*Sebastes melanops*, or *qwiq ma*) (Shannon Cowan 2012: personal communication). Sea mammals, e.g., northern fur seal (*Callorhinus ursinus*), Stellar sea lions (*Eumetopias jubata*), California sea lions (*Zalophus californianus*), and whales (esp. *Mysticeti*), were important for their presitge value and ranked after chum salmon and herring, although there was no dependence on them. Clams were said to be relatively less important to most Nuu-chah-nulth than to other NWC groups, and other shellfish were harvested mainly for dietary variety (Drucker 1951: 36). On the other hand, Sapir and Swadesh (1955: 41), reporting at approximately the same time, state that Barkley Sound groups consumed large quantities of both fresh and dried clams and mussels. This account is consistent with the still earlier chronicle of John Jewitt, a sailor held captive by Maquinna, chief of the Mowachaht Confederacy (a Nuu-chah-nulth social entity) at Nootka Sound in 1803–1805, who indicates that fresh and dried clams were eaten and traded by the Mowachaht (Jewitt 1896: 81, 86, 108, 109). Still earlier, Cook (1821: 292) reports that mussels were a common food of the Nootka Sound groups. The LIA was coming to an end at the time of Cooks voyage and Jewitt's captivity, so their reports correspond to indigenous cultural practices, environmental conditions and available resources of that time. The reports by Drucker and by Sapir and Swadesh were compiled well after first contact and derive primarily from informant memory and oral history. Salmon in general was regarded as the most important resource, and chum salmon in particular was prized because of its relative abundance compared to other species, the timing (late fall) of itheir arrival, and its suitability for winter storage (Drucker 1951: 36; Cannon and Yang 2006: 133). Chum salmon, and small numbers of coho and spring salmon, spawn in Toquart River and in most of the smaller streams along the coast of Toquaht territory (McDougall 1987: Appendix I; St. Claire 1991: 155–167) (Fig. 10.2).

Herring (*Clupea pallasii*) spawn on specific coastal beaches in March after schooling in protected inshore waters during the winter. Barkley Sound, and particularly Loudon Channel, adjacent to the sites under discussion here, is presently one of the most productive areas in Canada's Pacific herring fishery (Taylor 1983; Government of Canada 2006), and two of the three major Toquaht sites considered here sit immediately adjacent to herring spawning beaches (Hay and Kronlund 1987: Fig. 1; St. Claire 1991: Map 13). Long-term research has shown that herring recruitment and survival tend to be below average in warm years when predators such as hake and mackerel are abundant, thus reflecting an inverse relationship between SST and herring production (Fisheries and Oceans Canada 2011: 11). A finer level of analysis by Hay and Kronlund (1987: Tables 4, 5, 1190, 1193) showed that herring spawn and catch were negatively correlated with SST between 2 and 6 years previously, with the strongest correlation and greatest statistical probability at 4 years previous. Analysis of sediment cores from Effingham Inlet, which opens into Barkley Sound, showed that northern anchovy, Pacific herring, rockfish and hake (*Merluccius productus*) comprised over 90% of all fish scales in the sediment core samples (Wright et al. 2005: Table 2). Fine-grained analysis was not possible, and climatic forcing was not seen as affecting the fish assemblage directly. Nevertheless, a warm regime favourable to anchovy existed until ~1,600 BP after which the count of scales declined. Herring abundance, primarily juveniles using Effingham Inlet as a nursery, was more or less constant over the 4,000 years of the core, although cycling was noted consistent with current regime shifts and shorter term fluctuations (Wright et al. 2005: Fig. 4, 378–379). Of interest for the purposes of this chapter, though, is the impressionistic interpretation that can be made of their Fig. 4. The most recent date in this figure, 970 ± 135 cal BP, falls toward the end of the MCA, and consideration of the trajectories of each species depicted suggests that in the transition to the LIA dogfish abundance increased, rockfish decreased, hake decreased, anchovy decreased, and herring may have increased slightly. Conversely, during the preceding MCA, surfperch (*Embiotoca*

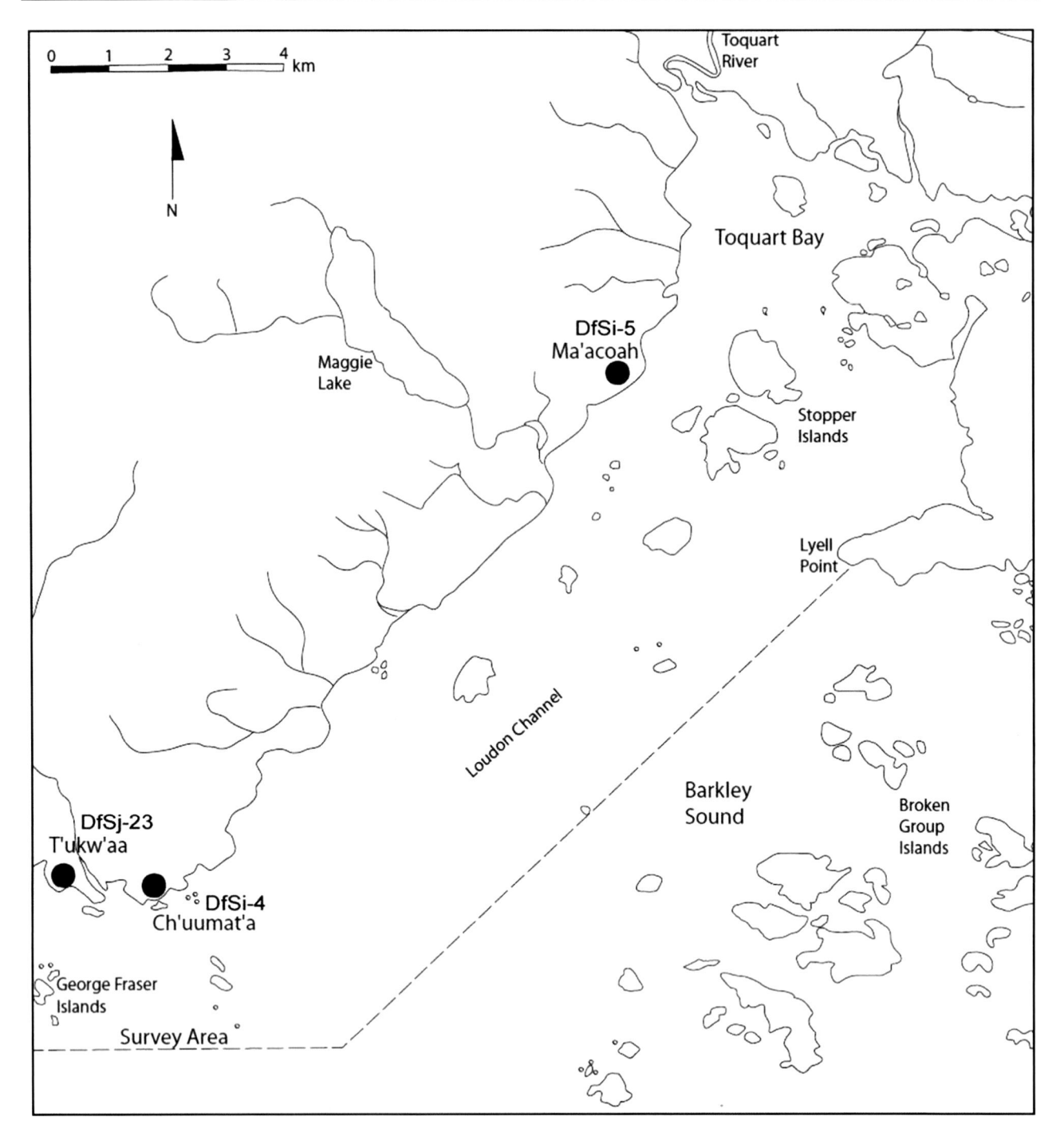

Fig. 10.2 Detail of the study area showing the location of T'ukw'aa Village (DfSj-23), adjacent Toquaht sites and geographic features (redrawn by Katherine Nichols and the author after McMillan and St. Claire 1991: Fig. 9, which was adapted from St. Claire 1991: Map 13)

lateralis) may have been somewhat more abundant, dogfish (*Squalus acanthius*) were not abundant, rockfish were more abundant, hake was more abundant, anchovy spiked briefly, and herring abundance was relatively low.

A halibut bank, clam beds, cod (native informant terminology) fishing locations, and sea mammal haul-outs and feeding areas were also found in Toquaht territory (St. Claire, 1991: 155–158). Before the extirpation of sea otters as a result of the maritime fur trade, kelp beds were more extensive than they are at present (Orchard 2009). All these higher trophic taxa depend on the zooplankton, phytoplankton and euphausids that are affected, short-term and long-term, by northeast Pacific oceanic regimes. They form the foundation of the food chains (Monks 1987) that were present under these regimes and that the Nuu-chah-nulth accessed for their subsistence and other purposes.

Theoretical Framework

The theoretical framework that is most useful for this chapter is optimal foraging theory (OFT) (see also Pilaar Birch and Miracle 2017), especially refinements to the prey choice model where small prey (fish in this chapter) are concerned. Small prey such as fish raise two important issues in OFT: the equation of body size with prey rank, and the access to prey. Although body size has often been used as a proxy for prey rank, the equation is not that simple (Stiner et al. 2000). Many characteristics of small prey, such as their mobility and their reproductive resilience (Stiner et al. 2000: 42) as well as the season in which they are sought and the technology used in their capture (Jones 2004: 313, 316; Bird et al. 2009: 22), influence their rank. It is therefore worth considering the characteristics of the main fish species taken by the Nuu-chah-nulth.

Rockfish (*Sebastes* spp.) inhabit both deep offshore waters and shallow inshore waters, and it is the latter that are of interest here. These species prefer kelp beds, rocky shores and reefs. They mature slowly, some taking as long as 15–20 years to reach reproductive maturity, and are known in some cases to live over a century. Consequently, they cannot be harvested intensively if viable breeding stocks of each species are to be maintained (Fisheries and Oceans Canada 2008). Chum salmon, by contrast, return from the ocean to their natal streams in October and November to spawn 3–5 years after birth, with most returning after 4 years. Similarly, herring return from the ocean 2–4 years after birth to school in protected inshore waters during the winter before spawning in early spring in the intertidal and sub-tidal zones of beaches that support aquatic vegetation. Thus both salmon and herring are characterized by short life spans and large population aggregates at narrowly defined times and locations, whereas rockfish are long-lived, widely distributed and consistently available.

The Nuu-chah-nulth used different technologies to take these taxa. Both rockfish and salmon could be taken by hook and line in inshore waters, but, while rockfish were available year-round, salmon were most abundant in late summer and fall, although trolling for spring salmon occurred earlier in the year. Return rates for this form of pursuit are unknown, but personal experience with this technology indicates that rockfish are easier to catch than salmon. Traps and weirs set in estuaries and rivers were far more productive for salmon and were commonly used by the Nuu-chah-nulth, and harpoons and leisters were also used in rivers (Drucker 1951: 16–21). Stone-walled tidal traps in the intertidal zone could be used to take small fish, and such traps were used to capture herring if they were constructed on the appropriate beaches. Herring were also raked and netted from canoes in open water when schools were sufficiently dense, and their roe was collected on branches submerged in the water at spawning sites (Drucker 1951: 23, 41–42; Sapir and Swadesh 1955: 30).

These major fish species therefore exhibit different characteristics with respect to their accessibility by humans, the technology required to access them, and their ability to withstand intensive harvest pressure. Hook and line fishing from canoes was the major means of taking rockfish, and this harvest strategy could be undertaken on a constant basis, weather permitting, up to the point at which the populations of this genus were able to sustain themselves. There is recent evidence, however, that this resource was sometimes over-exploited (McKechnie 2007). Warmer SST conditions of the MCA would have raised the "harvest quota" through expansion of kelp forests and northward migration of more species, thus greater numbers, of rockfish. Subsequent cooling and variation in SST would have reduced the availability of rockfish and their ability to withstand harvest pressure. Salmon and herring reproduction is more rapid, enabling these taxa to withstand intensive harvesting without population decline, but the conditions for their reproduction are not as optimal under warmer SST conditions as they are under cooler conditions. As well, they are only available for harvest at certain times of year and in certain places in quantities that exceed those of rockfish. Thus a shift to generally cooler conditions of the LIA promoted salmon and herring reproduction and their ability to withstand intensive harvest pressure, but more variable SST conditions meant increased risk to human predators in terms of accessibility.

One way to mitigate risk is through storage, and the Nuu-chah-nulth, along with every other Northwest Coast population, were known to store both salmon and herring in abundance. Indeed, Drucker (1951: 43) states that only fall salmon fishing and spring herring fishing involved all members of the local group, a mark of the importance of these two pursuits. It can be inferred, then, that both storage and the effort of the complete local group were risk mitigation strategies. The Nuu-chah-nulth stored halibut and "cod" (Cook 1821: 257; Sapir and Swadesh 1955: 40), and halibut, as well as other flatfish such as flounder and sole, are sometimes mentioned as fish for storage at other places along the coast (Croes 1995; Bowers and Moss 2001; Orchard 2009). Evidence of Pacific cod storage further north in Alaska is ambiguous at best (Partlow and Kopperl 2011: 212), but specific reports of rockfish storage, and archaeological evidence of it, are non-existent.

The effects on the Nuu-chah-nulth of these ecological shifts were likely considerable. Optimal foraging for small prey under conditions that favored rockfish and disfavored herring and salmon abundance, i.e., the MCA, could have two possible outcomes. The first is increased exploitative effort on the reduced numbers of herring and salmon, and the second is increased exploitative effort on an expanding, reliable resource base, i.e., rockfish. The first explanation would be supported in the archaeofaunal record by a consistent high level of salmon and herring production (NISP) relative to rockfish throughout deposits of both the MCA and the LIA. The second explanation would be supported by a greater emphasis (NISP) on rockfish during the MCA and a greater emphasis on salmon and herring during the LIA (see Carlson 2008: 105). Consequently, the substantive hypothesis I wish to test is this: the Nuu-chah-nulth shifted their fishing emphasis from rockfish to salmon and herring as a result of climatic change from the MCA to the LIA.

Zooarchaeology

The data I will use to detect the transition from the MCA to the LIA consist of the faunal sample from the T'ukw'aa Village Site (DfSj-23A) (Fig. 10.3). The site dates from $1{,}150 \pm 90$ ^{14}C BP (McMillan 1999: Table 4) into the twentieth century. This large village was the major Toquaht community, from which they take their name, but by the nineteenth century it was used only as a summer village at which halibut, cod, seals, sea lions and whales were taken (St. Claire 1991: 53; McMillan 1999: 66–67). The site was excavated in a series of 10 test units covering 38 m^2 located across the site so as to examine both sides of the stream that bisects the site and to sample both upper and lower terraces south of the stream. Four of these test units (44%) were selected from which to draw the faunal sample discussed here. Of these units, two 2×2 m units were selected because they produced radiocarbon dates, and one 2×2 m unit was sufficiently near a dated unit that, despite the complex and 2.75 m deep shell midden stratigraphy (McMillan 1999: 69), rough cross-dating of strata could be applied. The fourth unit was selected to provide information on the deposits north of the stream. The excavation was done in a system of natural stratigraphy and arbitrary 10 cm levels, and all excavated matrix was screened in the field through 1/4″ mesh (McMillan and St. Claire 1993: 103–109). A random sample of 40% of these arbitrary levels from each of the four units was selected (Monks 2000). In addition, several levels were arbitrarily selected if they contained a dated radiocarbon sample or if a large vertical gap between randomly selected levels was apparent. Thus, the total number of levels in the sample is between 40–45% of all levels in the four excavation units.

Some qualifications must be recognized with these data. The first is the location of the site. It is recorded as a summer fishing and sea mammal hunting site during the early twentieth century, a function it may have assumed by the mid-nineteenth century (McMillan 1999: 66). Previous to that, and since the time of its establishment, it is likely to have been the main Toquaht village. In ethnographic times it was not the fall and winter village site where one might reasonably expect to find the greatest amount of salmon and herring. That site is Ma'acoah (DfSi-5) located near the mouth of the Toquart River (see Fig. 10.2). Nevertheless, the faunal signature of the DfSj-23A is fully consistent with that of Ma'acoah (Monks 2006), the nearby sites of Ts'ishaa (DfSi-16) (Frederick and Crockford 2005; McKechnie, 2005; McMillan et al. 2008; Huu7ii McKechnie 2012: 165), and sites as far north as the Queen Charlotte Islands (Haida Gwaii) (Orchard and Clark 2005). No significant differences could be found between the sea mammals and the fish in the T'ukw'aa and Ma'acoah faunal assemblages to support a locational optimization hypothesis, although shellfish assemblages differentiated the two sites (Monks 2011). Another qualification lies in the excavation of the deposits. Units to be excavated were chosen arbitrarily and such disconnected excavation units that contain complex and deep stratigraphy are all but impossible to integrate stratigraphically. Any findings of the sample are therefore difficult to relate to the other site deposits. As well, use of arbitrary levels for this analysis ignores potentially meaningful natural stratification; there may therefore be "blurring" of the possible faunal distinctions between one natural stratum and another within an arbitrary level. Nevertheless, the broad tendencies in the assemblage of each excavation unit are still clear, and results from the excavation units can be cross-dated with each other because of the dates they contain and/or their physical proximity and faunal signatures. A third qualification of the data is the selection of the sample. The excavation units for analysis were chosen arbitrarily on the basis of the presence of radiocarbon dates contained therein, on the basis of proximity to such units, and with an eye to examining all major sections of the site. No representativeness of the entire site can be claimed by the sample, although the general outlines of the complete faunal assemblage are likely approximated. With these qualifications in mind, I now turn to analytical results of the sample.

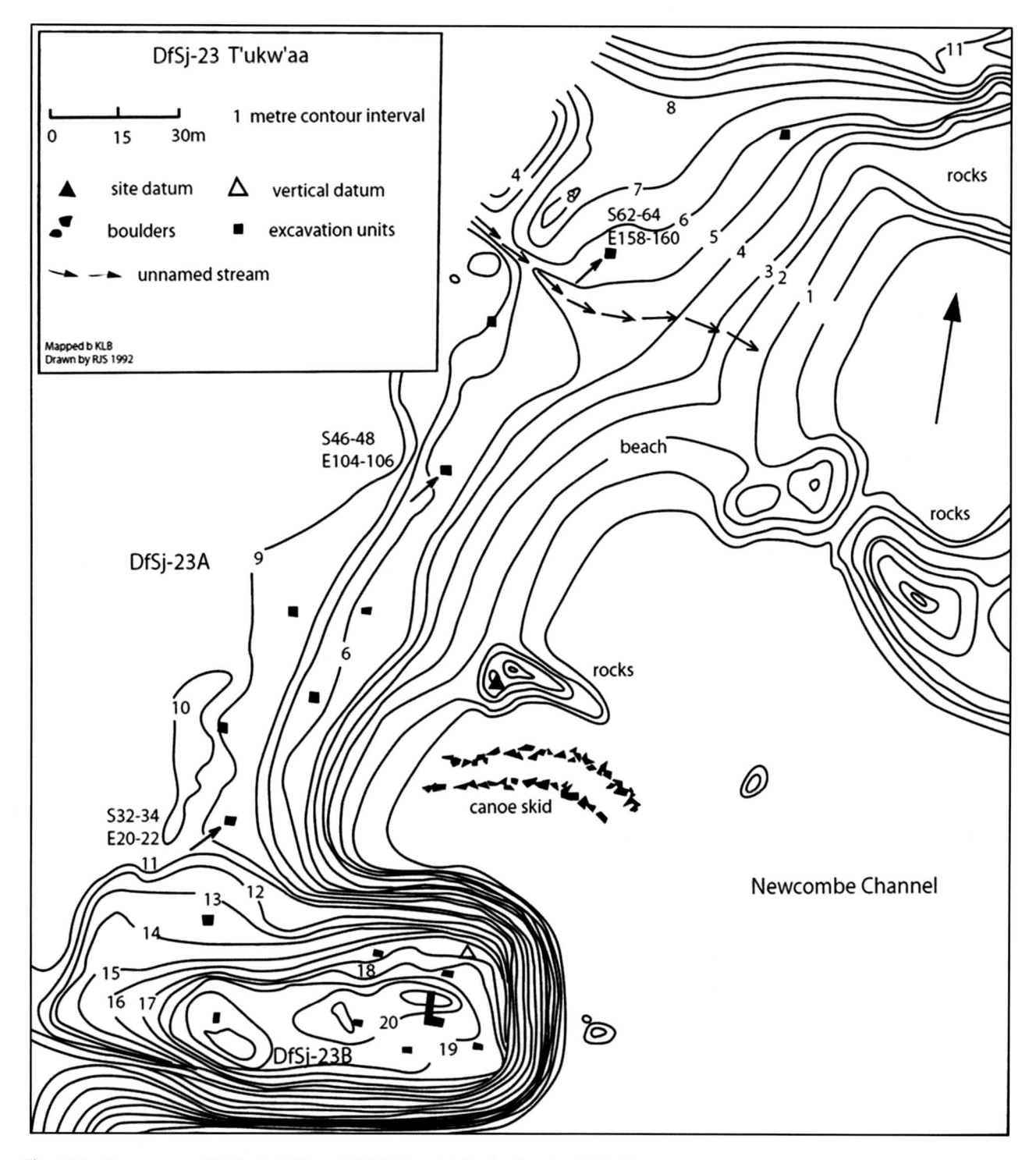

Fig. 10.3 Contour map of T'ukw'aa Village (DfSj-23A) and defensive location (DfSj-23B) showing locations of excavation units from which the analyzed sample was drawn (redrawn by the author after McMillan and St. Claire 1992: Fig. 12)

Analytic Results

The analytic results to date have emerged in an historical-chronological manner. That is, I came upon the issue of climatic change when I began looking at the vertical distribution of fish in relation to available dates. I noticed an odd phenomenon in which rockfish (*Sebastes*) NISP values crossed over with salmon (*Oncorhynchus* spp.) NISP in all four excavation units (Fig. 10.4). I had observed the predominance of salmon and herring over rockfish at the nearby site of Ma'acoah (DfSi-5) (Monks 2006: Fig. 5), and the same phenomenon had been observed elsewhere in Barkley Sound at Ma'acoah (Frederick and Crockford 2005; McKechnie 2005) and at Huu7ii (McKechnie 2012) as well as in Haida Gwaii (Orchard and Clark 2005). In all cases this cross-over seemed to occur around the end of the MCA and the onset of the LIA, hence the hypothesis that I test here. Figure 10.4 provisionally identifies the levels that are thought to belong to each climatic period in order to provide a framework for testing the hypothesis.

One would expect to see a higher abundance of rockfish (*Sebastes* spp.) in the MCA levels of the sample and a higher abundance of salmon (*Oncorhynchys* spp.) and herring (*Clupea pallassi*) in the LIA levels of the sample if the hypothesis is to receive support. Figure 10.4 shows the trajectories of rockfish, salmon and herring NISP values by sampled levels for the four excavation units in question. The chronological relationship between salmon and rockfish is most easily seen in Fig. 10.4a where rockfish are more abundant than salmon in levels 19 through 25 and where a calibrated date of A.D. 1,040–1,220 is found in level 23. The lowest level (25) shows the greatest difference between rockfish and salmon abundance and dates to the late MCA. The NISP trajectories of salmon and rockfish cross between levels 18 and 19, and by level 15, which contains a calibrated date of A.D. 1,260–1,210, salmon has become predominant. Figure 10.4b shows a similar pattern but less clearly. Level 25 was not included in the sample, but it provided a calibrated date of A.D. 760–990, i.e., around the beginning of the MCA. The same transition in predominance from rockfish to salmon is seen between levels 14 and 17 after which point salmon remains are clearly more abundant. Figure 10.4c, d show the same cross-over pattern, although dates are missing from those units.

The data depicted in Fig. 10.4 are far from conclusive, but they tend to support the hypothesis of Nuu-chah-nulth preference for salmon during the LIA and rockfish during the MCA. The tendency for salmon bones to be less common in lower levels is unlikely to be a taphonomic effect. Shell midden geochemistry provides an excellent matrix for bone preservation, and, at the nearby site of Ts'ishaa (DfSi-16), which has chronologically overlapping dates, fish bone survivorship has been shown not to be correlated with shell density. Data comparing attrition due to differences in bone density of rockfish and salmon are not available, but experience suggests that rockfish bones are denser, and therefore more durable, than salmon bones, especially those of the skull (Butler and Chatters 1994). A recent study of Pacific cod (*Gadus macrocephalus*) bone density, which is more like that of rockfish, indicates that bone density is relatively even across the skeleton (Smith et al. 2011: 48). Variation in abundance of fish bone is therefore taken to be the result of spatial and temporal human depositional factors (McKechnie 2005: 54, Fig. 12) for deposits spanning the past 4,500 years (Frederick and Crockford 2005: 173). Similarly, data from the other Toquaht site of Ch'uumat'a (DfSi-4) show that salmon NISP does not decrease with age over 4,000 years. The pattern depicted in Fig. 10.4, then, seems likely to reflect cultural behavior rather than diagenetic factors.

One would expect rockfish, if they were favored by MCA climate and marine conditions, to be larger, on average, during this period compared to the LIA. In addition, one would expect to find that the sizes of rockfish varied more during the MCA because there were more species present, each with its own typical size range, and because environmental conditions are thought to have been more favorable to rockfish during this period than during the later one.

Rockfish hyomandibulars were the most abundant element within the genus, so they were used to evaluate this expectation. Three different dimensions of each hyomandibular following the precedent of Orchard's (2003: Fig. 5.5) measurements 1–3. Measurements were taken using digital calipers, each measurement was taken three times, then the mean and standard deviation were calculated. Only when an element was undamaged in the critical areas for the measurement was that measurement taken. All measurements were taken on hyomandibulars from the sample of units and levels discussed throughout this paper. The Mann-Whitney two-sample test of difference in location of medians was used in order to avoid assumptions about sample population distributions, especially because two samples of measurements were small. All calculations were performed with correction for continuity. The null hypothesis stated that there was no difference in location of medians between the LIA and MCA populations; the substantive hypothesis stated that the median value of the LIA population was smaller than that of the MCA population. Consistent with an exploratory study, the chosen significance level for a one-tail test was $\alpha \leq 0.025$. Both exact and approximate probabilities of U_1 were calculated, and the results are shown in Table 10.1.

Measurement 1 shows that rockfish from the LIA are significantly different, and smaller, than those from the MCA. Variability in size, as judged by the standard deviation of the mean measurement, is also greater among MCA rockfish. Measurement 2 is not statistically significant, but the standard deviation of mean measurements is greater in the MCA. Thus, while there is no statistically significant difference in the ranked measurements between the MCA

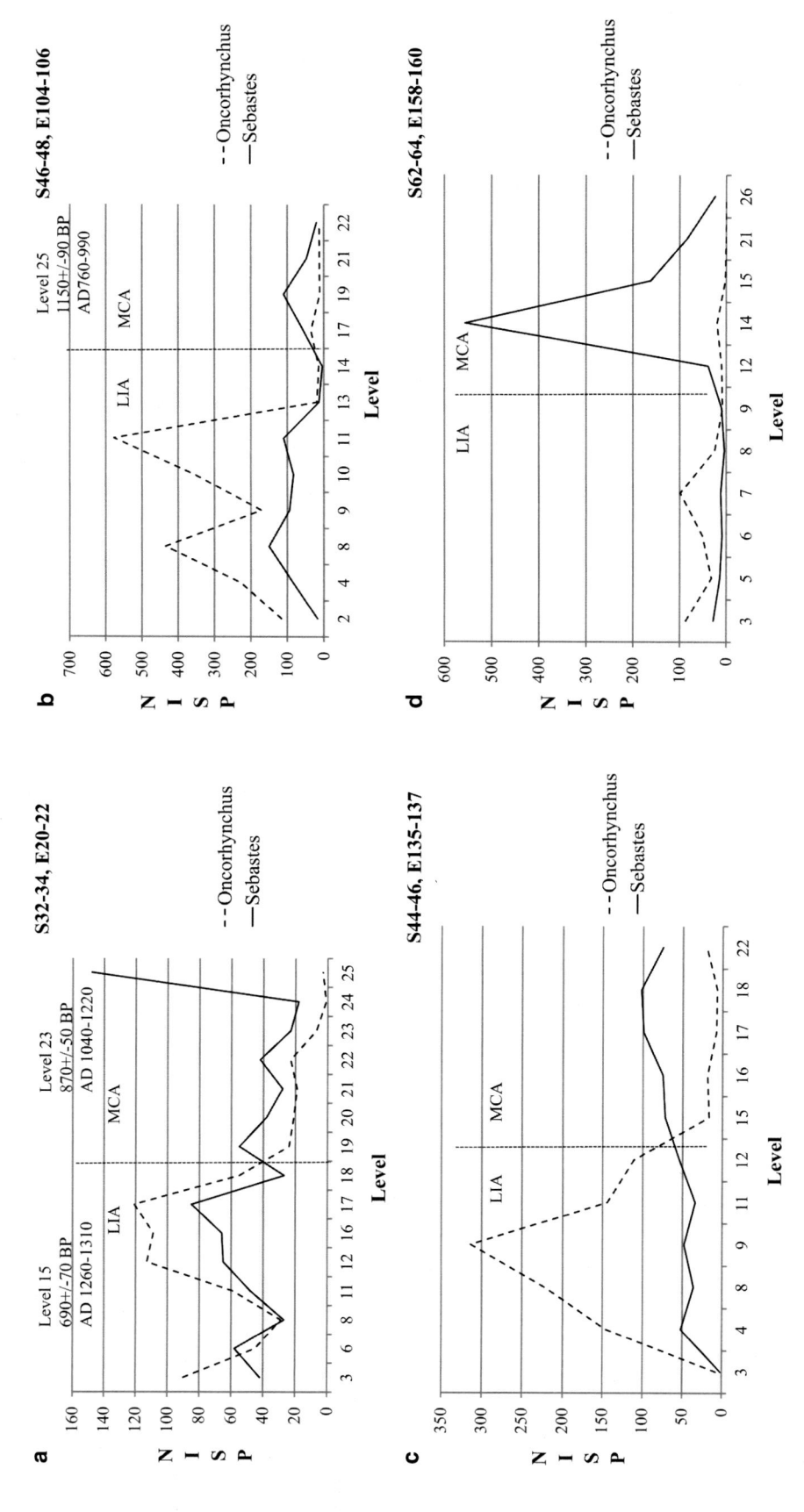

Fig. 10.4 Frequencies (NISP) of salmon and rockfish elements by level in four excavation units, DfSj-23A

Table 10.1 Rockfish Hyomandibular Measurements (mm) by Period, DfSj-23A

	LIA				MCA				M-W			
Measure	N	Mean	Median	σ	N	Mean	Median	σ	U_1	p_u (a)	p_u (b)	$p_u \leq 0.025$
1	30	15.46	14.95	4.5	20	20.15	18.27	5.54	152	0.00145	0.00174	yes
2	8	28.14	28.91	5.8	12	29.09	30.11	7.40	41	0.31194	0.30800	no
3	8	5.15	4.82	1.1	12	6.38	6.20	0.74	18	0.01007	0.01142	yes

a. exact probability
b. approximate probability

and LIA rockfish, there appears to be greater variation in absolute measurements during the MCA than during the LIA. Measurement 3 shows that there is a statistically significant difference between the two groups, with the MCA rockfish being larger. The standard deviations, though, suggest that there is less variation in MCA rockfish than in those from the LIA.

These results are indicative but not conclusive. The expectation of larger average size rockfish in the MCA is supported by two measurements out of three. The expectation of greater size variation in the MCA is also supported by two measurements out of three. Only measurement 1 meets both expectations, however. These results provide some support for the expectation of greater average size in the earlier of the two periods. It also points to the necessity for further examination of other rockfish elements, for measurement of salmon vertebrae, for consideration of the utility of using morphometrics on fish, and for the need to pursue other lines of inquiry, such as aDNA testing and collagen mass spectrometry, to determine which species were present in what quantities during each period. The results of this test also point to the need to consider other reasons, such as resource depression instead of, or in addition to, less favorable climatic conditions that might help explain decreases in average size of rockfish, decreased emphasis on them, and a growing preference for salmon during the LIA. Here it is worth noting that shifting subsistence evidence on Kodiak Island, Alaska, has been shown to be relatively unrelated to climate change and to salmon availability. Instead, other processes appear to drive the abundance of salmon remains in archaeological deposits of the last ∼550 years (West 2009: 234).

Is there independent zooarchaeological verification that a climatic shift actually occurred in the study area between the MCA and the LIA and that this shift took place as indicated in Fig. 10.4? If there was a climatic shift, it should be expressed in SST proxies. Specifically, such proxies should show that SST during the MCA were higher than those of the subsequent LIA and that the LIA temperatures were both lower and more variable than those of the MCA. The chosen proxy for this evaluation is $\delta^{16}O/^{18}O$ ratio of the carbonate fraction in shells of butter clams (*Saxidomus gigantea*) (see also Colonese et al. 2017). The sample of shells for analysis was selected from the overall sample described above. In order to attain sample independence, only one valve or valve fragment per sample level was chosen from the two units that contained radiocarbon dates and the one unit that could be stratigraphically cross-dated because of its proximity to a dated unit. In total, 40 samples were submitted and analyzed against a PDB standard using Scintillating Ion Mass Spectrometry (SIMS) in the Geology Department at the University of Manitoba. Two modern control valves were tested along with the archaeological samples and were used to verify the temperature calculations that were made according to Böhm et al. (2000). The results are presented in Table 10.2.

The median and mean temperature for all readings was 11.6°C. Calculation of mean temperatures for the MCA (N = 10) produced a value of 11.6°C and for the LIA (N = 28) a value of 12.0°C, which is the opposite of what one would expect if temperature were the sole defining characteristic of each climatic period. A Chi-square test of temperatures above and below the median = mean separated by period produced a value of zero, suggesting there was no difference between the populations of temperatures from the two periods. Similarly, a Mann-Whitney test of temperature ranks was not significant at $\alpha \leq 0.025$ with direction predicted (U = 163, U' = 117, p = 0.23). Thus, overall temperature differences did not distinguish the MCA from the LIA according to this proxy measure. However, consideration of the standard deviations and ranges of temperatures in each period was illuminating. The standard deviation for the MCA was σ = 0.49, and the range was 10.9–12.7°C, whereas the SD for the LIA was σ = 2.06 and the range was 2.6–14.0°C. The oxygen isotope proxy of SST therefore suggests that average inshore water temperature may have not changed between the two climatic periods, but consistency of water temperature within narrow limits characterized the MCA whereas highly variable water temperatures between widely separated limits characterized the LIA. Support for the separation of levels into those from the two climatic periods is provided by the oxygen isotope analysis, although that support does not come in the form of the expected drop in SST in the LIA. Instead, the isotope

Table 10.2 Oxygen isotope results of *Saxidomus gigantea* sample, DfSj-23A

Catalogue no.	Unit N/S	Unit E/W	Level	_18O (‰, VPDB)	_18O (‰, VSMOW)	Boehm et al. °C	MCA/LIA
DfSj-23A:7543	32–34	20–22	3	−1.432155241	29.43357684	14.30772617	LIA
DfSj-23A:7543	32–34	20–22	3	−1.304270654	29.56541434	13.74247629	LIA
DfSj-23A:7543	32–34	20–22	3	−0.754222153	30.13246484	11.31126192	LIA
DfSj-23A:7439	32–34	20–22	6	−1.035761279	29.84222334	12.55566485	LIA
DfSj-23A:7439	32–34	20–22	6	−1.294278385	29.57571547	13.69831046	LIA
DfSj-23A:7695	32–34	20–22	12	−1.024468829	29.85386484	12.50575222	LIA
DfSj-23A:7695	32–34	20–22	12	−1.331102288	29.53775334	13.86107211	LIA
DfSj-23A:7668	32–34	20–22	12	−0.590143815	30.30161484	10.58603566	LIA
DfSj-23A:7668	32–34	20–22	12	−0.782694571	30.10311234	11.43711	LIA
DfSj-23A:7591	32–34	20–22	16	−0.939437642	29.94152434	12.12991438	LIA
DfSj-23A:7591	32–34	20–22	16	−0.803059588	30.08211784	11.52712338	LIA
DfSj-23A:7702	32–34	20–22	16	−0.508104645	30.38618984	10.22342253	LIA
DfSj-23A:7590	32–34	20–22	17	−0.997154611	29.88202334	12.38502338	LIA
DfSj-23A:7590	32–34	20–22	17	−0.827960889	30.05644684	11.63718713	LIA
DfSj-23A:7626	32–34	20–22	17	−1.222149693	32.16992634	13.4	LIA
DfSj-23A:7215	32–34	20–22	25	−0.970822409	29.90916947	12.26863505	MWP
DfSj-23A:7215	32–34	20–22	25	−0.798040721	30.08729184	11.50493999	MWP
DfSj-23A:3973	46–48	104–106	8	−1.301346315	29.56842907	13.72955071	LIA
DfSj-23A:3973	46–48	104–106	8	−0.90279358	29.97930107	11.96794763	LIA
DfSj-23A:16024	46–48	104–106	10	−0.472876061	30.42250734	10.06771219	LIA
DfSj-23A:16024	46–48	104–106	10	−0.622187349	30.26858084	10.72766808	LIA
DfSj-23A:16037	46–48	104–106	10	−0.696022601	30.19246334	11.0540199	LIA
DfSj-23A:16037	46–48	104–106	10	−0.712076541	30.17591317	11.12497831	LIA
DfSj-23A:16075	46–48	104–106	11	−1.123591448	29.75167834	12.9438742	LIA
DfSj-23A:16075	46–48	104–106	11	−0.897838958	29.98440884	11.94604819	LIA
DfSj-23A:16074	46–48	104–106	11	−0.807885422	30.07714284	11.54845356	LIA
DfSj-23A:16074	46–48	104–106	11	−0.631935533	30.25853134	10.77075505	LIA
DfSj-23A:16352	46–48	104–106	21	−0.787423888	30.09823684	11.45801358	MWP
DfSj-23A:16352	46–48	104–106	21	−0.824389772	30.06012834	11.62140279	MWP
DfSj-23A:16370	46–48	104–106	22	−0.668129284	30.22121884	10.93073143	MWP
DfSj-23A:16370	46–48	104–106	22	−0.860776557	30.02261684	11.78223238	MWP
DfSj-23A:2901	44–46	135–137	8	−0.991170577	29.88819234	12.35857395	LIA
DfSj-23A:2901	44–46	135–137	8	−0.946097293	29.93465884	12.15935003	LIA
DfSj-23A:2916	44–46	135–137	9	−1.051203946	29.82630334	12.62392144	LIA
DfSj-23A:2916	44–46	135–137	9	−0.962601643	29.91764434	12.23229926	LIA
DfSj-23A:2916	44–46	135–137	9	−0.655389083	30.23435284	10.87441975	LIA
DfSj-23A:2978	44–46	135–137	15	−0.813772938	30.07107334	11.57447639	MWP
DfSj-23A:17008	44–46	135–137	18	−0.845816473	30.03803934	11.71610881	MWP
DfSj-23A:17008	44–46	135–137	18	−0.718317952	30.16947884	11.15256535	MWP
DfSj-23A:17084	44–46	135–137	22	−0.719283119	30.16848384	11.15683139	MWP
DfSj-23A:17084	44–46	135–137	22	−1.06066258	29.81655234	12.6657286	MWP

observations conform to the analysis by Nunn (2007) of greater variability in sea surface temperatures in the LIA compared to the MCA.

The reliability of paleoclimate estimation from stable oxygen isotope analysis has been challenged by Burchell et al. (2012) because of the amalgamated effects of spring snowmelt discharge and rainfall. Their findings, however, confirm that a distinct seasonal rhythm of temperature change throughout a given year can be clearly observed. One might then think of the stable isotope results in the present paper as the affective oxygen isotope environment in which the clams lived without necessarily accepting them as absolutely correct temperature. It is still the case, then, that the affective mean temperature of the water in the MCA was the same as in the LIA, and that MCA affective temperatures tended to be more stable than those of the LIA. Considering the discussion herein and in the larger literature of the changing currents and upwellings and downwellings in the northeast Pacific that are occasioned by El Niño/ENSO and PDO, in addition to the mixing effects of snowmelt and rainfall, the prospect for discovering the "real" SST and associated climate appear complicated and support the call

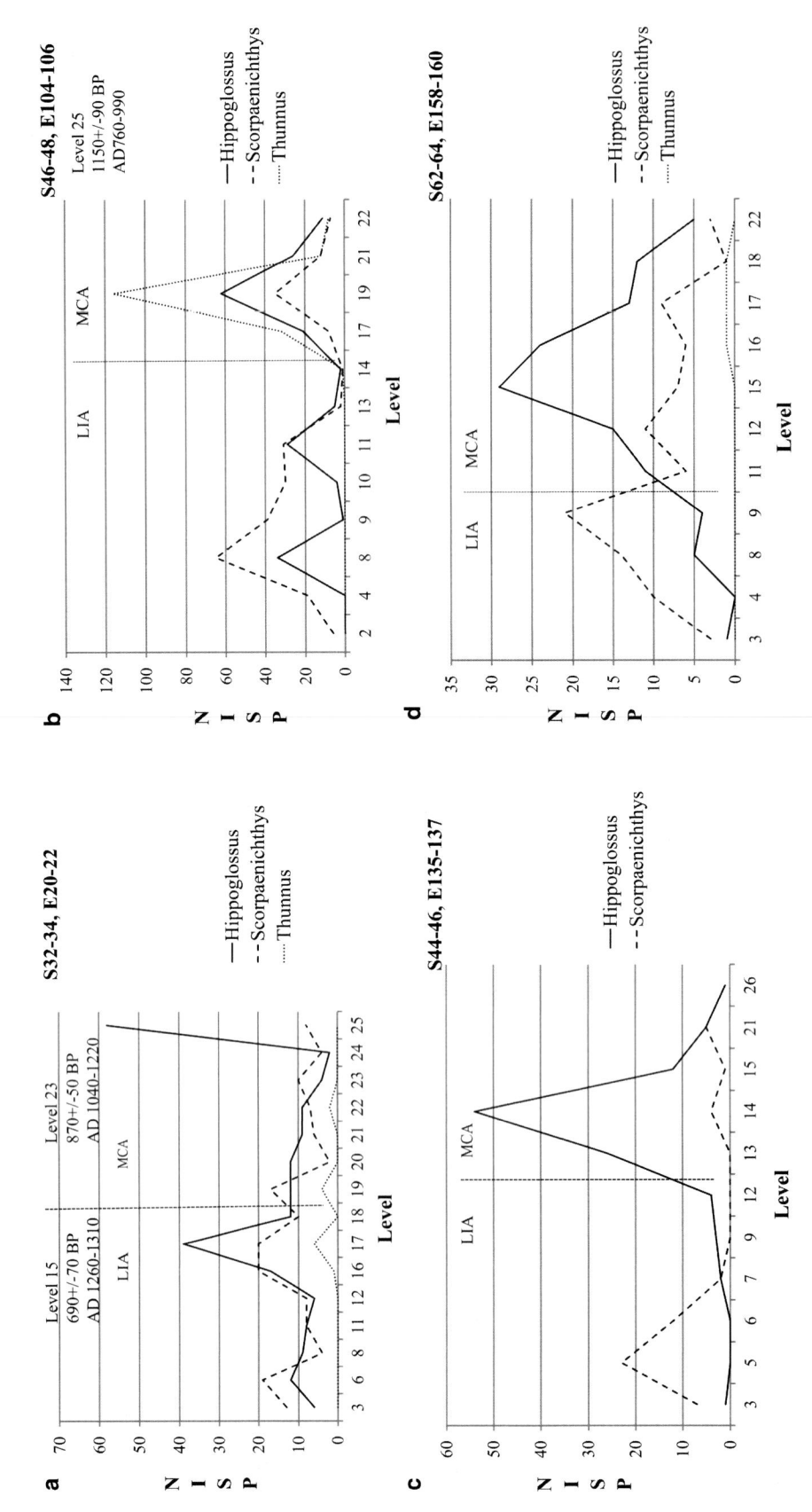

Fig. 10.5 Frequencies (NISP) of other major fish elements by level in four excavation units, DfSj-23A

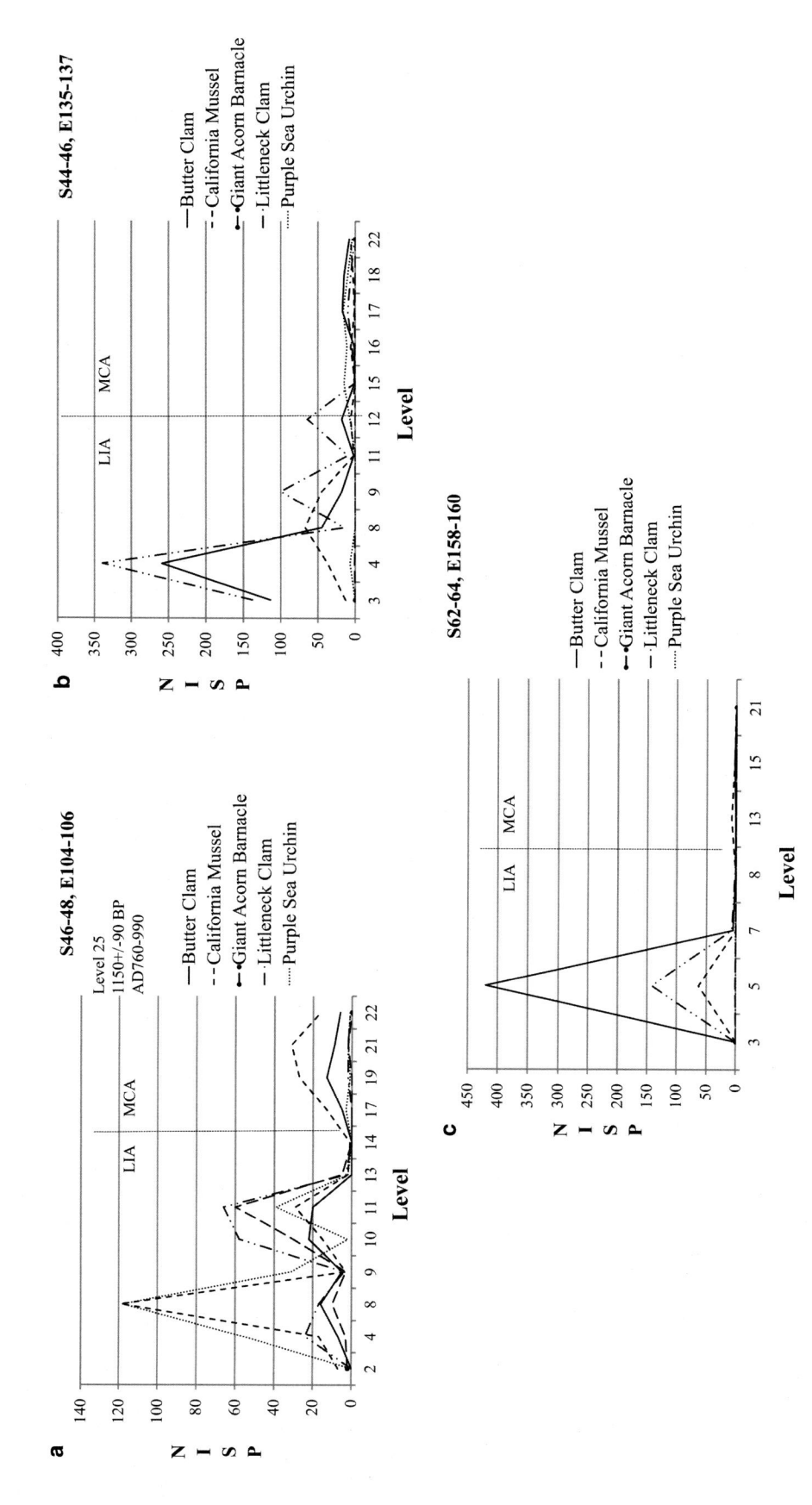

Fig. 10.6 Frequencies (NISP) of major Crustacea, Echinoidea and Bivalvia by level in three excavation units, DfSj-23A

by Burchell et al. (2012) for local calibration studies of stable oxygen isotopes prior to seasonality and affective paleoclimate studies.

Can support for the hypothesis that the climatic transition between the MCA and the LIA brought changes to the other fauna that were available to the Nuu-chah-nulth be found in other areas of the faunal assemblage? Among the fish, the trajectories of halibut, Bluefin tuna and cabezon are informative (Fig. 10.5). There is a tendency for halibut to be more abundant in MCA and early LIA levels and for cabezon be more abundant in late LIA levels. Their NISP trajectories cross one another the way salmon and rockfish do, only slightly later in time. Bluefin tuna, when it appears, is most common in MCA or early LIA levels. These patterns are least clear in E20-22 where halibut and cabezon follow roughly the same trajectories, except for the large numbers of halibut in level 25, and Bluefin tuna is found in small amounts during both climatic periods.

Examination of the sea mammal data did not reveal consistent patterning for any genus or family. This finding was inconsistent with what one might expect given the demonstrably different SST regimes of the two climatic periods. More extensive kelp beds should have harbored more sea otters, for example, but their numbers are small and inconsistent through time. Northern fur seal, the most abundant sea mammal (NISP = 1097), also did not display any consistent temporal tendencies between excavation units, possibly because of their dietary flexibility.

Mollusca should reflect differences in SST and Nuu-chah-nulth exploitation patterns. Generally warmer MCA temperatures and consequently greater risk of paralytic shellfish poisoning (PSP) would lead to the expectation of more seasonally restricted shellfish consumption during this period. Conversely, an increased number of cooler intervals during the LIA would likely lead to relatively high shellfish consumption. These expectations are most clearly borne out by the NISP counts of umbos from E135-137, E158-160 and E104-106 (Fig. 10.5a–c); all the umbos in E20-22 were found in LIA levels, so they are not depicted here. The data from these excavation units represent the molluscan taxa that are most economically important and, in the case of purple sea urchin, indicative of kelp bed exploitation. In all three excavation units, smaller NISP values are observed for each taxon in levels attributable to the MCA than to the LIA, and in E104-106, fewer taxa were collected during the MCA. The taxon that does not entirely conform to expectations is purple sea urchin in E104-106. Together with California mussel, it peaks in very late LIA levels.

Inspection of the bird data revealed no consistent patterning, likely because of their mobility and consequent ability to move to locations where food is available. Similarly, land mammals were hunted to such a limited extent that no reliable patterning could be discerned.

The fish, sea mammal and mollusk data taken together provide support for the hypothesis that subsistence patterns changed as a result of climatic change. The subsistence changes were not wholesale, however; a number of taxa were collected throughout the last 1,200 years, as one might expect. Nevertheless, the MCA subsistence strategy was characterized by emphasis on rockfish as opposed to salmon and halibut as opposed to cabezon, reduced quantity and variety of shellfish, acquisition of Bluefin tuna when it was available. There was also little emphasis on herring. The LIA subsistence strategy, in contrast, emphasized salmon over rockfish, cabezon over halibut, a late LIA emphasis on herring and extensive variety and quantity of shellfish. The graphs in Figs. 10.4, 10.5 and 10.6 require some interpretive caution, though. The main thing that I wish to emphasize by presenting them is the crossover phenomenon that seems to occur around the time of the MCA-LIA transition and the relative emphasis on taxa before and after that transition. The absolute amounts and fluctuations of taxa, however, likely reflect a variety of cultural, and perhaps natural, processes that lie beyond the scope of this paper. Intensity of harvest effort, shifting residential patterns, shifting patterns of site use that altered refuse disposal, occupation areas and areas of high traffic and consequent trampling of faunal remains, as well as short-term climatic fluctuations that affected taxonomic availability, are all possible causes for these observed fluctuations. All these processes are unlikely, though, to have affected the longer-term patterns that I wish to emphasize here.

Interpretation

Salmon would have benefited from colder intervals during the LIA and suffered from warmer ones that simultaneously favored rockfish. During the MCA, on the other hand, relatively consistent warm temperatures would have benefited rockfish and disadvantaged salmon. The opportunities that these sequential situations presented to the Nuu-chah-nulth are consistent with prey choice and patch choice models of optimal foraging theory. Both these models must be considered together in the present case because the environment is patchy but only certain resources are emphasized within chosen patches. The seasonal availability and accessibility of prey, the safety of their consumption and their size also bear directly on the coordinated choice of patches and prey species throughout the year under differing short and long-term climatic fluctuations.

MCA conditions appear to have favored a risk management strategy that emphasized consistent and extensive rockfish availability (numbers of species, extent of kelp beds, overall biomass) and supplemented this resource with other fish that were eaten fresh, as well as salmon and possibly small numbers of herring, which, along with halibut, may have been stored for winter consumption. Halibut would represent a large-package resource but one that could only be taken one at a time. Thus while it was consistently available and large (>100 kg would not be unusual in pre-European times), the hook and line technology required for its capture placed limits on the rate at which the resource could be acquired. Bluefin tuna, which can weigh 100–450 kg, was also a large-package resource that would have been favored by the relatively consistent, relatively warm MCA waters, although its size, speed and relatively infrequent appearance would have made it an opportunistic resource (see McMillan 1999: 140–143; Frederick and Crockford 2005: 194). Warmer water conditions would have promoted conditions for paralytic shellfish poisoning (PSP) and promoted avoidance of mollusk resources as a form of risk management. In conjunction with an emphasis on rockfish, halibut and tuna were exploited.

Under LIA conditions, decreased rockfish availability, in terms of numbers of taxa, size of patches and consequent reduction of overall biomass, and improved reproductive conditions for salmon during colder intervals, appears to have led to a risk assessment strategy by the Nuu-chah-nulth in which periodic (seasonal, annual, decadal) uncertainties in the abundances of these resources could be mitigated, and overall caloric production increased, by storage of small prey that were harvested *en masse*. Chum salmon spawning in a river channel primarily during October and November provide a seasonally restricted high density of small prey within a confined patch; they can therefore provide an increased return, relative to reduced rockfish availability, for the pursuit and handling times involved (moving to fishing stations, constructing and maintaining traps and weirs, building and maintaining drying racks, making stone and shell knives, cutting wood to smoke the processed fish, transporting the processed fish to the winter village). During times of abundance, i.e., under colder SST conditions, herring are present in great numbers during the winter when they school in northern Barkley Sound adjacent to spawning beaches. These schools can be accessed in suitable weather (absence of prevailing southwesterly storms) using canoes, herring rakes and hand nets (Drucker 1951: 23–25). The greatest abundance of herring, though, could be taken on spawning beaches for a short period in early spring. This event is also patchy and variable (seasonal, annual, decadal), but when they were available, they could provide a net increase over reduced rockfish productivity during the LIA. The temporally and spatially restricted super-abundance of this small prey has the capability to produce a net energetic return greater than continued primary reliance on rockfish. Herring fishing therefore justified the expenditure of time and effort to cut and set evergreen boughs and collect them with adhering egg masses later, rake and net fish as they spawned, build, maintain and operate stone-walled and wood-stake tidal traps, build racks and cut wood for smoking the fish and eggs, and transport the dried fish and eggs back to village sites. At the same time, increased fishing effort on cabezon in preference to halibut provided a medium-sized package that was far less dangerous and time consuming to acquire with highly similar technology. Cooler SST intervals during the LIA also reduced the likelihood of PSP, thus promoting an increased reliance on this low-risk, high-density small prey that could be taken with elementary technology (a pointed stick and a basket).

Cultural and Social Implications

The Nuu-chah-nulth appear to have adjusted their existing subsistence strategies in the face of shifting climatic conditions during the MCA and the LIA. Similar sets of resources were exploited during both periods, but the emphasis on certain resources changed in order to accommodate conditions as they arose at annual, decadal and centennial scales. These subsistence strategies, of course, were continuations of those that had been developed and practiced throughout the past 5,000 years and before. The evidence presented here suggests that the ethnographic and ethnohistoric descriptions of the Nuu-chah-nulth may be relatively recent developments in their overall cultural trajectory. The MCA conditions, at least, appear to have caused changes that were implemented in the decisions surrounding which subsistence resources to take, where they were to be taken, and when they were to be taken, i.e., seasonality and scheduling (see also Flannery 1968; Colonese et al. 2017; Pilaar Birch and Miracle 2017).

Scheduling, however, involves both decisions about when to acquire different resources and decisions about the task groups to acquire them. Thus social organization is implicated in the decisions around resource acquisition, and the resources that are taken will exert some influence on the size and composition of the task groups that are designed to take them. The task groups assigned to the acquisition of salmon and herring have already been noted. Whale hunting, on the other hand, is recorded as a prerogative of high-ranking men (Drucker 1951: 50), thus only a small percentage of the total labor budget, i.e., minimal risk investment, was allocated to this task (Monks in preparation).

If herring and salmon harvesting during fall at selected river channel patches and on specific spawning beaches were more common phenomena of the LIA, how did the resources that were emphasized during the MCA affect social organization? Similarly, if mollusk collection was not a large feature of MCA subsistence, how did that affect social organization? One feature of Nuu-chah-nulth ethnography and ethnohistory that seems likely to have persisted through both climatic periods (and before) is social ranking and patch ownership by descent groups. Ownership of salmon fishing stations on the Toquart River, clam beds, herring spawning beaches and foreshore kelp beds, for example, are likely to have been features of Nuu-chah-nulth culture and social organization during the MCA as well as in the more recent past. Resource patch ownership is only one aspect of social organization; allocation of labor to the exploitation of these patches is another, and control of stored resources is yet another. Rockfish acquisition was low-technology, relatively solitary (one or two people in a canoe with hand lines), and did not result in stored food (Carlson 2008: 106). Halibut fishing was similarly low-technology, but the task group consisted of one or more larger canoes paddled by adult men, probably from the same lineage. This pursuit may have resulted in a storable surplus that would be redistributed primarily within the households of the lineage. Small-scale salmon acquisition and minimal herring acquisition would have not required village-level task groups. Thus while lineage-level organization, resource patch ownership and surplus storage and redistribution were likely during the MCA, village-level task groups, mass harvesting of selected resources, storage and inter-household and inter-village redistribution seem more likely to have been features of the LIA that are consistent with existing ethnography and ethnohistory (see Cannon and Yang 2006; Monks and Orchard 2011).

Discussion

There are a lot of loose ends in this paper, and it would be a mistake to argue that the original hypothesis has been emphatically supported. Indeed, evidence presented here consists more of correlation than causation. There are interesting pieces of evidence that conform to the implications of the hypothesis, and they are sufficient to encourage further research on the role of climate change in Nuu-chah-nulth culture of the past. Much of the interpretation is speculative, and its conformity to optimal foraging theory is *post hoc*. Perhaps the chief virtue of this paper is the further questions that it raises, some of which are set out here.

The paucity of sea urchin remains during the MCA was surprising. If kelp beds were more extensive during the MCA, this resource should have been available in relatively great abundance. Were kelp beds not as extensive as expected? Were urchins not intensively sought? The availability and exploitation of urchins requires additional investigation, especially because of the implications for rockfish habitat that the answers to these questions may have.

Another puzzle is the apparent avoidance of mackerel. If SST was relatively consistently warmer than the LIA during the MCA, and if ENSO events tended to be more common, why would the Nuu-chah-nulth not exploit the abundant presence of mackerel? Lack of acquisition technology hardly seems like an insurmountable problem. They have a lot of fine bones, which may make them unpalatable to some, but that hardly seems like a barrier to consumption. Were they not susceptible to storage, so their irregular presence could not be counted on in the same way as salmon? Have faunal analysts failed to identify mackerel remains? Do mackerel remains easily fracture beyond recognition or decay quickly?

One might similarly ask why, if MCA conditions were as they are posited to be in this paper, warmer SST conditions would not have brought more sardines to the study area. Certainly they do not breed like herring and may have been less easily taken by traps and underwater egg collection technology, but their numbers seem likely to justify the effort of acquisition. Again, is it a failure of zooarchaeologists to recognize them. That seems unlikely considering how many independent researchers work on local faunal collections. Taphonomic factors may be at play, and optimal foraging decisions may also been at work. Perhaps the definition of "warm" and "cold" SST needs to be more fully refined for this area? More investigation is required here.

There are some larger considerations beyond fauna that might usefully be presented here by way of closing out this paper. One area of interest is the history of site proliferation and occupancy in the Toquaht traditional territory. The Toquaht have been claimed as the ancestral group for all the Barkley Sound Nuu-chah-nulth (Sproat 1868: 19; St. Claire 1991: 53). The earliest site occupancy of record is the site of Ch'uumat'a (DfSi-4) with a basal date of $4{,}000 \pm 40$ ^{14}C BP (McMillan 1999: Table 4). The earliest basal date of another site in Toquaht territory is $1{,}150 \pm 90$ BP at T'ukw'aa Village (DfSj-23A). This date approximates the onset of the MCA. The refuge site adjacent to the village (DfSj-23B) has a basal date of 780 ± 90 ^{14}C BP (McMillan 1999: 69, Table 4), which coincides approximately with the end of the MCA and the onset of the LIA and which is consistent with another defensive location at the southern entrance of Barkley Sound dated at approximately 700 BP (McMillan 1999: 82). The most recent dates from Ch'uumat'a (DfSi-4) are 720 ± 50 ^{14}C BP (back area) and 970 ± 60 ^{14}C BP (front area, although upper levels contain historic materials) (McMillan 1999: Table 4), about the time that the defensive site (DfSj-23B) came into use. Finally, the other major Toquaht site of Ma'acoah (DfSi-5) came into use at or after this time (McMillan 1999: Table 4), and became an important base for exploitation of the neighboring salmon streams, including the Toquart River, and it was also strategically located in relation to herring spawning beaches (St. Claire 1991: 163).

Rahn's (2002) nearest neighbour analysis of Toquaht Archaeological Project site survey data indicates that earlier sites, in what he called the "archaeological period", were located in relation to each other on the basis of social distance whereas later sites, in what he called the "ethnographic period", were located primarily in relation to subsistence resource locations. Taken together with the transition between rockfish and salmon predominance in the fish assemblages from the Toquaht sites, first identified by Streeter (2002), and the evidence presented in this paper, there seem to have been not only subsistence alterations but also demographic alterations and possibly social alterations that occurred in conjunction with changes in climate. Although they are beyond the scope of this paper, the effects of population growth, issues of site function, and much more and closer attention to indicators of climate change and their cultural and social implications are required in future.

This chapter set out to accomplish two aims. The first aim was to consider possible effects of the MCA-LIA transition on the environment of western Vancouver Island and its inhabitants, the Nuu-chah-nulth, as revealed through the analysis of zooarchaeological remains. The second aim, which is also that of this entire volume, was to advocate for a greater presence of archaeology and the time depth that it can bring to the study of animals (and plants) in the past (see Lyman 2006). Thus, in addition to archaeologists, this chapter and volume is intended to reach an audience that produces and consumes information relating to the temporal and spatial variation in the distribution of animal taxa in the face of environmental change resulting from climate change (e.g., Wintemberg 1919 cited in Lyman 2011; Magnell 2017; Rindel et al. 2017). Regardless of the domain in which one works, it is also important to recognize that humans were not hapless victims of climate, environment, and the availability of animals (and plants). Historical ecological processes must be recognized in which people played an active role in the distribution and variation of taxa over time and space. Improved fruitfulness of research in all disciplines that are involved in this common area of research and application will result from the adoption of this perspective.

Acknowledgments The Toquaht First Nation, especially the late Chief Bert Mack, generously gave permission and encouragement to excavate and analyze the data used herein. The Toquaht First Nation, the Nuu-chah-nulth Tribal Council, the Social Sciences and Humanities Research Council of Canada (grants # 410-03-0962 and 410-2006-1068), the British Columbia Heritage Trust, The University of Manitoba Social Sciences and Humanities Research Council (Research Grants Committee), and the University of Manitoba, Department of Anthropology, all provided generous financial support. Project co-directors Alan McMillan and Denis St. Claire invited me to examine the fauna and have been instrumental in facilitating all aspects of this work. Becky Wigen and Susan Crockford of Pacific Identifications and Megan Caldwell provided valuable assistance with fish identifications. I thank Ms. Linda Fitzpatrick and Dr. Richard Shelton of the Scottish Fisheries Museum for identifying "tusk" fish. Ms. Linda Gomez, Ms. Sabrina Crowley, Ms. Shannon Cowan and Ms. Kelly Johnson of the Nuu-chah-nulth Tribal Council provided assistance with Nuu-chah-nulth names for various fish species. I thank Elizabeth Reitz, Gary Coupland, Alan McMillan, Ken Ames and Nathan Mantua for their valuable comments on earlier drafts of this paper. All errors, omissions, and misinterpretations are solely my responsibility.

References

Ahlenius, H. (2007). *Temperatures Over Previous Centuries From Various Proxy Records*. Retrieved from Global Outlook for Ice and Snow: http://www.grida.no/publications/geo-ice-snow.

Ames, K. (1991). The archaeology of the longue druée: Temporal and spatial scale in the evolution of social complexity on the southern Northwest Coast. *Antiquity, 65*, 935–945.

Ames, K. (2003). The Northwest Coast. *Evolutionary Anthropology, 12*(1), 19–33.

Ames, K. (2005). Tempo and scale in the evolution of social complexity in western North America: Four case studies. In T. Pauketat & D. D. Loren (Eds.), *North American archaeology*. Malden, MA, USA: Blackwell Publishing.

Barclay, D., Wiles, G. C., & Calkin, P. E. (1999). A 1119 year tree-ring-width chronology from western Prince William Sound, southern Alaska. *Holocene, 9*(1), 79–84.

Bird, D., Bliege Bird, R., & Codding, B. (2009). In pursuit of mobile prey: Martu hunting strategies and archaeofaunal interpretation. *American Antiquity, 74*(1), 3–30.

Black, B., Boehlert, G. W., & Yoklavich, M. M. (2008). Establishing climate-growth relationships for Yelloweye Rockfish (*Sebastes ruberrimus*) in the northeast Pacific using a dendrochronological approach. *Fisheries Oceanography, 17*(5), 368–379.

Boas, F. (1966). *Kwakiutl ethnography*. Chicago, Illinois, USA: University of Chicago Press.

Böhm, F., Joachimski, M. M., Dullo, W.-C., Eisenhauer, A., Lehnert, H., Reitner, J., et al. (2000). Oxygen isotope fractionation in marine aragonite of coralline sponges. *Geochimica et Cosmochimica Acta, 64*(10), 1695–1703.

Bowers, P., & Moss, M. L. (2001). The North Point wet site and the subsistence importance of Pacific Cod on the northern Northwest Coast. In S. Gerlach & M.S. Murray (Eds.), *People and wildlife in northern North America: Essays in honor of R. Dale Guthrie* (pp. 159–177). Oxford: British Archaeological Reports, International Series, No. 944.

Bradley, R., Hughes, M., & Diaz, H. (2003). October 17). *Climate in medieval time. Science, 302*, 404–405.

Burchell, M., Cannon, A., Hallman, N., Schwarcz, H. P., & Schöne, B. R. (2012). Refining estimates for the season of shellfish collection on the Pacific Northwest Coast: Applying high-resolution stable oxygen isotope analysis and sclerochronology. *Archaeometry*. doi:10.1111/j.1475-4754.2012.00684.x.

Butler, V., & Chatters, J. C. (1994). The role of bone density in structuring prehistoric salmon bone density. *Journal of Archaeological Science, 21*(3), 413–424.

Campbell, C. (2002). Late Holocene lake sedimentology and climate change in southern Alberta, Canada. *Quaternary Research, 49*, 96–101.

Cannon, A. (1998). Contingency and agency in the growth of Northwest Coast maritime economies. *Arctic Anthropology, 35*(1), 57–67.

Cannon, A. (2002). Sacred power and seasonal settlement on the central Northwest Coast. In B. Fitzhugh & J. Habu (Eds.), *Beyond foraging and collecting: Evolutionary change in hunter-gatherer settlement systems* (pp. 311–338). New York: Kluwer/Plenum.

Cannon, A., & Yang, D. (2006). Early storage and sedentism on the Pacific Northwest Coast: Ancient DNA analysis of salmon remains from Namu. British Columbia. *American Antiquity, 71*(1), 123–140.

Carlson, R. (2008). The rise and fall of native Northwest Coast cultures. In Y. Gomez Coutouly (Ed.), *North Pacific Prehistory* (pp. 93–118). Madrid: The University Book.

Carré, M., Bentaleb, I., Fontugne, M., & Lavallée, D. (2005). Strong El Niño events during the Early Holocene: Stable isotope evidence from Peruvian shells. *The Holocene, 15*(1), 42–47.

Castro-Ortiz, J., & Lluch-Beldaa, D. (2007). Low frequency variability of fishing resources, climate and ocean. *Fisheries Research, 85*(1–2), 186–196.

Clague, J., & Turner, R. J. (2000). Climate change in British Columbia: Extending the boundaries of earth science. *Geoscience Canada, 27* (3), 111–120.

Colonese, A., Clemente, I., Gassiot, E., & Lopez-Saez, J. A. (2017). Assessing seasonal mollusc exploitation in tropical estuaries: Isotopic evidence from *Polymesoda sp.* (Bivalvia: Cyrenidae) shells from Pearl Lagoon (Nicaragua). In G. G. Monks (Ed.), *Climate Change And Human Responses: A Zooarchaeological Perspective* pp. (139–152). Dordrecht: Springer.

Cook, C. J. (1821). *The three voyages of Captain James Cook around the world* (Vol. VI). London: Longman, Hurst, Rees, Orme and Brown.

Crockford, S. (1994). New archaeological and ethnographic evidence of an extinct fishery for giant Bluefin Tuna (*Thunnus thynnus orientalis*) on the Pacific Northwest Coast of North America. In W. Van Neer (Ed.), *Fish exploitation in the past: Proceedings of the 7th Meeting of the ICAZ Fish Remains Working Group* (pp. 163–168). Tervuren: Annales du Musee Royal de l'Afrique Centrale, Sciences Zoologiques.

Crockford, S. (1997). Archaeological evidence of large Northern Bluefin Tuna, *Thunnus thynnus*, in coastal waters of British Columbia and northern Washington. *Fishery Bulletin, 95*, 11–24.

Croes, D. (1995). *The Hoko River archaeological site complex: The wet/dry site (34CA213), 3000–1700 B.P.* Pullman, WA: Washington State University Press.

Donald, L., & Mitchell, D. H. (1975). Some correlates of local group rank among the southern Kwakiutl. *Ethnology, 14*, 325–346.

Donald, L., & Mitchell, D. H. (1994). Nature and culture on the Northwest Coast of North America: The case of Wakashan salmon resources. In E. Burch & L. Ellana (Eds.), *Key issues in hunter-gatherer research* (pp. 95–117). Providence: Berg Publishers.

Drucker, P. (1951). *The northern and central Nootkan tribes* (Vol. Bulletin 144). Washington, D.C.: Smithsonian Institution, Bureau of American Ethnology.

El Niño and La Niña effects on Canada. (2010). Retrieved April 4, 2010, from http://www.ocgy.ubc.ca/projects/clim.pred/enso.canada.html.

Finney, B., Gregory-Evans, I., Douglas, M. S., & Smol, J. P. (2002). Fisheries productivity in the northeastern Pacific Ocean over the past 2,200 years. *Nature, 416*(April), 729–733.

Fisheries and Oceans Canada. (2008, 03 31). *Protecting British Columbia's rockfish (Brochure).* Retrieved January 17, 2012, from Fisheries and Oceans Canada: http://www.pac.dfo-mpo.gc.ca/publications/docs/rockfish-sebastes-eng.htm.

Fisheries and Oceans Canada. (2011, 05 20). *Herring spawn and catch records: Section 232, west Barkley Sound herring spawn records.* Retrieved January 15, 2012, from Fisheries and Oceans Canada: http://www.pac.dfo-mpo.gc.ca/science/species-especes/pelagic-pelagique/herring-hareng/herspawn/232fig-eng.htm.

Fisheries and Oceans Canada. (2011). *Stock assessment report on Pacific herring in British Columbia in 2011.* Science Advisory Report 2100/061, Science Advisory Secretariat. Retrieved from Fisheries and Oceans Canada: http://www.dfo-mpo.gc.ca/CSAS/Csas/publications/sar-as/2010/2010_064_e.pdf.

Flannery, K. (1968). Archaeological systems theory and early Mesoamerica. In B. Meggers (Ed.), *Anthropological archaeology in the Americas* (pp. 67–87). Washington, D.C.: Anthropological Society of Washington.

Francis, R., & Hare, S. R. (1994). Decadal-scale regime shifts in the large marine ecosystems of the north-east Pacific: A case for historical science. *Fisheries Oceanography, 3*(4), 279–291.

Frederick, S., & Crockford, S. (2005). Appendix D: Analysis of vertebrate fauna from Ts'ishaa Village, DfSi-16, Benson Island, B. C. In A. McMillan & D. E. St. Claire, *Ts'ishaa: Archaeology and ethnography of a Nuu-chah-nulth origin site in Barkley Sound,* (pp. 173–205). Burnaby, British Columbia: Archaeology Press, Simon Fraser University.

Gavin, D., Brubaker, L. B., & Lertzman, K. P. (2003). An 1800-year record of the spatial and temporal distribution of fire from the west coast of Vancouver Island. Canada. *Canadian Journal of Forestry Research, 33*(4), 573–586.

Gedalof, Z., Mantua, N. J., & Peterson, D. L. (2002). A multi-century perspective of variability in the Pacific Decadal Oscillation: New insights from tree rings and coral. Geophysical Research Letters, 29 (24), 57-1–57-4.

Graham, N., Ammann, C. M., Fleitmann, D., Cobb, K. M., & Luterbacher, J. (2011). Support for global climate reorganization during the "Medieval Climate Anomaly". *Climate Dynamics, 37*, 1217–1245.

Hare, E., Mantua, N. J., & Francis, R. C. (1999). Inverse production regimes: Alaska and west coast Pacific salmon. *Fisheries, 24*(1), 6–14.

Hay, D., & Kronlund, A. R. (1987). Factors affecting the distribution, abundance, and measurement of Pacific herring (*Clupea harengus pallasii*) spawn. *Canadian Journal of Fisheries and Aquatic Sciences, 44*(6), 1181–1194.

Hebda, R. (1982). Late-glacial and postglacial vegetation history at Bear Cove Bog, northeast Vancouver Island, British Columbia. *Canadian Journal of Botany, 61*, 3172–3192.

Jewitt, J. (1896). *The adventures of John R. Jewitt, only survivor of the ship Boston, during a captivity of nearly three years among the Indians of Nootka Sound in Vancouver Island.* London: Clement Wilson.

Jones, E. (2004). Dietary evenness, prey choice and human-environment interactions. *Journal of Archaeological Science, 31*, 307–317.

Kelly, R. (1995). *The foraging spectrum: Diversity in hunter-gatherer lifeways.* Washington, D.C.: Smithsonian Institution Press.

Kennett, D., & Kennett, J. P. (2000). Competitive and cooperative responses to climatic instability in coastal southern California. *American Antiquity, 65*(2), 379–395.

Koch, J., & Clague, J. (2011). Extensive glaciers in northwest North America during medieval time. *Climatic Change, 107*, 593–613.

Laird K. R., Cumming, F. F., Wunsam, S., Rusak, J. A., Oglesby, R. J., Fritz, S. C., et al. (2003). Lake sediments record large-scale shifts in moisture regimes across the northern prairies of North America during the past two millennia. *Proceedings of the National Academy of Sciences, 100*(5), 2483–2488.

Lamb, H. (1965). The Early Medieval Warm Epoch and its sequel. *Palaeogeography, 1*, 13–37.

Lertzman, K., Gavin, D., Hallett, D., Brubaker, L., Lepofsky, D., & Mathewes, R. (2002). Long-term fire regime estimated from soil charcoal in coastal temperate rainforests. *Conservation Ecology, 6*(2), 5.

Love, M., Yoklovich, M., & Thorsteinson, L. (2002). *The rockfishes of the northeast Pacific*. Berkeley, CA: University of California Press.

Luckman, B., & Wilson, R. (2005). Summer temperatures in the Canadian Rockies during the last millenium: A revised record. *Climate Dynamics, 24*(2–3), 131–144.

Luckman, B., Briffa, K. R., Jones, P., & Schweingruber, F. H. (1997). Tree-ring based reconstruction of summer temperatures at the Columbia Icefield, Alberta, Canada, A.D. 1073–1983. *Holocene, 7*(4), 375–389.

Lyman, R. (2006). Paleozoology in the service of conservation biology. *Evolutionary Anthropology, 15*(1), 11–19.

Lyman, R. (2011). A history of paleoecological research on sea otters, and pinnipeds of the eastern Pacific rim. In T. Braje & T. C. Rick (Eds.), *Human impacts on seals, sea lions and sea otters* (pp. 19–40). Berkeley and Los Angeles: University of California Press.

MacDonald, G., & Case, R. A. (2005). Variations in the Pacific Decadal Oscillation over the past millennium. *Geophysical Research Letters, 32*, L08703.

Magnell, O. (2017). Climate and wild game populations in south Scandinavia at the Holocene thermal maximum. In G. G. Monks (Ed.), *Climate change and human responses: A zooarchaeological perspective* (pp. 123–135). Dordrecht: Springer.

Mantua, N. (2002). Pacific-Decadal Oscillation (PDO). In M. MacCracken & J. S. Perry (Eds.), *Encyclopedia of global environmental change* (Vol. 1, pp. 592–594). Chichester: Wiley.

Mantua, N., Hare, S. R., Zhang, Y., Wallace, J. M., & Francis, R. C. (1997). A Pacific interdecadal climate oscillation with impacts on salmon production. *American Meteorological Society Bulletin, 78*, 1069–1079.

Matson, R. (2003). Introduction: The Northwest Coast in perspective. In R. Matson, G. Coupland, & Q. Mackie (Eds.), *Emerging from the mist: Studies in Northwest Coast culture history* (pp. 1–11). Vancouver: University of British Columbia Press.

Matson, R. (2006). *The coming of the salmon economy to Crescent Beach, B.C.* Toronto: Canadian Archaeological Association Annual Meeting, May 25, 2006.

McDougall, R. (1987). *Classification of British Columbia salmon stream escapements by species and subdistrict*. Fisheries and Oceans Canada, Planning and Economics Branch. Vancouver, B.C.: Canadian Manuscript Report of Fisheries and Aquatic Sciences No. 1870.

McKechnie, I. (2005). *Five thousand years of fishing at a shell midden in the Broken Group Islands, Barkley Sound, British Columbia*. Burnaby: Department of Archaeology, Simon Fraser University.

McKechnie, I. (2007). Investigating the complexities of sustainable fishing at a prehistoric village on western Vancouver Island, British Columbia, Canada. *Nature Conservation, 15*, 208–222.

McKechnie, I. (2012). Appendix B: Zooarchaeological analysis of the indigenous fishery at the Huu7ii big house and back terrace, Huu-ay-aht territory, Southwestern Vancouver Island. In A. McMillan, & D. St. Claire (Eds.), *Huu7ii: Household archaeology at a Nuu-chah-nulth village site in Barkley Sound* (pp. 154–186). Burnaby, B.C.: Archaeology Press, Simon Fraser University.

McKenzie, D. (1998). Vegetation and scale: Toward optimal models for the Pacific Northwest. *Northwest Science, 72*(Special Issue 1), 49–65.

McMillan, A. (1979). Archaeological evidence for aboriginal tuna fishing on western Vancouver Island. *Syesis, 12*, 117–119.

McMillan, A. (1999). *Since the time of the transformers: The ancient heritage of the Nuu-chah-nulth, Didiaht, and Makah*. Vancouver: University of British Columbia Press.

McMillan, A. D., & St. Claire, D. E., (1992). *The Toquaht Archaeological Project: Report on the 1992 Field Season*: Unpublished Report Submitted to the British Columbia Heritage Trust and Archaeology Branch, Victoria, and the Toquaht Band, Ucluelet.

McMillan, A., & St. Claire, D. E. (1993). *The Toquaht Archaeological Project: 1991*. British Columbia Heritage Trust, British Columbia Department of Tourism: Archaeology Branch, and Toquaht First Nation. Burnaby, B.C.: Simon Fraser University.

McMillan, A., & St. Claire, D. E. (2005). *Ts'ishaa: Archaeology and ethnography of a Nuu-chah-nulth origin site in Barkley Sound*. Burnaby: Archaeology Press, Simon Fraser University.

McMillan, A., McKechnie, I., St. Claire, D. E., & Frederick, S. G. (2008). Exploring variability in marine resource use on the Northwest Coast: A case study from Barkley Sound, Western Vancouver Island. *Canadian Journal of Archaeology, 32*(2), 214–238.

Minobe, S., & Mantua, N. (1999). Interdecadal modulation of interannual atmospheric and oceanic variability over the North Pacific. *Progress in Oceanography, 43*(2–4), 163–192.

Monks, G. G. (1987). Prey as bait: The Deep Bay example. *Canadian Journal of Archaeology, 11*, 119–142.

Monks, G. G. (2000). How much is enough? An approach to sampling ichthyofaunal assemblages. In T. Friesen (Ed.), *Studies in Canadian zooarchaeology: Papers in honour of Howard G. Savage* (Vol. 69, pp. 65–75). Ontario Archaeological Society.

Monks, G. G., McMillan, A. D., & St. Claire, D. E. (2001). Nuu-chah-nulth whaling: Archaeological insights into antiquity, species preferences, and cultural importance. *Arctic Anthropology, 38*(1), 60–81.

Monks, G. G. (2006). The fauna from Ma'acoah (DfSi-5), Vancouver Island, British Columbia: An interpretive summary. *Canadian Journal of Archaeology, 30*(2), 272–301.

Monks, G. G. (2011). Locational optimization and faunal remains in northern Barkley Sound, Western Vancouver Island, British Columbia. In M. Moss & A. Cannon (Eds.), *The archaeology of North Pacific fisheries* (pp. 129–148). Anchorage: University of Alaska Press.

Monks, G. G. (in preparation). Optimal foraging, costly signaling and Nuu-chah-nulth whaling. *Manuscript in preparation*.

Monks, G. G., & Orchard, T. J. (2011). Comment on Cannon and Yang: Early storage and sedentism on the Pacific Northwest Coast. *American Antiquity, 76*(3), 573–584.

Moser, H., Charter, R. L., Watson, W., Ambrose, D. A., Butler, J. L., Charter, S. R., et al. (2000). Abundance and distribution of rockfish (*Sebastes*) larvae in the southern California bight in relation to environmental conditions and fishery exploitation. *CalCOFI Report, 41*, 132–147.

Moss, M., Peteet, D. M., & Whitlock, C. (2007). Mid-Holocene climate and culture on the Northwest Coast of North America. In D. Anderson, K. Maasch, & D. H. Sandweiss (Eds.), *Climate change and cultural dynamics: A global perspective on Mid-Holocene transitions* (pp. 491–529). Amsterdam: Elsevier.

Mueter, F., Peterman, R. M., & Pyper, B. J. (2002). Opposite effects of ocean temperature on survival rates of 120 stocks of Pacific Salmon (*Oncorhynchus* spp.) in northern and southern Areas. *Canadian Journal of Fisheries and Aquatic Sciences, 59*(3), 456–463.

Nederbragt, A., & Thurow, J. W. (2001). A 6000 year varve record of Holocene climate in Saanich Inlet, British Columbia, from digital sediment colour analysis of ODP Leg 169S cores. *Marine Geology, 174*, 95–110.

North, G. (2006). *Surface temperature reconstructions for the last 2000 years*. National Research Council of the National Academies. Washington, D.C.: The National Academies Press.

Nunn, P. (2007). The A.D. 1300 event in the Pacific basin. *The Geographical Review, 97*(1), 1–23.

Nuu-chah-nulth Language. (n.d.). Retrieved June 7, 2012, from Nuu-chah-nulth Tribal Council: http://www.nuuchahnulth.org/language/sea_creatures.

Orchard, T. (2003). *An application of the linear regression technique for determining the length and weight of six fish taxa: The role of selected fish species in Aleut paleodiet* (Vol. S1172). Oxford: British Archaeological Reports, International Series.

Orchard, T. (2009). *Otters and urchins: Continuity and change in Haida economy during the Late Holocene and maritime fur trade period.* Oxford: British Archaeological Reports, International Series 2027.

Orchard, T., & Clark, T. N. (2005). Multidimensional scaling of Northwest Coast faunal assemblages: A case study from southern Haida Gwaii. *British Columbia. Canadian Journal of Archaeology, 29*(1), 88–112.

Partlow, M., & Kopperl, R. E. (2011). Processing the patterns: Elusive archaeofaunal signatures of cod storage on the North Pacific Coast. In M. Moss & A. Cannon (Eds.), *The archaeology of North Pacific fisheries* (pp. 195–220). Fairbanks: University of Alaska Press.

Pilaar Birch, S., & Miracle, P. (2017). Human response to climate change in the northern Adriatic during the Late Pleistocene and Early Holocene. In G. G. Monks (Ed.), *Climate change and human responses: A zooarchaeological perspective* (pp. 87–100). Dordrecht: Springer.

Pyper, B., Mueter, F. J., Peterman, R. M., Blackbourn, D. J., & Wood, C. C. (2002). Spatial covariation in survival rates of northeast Pacific chum salmon. *Transactions of the American Fisheries Society, 131*(3), 343–363.

Rahn, R. (2002). *Spatial analysis of archaeological sites in Barkley Sound, Vancouver Island.* M.A. thesis, University of Manitoba.

Rindel, D., Goni, R., Belardi, J., & Bourlot, T. (2017). Climatic changes and hunter-gatherer populations: Archaeozoological trends in southern Patagonia. In G. G. Monks (Ed.), *Climate Change and Human Responses: A Zooarchaeological Perspective* (pp. 153–172). Dordrecht: Springer.

Royer, T., Grosch, C., & Mysak, L. A. (2001). Interdecadal variability of northeast Pacific coastal freshwater and its implications on biological productivity. *Progress in Oceanography, 49*, 95–111.

Sandweiss, D., Maasch, K. A., Chai, F., Andrus, C. F., & Reitz, E. J. (2004). Geoarchaeological evidence for multidecadal natural climatic variability and ancient Peruvian fisheries. *Quaternary Research, 61*(1), 330–334.

Sapir, E., & Swadesh, M. (1955). *Native accounts of Nootka ethnography.* Bloomington: Indiana University, Research Center in Anthropology, Folklore and Linguistics.

Smith, R., Butler, V. L., Orwoll, S., & Wilson-Skogen, C. (2011). Pacific cod and salmon structural bone density. In M. Moss & A. Cannon (Eds.), *The archaeology of North Pacific fisheries* (pp. 45–56). Fairbanks: University of Alaska Press.

Sproat, G. (1868). *Scenes and studies of savage life.* London: Smith, Elder.

St. Claire, D. (1991). Barkley Sound tribal territories. In E. Arima, D. E. St. Claire, L. Clamhouse, J. Edgar, C. Jones, & J. Thomas, *Between Ports Alberni and Renfrew: Notes on West Coast peoples* (pp. 13–202). Ottawa: Canadian Museum of Civilization, Canadian Ethnology Service, Mercury Series Paper 121.

Stine, S. (1994). Extreme and persistent drought in California and Patagonia during the medieval time. *Nature, 369*(6481), 346–349.

Stiner, M., Munro, N. D., & Surovell, T. A. (2000). The tortoise and the hare: Small-game use, the broad-spectrum revolution, and Paleolithic demography. *Current Anthropology, 41*(1), 39–79.

Streeter, I. (2002). *Seasonal implications of rockfish exploitation in the Toquaht area, British Columbia.* M.A. thesis, University of Manitoba.

Strom, A., Francis, R. C., Mantua, N., Miles, E. L., & Peterson, D. L. (2004). North Pacific climate recorded in growth rings of geoduck clams: A new tool for paleoenvironmental reconstruction. *Geophysical Research Letters, 31*, 1–4.

Suttles, W. (1962). Variation in habitat and culture on the Northwest Coast. In *Proceedings of the 34th International Conference of Americanists* (pp. 522–537). Vienna.

Taylor, F. (1983). *The hydroacoustic assessment of herring distribution and abundance in Barkley Sound, February 18 to March 12, 1982.* Department of Fisheries and Oceans, Fisheries Research Branch, Pacific Biological Station. Nanaimo, B.C.: Canadian Technical report of Fisheries and Aquatic Sciences, No. 1197.

Tolimieri, N., & Levin, P. S. (2005). The roles of fishing and climate in the population dynamics of Boccacio rockfish. *Ecological Applications, 15*(2), 458–468.

Trouet, V., Esper, J., Graham, N. E., Baker, A., Scourse, J. D., & Franck, D. C. (2009). Persistent positive North Atlantic oscillation mode dominated the Medieval Climate Anomaly. *Science, 324*, 78–80.

U.S. Department of Commerce. (2011, August 31). *NOAA-NMFS-NWFSC TM-32: Chum Status Review.* Seattle. Retrieved August 31, 2011, from Chum Status Review: http://www.nwfsc.noaa.gov/publications/techmemos/tm32/t&flinks.html.

Viau, A. E, Gajewski, K., Fines, P., Atkinson, D. E., Sawada, M. C. (2002). Widespread evidence of 1500 yr climate variability in North America during the past 14,000 years. *Geology, 30*(5), 455–458.

Wang, Y., Flannigan, M., & Anderson, K. (2009). Correlations between forest fires in British Columbia, Canada, and sea surface temperatures of the Pacific Ocean. *Ecological Modelling, 221*(1), 122–129.

West, C. (2009). Kodiak Island's prehistoric fisheries: Human dietary response to climate change and resource availability. *Journal of Island and Coastal Archaeology 4(2)*, 223–239.

Whitlock, C., Higuera, P. E., McWethy, D. B., & Briles, C. E. (2010). Paleoecological perspectives on fire ecology: Revising the fire-regime concept. *The Open Ecology Journal, 3*, 6–23.

Wintemberg, W. (1919). Archaeology as an aid to zoology. *The Canadian Field Naturalist, 33*, 63–72.

Wright, C., Dallimore, A., Thomson, R. E., Patterson, R. T., & Ware, D. M. (2005). Late Holocene paleofish populations in Effingham Inlet, British Columbia, Canada. *Palaeogeography, Palaeoclimatology, Palaeoecology, 224*(4), 367–384.

Chapter 11
Biometry and Climate Change in Norse Greenland: The Effect of Climate on the Size and Shape of Domestic Mammals

Julia E.M. Cussans

Abstract This paper examines the climatic deterioration occurring in the 14th and 15th Centuries towards the end of the Norse Settlement in Greenland and its possible effects on the size and shape of domestic mammal (sheep and goat) bones. A review of biogeographical and nutritional factors affecting the size and shape of mammal bones is presented and used as a framework to predict potential changes in sheep bone size and shape at two sites from Norse Greenland; Gården under Sandet in the Western Settlement and Ø34 in the Eastern Settlement. The results are tentatively interpreted as indicating that bone growth was influenced both as a direct result of decreased temperature and as a result of a reduction in the vegetation productivity and hence animal nutrition. The negative effect of this on the human population is discussed.

Keywords Biogeography • Bone growth • Goats • Nutrition • Sheep • Temperature

Introduction

The Norse settlement in Greenland was short lived spanning nearly 500 years from the late 10th Century to the mid 15th Century. Deteriorating climate towards the end of the settlement period, whilst not being the sole cause of its demise (McGovern 2000) appears to have been a strong contributing factor (Buckland et al. 1996). Throughout the course of the Greenland settlements, adaptations to the sub-arctic environment were made, e.g., increased reliance on wild food resources compared to other Norse colonies (McGovern 1992); nevertheless, they maintained a strong pastoral element in their economy and raised livestock for meat, milk and wool (Enghoff 2003).

This paper examines the relationship between the growth of these livestock (particularly sheep but also goats) and the changes in climate occurring towards the end of the settlement. The interplay between biogeography, nutrition and bone growth is explored and used to determine how the deterioration in climate may have affected the size and shape of sheep and goat bones from two Norse farm sites in Greenland.

Hypothesis

In Greenland as the climate deteriorated towards the end of the settlement it seems likely that vegetation growth, or annual productivity pulse, would have become reduced, compared to earlier on in the settlement period, and grazing resources and hay production would have become depleted. Consequently, it is expected that over time there will be reduced availability of nutrition and hence reduced bone growth.

The hypothesis states that as environmental conditions become poorer, for example, through a shortened growing season or a poorer quality of vegetation, livestock animals will become smaller. This should be particularly manifested in breadth measurements which are preferentially diminished over length measurements through poor nutrition. According to this hypothesis, livestock bone size should diminish over the time of the settlement, particularly towards the very end of the occupation when climatic conditions were at their poorest. Two sites are examined; Gården under Sandet (GUS) in the Western Settlement and Ø34 in the Eastern Settlement. Although many other Norse sites in Greenland have provided domestic mammal bone assemblages and biometrical data (e.g., Degerbøl 1941), very few come from stratigraphically excavated deposits, i.e., they were excavated in the earlier part of the 20th century and hence cannot be examined in this study.

J.E.M. Cussans (✉)
Archaeological Solutions, 6 Brunel Business Court, Eastern Way, Bury St. Edmunds, Suffolk, IP32 7AJ, UK
e-mail: julia.cussans@ascontracts.co.uk

Norse Greenland

Settlement and Economy

According to the Icelandic scholar Arifroði Þorgilsson, Greenland (Fig. 11.1) was first discovered by the Norse around AD 982 (Arneborg 2003a) when Erik the Red was banished from Iceland for 3 years and decided to sail west. At the end of his banishment Erik returned to Iceland with tales of a newly discovered and fertile land that he named Greenland (Banks 1975). The following summer he and a group of followers returned to Greenland to settle (15 winters before Iceland adopted Christianity according to Islendingabok (Arneborg 2000, 2003a). As no farming communities were present in Greenland before this time they would have had to take all of their livestock with them. Settlers divided themselves between two main areas (Fig. 11.1): the Eastern Settlement near present day Igaliko (Garðar) and further north in the Western Settlement in the area of Nuuk (Godthåb) the present day capital.

Scientific evidence for the date of the Greenland landnám comes from paleoenvironmental studies. Edwards et al. (2008) found a swift decline in *Betlua glandulosa* pollen and concurrent increase in pollen from Poaceae and other herbs dating to AD 950–1020, which is in close agreement with the historical evidence for settlement. These changes in pollen represented a clearance of dense scrubland, possibly at least partly through burning and then replacement with grassland or steppe shortly after settlement (Edwards et al. 2008).

A great deal of archaeozoological interest has been paid to the settlements of Norse Greenland, with a large number of excavated bone assemblages; however the quality of these varies greatly (McGovern 1985, 1992). A large proportion of the evidence from these relates to the later part of the Norse occupation and only more recent excavations have been properly stratified and employed systematic sieving (McGovern 1985, 1992). McGovern (1985, 1992) collated much of the archaeozoological data and attempted to summarize the overall patterns in presence and distributions of animals at the Greenland Norse settlements. In terms of domesticates, cattle (*Bos taurus*) and caprines (*Ovis aries*; *Capra hircus*) were dominant; there was also a small presence of horse (*Equus cabalus*), dog (*Canis familiaris*) and pig (*Sus scrofa domesticus*). Cattle, horse and pigs were more common in the Eastern Settlement and dogs and caprines in the Western Settlement (McGovern 1985, 1992). Seal remains were very frequent, the most numerous being harp seals (*Pagophilus groenlandicus*). Ringed (*Phoca hispida*) and bearded (*Erignathus barbatus*) seals were present in small numbers, hooded seals (*Cystophora cristata*) were common in the Eastern Settlement and common seals (*Phoca vitulina*), although present at both settlements, were much more numerous in the Western settlement assemblages (McGovern 1985, 1992). Other wild resources included caribou (*Rangifer tarandus*), walrus (*Odobenus rosmarus*) and shellfish (Mollusca), all of which were more common at the Western Settlement. Whales (Cetacea) were more common in the Eastern Settlement, and birds and fish were found in very small numbers across both settlements (McGovern 1985, 1992). Bird species included Ptarmigan (*Lagopus* sp.), mallard (*Anas platyrhynchos*) and guillemot (*Uria* sp.), whilst other wild species included polar bear (*Ursus maritimus*), polar hare (*Lepus arcticus*) (Degerbøl 1943) and arctic fox (*Vulpes lagopus*) (Degerbøl 1934, 1941). The Greenlandic assemblages are unusual compared to others from the Norse North Atlantic as they show a heavy dependence on seals and caribou and very low quantities of fish bones, even at sites where sieving was employed during excavation; this is in contrast to elsewhere in the Norse North Atlantic where archaeozoological assemblages are dominated by domestic mammals and fish (McGovern 1992). There is also a relatively high frequency of goats (McGovern 1992). Age data for domestic mammals suggests that cattle were largely used for dairying and caprines were utilized for a mix of meat and secondary products (McGovern 1992). Information gained from stable isotope data (Arneborg et al. 1999), show that there was an increased marine component in the diet over the time of the Greenland Norse settlement from c. 20% marine input near to the beginning of settlement to an 80% input towards the end of the settlement.

It is clear from the high input of wild resources into the diet of the Norse Greenlanders that hunting played an important part in the economy particularly in the later settlement period. In fact the northern hunting grounds located around Disko Bay, known as the Norðrsetur, were a crucial part of their economy (Arneborg 2003a). These hunting grounds provided walrus hide for the production of rope, walrus and narwhal tusks, and polar bear furs, all of which could be traded for imports such as iron and timber (Arneborg 2000, 2003a).

The excellent preservation conditions found in Greenland have provided many textile remains, showing the importance of sheep for their wool (Berglund 2000). Other craft products were manufactured from wood, horn, soapstone (steatite), bone and iron (Berglund 2000). Cereal crops could not be grown in Norse Greenland as there was not enough consistently warm weather for them to complete their life cycle (Ross 1997: 18). However wild plant resources such as crowberry (*Empetrum nigrum*) and bilberry (*Vaccinium uliginosum*) were exploited (Buckland et al. 1996).

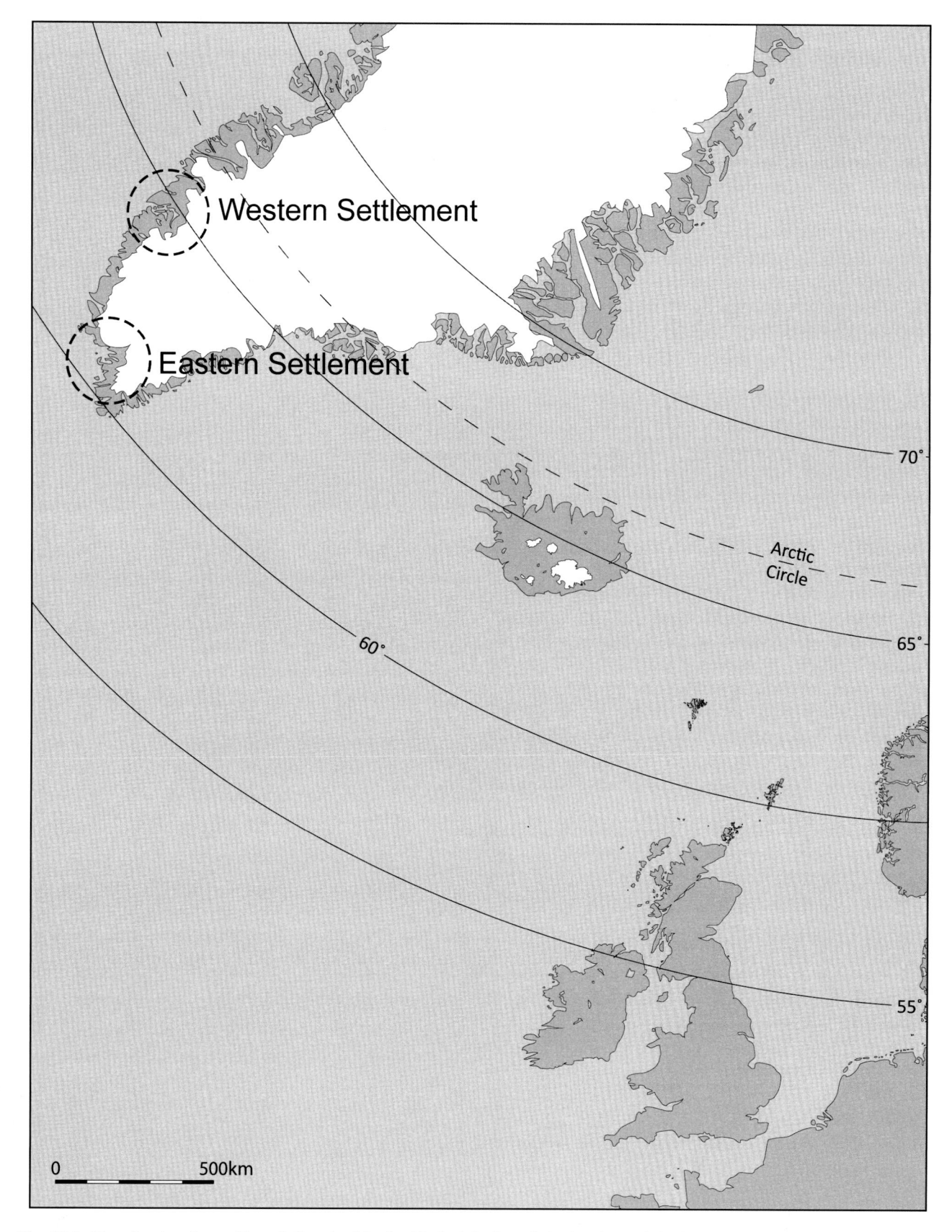

Fig. 11.1 Map showing the position of Greenland in the North Atlantic and the locations of the Eastern and Western Norse settlements (illustration by D. Bashford)

Changes in Climate and Environment, and Settlement Demise

The majority of the island of Greenland is encompassed by the Arctic Circle, resulting in the interior being covered by a vast icecap, the only ice-free land existing around the long coastline, made up of many islands and fjords. The Norse settlements lay within these ice-free areas and below the Arctic Circle. The land surrounding the fjords was covered by large areas of moorland and mountainside capable of supporting coarse vegetation whereas in the extreme southwest a more temperate climate allowed for the presence of extensive grasslands (Banks 1975).

Settlement took place during a period of relative warmth, with reduced storminess and sea ice, known as the Little Optimum (Bell and Walker 1992), which was greatly advantageous to the pastoral way of life and lasted until about 1,300 AD (Bell and Walker 1992) after which there was a period of distinct cooling. After flourishing for approximately 500 years, the Norse Greenlanders lost contact with mainland Europe and the settlement was later found to be abandoned (Arneborg 2003a). Historical sources (the *Greenland Description*) date the abandonment of the Western Settlement to c. 1360 (Arneborg 2003a). However the archaeological evidence would suggest a slightly later date; the farm at Gården under Sandet is thought to have been occupied until the 15th century (Berglund 2000). The last direct evidence of occupation of the Eastern Settlement is a marriage certificate dated to 1408; however it is thought the settlement survived until the middle of the 15th century (Arneborg 2003a).

What is uncertain about the abandonment of the settlement is the exact combination of reasons for its occurrence. Over the years many different explanations have been offered, the most enduring of which was the cooling in climate known as the Little Ice Age) (e.g., Bell and Walker 1992: 142). Early theories state that it simply became too cold for the Norse to survive (see McGovern 2000). A variety of other monocausal theories have also been offered, such as conflict with the Inuit or isolation from Europe (see McGovern 1981, 2000).

There are a number of reasons for these monocausal explanations being discredited, the most resounding relating to the cooling of the climate. The Norse were not the only group living in Greenland at the time of their extinction; Inuit were also present further to the north. Their population did not suffer when the climate cooled, pointing to the fact that there was more to the Norse extinction than cooling in climate alone (McGovern 2000).

Although climate change during the Little Ice Age cannot be named as the sole reason for the abandonment of the Norse settlement in Greenland it does appear to have made some contribution to the demise of settlement. It is this climate change and its possible effects on the domestic livestock being raised in Greenland that is the main focus of this paper. Detailed insights into climate change in Greenland (and further afield) have been gained from examination of ice core data.

Climate proxy data from the GISP2 ice core indicated that a large and abrupt climatic change occurred in the early part of the 15th Century. Specifically these were increased calcium concentrations, indicating increased soil erosion at this time, probably as a result of cooler climate and hence reduced vegetation cover (Buckland et al. 1996) and an increase in sea-salt sodium, an indicator of increased storminess (Mayewski et al. 1993). Additionally Dugmore et al. (2007) identified a key change in the cumulative deviation from the mean in North Atlantic storminess at around AD 1425 indicating that at this point past trends in weather patterns were poor indicators of future weather variations; a particularly problematic situation for farming communities reliant on knowledge of seasonal changes in planning their farming calendar. A similar reversal in sea-ice trends was also identified around AD 1450 (Dugmore et al. 2007). Similarly Monks (2017) also notes environmental unpredictability during the Little Ice Age in relation to salmon spawning conditions in western Canada, which would have a knock on effect on human populations fishing for these species.

The climatic changes examined so far, have all occurred in the earlier part of the 15th Century and are likely to have been too late to have impacted on the Norse livestock, particularly in the Western Settlement, as current evidence suggests this would have been abandoned in the very early 15th Century and the Eastern Settlement by the mid-15th Century. However examination of other proxy records (deuterium and oxygen isotopes) in the GISP2 ice core suggest that not only was the 14th Century on average colder than the 15th Century (Barlow et al. 1993) but that more prolonged periods of cold were noted in the 14th Century (Barlow 1994, from Buckland et al. 1996). From the examination of ice core data and historical records, Grove (2001) noted successions of cold years from 1308–1318, 1324–1329, 1343–1362 and 1380–1384. Perhaps of most significance to the current study though is Barlow's (1994, from Buckland et al. 1996) work looking at seasonal variations in ice core isotope records which noted the period 1343–1362 as having particularly low summer temperatures; this would have had a profound, detrimental effect on pasture growth and hay harvest.

It appears that the eventual demise of the settlement came from a variety of factors (McGovern 2000). Climate change would certainly have contributed, but factors such as erosion, the inability of the Norse to move around in the way that their Inuit counterparts could, a decrease in demand for trade goods such as walrus ivory and possibly poor political decisions made by those in charge (i.e., Bishops, all of whom were appointed from outside of Greenland) would have all played their part (McGovern 2000). Therefore, it appears that although some adaptations were made to the Greenlandic environmental conditions, e.g., increased exploitation of marine resources, these were not enough to offset the climatic changes of the Little Ice Age.

Biogeography and Bone Growth

Biogeography examines the relationship between an organism and its geographical situation, be it in terms of altitude, latitude, ambient temperature or available water and food resources. In turn each of these factors has a direct or indirect influence over an animal's growth. To examine the influence of specific habitats or environments on animal growth (and evolution) ecologists have developed a number of biogeographical rules; those particularly pertinent to this study are outlined below.

Allen's Rule

Allen's rule, also known as the proportion rule, states that animals living in cold climates have shorter extremities than those in warmer climates. This results in a reduction in body surface area and a consequent reduction in heat loss (Weaver and Ingram 1969; Hugget 2004). Weaver and Ingram (1969) investigated this rule and demonstrated that pigs raised at a low temperature (5°C) were much shorter in the limbs and snout than those from the same litter raised in considerably warmer conditions (35°C).

Bergmann's Rule

Bergmann's rule, or the size rule, is probably one of the best known of the biogeographical rules and has often been quoted in the past to explain size differences within mammalian species (e.g., Davis 1981; Weaver and Ingram 1969). The basic premise of the rule is that members of a particular genus or species will be larger in areas of cold habitat than members of the same genus/species living in a warmer climate (Hugget 2004). The reason for this being, once again, a matter of reducing heat loss in animals living in relatively cold habitats by reducing the surface area to volume ratio of the body, compared to those living in warmer conditions. A number of researchers have found that with increasing latitude and resulting decrease in temperature animals of the same or related species increase in size (e.g., Schrieder 1950). Conversely, Davis (1981) found that with an increase in temperature in Israel in the Late Pleistocene a whole array of mammal species decreased in size. Referring to this work Lister (1997) points out that this size change in mammals may not be directly related to temperature but rather to increased aridity, which in turn would be related to increased temperature, but probably more importantly would affect the quantity of available food. Magnell (2017) also noted a change in red deer size in Scandinavia in relation to temperature changes in the Holocene thermal maximum.

Guthrie's or Geist's Rule

Lister (1997) is by no means alone in casting doubt upon Bergmann's rule and its validity. Geist's (1987) paper "*Bergmann's rule is invalid*" makes a very clear stand. In this work Geist points out that although many researchers found an increase in body size with increasing latitude (and consequent decrease in temperature), some also found the opposite to be true (McNab 1971). Geist (1987) looked at data from a number of deer species and wolves, both of which have habitats spanning a wide range of latitudes including those around 60°N. The results showed that the animals increased in size with latitude up to around 60–65°N, at which point the trend towards greater size was reversed and animals started to get smaller with increasing latitude (Geist 1987).

Geist's (1987) explanation for this distribution of mammal sizes was that instead of body size being dependent on temperature it was actually dependent on available nutrition or more precisely the length of the annual productivity pulse, i.e., the quantity and quality of vegetative nutrition available during the peak growth season.

Guthrie's (1984) work, examining changes in mammal size between the Late Pleistocene and the Holocene, argued that during this time changes in the plant communities meant that in any one place resources were less varied and had shorter growing seasons. This in turn shortened the somatic (skeletal) growth season of herbivores which Guthrie believed was fundamental to dwarfing in the late glacial. Guthrie indicates that the growth season was critical to body size and "better accounts for Bergmann's law than any other factor" (Guthrie 1984: 269). Therefore both Guthrie (1984) and Geist (1987), believe that although mammalian body size does vary with latitude, the key factor in determining this is not temperature but available nutrition. Moreover this is not just the year-long average of available nutrients but the duration of the peak in quality growth resources, i.e., quality forage high in usable protein needed for somatic growth (Guthrie 1984).

Island Dwarfism

The biogeographical phenomenon of island dwarfism is of interest here, not because of land mass size but because of Greenland's limited plant resources. This phenomenon has been noted in a variety of species in a variety of places (e.g., Lister 1995, 1996). As mentioned above dwarfing has also taken place in continental regions during the Late Pleistocene. The reasons for both of these forms of dwarfing are essentially the same and relate to a shortage of resources (Marshall and Corruccini 1978; Guthrie 1984; Schüle 1993: 403). Large mammals have higher total rates of metabolism than do smaller mammals (McNab 1990: 15) and consequently need greater amounts of energy to sustain them; therefore in times and places of shortage small body size is of an advantage.

Summary of Biogeographical Effects

With the exception of Allen's rule, which does appear to be directly connected to environmental temperature, all of the other biogeographical effects appear to be linked with available nutrition. Available nutrition is limited by the annual productivity pulse which is affected by rainfall, angle of the sun and soil fertility, which are in turn are affected by latitude, altitude, geology and climate. All of these factors have an effect on the quality and quantity of vegetative nutrition. Therefore in areas where the annual productivity pulse is reduced, by whatever means, available nutrition and hence individual body size will also be reduced. The following section therefore examines the effects of reduced nutrition on bone growth in domestic mammals.

Nutrition and Bone Growth

Plentiful, good quality nutrition is required for animals to reach their full genetic potential (Pálsson and Vergés 1952a: 68f). This is a situation unlikely ever to have been achieved in Norse Greenland, which even at the height of the settlement period could still be described as marginal. Therefore how does nutrition, or more pertinently lack of it, affect the growth and hence bone size and shape of an animal, in particular that of domestic livestock? Growth and production performance in domestic livestock is very important to modern agriculture; a wide variety of experiments have been employed in the past to assess the effect of inadequate nutrition on sheep and other livestock. A small number of these studies have examined skeletal growth, most notable of which in relation to this research, is Pálsson and Vergés (1952a, b) work on the "*Effects of the plane of nutrition on growth and the development of carcass quality of lambs*". From this work a number of basic principles in (skeletal) growth relating to nutrition can be established. These ideas were also touched upon by Hammond (1932) and are as follows; the lower the level of nutrition an animal receives the less it will grow and the smaller its adult size will be. Later maturing bones (and body tissues) will be more greatly affected, in terms of growth retardation, by a low plane of nutrition than will more early maturing bones (or body tissues). Growth in bone thickness will be more greatly affected than growth in bone length and finally, the timing of shorter periods of poor nutrition will determine which bones are most greatly affected. Each of these factors is explored in more depth below. In terms of overall growth a variety of experiments have shown that animals that have higher planes of nutrition grow to a larger size whereas animals receiving poorer nourishment grow more slowly, mature later and cannot attain the proportions of well-fed animals Pálsson and Vergés (1952a), McCance et al. (1968).

Differential Effects of Poor Nutrition on the Skeleton

Growth of the mammalian body occurs in a wave starting at the extremities and gradually moving inwards towards the centre of the body. Therefore the skeleton does not mature all together at the same time, but in stages, starting at the head and feet and working inwards. In addition to this, skeletal tissue is one of the earlier maturing tissues in the body, muscle and fat maturing later in life. Dickerson and McCance (1961) found that in pigs fed on different nutritional planes, slaughtered at the same weight, but different ages, the undernourished pigs had larger (heavier) humeri than those pigs which had been well fed. This indicates that a greater proportion of the body weight of undernourished pigs was accounted for by the skeleton, which had been preferentially fed over other tissues given the limited available nutrition. Therefore in poorly nourished animals skeletal growth takes precedence over muscle and fat growth.

As well as the early maturing skeleton being preferentially fed over later maturing tissues, the skeleton itself is differentially affected by poor nutrition, and as with the different tissues it is the later maturing parts that are most affected. Pálsson and Vergés' (1952a, b) studies showed that the bones of the head and feet were less affected in terms of growth retardation than the upper limbs, ribs and vertebrae. The likely explanation for this phenomenon is that the more essential elements of the body develop first. The skull is important in protecting the brain, the skull and the lower jaw together are needed for eating (the acquisition of nutrients), and the feet are needed for moving around (finding food) and escaping predators. These elements are therefore given nutritive preference.

Effects of Poor Nutrition on Bone Shape

Not only does the nutritional level of an animal affect different parts of the skeleton preferentially it also differentially affects bone growth in different dimensions. Bone length is more early maturing than bone thickness (breadth) and therefore bone breadth is more greatly retarded by poor nutrition than bone length; hence bone shape is more greatly affected or altered than bone size (Pálsson and Vergés 1952a, b). Low nutritional plane lambs are relatively much skinnier than high plane lambs as described here: "The cannons of the Low-Plane lambs at 41 weeks though 10% shorter as an absolute figure than those of the High-Plane ones at the same age are, relative to their thickness, much longer, having 36% less weight: length ratio." (Pálsson and Vergés 1952a: 68). Measurements of bone length, minimum circumference and weight of bone per 10 cm length were measured for bones of the hind limb (Pálsson and Vergés 1952a) and these were compared between the high plane and low plane lambs. Least difference was found in the length measurements and the most difference was noted in the weight per 10 cm, indicating that bone thickness and possibly epiphysial breadth was most greatly affected by poor nutrition (Pálsson and Vergés 1952a).

More recent research, carried out by the English Heritage Sheep Project, has found some similar results in that bone breadth in males and castrates shows a relatively greater increase in size than bone length with increased nutrition (low quality pasture versus high quality pasture and additional hay for both) (Peter Popkin 2010: personal communication); conversely females show more significant increases in bone length than in bone breadth with increased nutrition.

Summary and Further Considerations of Nutrition and Bone Growth

From the above it can be seen that nutritional regime has a substantial effect on bone size and shape. Inadequate nutrition will reduce bone size, although not to the extent that overall body size will be reduced. It will also differentially affect later and earlier maturing body parts; later maturing parts being the most affected. Bone breadth will be more greatly retarded than will bone length (although this may be somewhat dependent on sex); additionally the timing of shorter periods of poor nutrition will most affect bones that should have a high growth intensity at the time of poor nutrition (Pálsson and Vergés 1952b).

Although studies such as Pálsson and Vergés (1952a, b) are excellent and yield a great deal of useful information there are some factors that need to be taken into consideration when using them to make inferences on archaeological data. Although some linear bone measurements are used, the majority of studies of this nature tend to use bone weight as a measure of bone size. There are several reasons why weight may not always be directly related to bone length or breadth i.e., bone weight may be affected by factors other than their external linear dimensions.

Dickerson and McCance (1961) found that in pigs undernourished animals had a thin and brittle bone cortex and that the marrow cavity was enlarged via the process of medullary erosion. In addition Pratt and McCance (1964) found that animals subjected to severe, prolonged poor nutrition had very thin and very dense cortical bone. These factors would not necessarily affect the external measurements of a bone but would most certainly affect its weight. In addition Dickerson and McCance's (1961) research also found that the bones of undernourished animals contained a much greater percentage of water than those of well-fed animals and conversely a much lower proportion of substances such as fat, collagen and calcium; this may also have an effect on bone weight. Nutritional stress occurring in pregnant and lactating ewes triggers the resorption of minerals from the skeleton, decreasing bone weight but not external measurements (Benzie et al. 1955).

Finally none of the papers discussed above examines fully adult animals (i.e., all bones fused); the oldest animals studied by Pálsson and Vergés (1952a, b) were 41 weeks old. They also showed that from three slaughtering ages of birth, 9 and 41 weeks, size differences between high and low nutritional plane lambs were least at birth, greatest at 9 weeks old and intermediate at 41 weeks; this suggests that some of the early shortfall in growth had been made up later in life. To what extent this catch up continues into adulthood is difficult to determine although it seems unlikely to be complete.

Sites and Phasing
GuS

The site of Gården under Sandet (GUS) in the Western Settlement (Fig. 11.2) is a typical farm settlement made up of a complex series of buildings forming a single farm, with a number of phases of modification and construction (Arneborg

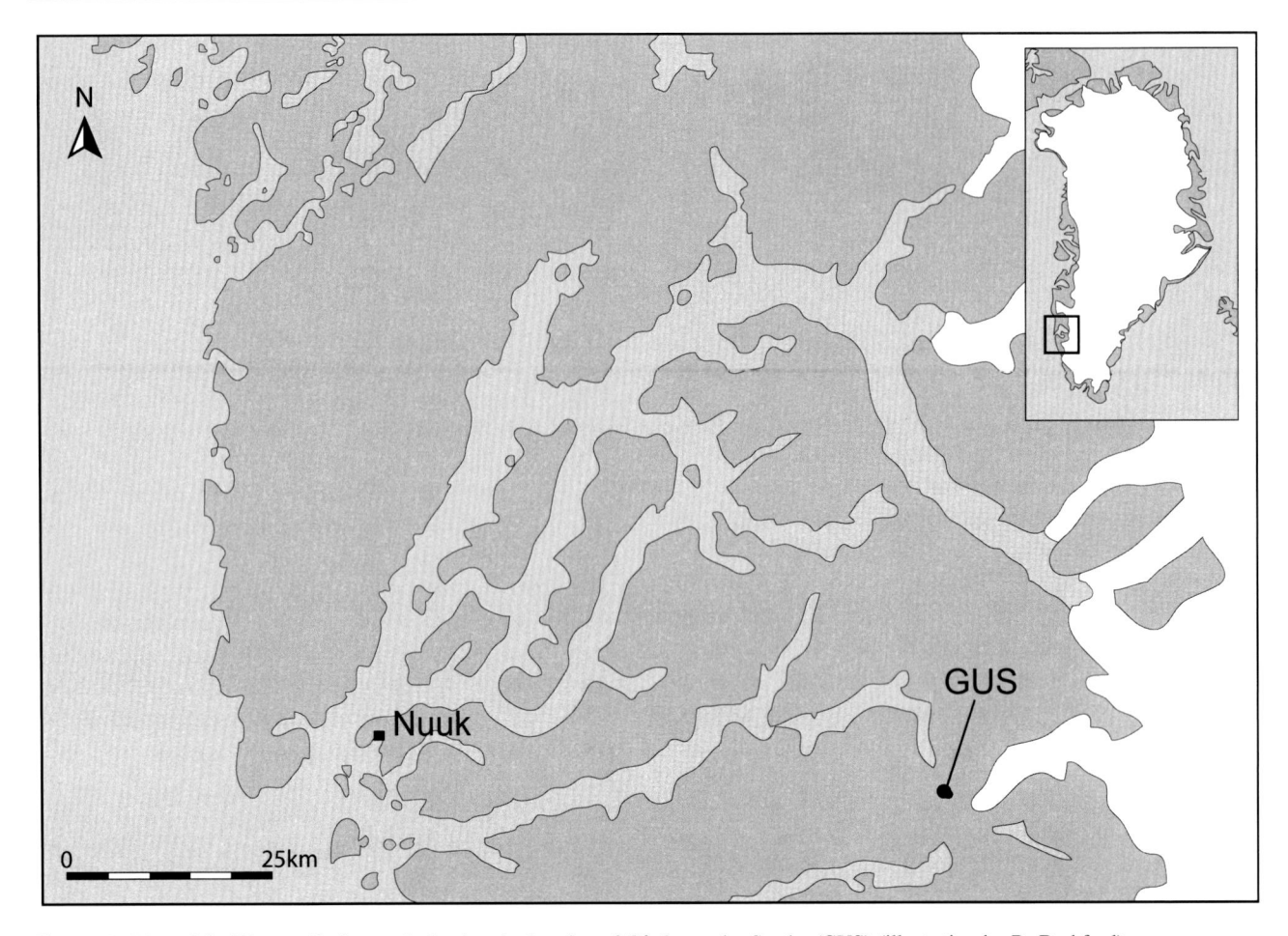

Fig. 11.2 Map of the Western Settlement indicating the location of Gården under Sandet (GUS) (illustration by D. Bashford)

2003b). In all, eight phases of construction were identified, with dates ranging from c. AD 1,000–1350. The animal bone finds from the site were split into three main phases for analysis, based on the construction phases. Phase 1 was dated to c. AD 1,000–1,150 (construction phases 1–3), Phase 2 dates were AD 1,150–1,300 (construction phases 4–6) and Phase 3 dated from AD 1,300 to AD 1,400, approximately the time of site abandonment (Arneborg 2003b).

Ø34

The site of Ø34 (Fig. 11.3.) is a middle sized farm, located near to Brattahlid, which consists of a group of 17 ruins including a small dwelling and a byre (Guldager et al. 2002: 53f). Excavations have focused on a midden deposit located next to the dwelling house (Nyegaard 1996). However most of the dating evidence has come from paleoenvironmental investigations. These showed evidence for peat cutting activity some time previous to AD 1020–1190, thought to be associated with the building of the farm, and evidence for site abandonment at around AD 1420–1630 (Schofield et al. 2008).

Data Collection

For the site of GUS bone measurement data were collected from two sources; the majority of the data were collected by the author at the Zoological Museum, University of Copenhagen and additional metapodial measurements were taken from the published animal bone report

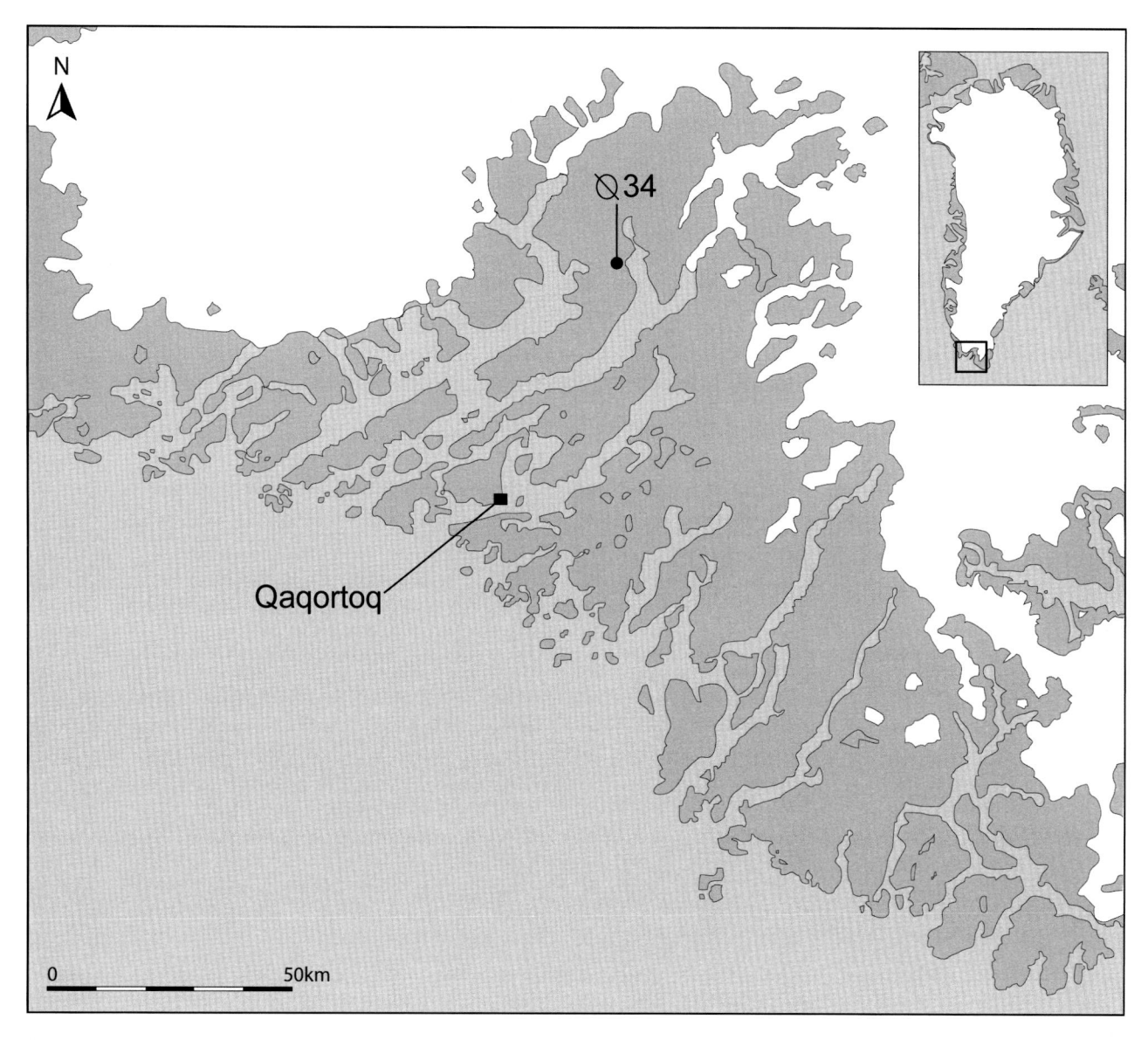

Fig. 11.3 Map of the Eastern Settlement showing the location of Ø34 (illustration by D. Bashford)

(Enghoff 2003). Measurements from the site Ø34 were kindly donated by Georg Nyegaard of the Greenland National Museum. All of the measurements (Table 11.1) were taken following the guidelines of von den Driesch (1976), recorded at a precision of 0.1 mm. Each measurement was taken at least three times and good agreement was usually found. During measurement recording, notes were also made where taphonomic or pathological factors may have had an effect on the bone dimension in question. In cases of severe damage no measurement was taken; however, if damage was slight the measurement was taken but a note was made of the nature and extent of the damage. These data were later excluded from any data analysis carried out but are available for later study if deemed necessary. The raw measurement data used in this study are provided in Table 11.1.

Table 11.1 Raw bone measurement data (mm) used in this study, by site, species, bone and phase/layer. Shaded cells indicate measurements used for log ratio analysis

Site	Species	Bone	Phase/layer	GL	GLl	GLm	Bp	SD	Dl	Bd	Source
GUS	Goat	AST	1		30.7	28.5			16.2	20.4	a
GUS	Goat	AST	1		31.3	28.9			16.4	20.3	a
GUS	Goat	AST	1		31.6	28.9			15.9	21.2	a
GUS	Goat	AST	2		28.4	26.6			14.7	18.5	a
GUS	Goat	AST	2		29.7	28.7			15.5	19.6	a
GUS	Goat	AST	2		29.9	28.7			15.3	19.2	a
GUS	Goat	AST	2		30	28.6			15.1	18.2	a
GUS	Goat	AST	3		24.4	23.7			12.7	16.3	a
GUS	Goat	AST	3		28.5	26.8			15.1	18.2	a
GUS	Goat	AST	3		28.5	27.7			15.1	18.3	a
GUS	Goat	AST	3		29.9	28.6			15.7	19.4	a
GUS	Goat	AST	3		30.5	29.2			15.4	18.9	a
GUS	Goat	AST	3		30.5	28.8			15	19.4	a
GUS	Goat	AST	3		30.6	28.4			16.1	20.7	a
GUS	Sheep	AST	1		24.9	24.9			14.4	17.2	a
GUS	Sheep	AST	1		25.4	24.5			14.9	16.9	a
GUS	Sheep	AST	1		26.2	24.5			14.6	17.4	a
GUS	Sheep	AST	1		26.6	26.1			15.4	17.3	a
GUS	Sheep	AST	1			25.2			14.9	17.8	a
GUS	Sheep	AST	2		26.5	26.1			15.4	17.6	a
GUS	Sheep	AST	2		28	27.6			16.3	18.9	a
GUS	Sheep	AST	2		28.1	26.6			16.3	19.1	a
GUS	Sheep	AST	2		29.6	28.2			16.5	18.8	a
GUS	Sheep	AST	3		24.3	23.4			13.9	16.3	a
GUS	Sheep	AST	3		24.9	24.2			14.5	16.7	a
GUS	Sheep	AST	3		25.3	24.7			14.8	17.2	a
GUS	Sheep	AST	3		25.8	25.1			14.9	18	a
GUS	Sheep	AST	3		26.5	25.6			15.5	18.5	a
GUS	Sheep	AST	3		27.1	27.2			15.8	19.4	a
GUS	Sheep	HUM	2							30.9	a
GUS	Sheep	HUM	3							30.4	a
GUS	Sheep	MTC	1	114.8			23.5	15.4		26.4	b
GUS	Sheep	MTC	2							23.4	b
GUS	Sheep	MTC	3	117.2			21.6	13.3		23.5	a
GUS	Sheep	MTC	3							23.6	a
GUS	Sheep	MTC	3							23.8	b
GUS	Sheep	MTC	3				21.2	13		24.4	b
GUS	Sheep	MTC	3							24.9	a
GUS	Sheep	MTC	3	127			24	14.7		26.6	b
GUS	Sheep	MTC	3							26.7	a
GUS	Sheep	MTT	1	128.2			18.6	11.4		21.9	b
GUS	Sheep	MTT	1							22.7	a
GUS	Sheep	MTT	1				20.3	11.9			b
GUS	Sheep	MTT	2	132.4			21.5	11.1		21.5	b
GUS	Sheep	MTT	2				19.5	11.7			b
GUS	Sheep	MTT	2				21.3	11.9			a
GUS	Sheep	MTT	3	115.9				11.4		20.7	b
GUS	Sheep	MTT	3							23.1	b
GUS	Sheep	MTT	3	135.7			20.7	12.5		24	a
GUS	Sheep	MTT	3							24.8	b
GUS	Sheep	PH1	2				11	9.5		10.8	a
GUS	Sheep	PH1	2				11.3	9.6		10.9	a
GUS	Sheep	RAD	3							27.5	a

(continued)

Table 11.1 (continued)

Site	Species	Bone	Phase/layer	GL	GLl	GLm	Bp	SD	Dl	Bd	Source
GUS	Sheep	RAD	3				32.6				a
GUS	Sheep	CAL	3	50.2							a
GUS	Sheep	FEM	3							36.8	a
Ø34	Sheep	TIB	0–10							24.5	c
Ø34	Sheep	TIB	10–20							21.7	c
Ø34	Sheep	TIB	10–20							23.2	c
Ø34	Sheep	TIB	10–20							23.4	c
Ø34	Sheep	TIB	10–20							23.8	c
Ø34	Sheep	TIB	10–20							24.0	c
Ø34	Sheep	TIB	10–20							24.3	c
Ø34	Sheep	TIB	10–20							25.9	c
Ø34	Sheep	TIB	20–30							23.7	c
Ø34	Sheep	TIB	20–30							25.7	c
Ø34	Sheep	TIB	30–40							21.7	c
Ø34	Sheep	TIB	30–40							23.2	c
Ø34	Sheep	TIB	30–40							24.8	c
Ø34	Sheep	TIB	30–40							25.1	c
Ø34	Sheep	TIB	30–40							26.3	c
Ø34	Sheep	TIB	30–40							26.7	c
Ø34	Sheep	TIB	30–40							26.8	c
Ø34	Sheep	TIB	40–50							23.9	c
Ø34	Sheep	TIB	40–50							24.3	c
Ø34	Sheep	TIB	40–50							24.9	c
Ø34	Sheep	TIB	40–50							25.4	c
Ø34	Sheep	TIB	40–50							26.5	c
Ø34	Sheep	TIB	40–50							26.8	c
Ø34	Sheep	TIB	50–60							23.8	c
Ø34	Sheep	TIB	50–60							24.2	c
Ø34	Sheep	TIB	50–60							25.8	c
Ø34	Sheep	TIB	50–60							26.1	c
Ø34	Sheep	TIB	60–70							24.1	c
Ø34	Sheep	TIB	60–70							26.0	c

Bone AST = astragalus, ATL = atlas, HUM = humerus, MTC = metacarpal, MTT = metatarsal, PH1 = phalanx 1, RAD = radius, CAL = calcaneus, FEM = femur, TIB = tibia. *Source* a = J. Cussans, b = Enghoff (2003), c = G. Nyegaard

Results
GUS, Western Settlement

From the site of Gården under Sandet sheep and goat bone measurements were examined across the three phases of the site; bones designated as sheep/goat were not included in this analysis. Great effort was made during the original archaeozoological analysis to distinguish sheep from goat where possible (Enghoff 2003) and therefore it is felt that the species identifications are fairly reliable.

The first analysis made was an examination of mean measurements of dimensions of the astragalus (greatest length of the lateral side (GLl), greatest length of the medial side (GLm), depth of the lateral side (DL), breadth of the distal end (Bd)), shown in Fig. 11.4. The figure illustrates both sheep and goat measurements and highlights a distinct difference between the two assemblages. Sheep astragalus measurements increase in size between Phase 1 and 2 and then decrease again between Phases 2 and 3. This appears to fit reasonably well with the hypothesis. An increase in size following settlement may well be due to the mild climate and expansion of grassland with the removal of wood and shrub land, bone size being reduced again as climate deteriorates in the later part of the settlement.

The measurements for goats however give a very different picture. These show a decrease in bone size between the

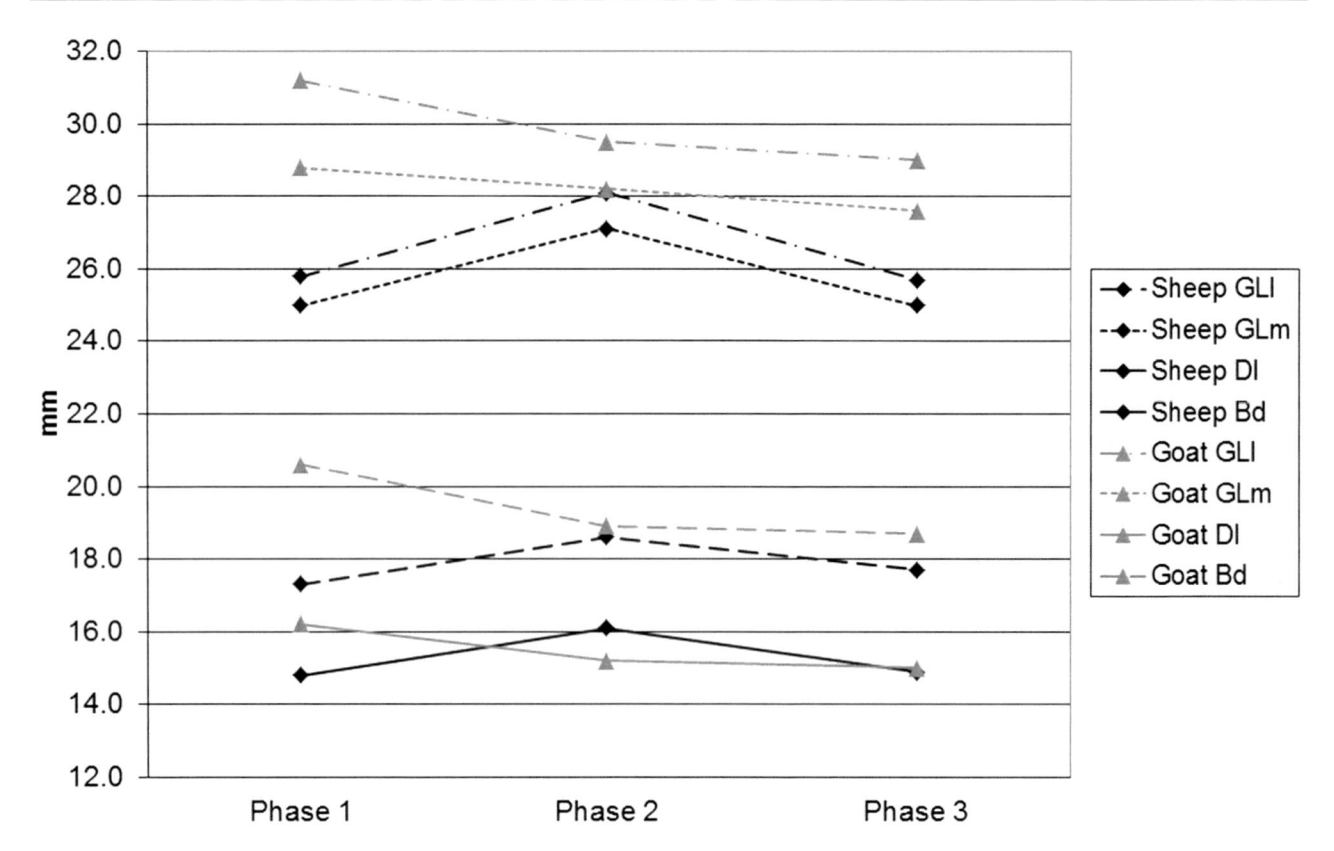

Fig. 11.4 Line graph showing mean sheep and goat astragalus dimensions (mm) from GUS, Western Settlement, Greenland. GLl = greatest length of the lateral side, GLm = greatest length of the medial side, DL = depth of the lateral side, Bd = breadth of the distal end

first two phases and then very little change in the last phase. There are a number of possible reasons why the size changes observed differ between sheep and goat and these will be discussed more fully below. However, it seems that the most likely reason for this difference is the change in vegetation occurring after settlement. As mentioned above, the main change that took place was the clearance of wood and shrub land (possibly greatly assisted by the presence of goats) and the expansion of grasslands. This would result in an increase in food availability for grazing sheep and a concurrent decrease in food availability for browsing goats. Interestingly there is evidence from element representation that during the first phase sheep and goats were allowed to wander some distance from the farm whereas later on they appeared to have been kept close by Enghoff (2003). Additionally, overall sheep seem to have been present in slightly larger numbers than goats and this is most pronounced in Phase 3 towards the end of settlement (Enghoff 2003); it is possible that this reduction in goat numbers was due to a lack of suitable habitat for them, first seen as a reduction in size in Phase 2.

Following the analysis of the astragalus means, a broader view of changes in sheep bone size was required. To achieve this, given the small sample sizes of the other bone dimensions available, a log ratio analysis (Meadow 1999) of all other measurable bones was carried out. This served to maximize the very small amount of data available and allowed some comparability of bones other than the astragalus between phases. Log ratio values were calculated using the following formula:

$Log\ ratio = log\ (archaeological\ measurement/standard\ measurement)$

When the log ratio value is zero it indicates that the archaeological measurement was the same as the standard; when the log-ratio value is positive it indicates that the archaeological measurement was larger than the standard; negative values indicate that the archaeological specimen was smaller than the standard. The standard used for the comparison of sheep measurements was a hypothetical standard created from published data in Clutton-Brock et al.'s (1990) study of the osteology of the Soay sheep. The standard measurements are listed in Table 11.2 and come from a group of Soay males. Although within Clutton-Brock et al.'s (1990) paper notation given by von den Driesch (1976) was not used, the diagrams provided allowed measurements to be selected that matched von den Driesch's (1976) criteria for specific bone dimensions and hence only directly comparable measurements were used.

Table 11.2 Standard measurements (mm) used for the calculation of log-ratio values for sheep measurements

Bone	Dimension (von den Driesch 1976)	Value
Humerus (7)	Bd	28.62
Radius (7)	Bd	27.64
Femur (7)	Bd	36.88
Calcaneus (7)	GL	52.22
Metacarpal (8)	GL	119.93
	Bd	23.70
Metatarsal (8)	GL	128.66
	SD	11.87
Phalanx 1 (16)	SD	9.23

Values are from a group of Soay males (Clutton-Brock et al. 1990) and represent mean measurements; figures in brackets after bone name show number in sample. Values for phalanx 1 are a mean of equal numbers of fore and hind limb bones

Due to the differential effects of poor nutrition on bone breadth and length, two figures were constructed to allow length and breadth dimensions to be examined separately. To avoid repetition of data the astragali were excluded from this analysis and likewise to avoid biasing toward more complete bone specimens no more than one length and breadth measurement from each bone was submitted for the log ratio analysis. Despite the pooling of data the sample is still very small. Log ratio values for bone lengths are shown in Fig. 11.5 and those for breadths in Fig. 11.6; mean log values for each phase are indicated. The mean log values shown here do not disagree with the data presented for the sheep astragalus (not included in these figures). They appear to indicate an increase in size between Phase 1 and 2 and then a decrease in length (but not necessarily breadth) in Phase 3. However as can be seen from the figures the sample size is so small statistically reliable interpretations cannot be drawn. Looking back at Fig. 11.4 the decrease in size seen in Phase 3 appears greater in terms of length measurements than breadth measurements. It appears then that there may be a greater change in bone lengths than in bone breadths and if this was so it seems unlikely that a decrease in nutrition could have been the only determining factor as, if this were the case, one may expect bone breadth to be more greatly affected than length; although as noted by Peter Popkin (2010: personal communication) this may be sex dependent. This is explored in more detail for the sheep astragali below.

Figure 11.7 is a scatterplot of astragalus length versus breadth and shows apparent size and shape differences between the three phases; annotations have been added with proposed theories for the changes in size and shape observed. There appears to be an overall size increase between Phase 1 and Phase 2, as noted above, which is postulated to be the result of improved nutrition. Such an increase in size may also be suggested to be the result of a change in the proportions of males and females present at the site, with potentially an increase in males between Phase 1 and 2. However Davis (2000) found only a 2% difference between male and female sheep astragalus measurements, whereas an 8% difference is seen between the means for Phases 1 and 2 here; indicating that sexual dimorphism is an unlikely cause of the observed size difference. The astragali for Phase 3 have a much more dispersed distribution and it is proposed that the cause of this may be twofold. Overall the Phase 3 astragali are relatively shorter (compared to their breadth) than the majority of those from the previous two phases. It is proposed that the reduction in bone length is not necessarily related to a reduction in nutritional intake but rather more directly a result of a drop in temperature. Allen's rule, discussed above, states that animals raised at lower temperatures will be shorter limbed than those of the same taxa or species raised at higher temperatures; a heat saving mechanism designed to reduce body surface area over which heat can be lost. These are however tentative interpretations given the small size of the data set.

Further examination of the astragalus scatterplot shows that three of the data points have breadths within the range of Phase 2 measurements and the other three have much lower values, similar to Phase 1 and smaller. It is proposed that these particularly small specimens come from the very latest period of occupation when decreasing nutrition starts to take effect on bone size. However, the fine stratigraphic detail needed to substantiate this is not currently available. An alternative explanation for the presence of these small specimens is that they belong to younger individuals; however during data collection care was taken to exclude astragali that appeared porous and not fully formed in order to minimize the inclusion of younger animals.

As mentioned above, the size changes observed in goats were quite different to those seen in sheep. To illustrate this further a scatterplot of goat astragalus length and breadth measurements (Fig. 11.8) was also made. This again shows the decrease in size between Phase 1 and 2 and the similar size of the Phase 2 and 3 animals. One exception is a very

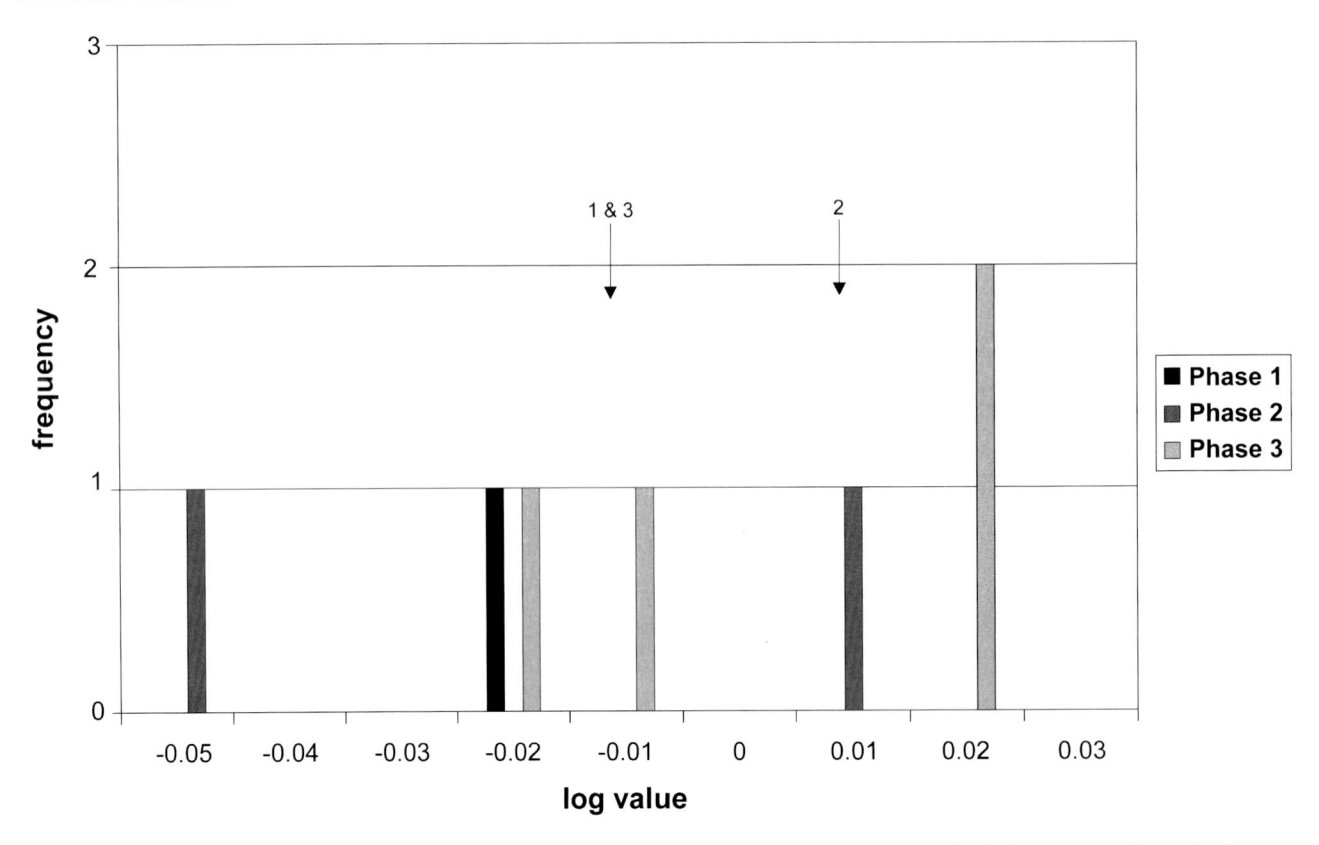

Fig. 11.5 Log-ratio values for sheep bone length dimensions from GUS, Western Settlement, Greenland. Mean values for each phase are indicated with arrows

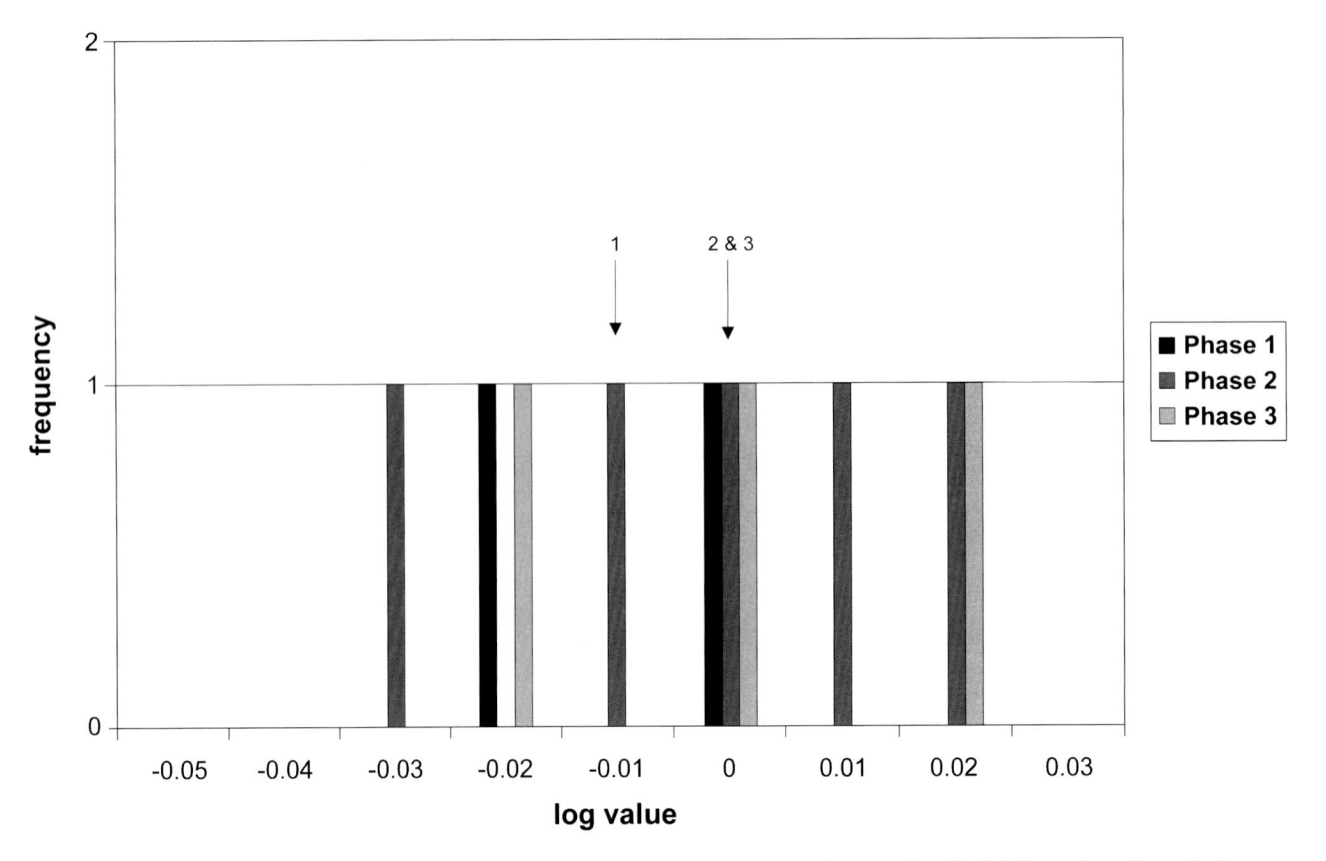

Fig. 11.6 Log-ratio values for sheep bone breadth dimensions from GUS, Western Settlement, Greenland. Mean values for each phase are indicated with arrows

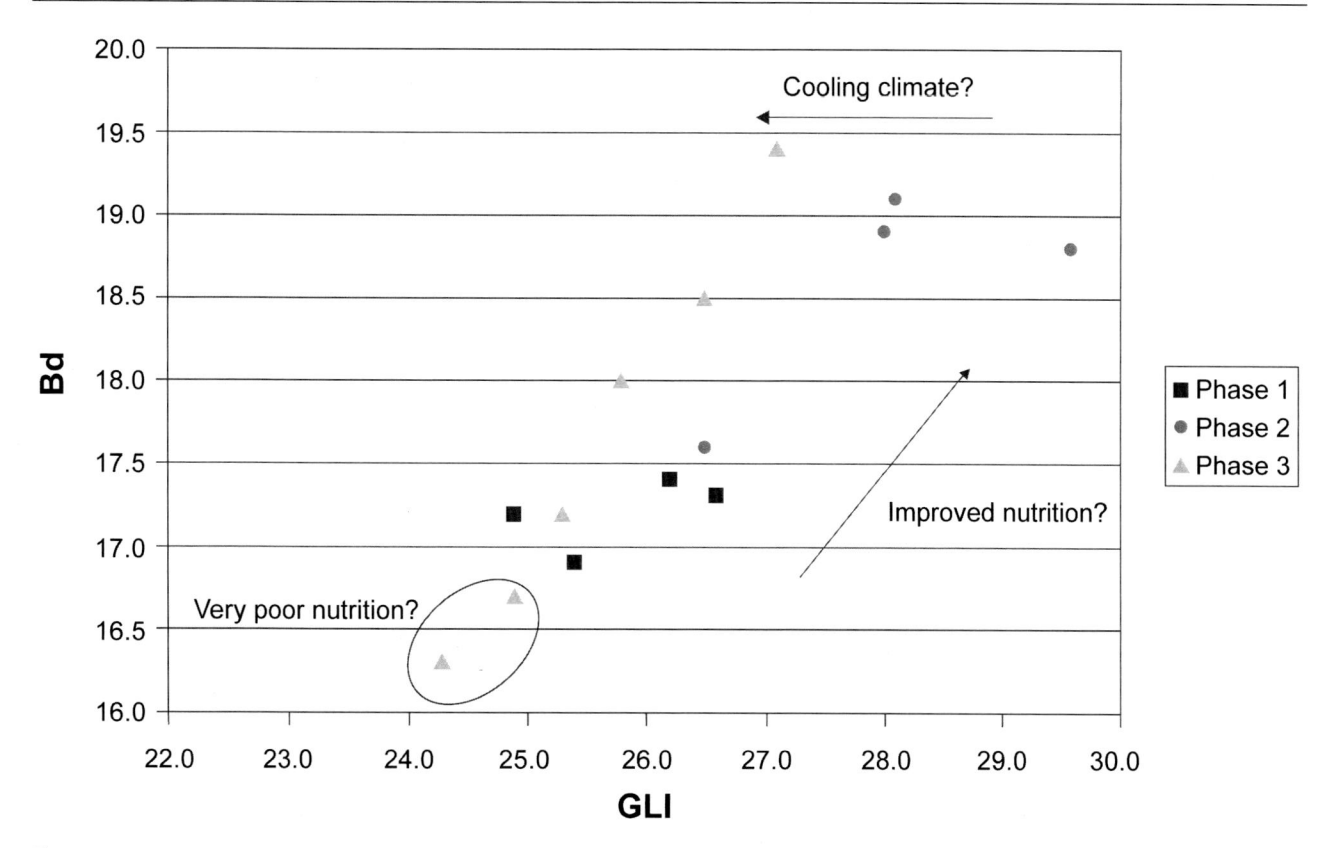

Fig. 11.7 Scatterplot of sheep astragalus greatest length (mm) versus distal breadth (mm) by phase for GUS, Western Settlement, Greenland. Annotations show possible explanations for size and shape changes

small astragalus plotting in the bottom left hand corner of the graph; it seems likely that this maybe a miss-identified sheep astragalus as it would be well placed with the other Phase 3 sheep astragalus plots. This figure does serve to further illustrate that the size changes taking place in goats are quite different to those taking place in sheep. Of particular interest is why no shortening of limb length is apparent in goats at the same time as it is seen for sheep, particularly if, as suggested, this is caused by a drop in temperature. One possible explanation is that due to their destructive effect on the local vegetation goats were, in the later phases, more restricted in their movements than sheep and possibly housed and fed indoors. This would clearly shelter them somewhat from a climatic drop in temperature. A complete goat skeleton from Phase 2 found inside a collapsed building (Enghoff 2003) suggests that goats may have been kept indoors some of the time at GUS. However based on the current data set such interpretations are only tentative and are somewhat at odds with Mainland's (2006) tooth micro-wear evidence that suggests goats were more likely to be kept outdoors over winter than sheep.

Ø34, Eastern Settlement

The second site, Ø34 in the Eastern Settlement, was only briefly examined. The only archaeological excavation carried out at this site to date was that of a midden deposit next to the farm buildings. Although no specific site phasing is currently available and AMS Radiocarbon dates are yet to be published for the midden itself it was thought feasible to carry out some preliminary analysis on the bone measurement data. Evidence for, and dating of, peat cutting activity from pollen analyses at the site indicates that the farm was likely occupied for the duration of the Norse settlement of Greenland (Schofield et al. 2008). Additionally, although not phased, the midden was excavated in spits and here these spits were used as analytical units, assuming (it should be noted here that this is the assumption of the present author, dating evidence suggests that there may be some disturbance in the midden sequence (Georg Nyegaard 2010: personal communication) that the lower spits relate to earlier in the occupation and higher spits relate to later in the occupation. Following these assumptions, tibia distal breadth measurements were examined and a box-and-whisker plot of these data is shown

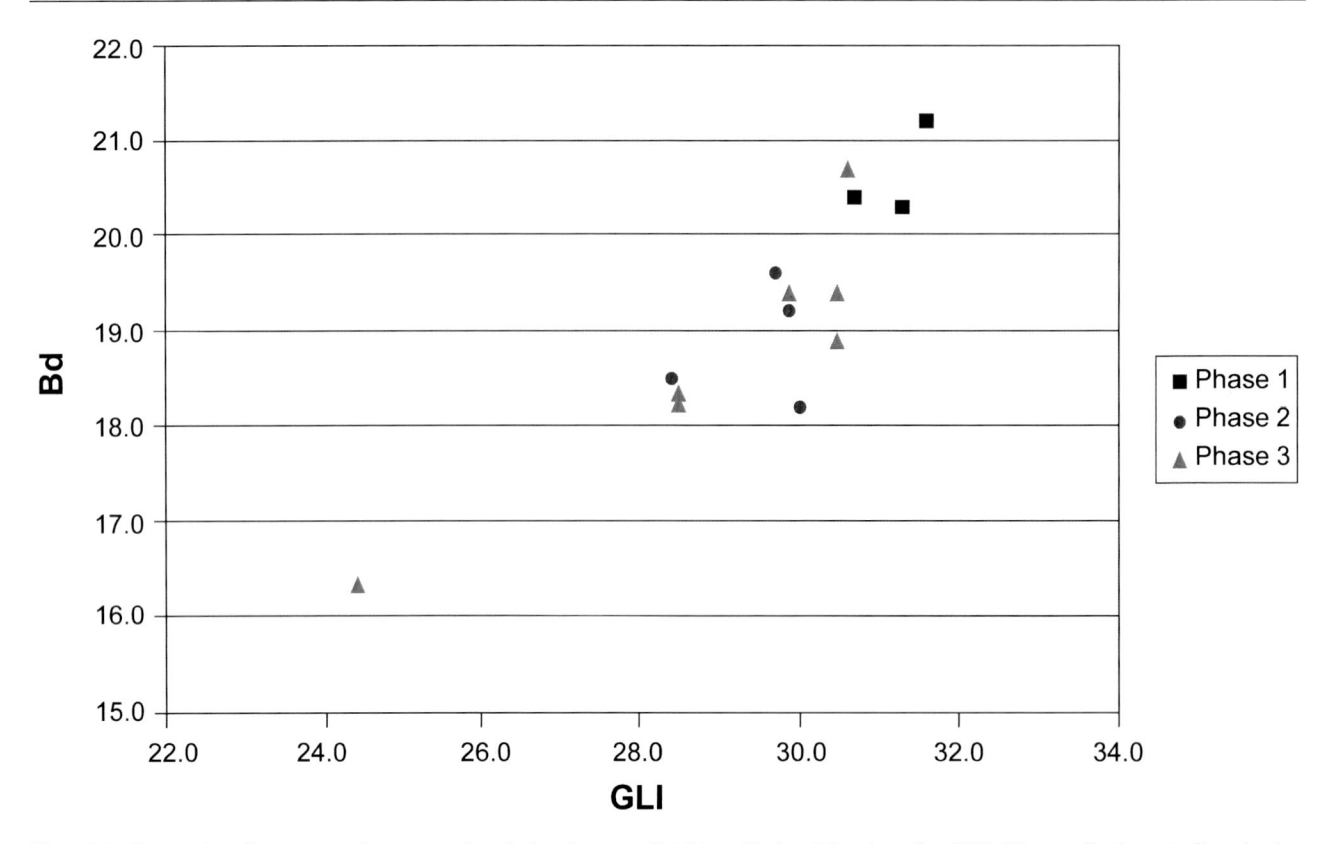

Fig. 11.8 Scatterplot of goat astragalus greatest length (mm) versus distal breadth (mm) by phase for GUS, Western Settlement, Greenland

in Fig. 11.9. This figure does seem to indicate that during the later period of midden formation, the sheep become reduced in size. If the author's assumptions regarding the phasing and dating of this midden could be substantiated, these data would support the hypothesis; however as mentioned above this may not be the case. Further investigation is necessary once a full dating sequence has been determined.

Discussion

It should be emphasized at this point that sample sizes were too small to allow for meaningful statistical significance testing of differences between the phases and that only a visual examination of the data could be made. As such any interpretations made here are necessarily tentative and are made in hope of encouraging further work in this area as larger, well stratified data sets become available. Changes in the size and shape of sheep bones observed for GUS in the Western Settlement may be attributable to environmental factors, although decreasing nutrition alone cannot explain all of the variation. It seems likely that the direct effects of decreasing temperature may have also played a significant role, possibly as a precursor to decreasing levels of nutrition.

However other factors that may cause the observed size changes must also be considered; in particular the sexual make-up of the population and possible changes in the economy over time. The current published data on the GUS animal bone assemblage (Enghoff 2003) makes little assessment of the sex of domestic livestock, sexually dimorphic bones such as pelves and metapodia not being particularly numerous; plots of metapodial measurements give a slight indication of bi-modality (Enghoff 2003). Age profiles were however created and used to infer that the main economic use of sheep and goats was as meat providers; other archaeological evidence indicates that wool was also an important product (Enghoff 2003). Cattle appear to be the main milk providers (Enghoff 2003). Although this gives an indication of the overall economy at GUS, the age profiles were not divided by phase, again probably due to the shortage of available data and therefore no indication can be gained of whether the main economic functions of the animals changed over time. If they were to have changed over

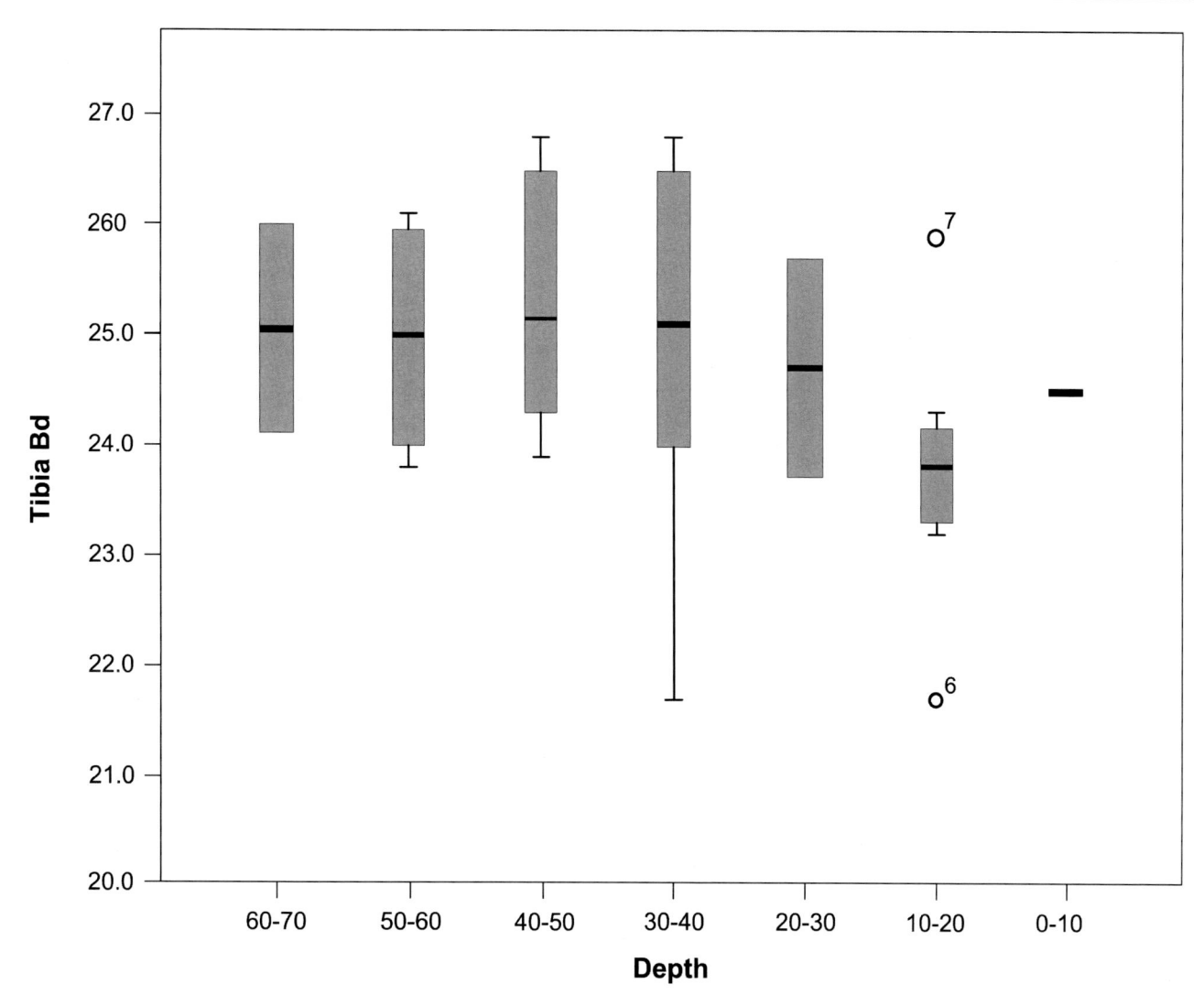

Fig. 11.9 Boxplot of sheep tibia distal breadth (mm) for Ø34, Eastern Settlement, Greenland. *Depth* refers to depth of excavation spit, number of cases: 0–10 = 1, 10–20 = 7, 20–30 = 2, 30–40 = 7, 40–50 = 6, 50–60 = 4, 60–70 = 2

the period of site occupation then it is also possible that the sexual make-up of the livestock populations would have varied over time which may have caused the observed size changes. For instance changes in bone length could be attributed to fluctuations in castration practices, castration tending to cause increased bone length (Davis 2000). However the main element examined here was the astragalus, a bone that shows low levels of sexual dimorphism in sheep and generally is agreed to be a good indicator of overall body size (Noddle 1979; Davis 2000). Therefore it is felt that the observed changes are most likely to be environmentally determined and related to changes in climate in terms of both the direct effects of temperature changes and the reduced availability of quality nutrition.

Buckland et al. (1996) recognized the vulnerable situation that the Norse society had created in Greenland by the beginning of the 14th Century and that any shortening of the growing season or reduction in pasture productivity was potentially devastating to their way of life. The importance of the hay harvest to the Norse Greenlanders is attested to by the widespread finds of fossil synanthropus beetle fauna distinctly associated with the storage of hay (Buckland et al. 1996). Another species linked with the storage of hay is the fly *Delia fabricii* that breeds in meadow grass (*Poa pratensis*) and is often found in samples from Norse farms in Greenland (Buckland et al. 1996). However its absence from the final floor layers of the farm of Nipaatsoq (V54) in the Western Settlement was interpreted as an indication that the hay yields had become poorer (i.e., less grass-rich) towards the end of the settlement (Buckland et al. 1996). If this is true across the Western Settlement then it would certainly have had an effect on the nutrition available for livestock growth. Further evidence for the poor state of grazing comes from analysis of tooth microwear; Mainland (2006) found evidence for high abrasive grazing (principally through soil ingestion) on sheep and goat teeth from GUS and other Norse sites in both the Western and Eastern settlements, indicative of high stocking levels and/or poor pasture management. However these indicators were not only present towards the end of the settlement period but at GUS were found to be present from at least AD 1150 (Mainland 2006).

Caution should be exercised when examining the data from Ø34 due to the assumptions that have been made regarding the dating of the site. However as they stand the data do appear to support the hypothesis of decreasing animal stature towards the end of settlement. Further discussion may be made once a firmer dating structure has been achieved.

What then are the impacts of reduced bone size in livestock animals for the human population? As noted above in situations of poor nutrition, skeletal growth is given precedence over the growth of other later maturing tissues, such as muscle and fat (Dickerson and McCance 1961). This would indicate that where nutrition is reduced the percentage decrease in the weight of soft, edible, tissues would be higher than the decrease in bone weight (or size). Therefore, where there are noticeable decreases in bone size, an animal is likely to have very little fat reserves and poor muscle mass, resulting in an extremely poor carcass (Pálsson and Vergés 1952a) that has a much decreased nutritional value for the humans that slaughter it, relative to an animal of more robust conformation. Even where no reduction in bone size could be detected it is possible that fat a muscle mass could have been reduced (see section on nutrition above). Reductions in the nutritional value of one of their key meat sources would have had severe consequences for the human population in Norse Greenland. Such situations are evidenced at several farms in the Western Settlement where in the final occupation layers the bones of dogs (previously used for hunting) were found disarticulated and butchered (Buckland et al. 1996); these were tentatively interpreted as having been killed and eaten by the final human inhabitants of the farms (McGovern et al. 1983; Buckland et al. 1996).

The differences seen between changes in sheep and goat astragali were particularly interesting. It is postulated that these were associated with the vegetation changes occurring after settlement that may have been favorable to sheep but not goats; by the final phase of the settlement sheep do appear to be the dominant species at GUS (Enghoff 2003). Orlove (2005) noted that sheep were likely to have been advantaged as the vegetation changed from birch and willow to a landscape more dominated by grasses and sedges. Differences in husbandry of the two species may be an alternative or additional cause for the variation. This may relate to the animals being kept on different parts of the landscape, being differently provisioned with winter fodder or shelter, or having a different sexual make-up to the sheep population. Mainland (2006) noted, from evidence of tooth microwear and season of death that goats were more likely to be kept outside over winter than sheep. This may be a possible cause for the decrease in size of goats between phases 1 and 2 as a lack of shelter in periods of harsh weather conditions can be detrimental to growth of livestock as a greater proportion of energy is expended in keeping warm (Smith 1964; McArthur 1991).

Conclusions and Further Work

From the results presented above the following sequence is tentatively proposed for changes in sheep skeletal growth at the site of GUS. After settlement, grassland increases and sheep grow larger given the greater quantity of food resources. As the climate deteriorates and temperatures drop, sheep limb length decreases to conserve body heat. Climatic deterioration also leads to a decrease in quality and quantity of pasture land and hay production and sheep from the latest period of occupation cannot grow to the same size as their predecessors.

It is felt that this study shows some indications that domestic mammal bone size and shape is partly influenced either directly or indirectly by climate. However care must be taken to account for other possible influencing factors such as sex and variations in husbandry practices. To further verify the proposals made in this study several additions and improvements could be made. Detailed examination of the sexual make-up of the sample and examination of any possible changes in husbandry over time would be greatly advantageous in affirming the climatic influence on changes in bone growth. The addition of more securely stratified bone assemblages would also allow for a wider study of the Norse Greenland settlement to be made; larger datasets are certainly needed if any real progress is to be made. The integration of such data with detailed analysis of vegetation and soil changes over the period of settlement would make for a much more robust appraisal of the association of bone size and shape with climate. Examination of factors such as bone cortical thickness may also provide more fruitful results. Overall the potential of biometry in zooarchaeology to enlighten on climate change should be further explored; this is emphasized by its use in other papers in this volume (e.g., Magnell 2017; Monks 2017).

Although the archaeological dataset presented here is small and comes with the inherent problems of small sample size, it is hoped that the theoretical framework and data presented have, to some extent, highlighted the potential of biometrical data to inform on climate and environmental change on both historical and archaeological timeframes. Well stratified midden deposits have the potential to provided datable sequences of material including both wild and domestic fauna, both of which may show reactions to local changes in climate and environment. As middens are inherently human constructs, the animal bones they contain can act as a proxy for the effects of climate and environmental changes observed in, for example, pollen or ice core data, on the humans populations that hunt or raise the animals found within these middens. The use of the log ratio method for pooling data is in some ways essential for increasing sample size but care must be taken not to blur the effect of other factors that influence bone growth and eventual size and shape such as sexual dimorphism and castration.

Acknowledgments I thank Greg Monks for giving me the opportunity to present this work at ICAZ2010 in Paris and the Paddy Coker Research Fund for funding my trip to Paris to attend the conference. I also thank SYNTHESIS for funding a research trip to the Zoological Museum in Copenhagen where a large part of the data used in this paper was collected, and Kim Aaris-Sørensen and Inge Bødker Enghoff of the Zoological Museum for their help and advice during my stay there. Georg Nyegaard is due thanks for donating his bone measurement data from Ø34 and for his advice on the site. Peter Popkin, Polydora Baker and Fay Worley kindly let me look at their results from the English Heritage Sheep Project and discuss them within this paper. Dan Bashford drew the map illustrations Figs. 11.1, 11.2 and 11.3 and Charlotte Davies provided further assistance with graphics. Julie Bond provided academic support throughout my Ph.D. project of which this research forms a small part. I thank Greg Monks, Ola Magnell and Jette Arneborg for their constructive comments on earlier versions of this paper. Finally, I thank the many others who provided help and support throughout my Ph.D. research.

References

Arneborg, J. (2000). Greenland and Europe. In W. W. Fitzhugh & E. I. Ward (Eds.), *Vikings, the North Atlantic saga* (pp. 304–317). Washington: Smithsonian Institute.

Arneborg, J. (2003a). Norse Greenland: Reflections on settlement and depopulation. In J. H. Barrett (Ed.), *Contact, continuity and collapse. The Norse colonisation of the North Atlantic* (pp. 163–181). Turnhout: Brepols.

Arneborg, J. (2003b). The archaeological background. In I. B. Enghoff (Ed.), *Hunting, fishing and animal husbandry at the farm beneath the sand, Western Greenland* (Vol. Meddelelser om Grønland, Man and Society 28, pp. 9–17). Copenhagen: Danish Polar Centre with SILA.

Arneborg, J., Heinemeier, J., Lynnerup, N., Nielsen, H. L., Rud, N., & Sveinbjörnsdóttir, Á. E. (1999). Change of diet of the Greenland Vikings determined from stable carbon isotope analysis and 14C dating of their bones. *Radiocarbon, 41*(2), 157–168.

Banks, M. (1975). *Greenland*. Newton Abbot: David & Charles Ltd.

Barlow, L. K. (1994). *Evaluation of seasonal to decadal scale deuterium and deuterium excess signals, GISP2 ice core, Summit, Greenland, AD 1270–1985*. Ph.D. Dissertation, University of Colorado.

Barlow, L. K., White, J. W. C., Johnson, S., Jouzel, J., & Grootes, P. M. (1993). Climate variability during the last 1000 years from delta Deuterium and delta ^{18}O in the GISP2 and GRIP deep ice cores. In *EOS, Transactions of the American Geophysical Union, fall meeting supplement, 1993* (pp. 118).

Bell, M., & Walker, M. J. C. (1992). *Late quaternary environmental change. Physical and human perspectives*. Harlow: Longman.

Benzie, D., Boyne, A. W., Dalgarno, A. C., Duckworth, J., Hill, R., & Walker, D. M. (1955). Studies of the skeleton of the sheep I. The effect of different levels of dietary calcium during pregnancy and lactation on individual bones. *Journal of Agricultural Science, 46*, 425–440.

Berglund, J. (2000). The Farm Beneath the Sand. In W. W. Fitzhugh & E. I. Ward (Eds.), *Vikings, the North Atlantic saga* (pp. 295–303). Washington: Smithsonian Institution Press.

Buckland, P. C., Amorosi, T., Barlow, A., Dugmore, A., Mayewski, P., McGovern, T. H., et al. (1996). Bioarchaeological and climatological evidence for the fate of Norse farmers in medieval Greenland. *Antiquity, 70*, 88–96.

Clutton-Brock, J., Dennis-Bryan, K., Armitage, P. L., & Jewell, P. A. (1990). Osteology of the Soay sheep. *Bulletin of the British Museum of Natural History (Zool), 56*(1), 1–56.

Davis, S. J. M. (1981). The effects of temperature change and domestication on the body size of Late Pleistocene to Holocene mammals of Israel. *Paleobiology, 7*(1), 101–114.

Davis, S. J. M. (2000). The effect of castration and age on the development of the Shetland Sheep skeleton and a metric comparison between bones of males, females and castrates. *Journal of Archaeological Science, 27*(5), 373–390.

Degerbøl, M. (1934). Animal bones from the Norse ruins at Brattahlid. *Meddelelser om Grønland, 88*(1), 149–155.

Degerbøl, M. (1941). The osseous material from Austmannadal and Tungmeralik, West Settlement. Greenland. *Meddelelser om Grønland, 89*(1), 345–354.

Degerbøl, M. (1943). Animal bones from the inland farms in the East Settlement. *Meddelelser om Grønland, 90*(1), 113–119.

Dickerson, J. W. T., & McCance, R. A. (1961). Severe under nutrition in growing and adult animals 8. The dimensions and chemistry of the long bones. *British Journal of Nutrition, 15*, 567–576.

Dugmore, A., Borthwick, D. M., Church, M., Dawson, A., Edwards, K. J., Keller, C., et al. (2007). The role of climate in settlement and landscape change in the North Atlantic islands: An assessment of cumulative deviations in high-resolution proxy climate records. *Human Ecology, 35*, 169–178.

Edwards, K. J., Schofield, J. E., & Mauquoy, D. (2008). High resolution paleoenvironmental and chronological investigations of Norse landnám at Tasiusaq, Eastern Settlement, Greenland. *Quaternary Research, 69*, 1–15.

Enghoff, I. B. (2003). *Hunting, fishing and animal husbandry at The Farm Beneath the Sand, western Greenland. An archaeozoological analysis of a Norse farm in the Western Settlement* (Vol. Meddelelser om Grønland, Man and Society 28). Copenhagen: Danish Polar Centre with SILA.

Geist, V. (1987). Bergmann's rule is invalid. *Canadian Journal of Zoology, 65*, 1035–1038.

Grove, J. M. (2001). The initiation of the "Little Ice Age" in regions round the North Atlantic. *Climatic Change, 48*, 53–82.

Guldager, O., Hansen, S. S., & Gleie, S. (2002). *Medieval farmsteads in Greenland. The Brattahlid region 1999–2000*. Copenhagen: The Danish Polar Centre.

Guthrie, R. D. (1984). Mosaics, allelochemicals and nutrients. An ecological theory of Late Pleistocene megafaunal extinctions. In P. S. Martin & R. G. Klein (Eds.), *Quaternary extinctions: A prehistoric revolution.* (pp. 259–298). Tuscon, Arizona: University of Arizona Press.

Hammond, J. (1932). *Growth and the development of mutton qualities in the sheep*. Edinburgh: Oliver and Boyd.

Hugget, R. J. (2004). *Fundamentals of biogeography* (2nd ed.). Abingdon: Routledge.

Lister, A. M. (1995). Sea-levels and the evolution of island endemics: The dwarf red deer of Jersey. In R. C. Preece (Ed.), *Island Britain: A Quaternary perspective* (Vol. 96, pp. 151–172, Geological Society Special Publication). Geological Society.

Lister, A. M. (1996). Dwarfing on island elephants and deer: Process in relation to time of isolation. *Symposium of the Zoological Society of London, 69*, 277–292.

Lister, A. M. (1997). The evolutionary response of vertebrates to Quaternary environmental change. In B. Huntley, W. Cramer, A. V. Morgan, H. C. Prentice, & J. R. Allen (Eds.), *Past and future rapid environmental changes: The spatial and evolutionary responses of terrestrial biota* (pp. 287–302). Berlin: Springer.

Magnell, O. (2017). Climate and wild game populations in south Scandinavia at Holocene Thermal Maximum. In G. Monks (Ed.) *Climate change and human responses: A zooarchaeological perspective* (pp. 123–135). Dordrecht: Springer.

Mainland, I. (2006). Pastures lost? A dental microwear study of ovicaprine diet and management in Norse Greenland. *Journal of Archaeological Science, 33*, 238–252.

Marshall, L. G., & Corruccini, R. S. (1978). Variability, evolutionary rates and allometry in dwarfing lineages. *Paleobiology, 4*(2), 101–119.

Mayewski, P. A., Meeker, L. D., Morrison, M. C., Twickler, M. S., Whitlow, S., Ferland, D. A., et al. (1993). Greenland ice core 'signal' characteristics: An expanded view of climate change. *Journal of Geophysical Research, 98*(D7), 12839–12847.

McArthur, A. J. (1991). Forestry and shelter for livestock. *Forest Ecology and Management, 45*, 93–107.

McCance, R. A., Owens, P. D. A., & Tonge, C. H. (1968). Severe under nutrition in growing and adult animals 18. The effects of rehabilitation on the teeth and jaws of pigs. *British Journal of Nutrition, 22*, 357–368.

McGovern, T. H. (1981). The economics of extinction in Norse Greenland. In T. M. L. Wigley, M. J. Ingram, & G. Farmer (Eds.), *Climate and history. Studies in past climates and their impact on man* (pp. 404–433). Cambridge: Cambridge University Press.

McGovern, T. H. (1985). Contributions to the palaeoeconomy of Norse Greenland. *Acta Archaeologica, 54*, 73–122.

McGovern, T. H. (1992). Bones, buildings and boundaries: Palaeoeconomic approaches to Norse Greenland. In C. D. Morris & D. J. Rackham (Eds.), *Norse and later settlement and subsistence in the North Atlantic* (pp. 193–230). Glasgow: University of Glasgow, Department of Archaeology.

McGovern, T. H. (2000). The demise of Norse Greenland. In W. W. Fitzhugh & E. I. Ward (Eds.), *Vikings, the North Atlantic saga* (pp. 327–339). Washington: Smithsonian Institution Press.

McGovern, T. H., Buckland, P. C., Savory, D., Sveinbjarnardottir, G., Andreason, C., & Skidmore, P. (1983). A study of the floral and faunal remains from two Norse farms in the Western Settlement, Greenland. *Arctic Anthropology, 20*, 93–120.

McNab, B. K. (1971). On the ecological significance of Bergmann's rule. *Ecology, 52*(5), 845–854.

McNab, B. K. (1990). The physiological significance of body size. In J. Damuth & B. J. MacFadden (Eds.), *Body size in mammalian palaeobiology. Estimation and biological implications*. Cambridge: Cambridge University Press.

Meadow, R. H. (1999). The use of size index scaling techniques for research on archaeozoological collections from the Middle East. In C. Becker, H. Manhart, J. Peters, & J. Schibler (Eds.), *Historia animalium ossibus. Festschrift für Angela von den Driesch* (pp. 285–300). Rahden/Westf: Verlag Marie Leidorf GmbH.

Monks, G. G. (2017). Evidence of changing climate and subsistence strategies among the Nuu-chah-nulth of Canada's west coast. In G. G. Monks (Ed.) *Climate change and human responses: A zooarchaeological perspective* (pp. 173–196). Dordrecht: Springer.

Noddle, B. A. (1979). A brief history of domestic animals in the Orkney Islands, Scotland, from the 4th Millennium B.C. to the 18th Century. In M. Kubasiewicz (Ed.), *Proceedings of the 3rd International Archaeozoology Conference, 23rd April 1978, Szczecin, Poland*. Szczecin: Szczecin Agricultural Academy.

Nyegaard, G. (1996). Investigation of a Norse refuse layer in Qorlortup Itinnera, South Greenland. In J. Berglund (Ed.), *Archaeological field work in the Northwest Territories, Canada, and Greenland in 1995* (Vol. Archaeology Report No. 17, pp. 71–73). Yellowknife: Prince of Wales Northern Heritage Centre.

Orlove, B. (2005). Human adaptation to climate change: A review of three historical cases and some general perspectives. *Environmental Science & Policy, 8*, 589–600.

Pállson, H., & Vergés, J. B. (1952a). Effects of the plane of nutrition on growth and the development of carcass quality in lambs. Part I. The effect of high and low planes of nutrition at different ages. *Journal of Agricultural Science, 42*(1), 1–92.

Pállson, H., & Vergés, J. B. (1952b). Effects of the plane of nutrition on growth and development of carcass quality in lambs. Part II. Effects on lambs of 30 lb. carcass weight. *Journal of Agricultural Science, 42*(2), 93–149.

Pratt, C. W. M., & McCance, R. A. (1964). Severe under nutrition in growing and adult animals 14. The shafts of the long bones in pigs. *British Journal of Nutrition, 18*, 613–624.

Ross, J. M. (1997). *A paleoethnobotanical investigation of garden under Sandet, a waterlogged Norse farm site. Western Settlement, Greenland (Kaiaallit Nunaata)*. M.A. Thesis, University of Alberta.

Schofield, J. E., Edwards, K. J., & Christensen, C. (2008). Environmental impacts around the time of Norse landnám in the Qorlortoq valley, Eastern Settlement, Greenland. *Journal of Archaeological Science, 35*, 1643–1657.

Schreider, E. (1950). Geographical distribution of the body-weight/body-surface ratio. *Nature, 165*, 286.

Schüle, W. (1993). Mammals, vegetation and the initial human settlement of the Mediterranean islands: A palaeoecological approach. *Journal of Biogeography, 20*, 399–412.

Smith, C. V. (1964). A quantitative relationship between environment, comfort and animal productivity. *Agriculture and Meteorology, 1*(4), 249–270.

von den Driesch, A. (1976). *A guide to the measurement of animal bones from archaeological sites*. (Vol. 1, Peabody Museum Bulletin). Harvard: Peabody Museum of Archaeology and Ethnology, Harvard University.

Weaver, M. E., & Ingram, D. L. (1969). Morphological changes in swine associated with environmental temperature. *Ecology, 50*(4), 710–713.

Part IV
Overview and Retrospective

Chapter 12
Zooarchaeology in the 21st Century: Comments on the Contributions

Daniel H. Sandweiss

Abstract In this chapter, I comment on all the papers in this volume. My discussion is organized around cross-cutting themes that appear in different chapters: collation, correlation, and causation; temporal, spatial, and analytic scale, new data on extinctions, the response to climatic change, archaeological data as climate proxies, and gaps in the record. Though recognizing areas that require reorientation and reconsideration, I conclude that the papers make important contributions to understanding human-climate-environment interaction as seen through the lens of zooarchaeology.

Keywords Causation • Scale • Extinctions • Climatic change • Climate proxies

Zooarchaeology (aka archaeozoology) has deep roots in the 19th century and a history of engagement with the burning questions of the day (Lyman 2005). Through time, those questions have multiplied and become more complex (Reitz and Wing 2008). The papers in this volume amply represent the multiplicity of contributions that 21st century studies of animal remains from archaeological sites offer to fields as diverse as anthropological archaeology, paleoclimatology, paleontology, and wildlife management.

In the interest of full disclosure, I was not present at the originating conference and did not participate in the discussions of the original papers. Also, I am at best a quasi-zooarchaeologist – I have worked with mollusks and with vertebrate zooarchaeologists, but I am really a geoarchaeologist focusing on climate change and maritime adaptations, mainly in coastal Peru. As an outsider, my comments examine six topics that particularly caught my attention and represent the lessons I took from the collection:

1. Collation, correlation, and causation.
2. Scale (temporal, spatial, and analytic).
3. New data on extinctions.
4. Lessons on the response to climatic change.
5. Archaeological remains as climate proxies.
6. Gaps in the record and open questions.

As Monks notes in his introduction (2017a), a principal focus of the papers is human-animal-environment interactions. This topic is well covered in the papers, with appropriately nuanced caution, and I do not address it here except as it impinges on the six categories listed above.

Collation, Correlation, and Causation

"Collation involves showing that events occurred simultaneously. Correlation, in formal terms, means demonstrating that events of interest co-vary in a statistically significant way. Causation means showing that outcomes are the necessary result of specific conditions and processes – that differences have occurred that make a difference."

Sandweiss and Quilter (2008:3)

This is not a new concern for archaeologists of any kind who wish to make interpretations about the past. Nevertheless, the degree of confidence in our understanding of past behavior of the climate, of animals, or of people depends on distinguishing between these different levels of relationship between data (however defined) and interpretation. Though most authors in this volume are admirably cautious about interpretation, few discuss causal or lower-order relations in detail. Even though it may appear to remove authority from the results, I urge the authors of these chapters (and others) to be explicit about interpretive uncertainties. This is particularly important in a volume "intended to bring [zooarchaeological studies] to the attention of natural historians (ecologists, climatologists, zoologists, botanists)" (Monks 2017a).

D.H. Sandweiss (✉)
Department of Anthropology, University of Maine, 5773 South Stevens Hall, Orono, ME 04469-5773, USA
e-mail: Dan_Sandweiss@umit.maine.edu

Monks (2017b) addresses issues of correlation and causation most explicitly in his discussion of climate change and Nuu-chah-nulth. He frames his study as a test of the hypothesis that "the Nuu-chah-nulth shifted their fishing emphasis from rockfish to salmon and herring as a result of climatic change from the MCA to the LIA". Despite suggestive results, enough uncertainty remains for Monks to recognize that "evidence presented here consists more of correlation than causation" (I would argue that collation is a more appropriate term, given the low number of cases).

Scale

Understanding the effects of temporal, spatial, and analytic scales is essential to interpreting the archaeological (and zooarchaeological) record. Time-averaging (e.g., Rollins et al. 1990:469) makes a difference – are we looking at time slices that cover a season? A year? A decade? A century? This issue is not addressed explicitly in the papers in this volume, except by Belmaker (2017) in passing, yet it is often critical in the interpretation of the interactions between humans, the biota, and the physical environment, as well as of other aspects of past behavior.

The last several decades of paleoclimatic research have demonstrated that abrupt change occurred in the past, sometimes on the order of magnitude of 10^{-1} years (e.g., Alley et al. 1993). Under current conditions of global change, we may cross thresholds that lead to abrupt change in the near future (e.g., Alley et al. 2003). This is an area in which zooarchaeological research on past periods of abrupt change could provide guidance to other disciplines, such as the target audience of "ecologists, climatologists, zoologists, botanists" mentioned by Monks in his introduction to this volume (2017a). Of course, understanding the effects of abrupt climatic change requires us to excavate sites and employ dating techniques that allow us to distinguish deposits at or near the scale of the change – a difficult but not impossible task. For instance, Elera et al. (1992) found molluscan bioindicators of one or a series of El Niño events in Puemape, a first millennium BC site in northern Peru; this abrupt event (less than 2 year duration) is associated with abandonment and reoccupation by people using a different pattern and style of cultural material (Elera 1993:251).

Spatial scale also influences what we learn from the zooarchaeological record. A case in point is Yacobaccio et al.'s (2017) recognition of the different settlement and animal use histories for the Chilean and Argentine sides of the southern altiplano of Andean South America. Rindel et al. (2017) employ an explicitly regional scale of analysis in their study of settlement logistics during the Medieval Climatic Anomaly in southern Patagonia and make a clear case that a microregional perspective would not answer the research questions about the seasonal and functional distribution of sites. They use multiple zooarchaeological analyses to assess and support an archaeological model for differential use of highland and lowland locales. Their conclusion that a regional scale is "appropriate…to monitor climate change and cultural change" makes sense, as long as we keep in mind that effective regional comparisons depend on detailed data from multiple individual sites throughout the region – the sort of data that Rindel et al. marshaled for their study. Perhaps it would be clearest to say that effective monitoring of both synchronic and diachronic patterns of culture and climate necessarily involves a dialogue between multiple spatial scales, from site components to regions, and with appropriate and comparable chronological control across the sites and periods of interest. This is hardly novel, but worth restating given the excellent example provided by Rindel et al.

Analytic scale (e.g., mesh size, sample size) and its role in archaeological interpretation are old and well-discussed topics in zooarchaeology (e.g., Grayson 1984; see Lyman 2005; Reitz and Wing 2008; cf. for instance Carlson 1996 on Bourque 1995). Many of the papers discuss sample size, mainly to point out the inadequacy of available or studied samples to give definitive results. I was surprised at the lack of explicit consideration of sample recovery techniques such mesh size, sampling strategy, and so on; indeed, there is little to no discussion of the processes (e.g., Sandweiss 1996:130–131) that produced the samples analyzed in the various papers. There are some partial exceptions: for instance, Belmaker (2017) notes that "[s]ince method of excavation highly affects both species richness and abundance, we can expect fauna richness to differ among sites with different collection protocols." Although she controls for under-sampling, she does not further discuss specific implications of the excavation methods that produced the sample for her meta-analysis. Rindel et al. (2017) mention in a footnote that their samples had been studied previously and "evaluated also from a taphonomic and site formational perspective; consequently, these discussions do not appear in" their paper. Ochoa and Piper (2017) summarize the stratigraphic context but do not discuss how the samples were selected and extracted. I understand that the papers are largely by zooarchaeologists who did not excavate the samples, but effective interpretation requires as complete a context as possible (for instance, see Reitz et al. 2015: 159–160 on the effects of mesh size on early coastal site samples from southern Peru).

New Data on Extinctions

Two of the papers in this volume provide new data on mammalian extinctions during the Late Quaternary. This research links current zooarchaeology to its 19th century origins (Lyman 2005). Ochoa and Piper provide data from two cave sites on Palawan Island in the Philippines that suggest the disappearance of three to four species including two deer species. The larger deer appears to have gone extinct in the Early Holocene while the smaller deer survived at least until the Metal Period (ca. 2000 BP). Scant remains of tiger suggest it was present only through the Early Holocene. Canid remains from Terminal Pleistocene and Early Holocene deposits at Ille Cave are morphometrically different from domestic dog remains in the Late Holocene deposits, suggesting that the earlier remains are dhole (*Cuon alpinus*), probably wild. The authors note that deer are extinct on the island now and not present in written records, and tiger and dhole are also extinct on Palawan now. Ochoa and Piper argue that a combination of natural (climatic and environmental change) and anthropogenic (hunting) pressures likely accounted for the local extinction of the larger deer, tiger, and dhole. For the later disappearance of the smaller deer, they suggest anthropogenic forces, since this species survived multiple cycles of climatic change previously. This is a case in which argument from absence has some validity, since the end result (local extinction of all four species) is known. The exact timing of disappearance remains uncertain; Ochoa and Piper provide comparative data from other sites in the broader region to support the chronology from Ille and Pasimbahan Caves, but greater confidence will require much larger data sets, as always with extinction studies. The authors helpfully conclude with suggestions for future research arising from this study.

The paper by Ferrusquia-Villafranca et al. (2017) also explores extinction. They compare Late Pleistocene and modern mammal inventories of southern North America (Mexico) to address the possible causes of one of the best-known mass extinctions. In their analysis, the cooler and moister climate of the end of the Pleistocene "made possible a biotic diversity unparalleled anywhere in Mexico at present", while warming and drying at the onset of the Holocene depleted and modified the composition of the biota. Climatic data come from multiple natural proxies, including pollen and soils. They also note that the habitat and diet requirements of the fauna provide information on the climate and environment that is generally in accord with the other proxies.

Although the availability of paleoenvironmental data varies in the different regions in the study, Ferrusquia-Villafranca et al. find a general collation across the broader region while also pointing to gaps that need to be filled (see final section). They see similar gaps in the early archaeological record of the study zone; while recognizing that the paucity of data make any interpretation suspect, they do suggest that the extinctions (local and general) were not caused by people. In support, they point to the millennia of co-existence between humans and some medium to large mammals (see also Lima-Ribeiro and Diniz-Filho 2013).

As with Ochoa and Piper (and any extinction study), the Ferrusquia-Villafranca et al. conclusions suffer from the "absence of evidence is not evidence of absence" problem (Lyman 1995). To their credit, they address this issue specifically and reject the Tyler-Faith and Surovell (2009) argument because it is based on postulating negative evidence (missing fossils) due to sampling error. In fact, both sides of this old argument are based on negative evidence, one side postulating that it exists and the other that it does not. Occam's razor suggests that, with caution, we should probably support Ferrusquia-Villafranca et al. and others, given that (as on Palawan), in the end the species in question did not survive to the present, at least in the study region, and must have gone locally or totally extinct sometime between the most recent fossil and the present.

Lessons on the Response to Climatic Change

It is increasingly easy and common to identify climatic change both directly in the archaeological record (see next section, below) and in natural proxies of relevance to past human groups. The papers in this collection demonstrate that zooarchaeologists have become increasingly adept at summarizing and incorporating multi-proxy climate data to inform their analyses; this is a strength of the volume. With the increased use of AMS for radiocarbon dates and more sophisticated sample collection and selection processes and pre-treatments (e.g., Taylor and Bar-Yosef 2014), control over chronology in natural and anthropogenic archives has improved. The advent of new dating techniques such as OSL (e.g., Waters et al. 2011; Huckleberry et al. 2012) and cosmogenic surface exposure-age dating (e.g., Bromley et al. 2011) permit the dating of site strata and climatically driven landscape features that previously could only be assigned relative ages. As a result of these advances, temporal collation between climate and people in the past has improved. Nevertheless, human nature intervenes between the demonstration of contemporaneity in cultural and climatic events and interpretation of the human response – in other words, in the (prehistoric) moment, people see and respond to external events such as climatic change through cultural lenses and with limited knowledge based on local observations. To interpret the human response to climatic change, we need to understand past cultural logic that would have conditioned responses to such stimuli – a difficult task, especially for pre-literate societies. We also need to consider what prehistoric people would have known about scale and trend in climatic change.

Given, then, that the human response to climatic and environmental change is always culturally nuanced (e.g., Sandweiss et al. 2001:605) and possibly unpredictable, it should not be a surprise that there *are* no big surprises in this volume's contributions to lessons on the human response to climatic change. However, extreme events do require some response, and papers in this volume provide useful examples.

Yacobaccio et al. (2017) show that the onset of aridity in the Mid-Holocene co-varied with change in patterns of animal acquisition on the southern Altiplano of Chile and Argentina. At the same time, they demonstrate the effects of historical trajectory and cultural idiosyncrasy by showing that the response to aridity differed on the eastern and western sides of this southern Altiplano region. While it is at first difficult to imagine how we might use adaptations from a time of hunter-gatherers with radically lower population densities than today, I would argue here for the value of basic research. Archaeologists have only begun to engage explicitly with sustainability science over the last decade (see discussion and examples in Hudson 2013). Indeed, this inter-discipline is only a few decades old and is still being defined (e.g., Kates 2011). Those interested in finding contemporary and future sustainability solutions may well see ways to use the successful strategies of the Mid-Holocene as well as other prehistoric periods, and it is important that we provide these examples.

Pilaar Birch and Miracle (2017) consider the relative merits of diet stress and increasing proximity of the shoreline as sea level rose (increasing resource availability) as explanations for the recent appearance of marine foods in Mid-Holocene strata in sites on the northeastern Adriatic, in particular in the Vela Špilja faunal collection. This collection only partially supports their expectations following optimal foraging theory, as they do not clearly see an intensification of carcass processing. Again, larger samples would help resolve the discrepancy one way or the other, as the authors readily admit. This paper contributes to the necessary and on-going dialogue between data and theory.

Magnell (2017) provides a different perspective by using the zooarchaeological record in Sweden to offer advice on wildlife management under conditions of climatic change. By looking at the response of wild game populations (as reflected in zooarchaeological collections) to the Holocene Thermal Maximum, he offers lessons from the great natural experiment of the Mid-Holocene that may be of relevance in the near future under conditions of global warming. Magnell is careful to point out that human behavior may also account for some of the changes in size and species composition; in other words, these are not entirely independent climatic proxies. In 50 years, we will probably know whether his predictions from the past (decrease in moose relative to deer, decrease in deer body size, and increase in wild boar size) are right.

Archaeological Remains as Climate Proxies

Archaeological sites often contain climate proxies that can be used to reconstruct regional paleoclimate (Sandweiss and Kelley 2012); Thomas McGovern (personal communication 2014) recently characterized archaeological sites as "long term distributed observing networks". Frequently, those proxies are zooarchaeological. Properly interpreted, such data are among the most important contributions we can make to our allied disciplines. Often, archaeological sites can provide a record where there are no nearby natural proxies. The Pacific coast of Peru is such a case. As a tropical coast, the nearest ice cores are inland and respond in part to Atlantic climate. As a cold water coast bathed by the Antarctic Humboldt Current, there are no corals. As a desert coast, there are no lakes. Offshore cores have provided some information but often lack the critical Mid-Holocene record. As a result, significant work on climatic issues such as Late Pleistocene to modern variability in El Niño frequency have come from archaeological sites, mainly from fish and mollusks (see Sandweiss et al. 2007; Sandweiss and Kelley 2012; cf. Carré et al. 2014 for alternative results also from archaeological proxies).

There is sometimes resistance to accepting archaeologically-derived climatic records; for instance, Cullen et al. (2000:379) referred to archaeological proxy data as "inherently subjective". As I have argued elsewhere, archaeological data are no more subjective or anecdotal than any other proxy (Sandweiss and Kelley 2012:385). All records respond to formation processes, whether they be natural, anthropogenic, or a combination of the two. Given the skepticism, however, it is incumbent on us to explain how we have accounted for these processes so that our results contribute to the broader study of past climate and therefore to the prediction of future climate.

Many of the papers in this volume contribute to reconstructing past climates using a variety of approaches involving animal remains. Some are more successful than others as a result of sample size and/or the nature of the proxy used. It is critically important that the archaeological proxies chosen be independent, that is, that they reflect climatic conditions directly, not through human intervention. Thus, presence of indicator species works, while arguments from absence are much weaker (see above, section on extinctions). Isotopic values from zooarchaeological remains are another source of independent information on climate used often in this volume. In all cases, the proxies must be calibrated by reference to modern samples, and a multi-proxy approach is more robust for archaeological remains just as it is for natural archives.

Cussans (2017) uses biological studies of the effects of temperature and nutrition on mammal body size to hypothesize that deteriorating climate during the Norse occupation of Greenland should be expressed in changes in bone growth due to temperature change, availability of nutrition (shorter growing season), or both. Measurements on sheep fit the hypothesis well, providing support for climatic models from natural proxies such as ice cores. Goat bones, however, have a different pattern. Cussans explores the possible reasons for the variation in growth between the two species; her discussion of the possible influence of cultural patterns and her recognition of inadequate sample size provide important caveats to the interpretation of the bone measurements. I believe that she has made a case for "the potential of biometry in zooarchaeology to enlighten on climate change" and support her call for continued work along this line.

Yacobaccio et al. (2017) begin by creating a consensus summary of Early to Mid-Holocene climate for the southern altiplano of Chile and Argentina using published studies of natural archives. By comparing the abundance of different species through time and across space, they make interesting suggestions concerning the impact of Mid-Holocene aridity events both on cultural practices concerning animal acquisition and on the actual distribution of prey populations. Because human behavior is a possible (or probable) explanation for some of the changes noted in the distribution of the remains by species, age, and size, these data cannot be used to reconstruct climate without running the risk of circular reasoning. However, Yacobaccio et al. (2017) also explore climatic nuances using C and N isotopes from archaeological camelid bones. The $\partial^{13}C$ data suggest that despite enhanced aridity in the Mid-Holocene, vegetational communities in the southern altiplano changed little in terms of composition and altitudinal range. This finding could be tested with pollen data from natural catchments, if available. The $\partial^{15}N$ show a change from the Early to the Mid Holocene that probably resulted from the effects of aridity on soils. Again, this might be tested by soil scientists if appropriately aged paleosols are available in the study region.

In their isotopic study of Late Holocene shell middens in Nicaragua, Colonese et al. (2017) use isotopic analysis of *Polymesoda* shells to assess climate and seasonality. Appropriately, they analyzed modern shells that lived under known conditions to provide baseline data for interpretation. They find that ^{18}O enrichment signals more saline conditions while ^{18}O depletion indicates less saline, more freshwater conditions. After analyzing a small sample of archaeological shells, Colonese et al. conclude that climatic conditions may have been different during the Late Holocene compared with today: "~ 2 k BP may have experienced different atmospheric-hydrological or nutrient conditions". In addition to useful information on seasonality, the results of this study support previous work indicating that isotopic analysis of a larger sample of *Polymesoda* could help resolve paleoclimate questions related to the seasonal migration of the Intertropical Convergence Zone (ITCZ). Further studies should include additional samples from different parts of the lagoon, including different depths, to test the suggestion that some of the variation is environmental rather than climatic. I also question the use of standard sized (~ 30 mm) shells, especially for seasonality studies. If growth rate is fairly consistent, shells of a standard size may have been harvested at the same season and therefore may not represent the full range of seasons of collection/occupation.

Belmaker (2017) addresses the hypothesis that climatic change caused the Neanderthal extinction in the Levant during MIS 4-3 using a meta-analysis of faunal data from archaeological sites in this region. She concludes that faunal differences between sites reflect the environmental mosaic and not climatically driven temporal variation, and therefore the climatic change evidenced in multiple natural proxy records was too low amplitude to be the driver of regional Neanderthal extinction. This paper introduces into zooarchaeology a new measure of climatic change: "*effective climate change*", based not on the absolute value of changes in physical variables such as temperature and precipitation, but rather (following Rahel's (1990) tiered response of biota to change) how effectively climatic change alters the biome in ways that affect the top predator (hominins). This is an approach worth pursuing elsewhere.

By Way of Conclusion: Gaps in the Record and Open Questions

One of the greatest strengths of this volume is the general and often explicit recognition that each study is just a step toward greater knowledge, with many gaps in the record and unresolved questions. This is an appropriate approach to any scientific endeavor, and all the more so for the paleo-disciplines in which samples are small, subject to known and unknown biases, unreplicable at first order, and observable only as trace evidence of past conditions, organisms, or behaviors.

Ferrusquia-Villafranca et al. (2017) most consistently recognize gaps in the record (of Late Pleistocene fauna, environment, climate, and human settlement). They also set an agenda to fill those gaps by pointing out the kinds of research needed. Pilaar Birch and Miracle and Ochoa and Piper also point explicitly to the necessity of future research to advance the goal of this volume to contribute to "understanding general patterns of human response to climate change through time" (Pilaar Birch and Miracle 2017). Similarly, Monks in his substantive chapter in this volume (2017b), raises a series of unanswered questions for continued investigation into the human eco-dynamics of ancient Nuu-chah-nulth culture: Why are sea urchins so scarce during the MCA (Medieval Climatic Anomaly)? Why were mackerel avoided (or were they)? Why aren't there more sardines? These questions, like others in the volume, strike at the core of human-climate-environment interaction as seen through the lens of zooarchaeology. At the outset of this chapter, I said that I would not address such interactions directly. This decision was not only due to my personal interests but also because the authors throughout the volume did a fine job in addressing this core question and in pointing the way towards improved understanding.

Acknowledgments I thank Greg Monks for inviting me to take on this fascinating task, and the three reviewers for very helpful suggestions and comments. However, responsibility for errors and opinions rests with me.

References

Alley, R. B., Meese, D., Shuman, C. A., Gow, A. J., Taylor, K., Ram, M., et al. (1993). Abrupt accumulation increase at the Younger Dryas termination in the GISP2 ice core. *Nature, 362*, 527–529.

Alley, R. B., Marotzke, J., Nordhaus, W. D., Overpeck, J. T., Peteet, D. M., Pielke, Jr., R.A., et al. (2003). Abrupt climate change. *Science, 299*, 2005–2010.

Belmaker, M. (2017). The southern Levant during the last glacial and zooarchaeological evidence for the effects of climatic-forcing on hominin population dynamics. In G. G. Monks (Ed.), *Climate change and human responses: A zooarchaeological perspective* (pp. 7–25). Dordrecht: Springer.

Bourque, B. J. (1995). *Diversity and complexity in prehistoric maritime societies: A Gulf of Maine perspective*. New York: Plenum Press.

Bromley, G. R. M., Hall, B. L., Schaefer, J. M., Winckler, G., Todd, C. E., & Rademaker, K. M. (2011). Glacier fluctuations in the southern Peruvian Andes during the late-glacial period, constrained with cosmogenic 3He. *Journal of Quaternary Science, 26*, 37–43.

Carlson, C. (1996). Review of "Diversity and complexity in prehistoric maritime societies: A Gulf of Maine perspective" by Bruce Bourque. *American Antiquity, 61*, 621–622.

Carré, M., Sachs, J. P., Purca, S., Schauer, A. J., Braconnot, P., Angeles Falcón, R., et al. (2014). Holocene history of ENSO variance and asymmetry in the eastern tropical Pacific. *Science, 345*, 1045–1048.

Colonese, A., Clemente, I., Gassiot, E., & López-Sáez, J. A. (2017). Oxygen isotope seasonality determinations of marsh clam shells from prehistoric shell middens in Nicaragua. In G. G. Monks (Ed.), *Climate change and human responses: A zooarchaeological perspective* (pp. 139–152). Dordrecht: Springer.

Cullen, H. M., deMenocal, P. B., Hemming, S., Hemming, G., Brown, F. H., Guilderson, T., et al. (2000). Climate change and the collapse of the Akkadian empire: Evidence from the deep sea. *Geology, 28*, 379–382.

Cussans, J. (2017). Biometry and climate change in Norse Greenland: The effect of climate on the size and shape of domestic mammals. In G. G. Monks (Ed.), *Climate change and human responses: A zooarchaeological perspective* (pp. 197–216). Dordrecht: Springer.

Elera, C. (1993). El Complejo Cultural Cupisnique: Antecedentes y desarrollo de su ideología religiosa. *Senri Ethnological Studies, 37*, 229–257.

Elera, C., Pinilla, J., & Vásquez, V. (1992). Bioindicadores zoológicos de eventos ENSO para el formativo medio y tardío de Puémape-Perú. *Pachacamac (Lima), 1*, 9–19.

Ferrusquia-Villafranca, I., Arroyo-Cabrales, J., Johnson, E., Ruiz-González, J., Martínez-Hernández, E., Gama-Castro, J., et al. (2017). Quaternary mammals, people, and climate change: A view from southern North America. In G. G. Monks (Ed.), *Climate change and human responses: A zooarchaeological perspective* (pp. 27–67). Dordrecht: Springer.

Grayson, D. K. (1984). *Quantitative zooarchaeology*. New York: Academic Press.

Huckleberry, G., Hayashida, F., & Johnson, J. (2012). New Insights into the evolution of an intervalley prehistoric irrigation canal system, north coastal Peru. *Geoarchaeology, 27*, 492–520.

Hudson, M. J. (2013). Navigating disciplinary challenges to global sustainability science: An archaeological model. *Documenta Praehistorica, 40*, 219–225.

Kates, R. W. (2011). What kind of a science is sustainability science? *Proceedings of the National Academy of Sciences, 108*, 19449–19450.

Lima-Ribeiro, M. S., & Diniz-Filho, J. A. F. (2013). American megafaunal extinctions and human arrival: Improved evaluation using a meta-analytical approach. *Quaternary International, 299*, 38–52.

Lyman, R. L. (1995). Determining when rare (zoo)archaeological phenomena are truly absent. *Journal of Archaeological Method and Theory, 2*, 369–424.

Lyman, R. L. (2005). Zooarchaeology. In H. D. G. Maschner & C. Chippendale (Eds.), *Handbook of archaeological methods* (pp. 835–870). Lanham, MD: AltaMira Press.

Magnell, O. (2017). Climate and wild game populations in south Scandinavia at the Holocene thermal maximum. In G. G. Monks (Ed.), *Climate change and human responses: A zooarchaeological perspective* (pp. 123–135). Dordrecht: Springer.

Monks, G. G. (2017a). Introduction. In G. G. Monks (Ed.), *Climate change and human responses: A zooarchaeological perspective* (pp. 1–4). Dordrecht: Springer.

Monks, G. G. (2017b). Evidence for changing climate and subsistence strategies among the Nuu-chah-nulth on Canada's west coast. In G. G. Monks (Ed.), *Climate change and human responses: A zooarchaeological perspective* (pp. 173–196). Dordrecht: Springer.

Ochoa, J., & Piper, P. (2017). Holocene large mammal extinctions in Palawan Island, Philippines. In G. G. Monks (Ed.), *Climate change and human responses: A zooarchaeological perspective* (pp. 69–86). Dordrecht: Springer.

Pilaar Birch, S., & Miracle, P. T. (2017). Human response to climate change in the northern Adriatic during the Late Pleistocene and Early Holocene. In G. G. Monks (Ed.), *Climate change and human responses: A zooarchaeological perspective* (pp. 87–100). Dordrecht: Springer.

Rahel, F. J. (1990). The hierarchical nature of community persistence: A problem of scale. *The American Naturalist, 136*(3), 328–344.

Reitz, E. J., & Wing, E. S. (2008). *Zooarchaeology* (2nd ed.). Cambridge: Cambridge University Press.

Reitz, E. J., deFrance, S. D., Sandweiss, D. H., & McInnis, H. E. (2015). Flexibility in southern Peru coastal economies: A Vertebrate Perspective on the Terminal Pleistocene/Holocene Transition. *Journal of Island and Coastal Archaeology, 10*(2), 155–183.

Rindel, D., Goñi, R., Bautista Belardi, J., & Bourlot, T. (2017). Climatic changes and hunter-gatherer populations: Archaeozoological trends in southern Patagonia. In G. G. Monks (Ed.), *Climate change and human responses: A zooarchaeological perspective* (pp. 153–172). Dordrecht: Springer.

Rollins, H. B., Sandweiss, D. H., & Rollins, J. C. (1990). Mollusks and Coastal Archaeology: A Review. In N. P. Lasca & J. D. Donahue (Eds.), *Decade of North American geology*, Centennial Special Volume 4, "Archaeological Geology of North America" (pp. 467–478). Boulder, CO: Geological Society of America.

Sandweiss, D. H. (1996). Environmental change and its consequences for human society on the central Andean coast: A malacological perspective. In E. J. Reitz, L. Newsom, & S. Scudder (Eds.), *Case studies in environmental archaeology* (pp. 127–146). New York: Plenum Publishing.

Sandweiss, D. H., & Kelley, A. R. (2012). Archaeological contributions to climate change research: The archaeological record as a paleoclimatic and paleoenvironmental archive. *Annual Review of Anthropology, 41*, 371–391.

Sandweiss, D. H., Maasch, K. A., Andrus, C. F. T., Reitz, E. J., Riedinger-Whitmore, M., Richardson III, J. B., et al. (2007). Mid-Holocene Climate and Culture Change in Coastal Peru. In D. G. Anderson, K. A. Maasch, & D. H. Sandweiss (Eds.), *Climatic change and cultural dynamics: A global perspective on Mid-Holocene transitions* (pp. 25–50). San Diego: Academic Press.

Sandweiss, D. H., Maasch, K. A., Burger, R. L., Richardson III, J. B., Rollins, H. B., & Clement, A. (2001). Variation in Holocene El Niño frequencies: Climate records and cultural consequences in ancient Peru. *Geology, 29*, 603–606.

Sandweiss, D. H., & Quilter, J. (2008). Climate, catastrophe, and culture in the ancient Americas. In D. H. Sandweiss & J. Quilter (Eds.), *El Niño, catastrophism, and culture change in cncient America* (pp. 1–11). Washington, DC: Dumbarton Oaks Research Library and Collection.

Taylor, R. E., & Bar-Yosef, O. (2014). *Radiocarbon dating: An archaeological perspective* (2nd ed.). Walnut Creek, CA: Left Coast Press Inc.

Tyler-Faith, J., & Surovell, T. A. (2009). Synchronous extinction of North America's Pleistocene mammals. *Proceedings of the National Academy of Sciences, 106*(49), 20641–20645.

Waters, M. R., Forman, S. L., Jennings, T. A., Nordt, L. C., Driese, S. G., Feinberg, J., et al. (2011). The Buttermilk Creek Complex and the origins of Clovis at the Debra L. Friedkin site, Texas. *Science, 331*, 1599–1603.

Yacobaccio, H. D., Morales, M., & Samec, C. (2017). Early to Middle Holocene climatic change and the use of animal resources by highland hunter-gatherers of the south-central Andes. In G. G. Monks (Ed.), *Climate change and human responses: A zooarchaeological perspective* (pp. 103–121). Dordrecht: Springer.

Index

Note: Page numbers followed by *f* and *t* indicate figures and tables respectively

A

Adriatic, 3, 87–89, 89*t*, 90–93, 95, 96, 98, 132, 222
Ageröd III, 124*f*, 126, 126*f*
Alaska, 57, 175–177, 181, 186
Alces
  alces ("elk" in Europe, "moose" in North America), 123, 125, 128*f*, 128*t*, 129*t*
Allen's rule (Proportion rule), 20, 201, 202
Altiplano, Puna, 103–105, 107, 220, 222, 223
  Argentinean, 103, 105, 111
  Chilean, 105, 111, 118, 220
Amud, 10*t*, 11*t*, 14*f*, 18, 21, 118
Andes, 3, 103, 105–108, 115, 156
Annual productivity pulse, 201
Anthropogenic impact, 73, 81. *See also* Human impact
Archaeofauna, 103, 182
Archaeozoology, 83, 124, 125, 133, 162, 219
Arctic, 1, 198, 200
Aridity, 19, 38, 103, 105, 107, 115, 118, 119, 201, 222, 223
Arlöv I, 124*f*, 126*f*, 127
Astragali, 127, 129, 131, 209, 214
Atlantic, 142, 144, 175, 198, 199*f*, 200, 222
Atmospheric circulation, 106, 107
Aurochs. *See Bos primigenius*
Axis
  calamianensis, 70, 77, 78, 78*t*, 80

B

Baltic Sea, 131
Barkley Sound, 173, 177, 179, 191, 192
Basaltic Plateaus, 153, 168
  Cardiel Chico, 154, 156, 156*f*
  Pampa del Asador, 154, 156, 156*f*, 158
  Strobel, 154, 156, 158
Bathymetry, 71, 72, 222
Bergmann's rule (Size rule), 201
Biodiversity, 27, 28, 70, 72
Biogeography, 27, 71, 197, 201
Bioindicator, 220
Bivalve. *See Polymesoda, Saxidomus*
Body size, 21, 28, 29–32*t*, 119, 123, 124, 127, 129, 132, 133, 181, 201–203, 213, 222, 223
Bone
  breadth, 203, 209, 210*f*
  density, 184
  growth, 197, 201–203, 214, 215, 223
  length, 202, 203, 209, 210*f*, 213

Boreal, 87, 89*t*, 90, 91, 127, 131, 141
Borneo, 70–73, 78*t*, 79*f*, 80, 82
Bos
  primigenius, 9, 125, 127*f*, 128*t*, 129*t*
British Columbia, 173, 175, 177, 193
Broad Spectrum, 87, 97, 98
Brown Bank, 139

C

C3 plants, 114, 115
C4 plants, 114, 115
Callorhinus
  ursinus, 179
Camelid. *See Lama, Vicugna*
  familiar groups, 113
  female groups with calves, 157
  guanaco, 108, 114*t*, 114, 115, 157
  guanaco social groups, 108
  male groups, 108, 157
  mixed groups, 157
Canid, 69, 76, 80, 221
Capra
  hircus, 198
  ibex, 9, 93
Capreolus
  capreolus, 9, 125, 128*t*, 129*t*, 129*f*, 130*f*, 131*t*
Carcass, 54, 92, 93, 95, 97, 158, 168, 202, 214
  cut marks, 165, 168
  marrow obtaining activities, 167
  percussion marks, 165
  perimetral marking, 162
  processing, 88, 94, 96, 158, 222
  transverse fracture, 162
Caribbean, 139, 140*f*, 140–142, 149, 150
Cascal de Flor de Pinos, 139
Causation, 192, 219, 220
Central America, 139, 140, 142–145, 147, 150
Cervus
  elaphus, 9, 32*t*, 48*t*, 93, 125, 128*t*, 129*t*, 129*f*, 130*f*, 131*t*
Chinchillas, 111
Ch'uumat'a (DfSi-4), 184, 192
Climate
  change, 1–3, 9–22, 27–29, 34, 53, 55, 57, 73, 87, 91, 98, 108, 123, 124, 131–133, 167, 175–178, 186, 192, 193, 200, 214, 219, 220, 223
  drought, 89*t*, 89, 176
  effective, 18

fluctuations, 7, 21, 71, 73, 190
forcing, 3, 7, 20, 21
global warming, 123, 133, 175
little ice age (LIA), 4, 173, 200, 201
little optimum, 200
medieval climatic anomaly (MCA), 4, 153, 173, 224
mediterranean, 3, 8, 13, 14, 17, 88, 91, 92, 98, 131
   hyper-mesic, 10*t*–12*t*, 13, 14, 16*f*, 18*f*
   mesic, 10*t*–12*t*, 13, 14, 16*f*, 18*f*
   xeric, 10*t*–12*t*, 13, 14, 16*f*, 18*f*
   primary productivity, 104, 105, 154
   southern westerlies, 153
   temperature, 91, 123, 131–133, 141, 154, 175–178, 186, 187, 190, 200–202, 209, 211–214, 223
   winds, 141
Clupea
   pallasii, 179
C:N ratios, 113, 114*t*
Coastal lagoon, 139, 142
Cod, 179–182, 184
Collation, 219–221
Colonization, 50, 69, 72, 73, 80, 153
Correlation, 13, 69, 77, 89*t*, 96, 97, 115, 118, 130*f*, 177, 179, 192, 219, 220
Croatia, 88
   Istria, 90, 91, 97
   Kvarner Gulf, 87, 88, 91
Culture
   ceramic, 140, 167
   change, 192
   colonization, 80
   engravings, 154
   extensification, 154
   grinding stones, 167
   human burials, 74, 154, 168
   hunting blinds, 154, 156, 162
   logistic, 220
   mobility, 97, 98
   organizational systems, 192
   paintings, 154
   projectile points, 54
   residential, 190
   rock art, 154, 156
   windbreaks, 162
Cuon
   alpinus, 30*t*, 41*t*, 50, 70, 80, 221
Current
   California, 178
   downwelling, 187
   Humboldt, 222
   subarctic current, 176
   upwelling, 177, 187

D
Dama
   dama, 133
   mesopotamica, 9, 20
Deer. *See Cervus, Dama*
Dendrochronological, 2
Desert, 8, 104, 105, 108, 222
Dhole. *See Cuon alpinus*
Diatom, 175, 176
Diet, 19–21, 29*t*–32*t*, 81, 87, 92, 94, 96–98, 105, 114, 115, 119, 132, 198, 221, 222
Diversification, 69, 72, 81, 87, 89*t*, 96, 97

E
Ecosystem, 1–3, 21, 28, 57, 69, 72, 73, 89, 92, 94, 98, 117–119, 123
El Niño, 175–177, 187, 220, 222
Embiotoca
   lateralis, 179, 180
Engraulis
   mordax, 177
ENSO, 106, 175, 176, 187, 192
Epipaleolithic, 11*t*, 12*t*, 16, 17*f*
Equus, 9, 32*t*, 47*t*, 53, 93, 198
Eumetopias
   jubatus, 179
Europe, 3, 7, 8, 87, 91, 97, 123, 132, 175, 200
   northern, 3, 123
   southern, 3, 87
Extinction, 7, 8, 20, 21, 28, 29, 50, 55–58, 70, 72, 73, 80–82, 108, 127, 200, 221, 223

F
Fallow deer. *See Dama*
Faunal turnover, 7, 70
FAUNMAP, 2
Fire Return Interval (FRI), 175
Formative Period, 149
Fossil fuel, 1
Fragmented landscape (Fragmentation of the habitat), 107, 108, 118

G
Gadus
   macrocephalus, 179, 184
Gården under Sandet (GUS), 197, 200, 203, 204*f*, 207
Gazella
   gazella, 9, 20, 20*f*
Geist's rule, 201
GISP2 ice core, 200
Glacial, 7, 8, 13, 15, 17, 28, 57, 69, 70*f*, 71, 72, 81, 89, 91–93, 156, 175, 176, 178, 201
   last maximum, 70*f*, 71
Global warming, 123, 124, 133, 175, 222. *See also* Climate change
Goats. *See Capra*
Greenland, 3, 4, 197, 198, 200, 205, 208*f*, 210–212*f*, 213, 214, 223
   eastern settlement, 4
   western settlement, 208*f*, 210–212*f*
Growth, 140, 142, 143, 143*f*, 144, 145, 145*f*, 146, 148, 149, 178, 193, 197, 200–203, 213–215, 223
Guthrie's rule, 201

H
Habitat, 7, 12, 113
   local, 8, 13, 15, 17, 22, 50, 73, 87, 92, 96, 98, 103, 113, 150
Haida Gwaii (Queen Charlotte Islands), 176, 178, 182, 184
Heinrich 5 event (H5), 8, 21
Hierarchical model, 17
Hippocamelus
   antisensis, 108
   bisulcus, 158
Historical ecology, 2
Holocene
   early Holocene, 3, 51, 54, 75, 76, 79–82, 87, 89–92, 96, 106, 107, 109*t*, 110*t*, 111, 112, 115, 118, 123, 124, 127, 131–133, 221
   late Holocene, 3, 75, 76, 79–81, 115, 139, 153, 154, 158, 159, 167–169, 175, 221, 223
   middle Holocene

middle Holocene I, 111, 112
middle Holocene II, 111, 112
thermal maximum (HTM), 123
Hominin, 3, 7, 18–21, 73, 223
Hornillos 2, 109t, 113, 114t, 114, 115, 118, 119
Human
early, 28, 55, 58, 87, 103, 158
impact, 2, 27, 58, 72. *See also* Anthropogenic impact
Humeri, 94, 127, 129, 131, 202
Hunter-Gatherers, 55, 103, 104, 107, 111, 119, 222
Hunting, 3, 21, 54, 55, 73, 77, 80, 81, 90, 96, 98, 107, 108, 118, 123, 125, 132, 133, 150, 154, 156, 156f, 157, 158, 161t, 162, 163, 164f, 167, 168, 182, 191, 198, 214, 221

I
Ille cave, 69, 73–75, 75f, 77, 78t, 80, 82f, 221
Individualistic hypothesis, 3
INPN, 2
Intensification, 87, 88, 94–98, 107, 154, 222
Intertropical Convergence Zone (ITCZ), 141, 146, 223
Isotope
$\delta^{13}C$, 96, 103, 113, 114t, 115f, 115, 132, 143, 144, 223
$\delta^{15}N$, 96, 103, 113, 114t, 115, 117–119, 223
$\delta^{18}O$, 91, 140, 143–145, 145f, 146t, 147, 148, 148f, 149, 149t, 150, 173, 186
stable, 3, 17, 20f, 95–97, 113, 132, 139, 142, 143, 187

J
Jaccard similarity Index, 12

K
Karoline, 139, 140, 140f, 142, 144, 147–150, 148f
KH-4, 140f, 142, 147, 148f, 150
Kebara, 8, 21

L
La Niña, 176, 177
Lacustrine Basins/Lake
Belgrano, 153, 156, 159t
Burmeister, 153
Cardiel, 153, 154, 156f, 158, 162
Parque Nacional Perito Moreno (PNPM), 154, 156f
Posadas, 153, 154, 156, 159, 162
Salitroso, 153, 154, 156, 156f, 158, 162
Tar, 156, 156f, 158, 159t, 160t, 162
Landnám, 198
Lama
glama, 114
guanicoe, 108, 114, 118, 157, 162
Latitude, 50, 104, 140, 201, 202
Lepus
europaeus, 93
Levant
southern, 3, 7–9, 13, 15–17, 19–21
Little Ice Age (LIA), 4, 173, 200, 201. *See also* Climate Change
Livestock, 197, 198, 200, 202, 212–214
Log ratio, 206t, 208, 209, 209t, 210t, 215
Lundby I, 124f, 125t, 126

M
Ma'acoah (DfSi-5), 182, 184, 192
Macromammals, 9
Mammalia
Artiodactyla, 32t, 47t, 109t, 110t
Antilocapridae, 32t, 47t
Bovidae, 32t, 47t
Camelidae, 28, 32t, 48t, 109t, 110t
Cervidae, 32t, 48t, 109t, 110t
Tayassuidae, 32t, 48t, 142
Carnivora, 28, 30t, 38, 41t, 51, 43, 109t, 110t
Canidae, 30t, 41t
Felidae, 28, 30t, 42t, 142
Hyaenidae, 30t, 42t
Mustelidae, 30t
Procyonidae, 30t, 42t
Ursidae, 28, 30t, 42t
Chiroptera, 28, 29t, 39t, 76
Antrozoidae, 29t, 39t
Emballonuridae, 29t, 39
Molossidae, 29t, 39
Mormoopidae, 30t, 40t
Natalidae, 30t, 40t
Phyllostomidae, 30t, 40t
Vespertilionidae, 30t, 41t
Didelphimorphia, 29t, 39t
Didelphidae, 29t, 39t
Insectivora *s.l.*
Soricidae, 29t, 39t
Lagomorpha, 32t, 46t
Leporidae, 32t, 46t
Liptoterna
Macrauchenidae, 32t, 49t
Notoungulata, 28
Toxodontidae, 32t, 49t
Perissodatyla
Equidae, 28, 32t, 47t
Tapiridae, 32t, 47t, 142
Primates, 8, 30t
Proboscidea, 28, 32t, 48t, 53
Elephantidae, 28, 32t, 48t
Gomphotheriidae, 28, 32t, 49t
Mammutidae, 28, 32t, 49t
Rodentia, 28, 30t, 43t
Castoridae, 30t, 43t
Cuniculidae, 30t, 43t
Dasyproctidae, 31t, 43t, 142
Erethizontidae, 31t, 43t
Geomyidae, 31t, 43t
Heteromyidae, 31t, 43t
Hydrochoeridae, 28, 44t
Muridae, 31t, 142
Sciuridae, 31t, 46t
Xenarthra *s.l.*, 29t, 39t, 111
Dasypodidae, 29t, 39t, 142
Glyptodontidae, 28
Megalonychidae, 28
Megatheriidae, 28
Mylodontidae, 28
Myrmecophagidae, 29t, 39t
Marine Isotope Stage (MIS)
MIS 4-3, 7, 8, 13, 15, 16, 223

MIS 6-3, 3, 19
Medieval Climatic Anomaly (MCA), 4, 153, 168, 173, 175–179, 181, 182, 184–187, 186t, 187t, 190–192, 224. *See also* Climate Change
Megafauna, 56, 57, 72, 73
Merluccius
  productus, 179
Mesoamerica, 3
Mesolithic, 87–98, 123, 125, 125t, 126t, 127, 128, 128f, 129f, 130f, 131, 132
Mesowear, 19, 20, 20f
Metal Period, 75, 79, 221
México
  Chiapas, 53, 54, 56f
    Ocozocuautla, 54
  Distrito Federal
    Peñón, 54
  Oaxaca, 54
    Guila Naquitz, 54
    Tlacolula, 54
  Puebla, 53, 54
    Valsequillo, 36t, 54
  Quintana Roo, 54
    Tulum, 54
  San Luis Potosí, 54
    Cedral, 35t, 54
  Sonora, 34, 38, 54
    Fin del Mundo, 54
  State of México, 54
    Santa Isabel Ixtapa, 54
    Tlapacoya, 36t, 54
    Tocuil, 54
Micromammals, 7–9, 11–13, 11t, 15, 18f, 19, 20
Microtus
  guentheri, 9, 19
Microwear, 19, 213, 214
Middle East, 3
Molle. *See Schinus marchandii*
Mollusc, 90, 91, 94, 98, 139, 140, 142, 144
Moose (North America. *See Alces alces*)
Morphotectonic Provinces, 33t, 34f, 35f, 38, 50f, 56f, 57
  Central and Eastern, 38, 58
    Central Plateau, 28, 38, 50, 51, 56f
    Gulf Coastal Plain (northern portion), 38, 53, 56f, 58
    Sierra Madre Oriental, 38, 50, 53, 55, 56f
  Northern
    Baja California Peninsula, 34, 53, 56f
    Chihuahua-Coahuila Plateaus and Ranges, 34, 53, 55
    Northwestern Plains and Sierras, 34, 53, 56f
    Sierra Madre Occidental, 34, 51, 53, 55, 56f
  Southern
    Gulf Coastal Plain (southern portion), 39, 53, 56, 58
    Sierra Madre de Chiapas, 53, 56f
    Sierra Madre del Sur, 51, 53, 55, 56f
    Yucatan Platform, 53, 56f
  Trans-Mexican Volcanic Belt, 28, 51, 55, 56f, 58
Mountain Gazelle. *See Gazella gazella*
Mousterian, 9
Mytilus
  californianus, 179

N
Nature conservation, 123–125, 133
Neanderthals, 7, 8, 13, 16, 19–22, 223

Nebekian, 11t, 12t, 13, 17f
New World, 3
Nicaragua, 3, 139–141, 150, 223
Non-metric multidimensional scaling (NMMDS), 7, 9, 11–14, 16f–18f
Norse, 3, 4, 197–200, 211, 213, 214, 223, 224
North American Land Mammal Ages
  Irvingtonian, 29, 34, 35t, 37t, 58
  Rancholabrean, 29, 34, 35f–38, 50–53, 55, 56, 58
Northwest Coast of North America (NWC), 3, 173, 179
Nugljanska, 90
Number of Identified Specimens per Taxon (NISP), 8, 9, 75, 82f, 94, 111, 112f, 125t, 126t, 128t, 129t, 157, 159t, 162, 163, 165f–167f, 182–185f, 188f–190
Nutrient Concentration Zones, 105
Nutrition
  plane, 28, 202, 203
Nuu-chah-nulth (Nootka), 173, 177–182, 184, 190–193, 220, 224

O
Ø34, 197, 204, 205f, 207t, 211, 213f, 214, 215
Occam's razor, 221
Old World, 3
Oncorhynchus
  keta, 177, 179, 183
  kisutch, 177
  nerka, 177
  tshawytscha, 177
Optimal Foraging Theory (OFT)
  patch choice, 190
  prey choice, 181
  small prey, 181, 191
Öresund Strait, 131

P
Pacific
  decadal oscillation (PDO), 175
Paleoclimatology, 219
Palaeoenvironment, 70–72, 106, 107, 140
Paleoenvironmental
  change, 106, 113, 123, 198. *See also* Climate change
  data (records), 104, 106, 221
  studies, 106, 198
Paleontology, 2, 219
Panthera
  tigris, 70, 79
Paleolithic
  middle "MP", 7, 12t, 16
  upper, 9, 12t, 15f, 18f, 88–91, 94, 95f, 95, 96
Palawan, 3, 69–73, 75, 76f, 77, 79–82, 221
Paralytic shellfish poisoning (PSP), 190, 191
Pasimbahan Cave, 70, 73, 74, 75f, 76f, 77, 81, 221
Pasture growth, 203, 213, 214
Patagonia, 3, 108, 153, 154, 157, 158, 162, 168, 169, 220
Pearl Lagoon
  Awas, 140f
  Haulover, 140f, 141
  Kukra Hill, 142
Peru, 106–108, 115, 177, 219, 220, 222
Philippines, 3, 69, 72, 73, 79f, 81, 82, 221
Pleistocene
  Late, 3, 7, 16, 27, 36, 37, 52t, 53–55, 56f, 57, 58, 73, 77, 82, 87, 89, 201, 221, 223
  megafauna, 73

Pollen, 8, 17, 18, 21, 51, 55, 89t, 91, 106, 123, 126, 132, 142, 175, 198, 211, 215, 221, 223
Polymesoda, Saxidomus
  arctata, 139, 140, 140f, 142, 143f, 145f, 146t, 146, 147, 150
  solida, 142
Principal Coordinates Analysis (PCA), 9, 12–14
Productivity, 8, 17, 21, 104, 105, 119, 154, 175, 176, 178, 191, 197, 201, 202, 213
Protective herding, 112, 114, 118, 119
Proxy, 1, 2, 19, 20, 91–93, 150, 173, 175–177, 181, 186, 200, 215, 222, 223
Puna, 3, 104, 104f, 105–108, 109t, 110t, 111, 112, 113f, 115, 117–119
Punta Masaya, 139
Pupićina, 90, 91, 96, 97

Q

Qafzeh, 8, 10t, 11t, 14f, 15f, 18
Quaternary, 21, 27–29, 57, 59, 69, 70, 73, 108, 221
  late, 59, 69, 70, 73, 108, 221
  Mexican Mammalian Database, 29

R

Red deer. See *Cervus elaphus*
Regional Scale, 118, 153, 157, 163, 167, 168, 220
  highlands, 118, 153, 157, 167, 168
  lowlands, 153, 157, 159t, 167, 168
  mesoregional, 167
  sand dunes, 154
  Santa Cruz Province, 158
  southern Patagonia, 153, 157, 158, 168, 220
  steppe, 153, 157, 167, 168
Resources, 17, 21, 28, 55, 76, 80, 81, 87, 94, 96–98, 103–105, 107, 108, 111, 113, 118, 119, 139, 140, 142, 149, 150, 154, 162, 165, 168, 176, 178, 179, 190–192, 197, 198, 201, 214
  structure, 108, 111, 118, 154
  extraction, 119
Risk, 92, 175, 176, 181, 190, 191, 223
Rodents, 19, 28, 38, 51, 76, 77, 111, 118, 158
Roe deer. See *Capreolus capreolus*
Rupicapra
  rupicapra, 93
Rusa
  marianna, 77, 78, 78t, 79f, 82

S

Sardinops
  sagax, 177
Saxidomus
  gigantea, 173, 186
Scale, 2, 92, 98, 103, 106, 107, 113, 118, 119, 123, 153, 156, 157, 163, 167, 168, 175, 219, 220, 222
Scandinavia, 3, 123, 124f, 127f, 128f, 129t, 129f–131f, 132, 133, 201
Scania, 123, 124f, 125, 126t, 127, 127f, 128, 128t, 128f, 129t, 129f, 131–133
Scheduling, 191
Schinus
  marchandii, 153
Scomber
  japonicus, 177
Sea
  level, 3, 69, 71–73, 81, 87, 88f, 89t, 89–92, 94, 96–98, 104, 131, 156, 176, 222
  surface temperature (SST), 141f, 187

Seasonality, 19, 89t, 91, 97, 104, 108, 139, 147, 158, 163, 190, 191, 223
Seasonal Mobility, 92, 95, 97–98, 168
Sebastes, 177–179, 181, 183, 184
  hyomandibular, 184, 186t
  paucispinus, 178
  ruberrimus, 178
Sediment cores, 17, 91, 175, 176, 179
Sheep, 4, 90, 91, 94, 197, 198, 202, 203, 206–207t, 207–209, 209t, 210f, 211f, 211, 212, 213f, 213–215, 223
Shell midden, 74, 81, 97, 139, 140, 140f, 142, 150, 182, 184, 223
Shifting baseline, 2
Sites (archaeological)
  caves, 17, 54, 70, 71, 73, 75f, 76f, 76, 80, 81, 94, 96, 156, 157, 158, 221
  kill, 92, 158, 175, 214
  open air, 156, 158, 163, 165
  rockshelters, 154, 156–158, 163
Sitetaia, 139
Sjöholmen, 124f, 126, 126f
Skeleton
  appendicular, 119, 186, 191
    carpal/tarsals, 160t, 161t, 162
    diaphysis, 114t, 158, 163
    epiphysis, 158, 163
    femur, 114t, 160t, 161t, 162, 207t, 209t
    humerus, 78t, 131t, 160t, 161t, 162, 207t, 209t
    metapodial, 94, 95, 114t, 160t, 161t, 162, 163, 168, 204, 212
    phalanges, 78, 79f, 97, 163, 167, 168
    radioulna, 160t, 161t
    scapula, 114t, 160t, 161t, 162, 163
    tibia, 78t, 160t, 161t, 162, 207t, 211, 213t
  axial
    crania, 163
    mandibles, 75, 94, 97, 114t, 160t, 161t, 163
    pelvis, 160t, 161t
    ribs, 162, 163, 202
    vertebrae, 94, 160t, 161t, 162, 163, 186, 202
  completeness
    forelegs, 97
    hind legs, 94, 97, 203, 209t
    lower limb bones, 162
    mid limb bones, 162
    upper limb bones, 163, 202
  growth, 197, 201–203, 214, 215
South America, 3, 21, 54, 55, 58, 105, 220
Southeast Asia
  island, 70f, 71, 73
  mainland, 71, 80
South Scandinavia, 123, 124, 124f, 127f, 128f, 129t, 129f, 130f, 131, 133
Squalus
  acanthius, 180
Statistic
  Jaccard similarity index, 12
  Kruskal statistic of stress, 11
  Kulczynski similarity index, 12
Steppe
  Herbaceous (*Pajonal*), 91, 105, 106, 115, 156
  mixed, 105, 115, 119
  Shrub (*Tolar*), 105–106, 153, 156, 168
Storage, 94, 179, 181, 191, 192, 213
Stress (Kruskal statistic), 11–13
Subsistence, 4, 76, 81, 87, 88, 92, 97, 103, 123, 149, 150, 157, 168, 173, 178, 180, 186, 190–193
Sunda, 70–72, 76

Sus
  scrofa, 9, 93, 125, 128*t*, 129*t*, 129*f*, 130*f*, 131*t*, 198
Sværdborg I, 124*f*, 125*t*, 126

T
Tabun B, C, D, 9, 10–12*t*, 13, 14*f*, 17*f*, 18*f*
Thynnus
  orientalis, 177
Toquaht (ethnic group), 173, 177, 180*f*, 182, 184, 192, 193
Toquart River, 177, 178, 192, 193
Time-averaging, 220
Trans-Mexican Volcanic Flora, 52*t*
Tree-Ring Dating (Dendrochronology), 123, 175, 176
Ts'ishaa (DfSi-16), 182, 184
T'ukw'aa, 180*f*, 182, 183*f*, 192
  T'ukw'aa defensive site (DfSj-23B), 183*f*, 192
  village site (DfSj-23A), 182

V
Vancouver Island, 173, 174*f*, 175–178, 193
Vegetation
  composition, 28, 119
  cover, 14, 200
Vela Špilja Lošinj, 88–89, 92, 96, 98
Vicugna
  vicugna, 108, 114, 118
Vizcachas, 111
Vulpes
  vulpes, 93, 198

W
Wetlands (*Vegas*), 105, 107, 108, 118, 119
Whales (Mysticeti), 179, 182, 198
Wild boar, 91, 94, 96, 97, 123, 125, 125*t*, 126*t*, 127, 128, 128*t*, 129*f*, 129*t*, 130*f*, 131–133, 131*t*, 222
Wildlife management, 2, 123–125, 133, 219, 222

Z
Zalophus
  californianus, 179
Zealand, 73, 123, 124*f*, 125, 125*t*, 126, 127, 127*f*, 128, 128*t*, 129*f*, 129*t*, 130*f*, 131*t*, 132, 133